Systems Neuroscience in Depression

Systems Neuroscience in Depression

Edited by

Thomas Frodl
Department of Psychiatry and Psychotherapy, Otto von Guericke University, Magdeburg, Germany; Department of Psychiatry, Trinity College Dublin, Ireland

AMSTERDAM • BOSTON • HEIDELBERG • LONDON
NEW YORK • OXFORD • PARIS • SAN DIEGO
SAN FRANCISCO • SINGAPORE • SYDNEY • TOKYO

Academic Press is an imprint of Elsevier

Academic Press is an imprint of Elsevier
125 London Wall, London EC2Y 5AS, UK
525 B Street, Suite 1800, San Diego, CA 92101-4495, USA
50 Hampshire Street, 5th Floor, Cambridge, MA 02139, USA
The Boulevard, Langford Lane, Kidlington, Oxford OX5 1GB, UK

British Library Cataloguing-in-Publication Data
A catalogue record for this book is available from the British Library

Library of Congress Cataloging-in-Publication Data
A catalog record for this book is available from the Library of Congress

ISBN: 978-0-12-802456-0

For information on all Academic Press publications
visit our website at https://www.elsevier.com/

Publisher: Mara Conner
Acquisition Editor: Melanie Tucker
Editorial Project Manager: Kathy Padilla
Production Project Manager: Chris Wortley
Designer: Mark Rogers

Typeset by TNQ Books and Journals
www.tnq.co.in

Contents

4. Anti-Inflammatory and Immune-Modulatory Therapeutic Approaches in Major Depression

Norbert Müller

5. Molecular Mechanisms of Depression

Artemis Varidaki, Hasan Mohammad and Eleanor T. Coffey

Contributors

Deanna M. Barch Program in Neuroscience, Washington University, St. Louis, MO, USA; Department of Psychiatry, Washington University, St. Louis, MO, USA; Department of Psychology, Washington University, St. Louis, MO, USA; Department of Radiology, Washington University, St. Louis, MO, USA

Christian F. Beckmann Department of Cognitive Neuroscience, Radboud University Medical Centre, Nijmegen, The Netherlands; Donders Institute for Brain, Cognition and Behaviour, Centre for Neuroscience, Nijmegen, The Netherlands; Oxford University Centre for Functional MRI of the Brain (FMRIB), Nuffield Department of Clinical Neurosciences, University of Oxford, Oxford, UK

Linda Booij Department of Psychology, Concordia University, Montreal, QC, Canada; Sainte-Justine Hospital Research Centre, University of Montreal, Montreal, QC, Canada; Department of Psychiatry, McGill University, Montreal, QC, Canada

Dante Cicchetti Institute of Child Development, University of Minnesota, Minneapolis, MN, USA

Eleanor T. Coffey Turku Centre for Biotechnology, Åbo Akademi University and the University of Turku, Turku, Finland

Marika V. Doucet Neuropsychopharmacology Research Group, School of Pharmacy and Pharmaceutical Sciences, Trinity College Institute of Neuroscience, Trinity College Dublin, Dublin, Ireland

Thomas Frodl Department of Psychiatry and Psychotherapy, Otto von Guericke University, Magdeburg, Germany; Department of Psychiatry, Trinity College Dublin, Ireland

Andrew Harkin Neuropsychopharmacology Research Group, School of Pharmacy and Pharmaceutical Sciences, Trinity College Institute of Neuroscience, Trinity College Dublin, Dublin, Ireland

Ulrich Hegerl Department of Psychiatry and Psychotherapy, University of Leipzig, Leipzig, Germany; Research Centre of the German Depression Foundation, Leipzig, Germany

Tilman Hensch Department of Psychiatry and Psychotherapy, University of Leipzig, Leipzig, Germany

Tomoya Hirota Department of Psychiatry, Division of Child and Adolescent Psychiatry, University of California San Francisco, San Francisco, CA, USA

Melissa L. Levesque Campbell Family Mental Health Research Institute, Centre for Addiction and Mental Health, Toronto, ON, Canada

Kevin J. Manning Department of Psychiatry, University of Connecticut Health Center, Farmington, CT, USA

Fiona McNicholas Department of Psychiatry, University College Dublin, Dublin, Ireland; Lucena Clinic, Dublin, Ireland; Department of Child Psychiatry, Our Lady's Children Hospital Crumlin, Dublin, Ireland

Gordana Milavić National and Specialist Services, Michael Rutter Centre, Maudsley Hospital, London, UK

Hasan Mohammad Turku Centre for Biotechnology, Åbo Akademi University and the University of Turku, Turku, Finland

Bernhard Mueller Department of Psychiatry and Psychotherapy, University Hospital Essen, Essen, Germany

Peter C. Mulders Department of Psychiatry, Radboud University Medical Centre, Nijmegen, The Netherlands; Department of Cognitive Neuroscience, Radboud University Medical Centre, Nijmegen, The Netherlands

Norbert Müller Department of Psychiatry and Psychotherapy, Ludwig-Maximilian University of Munich, Munich, Germany

Veronica O'Keane Department of Psychiatry, Trinity College, Dublin, Ireland

Eileen O'Toole Neuropsychopharmacology Research Group, School of Pharmacy and Pharmaceutical Sciences, Trinity College Institute of Neuroscience, Trinity College Dublin, Dublin, Ireland

David Pagliaccio Program in Neuroscience, Washington University, St. Louis, MO, USA

Christian Sander Department of Psychiatry and Psychotherapy, University of Leipzig, Leipzig, Germany; Research Centre of the German Depression Foundation, Leipzig, Germany

Eoin Sherwin Neuropsychopharmacology Research Group, School of Pharmacy and Pharmaceutical Sciences, Trinity College Institute of Neuroscience, Trinity College Dublin, Dublin, Ireland

Norbert Skokauskas Department of Psychiatry, Trinity College, Dublin, Ireland; Centre for Child and Adolescent Mental Health and Child Protection, Department of Neuroscience, Norges teknisk-naturvitenskapelige universitet NTNU, Trondheim, Norway

David C. Steffens Department of Psychiatry, University of Connecticut Health Center, Farmington, CT, USA

Moshe Szyf Department of Pharmacology, McGill University, Montreal, QC, Canada

Indira Tendolkar Donders Institute for Brain, Cognition and Behaviour, Nijmegen, The Netherlands; Department of Psychiatry, Radboud University Medical Centre, Nijmegen, The Netherlands; Department of Psychiatry and Psychotherapy, University Hospital Essen, Essen, Germany

Philip F. van Eijndhoven Department of Psychiatry, Radboud University Medical Centre, Nijmegen, The Netherlands; Department of Cognitive Neuroscience, Radboud University Medical Centre, Nijmegen, The Netherlands

Artemis Varidaki Turku Centre for Biotechnology, Åbo Akademi University and the University of Turku, Turku, Finland

Christina B. Young Department of Psychology, Northwestern University, Evanston, IL, USA; Donders Institute for Brain, Cognition and Behaviour, Nijmegen, The Netherlands; Department of Psychiatry, Radboud University Medical Centre, Nijmegen, The Netherlands

The Systems Neuroscience Approach

Thomas Frodl[1,2]
[1]*Department of Psychiatry and Psychotherapy, Otto von Guericke University, Magdeburg, Germany;* [2]*Department of Psychiatry, Trinity College Dublin, Ireland*

The current book gives an overview about the state in the field with regard to MDD. It reviews the literature from molecular and biological viewpoints and translates it in human experimental and clinical research.

DEVELOPMENT OF THE SYSTEMS PERSPECTIVE

Chronic social stress has been shown to induce glucocorticoid-mediated pyramidal dendrite retraction in the hippocampus and changes in dendrite arborization in the prefrontal cortex (Kole, Czeh, & Fuchs, 2004; Magarinos, McEwen, Flugge, & Fuchs, 1996; Wellman, 2001; Woolley, Gould, Frankfurt, & McEwen, 1990). Experimental studies have further shown that stress or cortisol administration may lead to atrophy of neurons in the hippocampus and to depressive-like states, (Duman, 2002) and therapy with antidepressants might reverse these changes (Santarelli et al., 2003).

Since the hippocampal structure was also found to be altered in depression, these stress-related changes seem to be associated with stress-related disorders like major depression (MacQueen & Frodl, 2011). Nowadays, we believe that depression is not associated with changes in single brain areas, but rather with changes in brain networks. Thus, there is evidence that specific neuronal circuits, particularly in the developing brain, are damaged by environmental stress-inducing changes in the hypothalamic-pituitary-adrenal (HPA) axis, inflammatory pathways, and affective and salience networks (Krishnan & Nestler, 2008).

INTRODUCTION TO MAJOR DEPRESSIVE DISORDER

Major depressive disorder (MDD) is one of the most prevalent and burdensome of all psychiatric illnesses associated with early stressors (Goetzel, Hawkins, Ozminkowski, & Wang, 2003; Murray & Lopez, 1996). The morbidity and mortality associated with MDD has a significant impact on individuals suffering from the illness as well as their families and society.

The economic costs, which include reduced work productivity and lost leisure opportunities, are estimated to be €15.5 billion in the United Kingdom and €100 billion in the United States (Greenberg et al., 2003; Thomas & Morris, 2003). About 15% of patients with depression commit suicide (Jamison, 2000). The World Health Organization's World Health Report (2001, http://www.who.int/whr/2001/en/) expects that depression will lead to the second highest health burden internationally and will be the primary cause of disability and death in developed countries by the year 2020.

The diagnosis is based on clinical observations and reports by patients, caregivers, parents, and teachers. This approach has served to improve diagnostic reliability in clinical practice and research. *Diagnostic and Statistical Manual of Mental Disorders, Fifth Edition* (DSM-5) and the International Classification of Diseases-10 (ICD-10), which include the main diagnostic categories, are the main classification systems used for the diagnosis of mental disorders. Thus, diagnostics in psychiatry are not following an etiological viewpoint like many other medical disciplines. Instead, they are following phenomenological viewpoints, which are based on symptoms, syndromes, and disease courses.

Unfortunately, the diagnostic schemes have not assimilated breakthroughs in neuroscience. For example, most genetic and neural circuit anomalies are associated with symptoms that are common to multiple DSM diagnostic categories. Diagnoses made of self-reported and observer-rated symptoms might therefore constrain advances in research on the biology of mental illness. Notwithstanding these difficulties, there is a general consensus that the biology of mental illness is insufficiently developed to support a classification scheme based on the integration of genetics, neuroscience, and psychopathology (see http://www.nimh.nih.gov/research-priorities/rdoc/index.shtml).

THERAPY

Psychological and pharmacological therapies are the most common forms of treatment. The most frequently applied form of psychotherapy is cognitive behavioral therapy (CBT), which is an effective therapy against depression. Here, the assumption that a person's mood is directly related to his or her patterns of thought is the basis for therapy. Negative, dysfunctional thinking affects a person's mood, behavior, and physical state. The goal of CBT is to help a person learn to recognize negative patterns of thought, evaluate their validity, and replace them with healthier ways of thinking and to change patterns of behavior.

Moreover, cognitive behavioral activation therapy is widely used and combines cognitive methods with positive activities and exercise, since exercise was found to have antidepressant effects on its own. Many other psychotherapy modules have been described as effective, including the Cognitive Behavioral Analysis System of Psychotherapy, interpersonal psychotherapy, and strategic short-term psychotherapy. In particular, when trauma-related interpersonal and personality development problems exist, more attention needs to be paid to these issues during psychotherapy. Understanding of the brain systems involved in depression is helpful not only for developing new antidepressant targets but also for the development of psychotherapies, e.g., those involving the interplay between body and mind.

Antidepressant medications improve depressive symptoms by modulating the monoamine neurotransmitter systems. The selective serotonin reuptake inhibitors (SSRI) class has become the most commonly prescribed (Ilyas & Moncrieff, 2012). Systems relevant for MDD are the serotonergic, noradrenergic, and dopaminergic systems as well as the glutamatergic and GABAergic system with known effects on depressive behavior.

There is evidence that the pathophysiology of depression involves low levels of serotonin, noradrenaline, and/or dopamine levels in the central nervous system (Carlsson, Corrodi, Fuxe, & Hokfelt, 1969). Depression involves dysfunctional glutamate signaling in the brain, which leads to impaired neuroplasticity (Duman, 2014). Moreover, depression is associated with hyperactivation of the cholinergic system and as a consequence decreased activity in the noradrenergic system (Drevets, Zarate, & Furey, 2013). Furthermore, reduced GABA neurotransmission in cortical circuits was also detected (Pehrson & Sanchez, 2015).

Tricyclic antidepressants (TCA) and monoamine oxidase (MAO) inhibitors were discovered first. SSRIs and serotonin and noradrenaline reuptake inhibitors (SNRIs) were later conceived mainly to improve the tolerability of TCAs. Since the discovery of SSRIs and serotonin and noradrenaline reuptake inhibitors (SNRIs), one strategy in antidepressant research has been to expand on the monoaminergic mechanisms by targeting monoaminergic receptors or additional transporters in one molecule. The main objectives have been to improve efficacy and/or reduce the time to onset the therapeutic effect of SSRIs and SNRIs. The interest in glutamate targets precipitated from the spectacular clinical finding that intravenous infusion of the *N*-methyl-D-aspartate (NMDA) receptor antagonist ketamine can produce an immediate antidepressant effect in patients with treatment-resistant depression (Berman et al., 2000). Several other glutamate targets have been defined, some of which have been tested in the clinic.

Brain stimulation therapies involve activating the brain directly with electricity, magnets, or implants to treat depression and other disorders. Electroconvulsive therapy is the most researched stimulation therapy, has the longest history of use, and has strong evidence for therapy-resistant depression. Other stimulation therapies like vagus nerve stimulation, repetitive transcranial magnetic stimulation (rTMS), and magnetic seizure therapy have more regional and local effects and are thought to have less side effects, e.g., on cognitive functioning. The US Food and Drug Administration approved rTMS as a therapy for treatment-resistant depression and vagus nerve stimulation for use in treating major depression in certain circumstances—if the illness has lasted 2 years or more, if it is severe or recurrent, and if the depression has not eased after trying at least four other treatments. Within a systems neuroscience perspective it is thought that when one key area in a brain system is influenced it has effects on the whole system as well as other interrelated brain systems.

Deep brain stimulation (DBS), which was first developed as a treatment for Parkinson disease to reduce tremors, stiffness, walking problems, and uncontrollable movements, has been studied as a treatment for depression. DBS is available on an experimental basis only.

INVOLVED SYSTEMS

HPA Axis and the Immune System

Many research studies in the field of immunology and psychiatry have reported the association between immune system changes detected in the peripheral blood or tissue or in cerebrospinal fluid and related changes in the brain in terms of imaging findings or postmortem findings. Several findings have been reported regarding the link between immune system changes, psychological abnormalities, and impairment in other systems such as energy metabolism (Myint & Kim, 2014). Moreover, changes to immune molecules in the brain are also reported to be associated with other neurochemical changes, such as serotonergic, noradrenergic, GABAergic, glutamatergic neurotransmissions, and stress hormones. Some studies have reported metabolisms such as tryptophan/kynurenine metabolism and tyrosine metabolism as the link between changes in the immune system and other neurochemicals. These results have provided the missing links between the glial—neuronal network and opportunities to develop new therapeutics and biomarkers in the field of psychiatry (Myint & Kim, 2014).

Furthermore, it was shown that both physiological and psychological stress can induce an increased production of proinflammatory mediators that can stimulate tryptophan catabolism in the brain (Myint, Schwarz, & Muller, 2012). Cytokines modulate the balance between humoral and cell-based immune responses and regulate the maturation, growth, and responsiveness of cells. They are known to have central effects, either crossing the blood—brain barrier or through the transmission of signals across the vagus nerve (Dantzer, O'connor, Freund, Johnson, & Kelley, 2008). In this regard, it has been suggested that cytokines could influence reward processing and promote the increase of depressive symptoms through altered activity of dopaminergic and serotonergic systems (Ikemoto & Panksepp, 1999; Ressler & Nemeroff, 2000). Interestingly, proinflammatory cytokines can contribute to glutamate toxicity, also stimulating microglial cells to release interleukin (IL)-1 and tumor necrosis factor alpha (TNF-α) (reviewed in McNally, Bhagwagar, & Hannestad, 2008).

In MDD, activity of the tryptophan/kynurenine pathway has been shown to modulate inflammatory processes, with kynurenine being directly associated with depressive symptoms and/or cognitive loss (Oxenkrug, 2007). Products of the "kynurenine" pathway include NMDA agonists (e.g., quinolinic and picolinic acids) (Jhamandas, Boegman, Beninger, Miranda, & Lipic, 2000) and antagonists as well as free radical generators, e.g., 3-hydroxykynurenine and 3-hydroxyanthranilic acids (Forrest et al., 2004; Thomas & Stocker, 1999).

NEUROIMAGING AND NEUROIMAGING GENETICS IN MDD

Neuroimaging techniques are important in order to characterize neural circuits built from different areas in the brain and relevant for certain brain functions. Here structural, functional, and diffusion imaging in MDD will be briefly reviewed.

Structural Imaging in MDD

Studies in humans strengthen the evidence that MDD is associated with structural changes. Many structural imaging studies have reported that the hippocampus is small in patients with MDD. A recent meta-analysis of hippocampal volumes in patients with MDD confirmed that patients had hippocampal volumes that were approximately 4–6% smaller than matched control subjects in the left and right hippocampus. The analysis included 1167 patients and 1088 control subjects across a wide range of ages from pediatric to geriatric populations (McKinnon, Yucel, Nazarov, & MacQueen, 2009). Conclusions from this meta-analysis were consistent with the findings of earlier meta-analyses of hippocampal volume in patients with MDD (Campbell, Marriott, Nahmias, & MacQueen, 2004; Videbech & Ravnkilde, 2004). A meta-analysis over 226 studies found areas affected in MDD to be the hippocampus, basal ganglia, and orbitofrontal cortex, in particular the gyrus rectus (Kempton et al., 2011). The above-mentioned associations between glucocorticoids and stress and neuronal damage in the hippocampus indicate that the neurotoxic effects of glucocorticoids on the hippocampus can be visualized in terms of overall volume changes. Evidence from neuroimaging, neuropathological, and lesion analysis studies further implicates limbic-cortical-striatal-pallidal-thalamic circuits, including orbitomedial prefrontal cortex, amygdala, ventromedial striatum, mediodorsal, and midline thalamic nuclei and ventral pallidum in the pathophysiology of mood disorders (Miller, SaintMarie, Breier, & Swerdlow, 2010).

In line with the glucocorticoid cascade theory are some longitudinal imaging studies. In a longitudinal study on 30 patients with MDD and 30 healthy controls it was demonstrated that a negative clinical outcome with more relapses and a chronic course during a 3-year follow-up was associated with hippocampal, amygdala, anterior cingulate cortex, and dorsomedial prefrontal cortex volume decline (Frodl, Koutsouleris, et al., 2008). In a long-term follow-up study it was evaluated whether any possible difference in hippocampal volume and brain structure between depressed in-patients and healthy controls at inclusion disappeared over an 11-year period when the patients were in remission. At baseline, patients had smaller volumes in the right and left superior and middle temporal gyri, medulla, and body of the right hippocampus. At follow-up, there were no significant local brain differences between patients and controls. In a group of 19 patients and 19 controls who were investigated at baseline and follow-up, no significant hippocampal volume differences were detected (Ahdidan et al., 2011). While the cross-sectional parts of the study are well powered, for the longitudinal part of the study a larger sample would have been desirable, since a sample of 19 seems to not have enough power to detect small changes in brain structure. Moreover, during successful treatment brain structures like the left inferior frontal cortex, right fusiform gyrus, and right cerebellum increased in size (Lai & Hsu, 2011). On the other hand, smaller hippocampal volumes were also found to be predictive of a poor clinical outcome in 1- and 3-year follow-up studies and also for response to a course of antidepressant therapy (Frodl et al., 2004; MacQueen & Frodl, 2011). Therefore, a predisposition to depression might be associated with smaller hippocampal volumes, which might further decline during the course of a chronic depression but normalize during remission. However, this is speculative at this stage and needs to be investigated in studies with longitudinal designs.

Functional Neuroimaging

Functional magnetic resonance imaging (fMRI) studies examining neural responses to emotional stimuli in patients with major depression indicated increased responses in the amygdala, anterior cingulum (ACC), fusiform gyrus, putamen, and prefrontal cortical regions (Frodl et al., 2007; Fu et al., 2004; Surguladze et al., 2005). Since many researchers assume that the depressive syndrome might arise from abnormal interactions between brain regions, functional neuroimaging studies have examined the connectivity of the neural network. With respect to connectivity, a study in 15 unmedicated patients with major depression and 15 healthy volunteers found decreased correlations between ACC and limbic regions, which is consistent with the hypothesis that decreased cortical regulation of limbic activation in response to negative stimuli may be present in depression (Anand et al., 2005). Again the amygdala was not only negatively coupled with the ACC but also positively coupled bilaterally with medial temporal and ventral occipital regions in 19 unmedicated patients with major depression and 19 healthy volunteers (Chen et al., 2008). Studies on functional connectivity in patients with major depression receiving antidepressant medication achieved varied results. The results indicated that a neural network consisting of the cingulate region, prefrontal cortical regions, amygdala, and subcortical regions may play key roles in MDD: compared to healthy controls, patients with depression showed increased functional connectivity between the amygdala, hippocampus, and caudate–putamen regions during emotion processing (Hamilton & Gotlib, 2008) but significantly reduced amygdala–prefrontal connectivity (Dannlowski et al., 2009). Uncoupling of the prefrontal cortex and gyrus cinguli was found in 14 patients with MDD and 14 healthy controls during a verbal working memory task (Vasic, Walter, Sambataro, & Wolf, 2008).

In 25 drug-free patients with MDD as compared to 15 healthy controls, dorsal anterior cingulate cortex, precuneus, and cerebellum activity showed less connectivity with the orbitofrontal cortex (OFC) in patients than in controls, while functional connectivity between the OFC and the right dorsolateral prefrontal cortex, right inferior frontal operculum, and left motor areas was increased in patients as compared to healthy controls (Frodl, Bokde, et al., 2010).

Interest is growing in the use of resting state fMRI, which does not require the use of a task and which has become a popular means of complementing the results of task-based fMRI studies. Resting state fMRI allows for the examination of large-scale neural systems that exhibit spontaneous synchronous fluctuations during goal-directed and nongoal-directed behavior (Castellanos & Proal, 2012). These low-frequency (<0.1 Hz) spontaneous fluctuations in blood oxygenated level-dependent (BOLD) signals correlate with interactions between adjacent and nonadjacent brain areas that form spatially distributed networks of brain function (Raichle et al., 2001). Functional connectivity is the observed correlation in spontaneous neural activity between brain areas at rest (Deco & Corbetta, 2011). Several resting state studies have found increased resting state functional connectivity in the cognitive control network (Sheline et al., 2009; Zhou et al., 2010), increased connectivity in the default mode network (Grimm et al., 2009; Sheline et al., 2010; Zhou et al., 2010), and increased functional connectivity in the affective network (Sheline et al., 2010).

Resting state fMRI demonstrated that subgenual cingulate and thalamic functional connectivity were significantly increased in 20 patients with MDD as compared to 20 healthy controls (Greicius, Supekar, Menon, & Dougherty, 2009).

Diffusion Imaging in MDD

Structural neuroimaging studies in MDD show volume reductions, and functional imaging studies indicate disconnectivity in the limbic and frontal brain regions. Whether or not white matter fiber bundles between limbic and frontal brain regions are altered can be investigated with diffusion tensor imaging (DTI).

Magnetic resonance DTI is a novel neuroimaging technique that can evaluate both the orientation and the diffusion characteristics of white matter tracts in vivo (Sexton, Mackay, & Ebmeier, 2009). DTI represents the forefront of neuroimaging techniques in the characterization of microstructural alterations (Frodl, Carballedo, et al., 2012). These findings appear to be heterogeneous, and therefore we conducted a meta-analysis of seven available studies, including 188 patients with MDD and 221 healthy controls. Patients with depression showed decreased white matter fractional anisotropy (FA) values in the superior longitudinal fasciculus (SLF) and increased FA values in the fronto-occipital fasciculus as compared to controls. In conclusion, the meta-analysis revealed a significant reduction in FA values in the left SLF, which may ultimately play an important role in the pathology of depression. More research in larger samples is needed to track changes during the disease course using DTI (Murphy & Frodl, 2011). Another meta-analysis of DTI studies of patients with MDD consistently identified decreased fractional anisotropy in the white matter fascicles connecting the prefrontal cortex within the frontal, temporal, and occipital lobes as well as the amygdala and hippocampus (Liao et al., 2013).

Some studies were carried out to investigate certain clinical characteristics. With respect to remission status in MDD, findings are inconclusive. Interestingly, remitted MDD patients revealed decreased mean diffusivity and increased fractional anisotropy within the left amygdala, which may be interpreted as a greater cell density and increased number of fibers, respectively. This last notion was supported by probabilistic tractography results, which revealed increased connectivity from the left amygdala to the hippocampus, cerebellum, and brain stem in remitted as compared to nonremitted patients (Arnold et al., 2012). A study comparing 13 patients with late-life depression to matched healthy controls found a reduction in FA in both the frontal and temporal lobes of depressed patients (Nobuhara et al., 2006). In addition, an inverse relationship was established between FA values and symptom severity (Nobuhara et al., 2006).

Importantly, reductions in FA have also been associated with early life adversity in the form of disrupted maternal–infant attachment and correlate with an increased risk of both anxiety and depression (Coplan et al., 2010). A study comparing 12 maternally deprived adult male macaques to nine normally reared controls found significant reductions in FA in the anterior limb of the internal capsule in the maternally deprived macaques (Coplan et al., 2010). This is another brain region that is important in emotional regulation and is involved in the medial and the basolateral limbic circuits (Coplan et al., 2010). Thus, disruption of this region may alter functional connectivity between the frontal and temporal lobes, conferring an increased risk of MDD (Coplan et al., 2010). Another study demonstrating the microstructural implications of early life adversity found significantly reduced FA within the genu of the corpus callosum among those exposed to high levels of early life stress (Paul et al., 2008). We have demonstrated that first-degree relatives of patients with MDD who managed to stay healthy showed greater fractional anisotropy than controls in the body and splenium of the corpus callosum, inferior fronto-occipital fasciculus, superior longitudinal fasciculus, and fornix, suggesting adaptive and protective mechanisms (Frodl et al., 2012).

Imaging Genetics

In functional models of MDD, overactivity in limbic areas like the amygdala, anterior cingulate cortex, and hippocampus is not adequately controlled by prefrontal areas (Mayberg et al., 1999). A meta-analysis of functional MRI studies support the involvement of these brain regions in MDD (Graham et al., 2013). A review about the impact of genetic variations on brain function concluded that genetic variants are associated with brain function in MDD (Scharinger, Rabl, Sitte, & Pezawas, 2010). Most evidence comes from genes involved in the regulation of the 5-HT system, especially with regard to the 5-HT transporter gene (SLC6A4). For instance, short allele carriers of the promoter polymorphism of the serotonin transporter gene commonly known as 5-HTTLPR have increased amygdala reactivity to masked emotional faces in 35 patients with MDD (Dannlowski et al., 2007).

Vulnerability models suggest that within the etiology of depression, environmental (e.g., stress) factors and genetic factors are supposed to interact, thereby altering the brain structure and function. In a first step it has been shown that genetic polymorphisms of the serotonin transporter gene affect hippocampal structure (Frodl, Zill, et al., 2008) and amygdala brain function (Hariri et al., 2002). We demonstrated that a genetic polymorphism in the promoter region of the serotonin transporter gene (5-HTTLPR) interacts and influences hippocampal volumes measured with MRI (Frodl, Reinhold, Koutsouleris, Donohoe, et al., 2010). Moreover, in an independent sample, hippocampal volume changes, which are related to a depressive vulnerability due to changes in neurocognitive processes (Gerritsen et al., 2012) and plasticity (MacQueen & Frodl, 2011), are caused by interactive effects of childhood adversity and genetic variation (Everaerd et al., 2012). It was further demonstrated—on top of the well-known finding of an increased risk of developing disorders like MDD due to childhood maltreatment—that childhood maltreatment affects the brain structure with respect to regional gray matter volume (Chaney et al., 2014; Frodl, Reinhold, Koutsouleris, Reiser, & Meisenzahl, 2010) and white matter tractography (Ugwu, Amico, Carballedo, Fagan, & Frodl, 2014). As variations in the mineralocorticoid receptor gene (NR3C2) have been related to a higher risk for depression, it was investigated whether NR3C2 variance is related to negative memory bias, an established endophenotype for depression, in healthy participants (Vogel et al., 2014). Furthermore, the influence of life adversity on the association between NR3C2 and negative memory bias was demonstrated. A set-based analysis was used to simultaneously test all measured variations in NR3C2 for an association with negative memory bias in 483 participants and an interaction with life adversity. To further specify this interaction, the sample was split into low and high live adversity groups and the analyses was repeated separately in both groups. NR3C2 variance was associated with negative memory bias, especially in the high life adversity group. Additionally, a functional polymorphism (rs5534) related to negative memory bias demonstrated a gene by life adversity interaction. Variations in NR3C2 are associated with negative memory bias, and this relationship appears to be influenced by life adversity (Vogel et al., 2014).

Studies showed that genes relevant for depression, like the serotonin transporter and glucocorticoid genes, and early life stress interact and influence the development of depression and affect brain structure and function (Bermingham et al., 2012; Frodl, Reinhold, Koutsouleris, Donohoe, et al., 2010). The mechanisms through which these interactive effects seem to be mediated seem to be epigenetics like methylation of DNA. The study of epigenetics in depression is at its start. In healthy subjects, an association was found between methylation of the serotonin transporter polymorphism and brain serotonin synthesis measured with *positron emission tomography* (Wang et al., 2012). Subjects committing suicide and having a history of early life adversity had higher methylation of glucocorticoid receptor genes as compared to subjects committing suicide without a history of early life adversity. Interestingly, this higher methylation was associated with a decreased expression of glucocorticoid receptor proteins in the hippocampus (McGowan et al., 2009).

Longitudinal Imaging Changes

It is well established from cross-sectional studies that patients with MDD have impaired glucocorticoid receptor functioning, increased circulating concentrations of inflammatory cytokines and C-reactive protein (CRP), and reduced hippocampal volumes, in particular in the hippocampal subfields CA1, CA2-CA3, and subiculum (Frodl & O'Keane, 2013; MacQueen & Frodl, 2011). Experimental studies in animals demonstrate that neuroplastic changes occur during the onset of depressive-like states, indicating that dysfunction of neuronal plasticity might potentially contribute to the pathophysiology of MDD (Duman, 2002). In animal models, glucocorticoid administration induces increases in neural cell death, reductions in brain-derived neurotrophic factor (BDNF), atrophy of neuronal processes, a reduction of hippocampal neurogenesis, and ultimately a reduction in hippocampal volume (Sapolsky, 2002). On the other hand, antidepressant therapy seems to increase hippocampal neurogenesis in rats (Santarelli et al., 2003) and restore homeostasis of the inflammatory and HPA axis systems. Glucocorticoids, via a process of repression of gene transcription, inhibit the key target genes in the inflammatory system such as IL-1β, IL-6, TNF α, and cyclo-oxygenase-2 (COX-2) (De Bosscher & Haegeman, 2009).

These experimental findings are being translated into clinical investigations. The first longitudinal long-term structural neuroimaging studies in humans with MDD found that the hippocampus, prefrontal cortex, and anterior cortex cinguli decline in patients developing a chronic depression (Frodl, Koutsouleris, et al., 2008), which seems to be related to neuroplastic processes. Moreover, we demonstrated that a smaller total hippocampal volume was related to a poor clinical outcome and that hippocampal volumes increased in patients with MDD under continuous therapy with antidepressants (Frodl et al., 2004). In a pilot study, 27 patients with treatment-resistant MDD were investigated at baseline and at follow-up, when they either achieved 6 months of sustained remission or 12 months of failure to remit. The 12 patients who achieved sustained 6-month remission demonstrated a significant mean increase in whole-brain volume in the right orbitofrontal cortex and the right inferior temporal gyrus as compared to nonremitted patients (Phillips, Batten, Aldosary, Tremblay, & Blier, 2012). Similar to our longitudinal study (Frodl, Koutsouleris, et al., 2008) nonremitters showed decreased brain volumes, although here in the white matter volume in the left anterior limb of the internal capsule. However, these previous studies only investigated brain structure and had limitations of a small sample size, so novel studies are necessary that also investigate brain function and structural connectivity using DTI and explore relationships to the stress hormone system and inflammation.

A few longitudinal antidepressant trials have previously investigated BOLD-responses due to emotional stimuli processing using fMRI. We found a significant decrease of BOLD-responses in the hippocampus, basal ganglia, thalamus, and cerebellum of venlafaxine (serotonin and noradrenaline reuptake inhibitor)-treated patients and a significant increase in BOLD-responses in the middle gyrus cinguli and supplementary motor area of mirtazapine (selective serotonin and norepinephrine reuptake inhibitor)-treated patients, indicating different locations of actions for different treatment strategies (Frodl, Scheuerecker, et al., 2011). A meta-analysis summarized the results of nine fMRI studies and showed that antidepressant treatment resulted in an increase of BOLD-responses of dorsolateral, dorsomedial, and ventrolateral prefrontal cortices, whereas subcortical regions and those related to emotion processing like amygdala, hippocampus, parahippocampus, cingulate cortex, orbitofrontal cortex, insula, and precuneus decreased in BOLD-responses (Delaveau et al., 2011). However, to our knowledge longitudinal studies using fMRI investigating the long-term effects of MDD are missing.

OUTLINE OF *SYSTEMS NEUROSCIENCE IN DEPRESSION*

The book starts with brain development, early life changes, and how early life stressors alter brain function and structure. In the first chapter, D. Cichetti and colleagues examine how the principles of a multilevel developmental psychopathology systems perspective provide insight into understanding the etiology, course, and sequelae of depressive disorders. Here depression is of particular interest to developmental psychopathologists because of the complex interplay of psychological (affective, cognitive, interpersonal) and biological (genetic, neurobiological, neurophysiological) systems that are involved. The knowledge about normative development and functioning will provide a background to characterize deviations evident among those with depression. In addition, pathways to resilience are discussed that represent competent adaptation despite the experience of significant adversity, and approaches for prevention are examined.

In the second chapter, D. Pagliaccio and D. Barch focus on the role of stress in the development of major depression and highlight within an integrated view other possible factors influencing the association between stress and depression. It is well known that the experience of stressful life events, comorbidity, genetic risk, and low social support can all contribute to ineffective treatment and treatment resistance. As such, understanding the role of factors such as stress in the etiological mechanisms underlying depression may be key to improved treatment and diagnostics. In the chapter, literature on the link between stressors and depression as well as potential mediators of this relationship is reviewed. The authors discuss a mechanism implicit in much of the literature, namely that childhood stress leads to alterations in cortisol function, which in turn leads to alterations in the brain that contribute to the risk for MDD.

Ultimately this leads to a chapter (T. Hirota, G. Milavić, F. McNicholas, T. Frodl, and N. Skokauskas) reviewing knowledge about child and youth depression and the current state of system neuroscience within the discipline. The development timeline of brain regions is discussed in terms of structural and functional brain changes in children and youth with depression. There is a clear trend that despite a growing body of research into the neurobiological correlates of depression, the evidence in children and adolescents is limited to the extent that no firm conclusions may yet be drawn.

Three chapters explore the underlying molecular backgrounds including genetics and epigenetics of depression. It is well investigated that long-term stress exposure leads to dendritic atrophy and atypical synaptic connections in regions controlling emotions, like the amygdala, prefrontal cortex, and hippocampus, and contributes to the development of depression.

A. Varidaki, H. Mohammad, and E.T. Coffey describe important findings from animal models of depression and postmortem tissues of subjects with mood disorders. They highlight how deregulated BDNF and NMDA disturb cellular signaling and dendritic spine plasticity in depression. Downstream effectors, including small GTPases and synaptic proteins, and the mechanisms whereby they regulate spine plasticity are reviewed. Also, the function of protein kinases LIMK, mTOR, MAPK, CamKII, SGK1, and GSK-3 in the pathophysiology of depression is discussed.

The book then moves toward an understanding of molecular and biological alterations affecting neural systems in depression and its rational toward developing new targets for antidepressant treatment. Here, E. O'Toole, M.V. Doucet, E. Sherwin, and A. Harkin review the glutamatergic system and its relevance for MDD. The NMDA-R antagonist ketamine has repeatedly shown rapid and sustained antidepressant effects following a single dose in a number of clinical studies. Since the adverse effects associated with ketamine limit its wider use in the clinic, investigations are instead aimed at developing "cleaner" ketamine compounds that could demonstrate a fast onset of action and sustained efficacy, without incurring adverse effects. In this regard, targets located downstream of the NMDA-R, such as nitric oxide synthase, may represent novel avenues for treating neuropsychiatric disorders. Here, the authors review the rationale for the development of glutamate and nitric oxide-modulating drugs for treating depression and highlight findings in the field, which provide a basis for ongoing drug development.

Since MDD is a heterogeneous disorder, in which many neurobiological systems have been shown to be involved, it is important to also consider genetic and environmental factors together that are underlying physiological mechanisms. In the chapter by M. Levesque, M. Szyf, and L. Booij, the gene by environment interaction model is described, and evidence that early adverse exposures are associated with epigenetic alterations in specific physiological systems relevant to MDD, including the serotonin system, the HPA axis, BDNF, and the immune system, is outlined. The authors also discuss the stability of epigenetic mechanisms and the methodological challenges encountered when studying epigenetic mechanisms in humans.

The inflammatory system was also found to be altered in patients with MDD as compared to healthy controls. So, high levels of several proinflammatory components of the immune system such as IL-6, CRP, TNF-α, or neopterin in patients suffering from MDD point to the involvement of an inflammatory process in the pathophysiology of MDD. The chapter by N. Müller discusses the direct and indirect effects of cytokines on the neurotransmitter storage and release mediated by microglia cells and astrocytes. It also highlights the importance of tryptophan/kynurenine metabolism and its actions on the glutamatergic neurotransmission. Some first studies showing the therapeutic benefits of anti-inflammatory medications, such as the COX-2 inhibitors and TNF-α antagonists, and the anti-inflammatory and immunomodulatory intrinsic effects of antidepressants are reviewed as well.

The next three chapters highlight the importance of neuroimaging in depression research. Advances in neuroimaging allow us to investigate depression from the perspective of large-scale brain networks, their structures, and their complex interactions.

P.C. Mulders, P.F. van Eijndhoven, and C.F. Beckmann describe different methods of unraveling the functional connections in the brain. This chapter provides an understanding of the associated tools and techniques for network identification that are crucial to anyone with an interest in depression's underlying pathophysiology at the systems level. The authors discuss the most common methods for network characterization, the methodological background, and a summary of the current understanding of network dysfunction in depression.

C. Young, B. Mueller, and I. Tendolkar highlight the relationship between common symptoms of MDD with activity and connectivity of neural systems. This chapter deals with backward translation from clinically and neuropsychologically classified functional systems to neural networks. They link (1) mood-congruent processing, (2) attention biases toward negative stimuli, (3) memory biases for negative material, (4) long-term memory deficits, (5) rumination, and (6) anhedonia to their associated brain networks.

T. Frodl and V. O'Keane summarize the associations between brain systems measured with imaging and biological markers of stress in MDD. This chapter highlights the associations between brain structure and function as well as HPA axis dysregulation. The role of inflammatory markers within these associations is further discussed, and changes in neurotrophic and the neurotransmitter systems such as the serotonergic system are described. Moreover, integrated models considering biological and environmental factors are developed.

K.J. Manning and D.C. Steffens provide an individual overview of the cognitive control, default mode, and salience networks as observed in late-life depression (LLD) and discuss how dysfunction in multiple networks contributes to common behavioral syndromes in LLD, including apathy, anxious-depression, and suicidality. Three prominent functional networks in LLD have been identified: a cognitive control network, active during cognitively demanding tasks; a default mode network, active during resting state and inhibited during cognitively demanding activity that is involved in internal mentation; and a salience network, relevant to attending to survival-relevant events in the environment.

The brain arousal system fundamentally impacts human behavior, and arousal has been implemented as a basic dimension of mental diseases in the Research Domain Criteria project of the National Institute of Mental Health. Brain arousal can be investigated using electroencephalography (EEG) techniques. The chapter by U. Hegerl, C. Sander, and T. Hensch provides an overview of brain arousal regulation in affective disorders. Initially, theoretical models and common means of assessments will be discussed, and the Vigilance Algorithm Leipzig, an EEG-based assessment of brain arousal regulation, will be presented. Afterward, the arousal regulation model will be described in detail, which explains several clinical phenomena and predicts the response to treatment based on the respective brain arousal disturbance.

REFERENCES

Ahdidan, J., Hviid, L. B., Chakravarty, M. M., Ravnkilde, B., Rosenberg, R., Rodell, A., … Videbech, P. (2011). Longitudinal MR study of brain structure and hippocampus volume in major depressive disorder. *Acta Psychiatrica Scandinavica, 123*, 211–219.

Anand, A., Li, Y., Wang, Y., Wu, J., Gao, S., Bukhari, L., … Lowe, M. J. (2005). Activity and connectivity of brain mood regulating circuit in depression: a functional magnetic resonance study. *Biological Psychiatry, 57*, 1079–1088.

Arnold, J. F., Zwiers, M. P., Fitzgerald, D. A., Van Eijndhoven, P., Becker, E. S., Rinck, M., … Tendolkar, I. (2012). Fronto-limbic microstructure and structural connectivity in remission from major depression. *Psychiatry Research, 204*, 40–48.

Berman, R. M., Cappiello, A., Anand, A., Oren, D. A., Heninger, G. R., Charney, D. S., & Krystal, J. H. (2000). Antidepressant effects of ketamine in depressed patients. *Biological Psychiatry, 47*, 351–354.

Bermingham, R., Carballedo, A., Lisiecka, D., Fagan, A., Morris, D., Fahey, C., … Frodl, T. (2012). Effect of genetic variant in BICC1 on functional and structural brain changes in depression. *Neuropsychopharmacology, 37*, 2855–2862.

Campbell, S., Marriott, M., Nahmias, C., & MacQueen, G. M. (2004). Lower hippocampal volume in patients suffering from depression: a meta-analysis. *American Journal of Psychiatry, 161*, 598–607.

Carlsson, A., Corrodi, H., Fuxe, K., & Hokfelt, T. (1969). Effect of antidepressant drugs on the depletion of intraneuronal brain 5-hydroxytryptamine stores caused by 4-methyl-alpha-ethyl-meta-tyramine. *European Journal of Pharmacology, 5*, 357–366.

Castellanos, F. X., & Proal, E. (2012). Large-scale brain systems in ADHD: beyond the prefrontal-striatal model. *Trends in Cognitive Sciences, 16*, 17–26.

Chaney, A., Carballedo, A., Amico, F., Fagan, A., Skokauskas, N., Meaney, J., & Frodl, T. (2014). Effect of childhood maltreatment on brain structure in adult patients with major depressive disorder and healthy participants. *Journal of Psychiatry and Neuroscience, 39*, 50–59.

Chen, C. H., Suckling, J., Ooi, C., Fu, C. H., Williams, S. C., Walsh, N. D., … Bullmore, E. (2008). Functional coupling of the amygdala in depressed patients treated with antidepressant medication. *Neuropsychopharmacology, 33*, 1909–1918.

Coplan, J., Abdallah, C., Tang, C., Mathew, S., Martinez, J., Hof, P., … Pantol, G. (2010). The role of early life stress in development of the anterior limb of the internal capsule in non-human primates. *Neuroscience Letters, 480*(2), 93–96.

Dannlowski, U., Ohrmann, P., Bauer, J., Deckert, J., Hohoff, C., Kugel, H., … Baune, B. T. (2007). 5-HTTLPR biases amygdala activity in response to masked facial expressions in major depression. *Neuropsychopharmacology, 33*, 418–424.

Dannlowski, U., Ohrmann, P., Konrad, C., Domschke, K., Bauer, J., Kugel, H., … Suslow, T. (2009). Reduced amygdala-prefrontal coupling in major depression: association with MAOA genotype and illness severity. *International Journal of Neuropsychopharmacology, 12*, 11–22.

Dantzer, R., O'connor, J. C., Freund, G. G., Johnson, R. W., & Kelley, K. W. (2008). From inflammation to sickness and depression: when the immune system subjugates the brain. *Nature Reviews Neuroscience, 9*, 46–56.

De Bosscher, K., & Haegeman, G. (2009). Minireview: latest perspectives on antiinflammatory actions of glucocorticoids. *Molecular Endocrinology, 23*, 281–291.

Deco, G., & Corbetta, M. (2011). The dynamical balance of the brain at rest. *Neuroscientist, 17*, 107–123.

Delaveau, P., Jabourian, M., Lemogne, C., Guionnet, S., Bergouignan, L., & Fossati, P. (2011). Brain effects of antidepressants in major depression: a meta-analysis of emotional processing studies. *Journal of Affective Disorders, 130*, 66–74.

Draganski, B., Gaser, C., Kempermann, G., Kuhn, H. G., Winkler, J., Buchel, C., & May, A. (2006). Temporal and spatial dynamics of brain structure changes during extensive learning. *Journal of Neuroscience, 26*, 6314–6317.

Drevets, W. C., Zarate, C. A., Jr., & Furey, M. L. (2013). Antidepressant effects of the muscarinic cholinergic receptor antagonist scopolamine: a review. *Biological Psychiatry, 73*, 1156–1163.

Duman, R. S. (2002). Pathophysiology of depression: the concept of synaptic plasticity. *European Psychiatry, 17*(Suppl. 3), 306–310.

Duman, R. S. (2014). Pathophysiology of depression and innovative treatments: remodeling glutamatergic synaptic connections. *Dialogues in Clinical Neuroscience, 16*, 11–27.

Eichenbaum, H., & Harris, K. (2000). Toying with memory in the hippocampus. *Nature Neuroscience, 3*, 205–206.

Everaerd, D., Gerritsen, L., Rijpkema, M., Frodl, T., Van Oostrom, I., Franke, B., … Tendolkar, I. (2012). Sex modulates the interactive effect of the serotonin transporter gene polymorphism and childhood adversity on hippocampal volume. *Neuropsychopharmacology, 37*, 1848–1855.

Feng, R., Rampon, C., Tang, Y. P., Shrom, D., Jin, J., Kyin, M., … Tsien, J. Z. (2001). Deficient neurogenesis in forebrain-specific presenilin-1 knockout mice is associated with reduced clearance of hippocampal memory traces. *Neuron, 32*, 911–926.

Forrest, C. M., Mackay, G. M., Stoy, N., Egerton, M., Christofides, J., Stone, T. W., & Darlington, L. G. (2004). Tryptophan loading induces oxidative stress. *Free Radical Research, 38*, 1167–1171.

Frodl, T., Bokde, A. L., Scheuerecker, J., Lisiecka, D., Schoepf, V., Hampel, H., … Meisenzahl, E. (2010). Functional connectivity bias of the orbitofrontal cortex in drug-free patients with major depression. *Biological Psychiatry, 67*, 161–167.

Frodl, T., Carballedo, A., Fagan, A. J., Lisiecka, D., Ferguson, Y., & Meaney, J. F. (2012). Effects of early-life adversity on white matter diffusivity changes in patients at risk for major depression. *Journal of Psychiatry and Neuroscience, 37*, 37–45.

Frodl, T. S., Koutsouleris, N., Bottlender, R., Born, C., Jäger, M., Scupin, I., … Meisenzahl, E. M. (2008). Depression-related variation in brain morphology over 3 years: effects of stress? *Archives of General Psychiatry, 65*, 1156–1165.

Frodl, T., Meisenzahl, E. M., Zetzsche, T., Hohne, T., Banac, S., Schorr, C., … Moller, H. J. (2004). Hippocampal and amygdala changes in patients with major depressive disorder and healthy controls during a 1-year follow-up. *Journal of Clinical Psychiatry, 65*, 492–499.

Frodl, T., & O'Keane, V. (2013). How does the brain deal with cumulative stress? A review with focus on developmental stress, HPA axis function and hippocampal structure in humans. *Neurobiology of Disease, 52*, 24–37.

Frodl, T., Reinhold, E., Koutsouleris, N., Donohoe, G., Bondy, B., Reiser, M., … Meisenzahl, E. M. (2010). Childhood stress, serotonin transporter gene and brain structures in major depression. *Neuropsychopharmacology, 35*, 1383–1390.

Frodl, T., Reinhold, E., Koutsouleris, N., Reiser, M., & Meisenzahl, E. M. (2010). Interaction of childhood stress with hippocampus and prefrontal cortex volume reduction in major depression. *Journal of Psychiatric Research, 44*, 799–807.

Frodl, T., Scheuerecker, J., Albrecht, J., Kleemann, A. M., Muller-Schunk, S., Koutsouleris, N., … Meisenzahl, E. (2007). Neuronal correlates of emotional processing in patients with major depression. *World Journal of Biological Psychiatry*, 1–7.

Frodl, T., Scheuerecker, J., Schoepf, V., Linn, J., Koutsouleris, N., Bokde, A. L., ... Meisenzahl, E. (2011). Different effects of mirtazapine and venlafaxine on brain activation: an open randomized controlled fMRI study. *Journal of Clinical Psychiatry, 72*, 448–457.

Frodl, T., Zill, P., Baghai, T., Schule, C., Rupprecht, R., Zetzsche, T., ... Meisenzahl, E. M. (2008). Reduced hippocampal volumes associated with the long variant of the tri- and diallelic serotonin transporter polymorphism in major depression. *American Journal of Medical Genetics Part B: Neuropsychiatric Genetics, 147B*, 1003–1007.

Fu, C. H., Williams, S. C., Cleare, A. J., Brammer, M. J., Walsh, N. D., Kim, J., ... Bullmore, E. T. (2004). Attenuation of the neural response to sad faces in major depression by antidepressant treatment: a prospective, event-related functional magnetic resonance imaging study. *Archives of General Psychiatry, 61*, 877–889.

Gerritsen, L., Tendolkar, I., Franke, B., Vasquez, A. A., Kooijman, S., Buitelaar, J., ... Rijpkema, M. (2012). BDNF Val66Met genotype modulates the effect of childhood adversity on subgenual anterior cingulate cortex volume in healthy subjects. *Molecular Psychiatry, 17*, 597–603.

Goetzel, R. Z., Hawkins, K., Ozminkowski, R. J., & Wang, S. (2003). The health and productivity cost burden of the "top 10" physical and mental health conditions affecting six large U.S. employers in 1999. *Journal of Occupational and Environmental Medicine, 45*, 5–14.

Graham, J., Salimi-Khorshidi, G., Hagan, C., Walsh, N., Goodyer, I., Lennox, B., & Suckling, J. (2013). Meta-analytic evidence for neuroimaging models of depression: state or trait? *Journal of Affective Disorders, 151*, 423–431.

Greenberg, P. E., Kessler, R. C., Birnbaum, H. G., Leong, S. A., Lowe, S. W., Berglund, P. A., & Corey-Lisle, P. K. (2003). The economic burden of depression in the United States: how did it change between 1990 and 2000? *Journal of Clinical Psychiatry, 64*, 1465–1475.

Greicius, M. D., Supekar, K., Menon, V., & Dougherty, R. F. (2009). Resting-state functional connectivity reflects structural connectivity in the default mode network. *Cerebral Cortex, 19*, 72–78.

Grimm, S., Boesiger, P., Beck, J., Schuepbach, D., Bermpohl, F., Walter, M., ... Northoff, G. (2009). Altered negative BOLD responses in the default-mode network during emotion processing in depressed subjects. *Neuropsychopharmacology, 34*, 932–943.

Hamilton, J. P., & Gotlib, I. H. (2008). Neural substrates of increased memory sensitivity for negative stimuli in major depression. *Biological Psychiatry, 63*, 1155–1162.

Hariri, A. R., Mattay, V. S., Tessitore, A., Kolachana, B., Fera, F., Goldman, D., ... Weinberger, D. R. (2002). Serotonin transporter genetic variation and the response of the human amygdala. *Science, 297*, 400–403.

Ikemoto, S., & Panksepp, J. (1999). The role of nucleus accumbens dopamine in motivated behavior: a unifying interpretation with special reference to reward-seeking. *Brain Research. Brain Research Reviews, 31*, 6–41.

Ilyas, S., & Moncrieff, J. (2012). Trends in prescriptions and costs of drugs for mental disorders in England, 1998-2010. *British Journal of Psychiatry, 200*, 393–398.

Jamison, K. R. (2000). Suicide and bipolar disorder. *Journal of Clinical Psychiatry, 61*(Suppl. 9), 47–51.

Jhamandas, K. H., Boegman, R. J., Beninger, R. J., Miranda, A. F., & Lipic, K. A. (2000). Excitotoxicity of quinolinic acid: modulation by endogenous antagonists. *Neurotoxicity Research, 2*, 139–155.

Kempton, M. J., Salvador, Z., Munafo, M. R., Geddes, J. R., Simmons, A., Frangou, S., & Williams, S. C. (2011). Structural neuroimaging studies in major depressive disorder. Meta-analysis and comparison with bipolar disorder. *Archives of General Psychiatry, 68*, 675–690.

Kole, M. H., Czeh, B., & Fuchs, E. (2004). Homeostatic maintenance in excitability of tree shrew hippocampal CA3 pyramidal neurons after chronic stress. *Hippocampus, 14*, 742–751.

Krishnan, V., & Nestler, E. J. (2008). The molecular neurobiology of depression. *Nature, 455*, 894–902.

Lai, C. H., & Hsu, Y. Y. (2011). A subtle grey-matter increase in first-episode, drug-naive major depressive disorder with panic disorder after 6 weeks' duloxetine therapy. *International Journal of Neuropsychopharmacology, 14*, 225–235.

Liao, Y., Huang, X., Wu, Q., Yang, C., Kuang, W., Du, M., ... Gong, Q. (2013). Is depression a disconnection syndrome? Meta-analysis of diffusion tensor imaging studies in patients with MDD. *Journal of Psychiatry and Neuroscience, 38*, 49–56.

Lisman, J. E. (1999). Relating hippocampal circuitry to function: recall of memory sequences by reciprocal dentate-CA3 interactions. *Neuron, 22*, 233–242.

MacQueen, G., & Frodl, T. (2011). The hippocampus in major depression: evidence for the convergence of the bench and bedside in psychiatric research? *Molecular Psychiatry, 16*, 252–264.

Magarinos, A. M., McEwen, B. S., Flugge, G., & Fuchs, E. (1996). Chronic psychosocial stress causes apical dendritic atrophy of hippocampal CA3 pyramidal neurons in subordinate tree shrews. *Journal of Neuroscience, 16*, 3534–3540.

Mayberg, H. S., Liotti, M., Brannan, S. K., McGinnis, S., Mahurin, R. K., Jerabek, P. A., ... Fox, P. T. (1999). Reciprocal limbic-cortical function and negative mood: converging PET findings in depression and normal sadness. *American Journal of Psychiatry, 156*, 675–682.

McGowan, P. O., Sasaki, A., D'alessio, A. C., Dymov, S., Labonte, B., Szyf, M., ... Meaney, M. J. (2009). Epigenetic regulation of the glucocorticoid receptor in human brain associates with childhood abuse. *Nature Neuroscience, 12*, 342–348.

McKinnon, M. C., Yucel, K., Nazarov, A., & MacQueen, G. M. (2009). A meta-analysis examining clinical predictors of hippocampal volume in patients with major depressive disorder. *Journal of Psychiatry and Neuroscience, 34*, 41–54.

McNally, L., Bhagwagar, Z., & Hannestad, J. (2008). Inflammation, glutamate, and glia in depression: a literature review. *CNS Spectrums, 13*, 501–510.

Miller, E. J., Saint Marie, L. R., Breier, M. R., & Swerdlow, N. R. (2010). Pathways from the ventral hippocampus and caudal amygdala to forebrain regions that regulate sensorimotor gating in the rat. *Neuroscience, 165*, 601–611.

Murphy, M. L., & Frodl, T. (2011). Meta-analysis of diffusion tensor imaging studies shows altered fractional anisotropy occurring in distinct brain areas in association with depression. *Biology of Mood and Anxiety Disorders, 1*, 3.

Murray, C. J., & Lopez, A. D. (1996). Evidence-based health policy—lessons from the Global Burden of Disease Study. *Science, 274*, 740–743.

Myint, A. M., & Kim, Y. K. (2014). Network beyond IDO in psychiatric disorders: revisiting neurodegeneration hypothesis. *Progress in Neuropsychopharmacology and Biological Psychiatry, 48*, 304–313.

Myint, A. M., Schwarz, M. J., & Muller, N. (2012). The role of the kynurenine metabolism in major depression. *Journal of Neural Transmission, 119*, 245–251.

Nobuhara, K., Okugawa, G., Sugimoto, T., Minami, T., Tamagaki, C., Takase, K., ... Kinoshita, T. (2006). Frontal white matter anisotropy and symptom severity of late-life depression: a magnetic resonance diffusion tensor imaging study. *Journal of Neurology Neurosurgery and Psychiatry, 77*, 120.

Olesen, P. J., Westerberg, H., & Klingberg, T. (2004). Increased prefrontal and parietal activity after training of working memory. *Nature Neuroscience, 7*, 75–79.

Oxenkrug, G. F. (2007). Genetic and hormonal regulation of tryptophan kynurenine metabolism: implications for vascular cognitive impairment, major depressive disorder, and aging. *Annals of the New York Academy of Sciences, 1122*, 35–49.

Paul, R., Henry, L., Grieve, S., Guilmette, T., Niaura, R., Bryant, R., ... Cohen, R. (2008). The relationship between early life stress and microstructural integrity of the corpus callosum in a non-clinical population. *Neuropsychiatric Disease and Treatment, 4*, 193.

Pehrson, A. L., & Sanchez, C. (2015). Altered gamma-aminobutyric acid neurotransmission in major depressive disorder: a critical review of the supporting evidence and the influence of serotonergic antidepressants. *Drug Design, Development and Therapy, 9*, 603–624.

Phillips, J. L., Batten, L. A., Aldosary, F., Tremblay, P., & Blier, P. (2012). Brain-volume increase with sustained remission in patients with treatment-resistant unipolar depression. *Journal of Clinical Psychiatry, 73*, 625–631.

Raichle, M. E., Macleod, A. M., Snyder, A. Z., Powers, W. J., Gusnard, D. A., & Shulman, G. L. (2001). A default mode of brain function. *Proceedings of the National Academy of Sciences of the United States of America, 98*, 676–682.

Ressler, K. J., & Nemeroff, C. B. (2000). Role of serotonergic and noradrenergic systems in the pathophysiology of depression and anxiety disorders. *Depression and Anxiety, 12*(Suppl. 1), 2–19.

Santarelli, L., Saxe, M., Gross, C., Surget, A., Battaglia, F., Dulawa, S., ... Hen, R. (2003). Requirement of hippocampal neurogenesis for the behavioral effects of antidepressants. *Science, 301*, 805–809.

Sapolsky, R. M. (2002). Chickens, eggs and hippocampal atrophy. *Nature Neuroscience, 5*, 1111–1113.

Scharinger, C., Rabl, U., Sitte, H. H., & Pezawas, L. (2010). Imaging genetics of mood disorders. *NeuroImage, 53*, 810–821.

Sexton, C., Mackay, C., & Ebmeier, K. (2009). A systematic review of diffusion tensor imaging studies in affective disorders. *Biological Psychiatry, 66*, 814–823.

Sheline, Y. I., Barch, D. M., Price, J. L., Rundle, M. M., Vaishnavi, S. N., Snyder, A. Z., ... Raichle, M. E. (2009). The default mode network and self-referential processes in depression. *Proceedings of the National Academy of Sciences of the United States of America, 106*, 1942–1947.

Sheline, Y. I., Price, J. L., Yan, Z., & Mintun, M. A. (2010). Resting-state functional MRI in depression unmasks increased connectivity between networks via the dorsal nexus. *Proceedings of the National Academy of Sciences of the United States of America, 107*, 11020–11025.

Surguladze, S., Brammer, M. J., Keedwell, P., Giampietro, V., Young, A. W., Travis, M. J., ... Phillips, M. L. (2005). A differential pattern of neural response toward sad versus happy facial expressions in major depressive disorder. *Biological Psychiatry, 57*, 201–209.

Thomas, C. M., & Morris, S. (2003). Cost of depression among adults in England in 2000. *British Journal of Psychiatry, 183*, 514–519.

Thomas, S. R., & Stocker, R. (1999). Redox reactions related to indoleamine 2,3-dioxygenase and tryptophan metabolism along the kynurenine pathway. *Redox Report, 4*, 199–220.

Ugwu, I. D., Amico, F., Carballedo, A., Fagan, A. J., & Frodl, T. (2014). Childhood adversity, depression, age and gender effects on white matter microstructure: a DTI study. *Brain Structure and Function, 220*(4).

Vasic, N., Walter, H., Sambataro, F., & Wolf, R. C. (2008). Aberrant functional connectivity of dorsolateral prefrontal and cingulate networks in patients with major depression during working memory processing. *Psychologie Medicale*, 1–11.

Videbech, P., & Ravnkilde, B. (2004). Hippocampal volume and depression: a meta-analysis of MRI studies. *American Journal of Psychiatry, 161*, 1957–1966.

Vogel, S., Gerritsen, L., Van Oostrom, I., Arias-Vasquez, A., Rijpkema, M., Joels, M., ... Fernandez, G. (2014). Linking genetic variants of the mineralocorticoid receptor and negative memory bias: interaction with prior life adversity. *Psychoneuroendocrinology, 40*, 181–190.

Wang, D., Szyf, M., Benkelfat, C., Provencal, N., Turecki, G., Caramaschi, D., ... Booij, L. (2012). Peripheral SLC6A4 DNA methylation is associated with in vivo measures of human brain serotonin synthesis and childhood physical aggression. *PLoS One, 7*, e39501.

Wellman, C. L. (2001). Dendritic reorganization in pyramidal neurons in medial prefrontal cortex after chronic corticosterone administration. *Journal of Neurobiology, 49*, 245–253.

Woolley, C. S., Gould, E., Frankfurt, M., & McEwen, B. S. (1990). Naturally occurring fluctuation in dendritic spine density on adult hippocampal pyramidal neurons. *Journal of Neuroscience, 10*, 4035–4039.

Zhou, Y., Yu, C., Zheng, H., Liu, Y., Song, M., Qin, W., ... Jiang, T. (2010). Increased neural resources recruitment in the intrinsic organization in major depression. *Journal of Affective Disorders, 121*, 220–230.

Part I

Developmental Aspects

Chapter 1

A Multilevel Developmental Psychopathology Systems Perspective on Depression

Dante Cicchetti
Institute of Child Development, University of Minnesota, Minneapolis, MN, USA

In this chapter, it is proposed that the principles inherent to a developmental psychopathology perspective can serve to elucidate the understanding of a systems approach to depression across the life course. A developmental psychopathology viewpoint espouses the conviction that comprehending the genesis (i.e., origins) and epigenesis (i.e., the development of new, different competencies across each stage of the life span) of depressive disorders in their full complexity necessitates that we possess a full comprehension of the organization and integration of diverse biological, psychological, and social systems at multiple levels of analysis within individuals across different contexts and varying developmental periods (Cicchetti & Blender, 2006; Cicchetti, Rogosch, & Toth, 1997; Cicchetti & Schneider-Rosen, 1986; Cicchetti & Toth, 1995).

Developmental psychopathology represents a movement toward comprehending the causes and determinants, pathways, sequelae, and prevention and treatment of mental disorders through its synthesis of knowledge from multiple disciplines. The undergirding developmental orientation impels researchers to pose new questions about the phenomena they study. For example, with regard to depression, it becomes necessary to move beyond identifying features that differentiate children, adolescents, and adults who have and who do not have depression (e.g., affect dysregulation; attributional distortions) to articulating how such differences have evolved developmentally within a multilevel and dynamic social ecology. Likewise, rather than being concerned with merely describing the symptoms of depressed children, adolescents, and adults, the emphasis shifts to ascertain how similar and different biological and psychological organizations contribute to the expression of depressive outcomes at each specific developmental level. Because psychopathology unfolds over time

Systems Neuroscience in Depression. http://dx.doi.org/10.1016/B978-0-12-802456-0.00001-7

in a dynamically developing organism, the adoption of a developmental perspective is critical in order to comprehend the processes underlying individual pathways to adaptive and maladaptive outcomes in persons with depressive disorders.

Although abnormalities in the broad domains of genetics, neurobiology, cognition, emotion, and interpersonal relations are present to varying degrees among individuals with depression, these diverse areas do not exist in isolation. Rather, they are complexly interrelated and mutually interdependent. Consequently, it is essential for researchers to strive to comprehend the interrelations among the biological, psychological, and social systems in order to delineate the nature of depression, including the discovery of ways in which the organization and integration of these systems may promote resilient functioning (Charney, 2004; Curtis & Cicchetti, 2003). Relatedly, because there are myriad risk factors associated with depression and its comorbid forms of psychopathology, it is critical for researchers and clinicians to acquire a firm grasp of the multilevel biological and psychological processes and mechanisms that contribute to the emergences, maintenance, and recurrence of depression. Because of the continuities and divergences from normal functioning that are manifested in depression, empirical research on pathways to depression as well as prospective longitudinal investigations of its developmental course and sequelae hold promise for advancing understanding of the relation between normality and psychopathology.

In this chapter, I begin by explicating why a developmental psychopathology perspective can be usefully applied toward enhancing the understanding of depression. Next, I discuss the parameters of developmental psychopathology, including the core principles of the discipline. Throughout, aspects of a developmental psychopathology approach that are especially relevant to the prevention and treatment of depression are highlighted.

WHAT IS DEVELOPMENTAL PSYCHOPATHOLOGY?

The integrative nature of a developmental approach to psychopathology was articulated by Eisenberg (1977), who stated that development "constitutes the crucial link between genetic determinants and environmental variables and between psychogenic causes" (p. 225). Development thus encompasses "not only the roots of behavior in prior maturation as well as the residual of earlier stimulation, both internal and external, but also the modulations of that behavior by the social fields of the experienced present" (p. 225). Not surprisingly, given the intimate link between the study of normality and psychopathology, similar depictions of normative developmental processes have been espoused in the literature.

Whereas the term "developmental psychopathology" has frequently been equated with the study of mental disorders among children and youths, this perspective encompasses a much broader approach to studying development, normal and abnormal, across the life span (Cicchetti, 1990, 1993). A developmental analysis is necessary for tracing the roots, etiology, and nature of maladaptation so that interventions may be sensitively timed and guided as well as developmentally appropriate (Toth & Cicchetti, 1999). Moreover, a developmental perspective will prove useful for uncovering the compensatory mechanisms, both biological and psychological, that may be used in the face of significant adversity (Cicchetti & Curtis, 2006).

Developmental psychopathology is an integrative scientific discipline that strives to unify, within a life span framework, contributions from multiple fields of inquiry with the goal of understanding the mutual interplay between psychopathology and normative adaptation (Cicchetti, 1990, 1993; Sroufe, 1990). A developmental analysis presupposes change and novelty, highlights the critical role of timing in the organization of behavior, underscores multiple determinants, and cautions against expecting invariant relations between causes and outcomes. A developmental analysis is as applicable to the study of the gene or cell as it is to the investigation of the individual, family, or society (Cicchetti & Pogge-Hesse, 1982; Werner & Kaplan, 1963).

Developmental psychopathologists seek to engage in a comprehensive evaluation of biological, psychological, and social processes and to ascertain how the transaction among these multiple levels of analysis may influence individual differences, the continuity or discontinuity of adaptive or maladaptive behavioral patterns, and the pathways by which normal and pathological outcomes may be achieved (Cicchetti & Schneider-Rosen, 1986). In practice, this entails comprehension of and appreciation for the developmental transformations and reorganizations that occur over time; an analysis of the risk and protective factors and mechanisms operating within and outside the individual and his or her environment over the course of development; the investigation of how emergent functions, competencies, and developmental tasks modify the expression of a disorder or lead to new symptoms and difficulties; and the recognition that a particular stressor or set of stressful circumstances may eventuate in different biological and psychological difficulties, depending on when in the developmental period the stress occurs.

In one of the initial statements concerning the goals of developmental psychopathology, Cicchetti (1990) remarked, "Developmental psychopathology should bridge fields of study, span the life cycle, and aid in the discovery of important new truths about the processes underlying adaptation and maladaptation, as well as the best means of preventing or ameliorating psychopathology." Cicchetti further commented, "This discipline should contribute greatly to reducing the dualisms that exist between the clinical study of and research into childhood and adult disorders, between the behavioral and biological sciences, between developmental psychology and psychopathology, and between basic and applied science" (p. 20). More recently, Kaplan, Spittel, and Spotts (2013) asserted that scientific inquiry could be advanced by blurring research boundaries between scientific disciplines.

The field of developmental psychopathology transcends traditional disciplinary boundaries and provides fertile ground for moving beyond descriptive facts to process-level understanding of adaptive and maladaptive as well as normal and abnormal trajectories of individual development. Research conducted within a developmental psychopathology framework may challenge assumptions about what constitutes health or pathology and may redefine the manner in which the mental health community operationalizes, assesses, classifies, communicates about, and treats the adjustment problems and functional impairments of infants, children, adolescents, and adults (Cicchetti & Toth, 1998). Thus, one of developmental psychopathology's potential contributions lies in the heuristic power it holds for translating facts into knowledge, understanding, and practical application (Cicchetti & Toth, 2000, 2006). Accordingly, such a developmental perspective may aid in the prevention and reduction of the individual and societal burden of mental disorder; alleviate the onus of suffering that mental illness engenders in individuals, their families, and the communities in which they reside; and contribute toward eliminating the stigma commonly associated with the presence of mental disorder.

DEVELOPMENTAL ANALYSIS

There are two interrelated goals inherent to a developmental analysis. First, a developmental analysis strives to investigate the specific evolving biological and psychological systems that are characteristic of individuals at varying developmental stages across the life span. This requires formulating questions about a phenomenon in terms of what capacities are characteristic of an individual during a particular developmental period and how a given process or mechanism becomes manifested in view of those developmental capacities and attainments of the individual. Age-appropriate limitations in children's cognitive, emotional, and social development as well as in neurobiological and hormonal development may make the expression of specific depressive symptoms beyond their capabilities. Thus, the delineation of those characteristics relevant to the overt manifestation of depression at different ages can probably only be accomplished by means of longitudinal prospective studies that measure skills and capacities in a variety of biological and psychological domains. Consequently, to comprehend depression, researchers must consider developmental variations in cognitive, social cognitive, and emotional capacities, in addition to other psychological and biological domains of functioning, to ascertain how particular outcomes—normal, psychopathological, or resilient—are exhibited during varying periods of development. One would not predict that developmental variations in internal cognitive structures would enable individuals with depression at different ages to use similar strategies to interpret, express, or defend against their affective experiences or internal emotional states. Likewise, cognitive difficulties associated with depression can lead to impairments in regulatory processes that affect and are affected by attention networks and executive functions. Thus, a developmental analysis is needed to highlight the processes most likely to contribute to vulnerabilities or strengths at each developmental level in persons with depression.

Second, a developmental analysis seeks to examine the prior sequences of adaptation or maladaptation in development that have contributed to an outcome in a particular developmental period. In order to achieve this goal, it is essential that the current status of any individual's functioning be examined in the context of how that status was attained across the course of development. For example, given the multiplicity of biological and psychological processes affected by depression, directing attention to examining early developmental functioning (i.e., prior development) that may be theoretically related to later appearing depression organizations may prove to be very fruitful (Cicchetti & Sroufe, 2000; Cicchetti & Tucker, 1994). Accordingly, to obtain an understanding of the abnormalities in emotion regulation, difficulties in interpersonal relations, or the core negative attributions about the self that often exist in depression, researchers may begin by investigating the early development of these features, their developmental course, and their interrelations with other psychological and biological systems of the individual.

NORMAL AND ABNORMAL DEVELOPMENT

The field of developmental psychopathology is concerned with expanding its knowledge base by focusing on the extremes of adaptation and nonnormative processes of development rather than on central tendencies and uniformities in normative processes of growth and development emphasized in classic developmental psychology. As such, developmental psychopathology underscores and highlights the dialectic between normal and abnormal development (Cicchetti, 1984; Cicchetti & Toth, 2009; Rutter, 1986; Rutter & Garmezy, 1983; Sroufe, 1990, 2013). By virtue of its emphasis on comparing and contrasting abnormal development with normative developmental patterns, and investigating the similarities as well as the differences between normality and psychopathology, the strengths and weaknesses associated with atypical development are underscored (Cicchetti, 1993; Karmiloff-Smith, 2007).

The central focus of developmental psychopathology is the elucidation of developmental processes and how they function as indicated and elaborated by the examination of extremes in developmental outcome. Such extremes contribute to the understanding of developmental processes. Research in the field of developmental psychopathology is not limited to the investigation of mental disorders. Scientists working in the discipline of developmental psychopathology are interested in examining the entire range of developmental processes and functioning. Not only are the disordered extremes the subject of study, but also the subclinical range of functioning is viewed as being important to the goal of understanding the organization of normal and abnormal development. Individuals in the subclinical range of adaptation (e.g., children with dysthymic moods) may be vulnerable to the subsequent emergence of psychopathology (e.g., the onset of depressive disorders). The investigation of processes that contribute to the later emergence of a disorder, such as depression, as well as processes that mitigate against disordered outcomes provides further insight into the full range of developmental phenomena.

Developmental psychopathology is especially applicable to the investigation of transitional turning points in development across the life span. This is due to its acknowledgment that disorders may appear for the first time in later life and because of its advocacy for the examination of the course of disorders once manifest, including their phases and sequelae (Zigler & Glick, 1986).

RESEARCH APPROACHES WITHIN DEVELOPMENTAL PSYCHOPATHOLOGY

The nature of the developmental processes elucidates a clear perspective on how to conceptualize empirical research on the origins and course of later emerging psychopathology. Researchers conducting investigations aimed at identifying early precursors of later emerging depression face numerous conceptual and methodological challenges. Because of developmental changes in neurobiological and physiological systems as well as parallel developments in cognitive, social cognitive, socioemotional, and representational systems, investigators cannot presume phenotypic similarity between early precursors and later impairments. Consequently, studies of the early precursors of later psychopathology should conceptualize and measure features of early development that are theoretically related but not necessarily behaviorally identical to the emergence of subsequent depression.

Given the importance of a life span view of developmental processes and an interest in delineating how prior development influences later development, a major issue in developmental psychopathology involves how to determine continuity in the quality of adaptation across developmental time. Crucial to this concept is a recognition that the same behaviors in different developmental periods may represent quite different levels of adaptation. Behaviors indicating competence within a developmental period may indicate incompetence when evidenced within subsequent developmental periods. Normative behaviors early in development may indicate maladaptation when exhibited later in development. Thus, the manifestation of competence in different developmental periods is rarely indicated by isomorphism in behavioral presentation (i.e., homotypic continuity—see Kagan, 1971).

Additionally, it must be recognized that the same function in an organized behavioral system can be fulfilled by two dissimilar behaviors, whereas the same kind of behavior may serve two different functions (Werner & Kaplan, 1963) and may also play different roles in different systems. As a result, it is especially important to distinguish between similarities and differences in higher order organization of symptomology (molar level) and component behavioral manifestation of symptomology (molecular level) during different developmental periods. The reorganization of biological and psychological systems that takes place at each level of development means researchers could not expect to see, for any symptom, behavioral isomorphism at the molecular level, even if there is isomorphism at the molar level. For example, individuals who experience recurrent depression during the transition from preoperational to concrete operational thought may display excessive and inappropriate guilt, a loss of self-esteem, and a decrease in activity throughout the episodes. Consequently, at a molar level, the depressive symptoms at the later period (i.e., concrete operational) will be isomorphic to those of the earlier period (i.e., preoperational). Nonetheless, the particular manifestation of the guilt feelings, loss of self-esteem, and psychomotor retardation may change and develop during the transition, when the child's cognitive, representational, socioemotional, and behavioral competencies undergo a rather radical development across these developmental periods. In this way, there may be noteworthy differences at the molecular level.

Because development typically involves the organization through integration of previously differentiated behaviors, we can predict that the expression of depression may be characterized by molar continuities but additionally by molecular discontinuities and changes. At the molar level, continuity will be preserved by an orderly development in the organization of behaviors; however, at the molecular level, the behaviors that are present at different periods may vary, but the meaning may remain coherent (i.e., heterotypic continuity). The study of the development of depression over the life course is likely to be fruitful and to reveal the relationship between pathological processes and normal development only if the behavior of individuals with an affective disorder is examined simultaneous at the molar and molecular levels.

Furthermore, examining the course of adaptation once an episode of depression has remitted would benefit from the utilization of a developmental perspective. For example, the examination of the functioning of characteristics of individuals previously diagnosed with depression who have returned to a nondisordered condition would provide additional valuable information about depressive disorders. It may be possible to identify core characteristics of functioning that remain stable but that no longer give rise to depression because of compensatory factors in the environment, within the individual, or through gene × environment (G×E) interactions that promote resilient adaptations (Caspi et al., 2003, 2010; Cicchetti & Blender, 2006; Karg, Burmeister, Shedden, & Sen, 2011). It is conceivable that research such as this might reveal that certain functioning characteristics that were causally relevant to depression in an earlier environment have become positively adaptive in a new context (Belsky & Pluess, 2009).

It also may be erroneous to assume that normalized behavior necessarily reflects improvements in processes that were once causal to the development of depression. Accordingly, a developmental psychopathology perspective encourages us to remain open to the possibility that at least some of the characteristics typically viewed as functioning deficits in fact may be neutral or even advantageous. Stated differently, they may translate into assets or deficits, depending on other biological and psychological characteristics of the individual or the environment.

PRINCIPLES OF DEVELOPMENTAL PSYCHOPATHOLOGY

In this section, the major principles that are central to elucidating the understanding of both normal and atypical patterns of development are discussed, and their relevance to the study of depression is highlighted. It is asserted that the incorporation of these principles into the design and implementation of longitudinal investigations from their inception will proffer a powerful framework for guiding and informing the future research agenda on the causes, course, sequelae, and treatment of depressive disorders.

The Mutual Interplay between Normal and Abnormal Development

A focus on the boundary between normal and abnormal development is central to a developmental psychopathology analysis (Cicchetti, 1984, 1993, 2006; Cicchetti & Cohen, 1995). Such a perspective emphasizes not only how knowledge from the study of normal development can inform the study of high-risk conditions and psychopathology but also how the investigation of risk and pathology can enhance our comprehension of normal development.

The study of depression from a developmental perspective can make many significant contributions to theories of normal development, primarily by contributing greater precision to existing theory and by forcing us to examine theories of development critically in relation to our knowledge about psychopathology. The results of such empirical and theoretical investigations may be the description of alternative developmental pathways that lead to the same or different outcomes of the developmental sequence and a weighting of the respective roles of biological, social, emotional, and cognitive factors in mental growth. Furthermore, before one is capable of identifying deviances that exist in a system, one must possess an accurate description of the system itself. Only when we understand the total ongoing development of normal systems can we fully comprehend developmental deviations as adaptational irregularities of those systems (von Bertalanffy, 1968). Because developmental change may be rapid or gradual, it is necessary to consider normative trends of developing skills in the social, emotional, and cognitive domains so as to be in a better position to evaluate deviation or maladjustment. In addition, it is critical to consider intraindividual variation in the overt manifestation of an episode of depression and individual protective factors or stressors that may inhibit or potentiate depressive illness.

Thus, the application of knowledge of normal biological, cognitive/social, cognitive, representational, and socioemotional development to the understanding of depressive illness results in an articulation of how components of individual functioning in persons with depression contribute to their symptomatic presentation. For example, many internal processes implicated in existing theories of depression do not exist in isolation. Deficits in neurobiological, neurochemical, social cognitive, emotion regulatory, parent–child attachment, impulse control, executive functions, neuropsychological development and functioning, and other systems tend to covary significantly in children and adults with depression.

This covariance, in turn, often renders difficult the important task of disentangling causal processes (Richters, 1997). In some instances, suspected causal processes actually may be the products of other covarying systems and only spuriously related to depression. In other cases, a process may indeed influence depressive behavior; however, the nature and extent of its causal influence may be masked or clouded by the influence of other interacting systems.

One strategy that could be used to help disentangle causal influences among multiple, interactive systems would be to identify and examine the functioning of individuals with depression who possess particular functioning deficits but not others. Moreover, the examination of aberrations of the biological, cognitive, social cognitive, socioemotional, and other biological and psychological domains in individuals with depression contributes to a more complete comprehension of how these systems function in normal development.

The Importance of a Life Span Perspective

Development extends throughout the entire course of life, and adaptive and maladaptive processes emerge over the life span. From infancy through senescence, each period of life has its own developmental agenda and contributes in a unique fashion to the past, present, and future organization of individual development, normal or abnormal. Thus, individuals with a mood disorder, such as depression, may move between pathological and nonpathological forms of functioning. Moreover, even in the midst of a disordered period, individuals may display adaptive as well as maladaptive processes so that it becomes possible to delimit the presence, nature, and boundaries of underlying psychopathology.

With respect to the emergence of psychopathology, all periods of life are consequential in that the developmental processes may undergo a pernicious turn toward psychiatric disorder at any phase. Many disorders have several distinct phases. The factors that are associated with the onset of a disorder may be very different from those that are associated with the cessation of a disorder or with its repeated occurrence. For example, a positive family history of depression is strongly associated with a higher risk of depression onset.

In contrast, to the often dichotomous world of mental disorder/nondisorder depicted in the extant literature, a developmental psychopathology perspective recognizes that normality often fades into abnormality. Thus, because individuals with depression can have extended periods of normal functioning and also can move into a disordered period unexpectedly, being cognizant of the boundary between normal and atypical functioning is particularly relevant for a person with a mood disorder. For example, it is quite likely that during an early episode, individuals with depression may not recognize that they are in an illness phase. Therefore, strategies for helping them to detect signals of deteriorating functioning during the wellness state is critically important. Family members, friends, and significant others can also be enlisted and may be helpful in the "detection" process.

Moreover, in developmental psychopathology, "adaptive" and "maladaptive" may assume differing definitions, depending on whether one's time referent is immediate circumstances or long-term development, and processes within the individual can be characterized as having shades or degrees of psychopathology. With respect to depressive illness, such a life span perspective suggests that, even when recurrent depression has occurred, future remission and more adaptive functioning are possible (Cicchetti & Toth, 1995).

Rutter (1996) has conjectured that key life "turning points" may be times when the presence of protective mechanisms are especially likely to help individuals redirect themselves from a risk trajectory onto a more adaptive developmental pathway. Likewise, Toth and Cicchetti (1999) have suggested that these periods of developmental transition may also afford opportunities when individuals are most amenable to profiting from therapeutic interventions. Whereas change in functioning remains possible at each transition turning point in development, prior adaptation does place constraints on subsequent adaptation. In particular, the longer an individual continues along a maladaptive pathway, the more difficult it is to reclaim a normal developmental trajectory. Furthermore, recovery of function to an adaptive level of developmental organization is more likely to occur after a period of pathology if the level of organization before the breakdown was a competent and adaptive one.

Developmental Pathways: Diversity in Process and Outcome

Since the emergence of developmental psychopathology as an interdisciplinary science, diversity in process and outcome has been among the hallmarks of its perspective. As Sroufe (1990, p. 335) has asserted, "One of the principal tasks of developmental psychopathology is to define families of developmental pathways, some of which are associated with psychopathology with high probability, others with low probability." Even before a psychiatric disorder emerges, certain pathways signify adaptational failures that probabilistically forebode subsequent psychopathology. An example comes from a 40-year follow-up of children who showed mild or moderate externalizing behavior, as rated by teachers when they were aged 13 to 15 years. By middle adulthood, these children showed greater rates of alcohol abuse, marital failure, occupational impairment, and psychiatric disorder than comparison children rated low in externalizing behavior (Colman et al., 2009).

It is expected that (1) there are multiple contributors to depressive outcomes in any individual, (2) the contributors vary between individuals with depression, (3) there is heterogeneity among persons with depression in the features of their biological and psychological disturbances and underlying dysfunctions, and (4) there are numerous pathways to depression. Moreover, it is believed that there is heterogeneity among individuals who possess many of the risk factors for depression but who do not develop the disorder. In this regard, the principles of equifinality and multifinality, derived from general systems theory (Cicchetti & Rogosch, 1996; von Bertalanffy, 1968), are germane.

Equifinality refers to the observation that a diversity of paths may lead to the same outcome. This alerts us to the possibility that a variety of developmental progressions may eventuate in depression rather than positing a singular primary pathway to disorder. In contrast, multifinality suggests that any one component may function differently, depending on the organization of the system in which it operates (Cicchetti & Rogosch, 1996; Wilden, 1980). Multifinality states that the effect on functioning of any one component's value may vary in a different system; thus, the same risk factor or starting point may eventuate in a wide dispersion of outcomes. Actual effects will depend on the conditions set by the values of additional components with which it is structurally linked. Consequently, the pathology or health of the system must be identified in terms of how adequately its essential functions are maintained. Stated differently, a particular adverse event should not necessarily be seen as leading to the same psychopathological or nonpsychopathological outcome in every individual with depression. Likewise, individuals with depression may begin on the same major pathway and, as a function of their subsequent "choices," exhibit very different patterns of adaptation or maladaptation (Cicchetti & Tucker, 1994; Sroufe, Egeland, & Kreutzer, 1990).

Because of the diversity in processes and outcomes that characterize development, the developmental psychopathology approach to depression does not proffer a simple unitary etiological explanation. Although commonalities in pathways in different clusters of persons with depression may be delineated, it is also possible that depression is not the only outcome associated with each pathway. Although pathways may be discovered that are specific to depression in some individuals, there are also likely to be a range of dysfunctions and comorbid dysfunctions and disorders (e.g., anxiety disorders, conduct disorders, substance abuse disorders, personality disorders), of which an affective disorder (e.g., depression) may be one. Thus, the empirical investigation of depression must be conceptualized within a larger body of inquiry into the developmental patterns promoting adjustment difficulties and psychopathology.

A pathways approach builds on knowledge gained from variable-oriented studies; however, attention is shifted to exploring the common and the uncommon outcomes as well as alternative routes by which outcomes are achieved by different individuals (Cicchetti & Natsuaki, 2014). Thus, what might be considered error variance at the group level must be critically examined for understanding diversity in process and outcome. The emphasis on person-centered observation highlights the transition from a focus on variable to a focus on individual, and this transition is essential for demonstrating equifinality and multifinality in the developmental course.

The growing knowledge that subgroups of individuals manifesting similar problems arrived at them from different beginnings and that the same risk factors may be associated with different outcomes has proved to be critical not only because it has the potential to bring about important refinements in the diagnostic classification of mental disorders but also because it calls attention to the importance of continuing to conduct process-oriented studies (Bergman & Magnusson, 1997; von Eye & Bergman, 2003). The examination of patterns of commonality within relatively homogeneous subgroups of individuals and concomitant similarity in profiles of contributory processes becomes an important data analytic strategy. Moreover, the need to examine the totality of attributes, psychopathological conditions, and risk and protective processes in the context of each other rather than in insolation is seen as crucial for understanding the course of development taken by individuals. For example, the presence of depression in a child, adolescent, or adult would have different developmental implications, depending on whether it occurs alone or in conjunction with other types of psychopathology. The meaning of any one attribute, process, or psychopathological condition needs to be considered in light of the complex matrix of individual characteristics, experiences, and social contextual influences involved; the timing of events and experiences; and the developmental history of the individual.

This attention to diversity in origins, processes, and outcomes in understanding developmental pathways does not suggest that prediction is futile as a result of many potential individual patterns of adaptation. There are constraints on how much diversity is possible, and not all outcomes are equally likely (Cicchetti & Tucker, 1994; Sroufe et al., 1990). Nonetheless, the appreciation of equifinality and multifinality in development encourages theorists and researchers to entertain more complex and varied approaches to how they conceptualize and investigate development and psychopathology. Researchers on depression should increasingly strive to demonstrate the multiplicity of processes and outcomes that may be articulated at the individual, person-oriented level within existing longitudinal data sets. Ultimately, future endeavors must conceptualize and design research on depression at the outset with these differential pathways concepts as a foundation. In so doing, progress achieving the unique goals of developmental psychopathology—to explain the development of individual patterns of adaptation and maladaptation—will be realized (Sroufe & Rutter, 1984).

Individuals Play an Active Role in Their Own Development

There has been a growing recognition of the role of the developing person as a processor of his or her experiences. The environment does not simply create an individual's experience; rather, individuals also choose and create their experiences and their own environments in a changing world (Scarr & McCartney, 1983). Individuals select, integrate, and actively affect their own development and the environment in a dynamic fashion (Cicchetti & Tucker, 1994; Wachs & Plomin, 1991). The principle of contextualism conceptualizes developmental processes as the ongoing interaction between an active, changing individual and a continuously unfolding, dynamic context (Cicchetti & Aber, 1998). Thus, maladaptation and psychopathology are considered to be products of the transaction among an individual's intraorganismic characteristics, adaptational history, and the current context (Boyce et al., 1998).

Various difficulties will constitute different meanings for an individual, depending on cultural considerations (Garcia-Coll, Akerman, & Cicchetti, 2000) as well as an individual's experiential history and current level of psychological and biological functioning. The integration of the experience, in turn, will affect the adaptation or maladaptation that ensues. Moreover, we now know that social contexts exert effects not only on psychological processes but also on biological structures, functions, and processes (Boyce et al., 1998; Cicchetti, 2002; Cicchetti & Tucker, 1994; DeBellis, 2001; Eisenberg, 1995; Nelson & Bloom, 1997). For example, persons at risk for developing depression who experience traumatic environmental adversity will possess a greater likelihood that their genetic vulnerability will get expressed and that the neural circuitry associated with aspects of depression will be activated.

Multilevels of Analysis

A "systems view" conceives development as being hierarchically organized into multilevels that mutually influence each other (Cicchetti & Natsuaki, 2014; Gottlieb, 1992; Grillner, Kozlov, & Kotaleski, 2005; Thelen & Smith, 1998). "Top-down" as well as "bottom-up" bidirectional effects are theorized to occur among the various levels. Accordingly, genetic activity ↔ neural activity ↔ behavior ↔ environment can serve as a schematic representation of this systems view. These bidirectional effects among levels of the system result in a probabilistic conceptualization of epigenetic development in all individuals, including those with a mental illness, such as persons with a depressive disorder (Cicchetti & Tucker, 1994; Gottlieb, 1992). The probabilistic epigenesis perspective thus implies that individuals are neither unaffected by earlier experiences nor immutably controlled by them. A change in developmental course is thought to be possible as a result of new experiences, reciprocal interactions between levels of the developing person, and the individual's active self-organizing strivings for adaptation (see also Cicchetti & Tucker, 1994). Thus, epigenesis is viewed as probabilistic rather than predetermined, with the bidirectional and transactional nature of genetic, neural, behavioral, and environmental influences over the life course capturing the essence of probabilistic experiences. Because development is a dynamic process, assertions about causality must include a temporal dimension that specifies and describes when the experience or coactions occurred (Gottlieb & Halpern, 2002).

Different levels of analysis—genetic, biological, social, psychological, familial, or cultural—constrain other levels (Cicchetti & Cannon, 1999). As scientists investigating depressive disorders learn more about multiple levels on analysis, researchers conducting their work at each level will need to develop theories that are consistent across all levels. When scientists in different disciplines function in isolation, they run the risk of formulating theories that will ultimately prove to be incorrect because vital information from other disciplines has either been ignored or is unknown. Just as is the case in systems neuroscience (Cowan, Harter, & Kandel, 2000), it is critical that there be an integrative framework that incorporates all levels of analysis about complex systems in the development of depression.

It is now widely understood that individual risk factors seldom are powerful enough to exert sufficient influence to result in psychopathology (Sameroff & Chandler, 1975). Moreover, when they appear to have such effects, it is highly likely that they are surrogates for multiple, unobserved influences. Much more commonly, adequate prediction of either disturbance or resilience necessitates the consideration of multiple risk and protective factors and their interplay. Moreover, the consequences of any risk factor depend on myriad other aspects embedded in the developmental context.

To comprehend depression in its full complexity, all levels of analysis must be examined and integrated (Frodl & O'Keane, 2013; Grillner et al., 2005). In a seminal article on the biology of emotion, Schildkraut and Kety (1967) pointed out that few studies had examined the concurrent links between the psychological and biological aspects of normal and abnormal emotion. As they noted, "The interactions between environmental determinants of affect, various physiological factors, and the complexity of psychological determinants, including cognitive factors derived from the individual's remote and immediate past experiences, have received only limited study under adequately controlled conditions." They further stated, "it is not likely that changes in the metabolism of the biogenic amines alone will account for the complex phenomena of normal or pathological affect…any comprehensive formulation for the physiology of affective state will have to include many other concomitant biochemical, physiological, and psychological factors" (Schildkraut & Kety, 1967, p. 28).

Cross-fertilization of neuroscience with psychology will result in major advances in our comprehension of mood disorders, especially if a developmental perspective to these collaborative ventures is adopted by investigators. In this regard, we must underscore a fundamental maxim that must characterize any interdisciplinary research: Adequate and state-of-the-art assessments must be made of each domain—biological, psychological, or environmental—that is examined. Moreover, the problem is compounded when we add the future requirement that the measurement strategy, design of the study, and the actual measures all must reflect a sensitivity to a variety of cultural and developmental issues (Causadias, 2013). Although the challenges are great in the quest to arrive at a sufficiently integrative approach to understanding the depressive disorder, a developmental psychopathology approach holds great promise for elucidating necessary questions and suggesting strategies to apply to such an undertaking.

It is now clear that investigations conducted at a single level of analysis (e.g., molecular, genetic, or cognitive) will not suffice to unravel how the brain works in terms of neurobiology (Grillner et al., 2005).

Resilience

Developmental psychopathologists are as interested in individuals at a high risk for the development of psychopathology who do not manifest it over time as they are in individuals who develop an actual mental disorder (Cicchetti, 2010; Cicchetti & Garmezy, 1993; Luthar, 2006; Masten, 2001, 2014). Moreover, researchers in developmental psychopathology emphasize the importance of understanding the functioning of individuals who, after having diverged onto deviant developmental pathways, resume normal functioning and achieve adequate adaptation (Cicchetti & Rogosch, 1997; Masten, Best, & Garmezy, 1990).

Resilience has been operationalized as the individual's capacity for adapting successfully and functioning competently despite experiencing chronic adversity of after exposure to prolonged or severe trauma (Luthar, Cicchetti, & Becker, 2000). Resilience is a dynamic developmental process; it is multidimensional in nature, exemplified by finding that individuals who are at a high risk for or who have a mental disorder may manifest competence in some domains and contexts, whereas they may exhibit problems in others.

Research on the determinants of resilience also highlights the need to examine functioning across multiple domains of development. Furthermore, the ability to function in a resilient fashion in the presence of biological, psychological, environmental, and sociocultural disadvantage may be achieved through the use of developmental pathways that are less typical than those negotiated in usual circumstances. Thus, an important question for researchers to address is whether the employment of alternative pathways to attaining competence renders individuals more vulnerable to manifesting delays or deviations in development. Although only prospective longitudinal investigations can fully address this issue, it is critical to ascertain whether these individuals are more prone to developing maladaptation or psychopathology in later life. Given the nonstatic nature of the construct, we do not expect children identified as resilient to be immune to declines in functioning at each subsequent developmental period.

Investigations aimed at discovering the processes leading to resilient outcomes and on the processes underlying recovery of adaptive function offer great promise as an avenue for facilitating the development of prevention and intervention strategies (Luthar et al., 2000; Toth & Cicchetti, 1999). Through the examination of the proximal and distal processes and mechanisms that contribute to positive adaptation in situations that more typically eventuate in maladaptation, researchers and clinicians will be better prepared to devise ways of promoting competent outcomes in individuals at a high risk for developing a depressive disorder (Luthar & Cicchetti, 2000; Toth, Rogosch, Manly, & Cicchetti, 2006).

Despite the attention paid to discovering the processes through which individuals at a high risk do not develop maladaptively, the empirical study of resilience has focused primarily on detecting the psychosocial determinants of the phenomenon (Charney, 2004; Curtis & Cicchetti, 2003). For research on resilience to grow in ways that are commensurate with the complexity inherent to the construct, efforts to understand underlying processes will be facilitated by the increased implementation of multidisciplinary investigations designed within a developmental psychopathology framework. Research of this nature would entail a consideration of biological, psychological, and environmental/contextual processes from which varied pathways to resilience (equifinality) might eventuate as well as those that result in diverse outcomes among individuals who have achieved resilient functioning (multifinality; see Cicchetti & Curtis, 2007). Along these lines, the investigation of multiple aspects of the processes underlying resilience can shed light on the nature of the interrelation among various developmental domains in individuals with depression. For example, how do cognition, affect, and neurobiological growth relate to one another at various developmental periods? When an advance or a lag occurs in one biological or psychological system, what are the consequences for other systems?

It is important that these issues receive focused attention from researchers, because the presence of capacities of one of these systems may be a necessary condition for the development or exercise of capacities of another system. For example, certain cognitive skills may be necessary for the development of particular affective expressions and experiences (Hesse & Cicchetti, 1982). Lags in these systems may then result in compensatory development, which may, in some instances, leave the child vulnerable to psychopathology. Over time, difficulty in the organization of one biological or psychological system may tend to promote difficulty in the way in which other systems are organized as hierarchical integration between the separate systems occurs. The organization of the individual may then appear to consist of poorly integrated component systems. As the converse of the effects of early competence, early incompetence will tend to promote later incompetence because the individual arrives at successive developmental stages with less than optimal resources available for responding to the challenges of that period. Again, however, this progression is not inevitable but probabilistic. Changes in the internal or external environment may lead to improvements in the ability to grapple with developmental challenges, resulting in a redirection in the developmental course.

The role of biological factors in resilience is suggested by evidence on neurobiological and neuroendocrine function in relation to stress regulation and reactivity (Heim & Binder, 2012), by behavioral genetics research on nonshared environmental effects, and by molecular genetics research that may reveal the genetic elements that serve a protective function for individuals experiencing significant adversity (Cicchetti & Blender, 2006; Cicchetti & Rogosch, 2012). To provide an example gleaned from the field of molecular genetics, research suggests that it is conceivable that the gene encoding high monoamine oxidase A (MAOA) activity and the l/l genotype of the serotonin transporter gene (5-HTT) may confer protection against the development of antisocial behavior in males who have been maltreated and against the development of depression in individuals who have been maltreated, respectively (Caspi et al., 2002, 2003; Cutuli, Raby, Cicchetti, Englund, & Egeland, 2012).

Children who develop in a resilient fashion despite having experienced significant adversity play an active role in constructing, seeking, and receiving the experiences that are developmentally appropriate for them. To date, research investigations that search for mechanisms of G×E interaction have rarely addressed the role that genetic factors may play in influencing how children who are developing in a resilient fashion have actively transformed their social environment (known as evocative gene–environment correlations; Rende & Waldman, 2006; Scarr & McCartney, 1983). An investigation that accomplishes this was conducted by Frodl et al. (2015). They found that individuals who were developing in a healthy fashion despite high methylation of 5-HTTLPR (found to be related to adversity) manifested a higher functional MRI that could conceivably be a strategy used for resilient adaptation (Frodl et al., 2015).

At the neurobiological level, different areas of the brain may attempt to compensate; on another level, individuals may seek out new experiences in areas where they have strength (Black, Jones, Nelson, & Greenough, 1998; Cicchetti & Tucker, 1994). The effects of social experiences, such as child abuse and neglect, on brain biochemistry and microstructure may be either pathological or adaptive. With respect to the experience of depression, depending on how the individual interprets and responds to depression, as well as the genetic elements that are expressed, the effects either may be pathological (the typical outcome) or may not preclude normative development (a resilient outcome; Gibb, Beevers, & McGeary, 2013). Thus, neither early neurobiological anomalies nor aberrant experience should be considered as determining the ultimate fate of the individual with a depressive disorder (the notion of probabilistic epigenesis).

A multilevel approach to resilience also affords an additional avenue for examining the biological and social constraints that may operate on aspects of the developmental process throughout the life course. Moreover, through investigating the multiple determinants of resilient adaptation, we are in a position to discover the range and variability in individuals' attempts to respond adaptively to challenges and ill fortune.

Prevention and Intervention

The major objective of the field of prevention science is to intervene in the course of development in order to reduce or eliminate the emergence of maladaptation and mental disorder as well as to promote resilient adaptation in individuals at a high risk for psychopathology (Ialongo et al., 2006). To achieve this laudable goal, it is essential that prevention scientists possess a complex, multilevel understanding of the course of normality to formulate an in-depth portrayal of how deviations in normal developmental processes can eventuate in maladaptation and mental disorder. Because of its focus on the mutual interplay between the investigation of normal and abnormal development, the field of developmental psychopathology is well poised to provide the theoretical foundation for prevention and intervention initiatives.

Developmental psychopathologists believe that efforts to prevent the emergence of psychopathology or to ameliorate its effects can also be informative for understanding processes involved in psychopathological development (Hinshaw, 2002; Kellam & Rebok, 1992). For example, if the course of development is altered as a result of the implementation of randomized controlled prevention trials and the risk for negative outcomes is reduced, then prevention research helps to specify processes that are involved in the emergence of psychopathology or other maladaptive developmental outcomes (Ialongo et al., 2006). As a consequence, if randomized controlled prevention trials examine mechanisms of intervention action, then they can be conceptualized as true experiments in modifying the developmental course, thereby providing insight into the etiology and pathogenesis of disordered outcomes (Cicchetti & Hinshaw, 2002; Hinshaw, 2002; Howe, Reiss, & Yuh, 2002; Kellam & Rebok, 1992). Thus, prevention research not only leads to support or lack of support for theoretical formulations accounting for the development of psychopathology but also can contribute to the knowledge base of strategies that can be implemented to reduce psychopathology and promote positive adaptation. Knowledge of developmental norms, appreciation of how developmental levels may vary within the same age group, sensitivity to the changing meaning that problems have at different developmental levels, attention to the effects of developmental transitions and reorganizations, and understanding of the factors that are essential features to incorporate into the design and implementation of preventive interventions all may serve to enhance the potential for optimal intervention efficacy (Noam, 1992; Toth & Cicchetti, 1999).

For example, at what phases of development will psychosocial interventions have a maximal preventive effect among children at risk for depression and by what mechanisms? Basic neurobiological research could inform our understanding of when children at risk for depression develop facial emotion recognition errors (e.g., viewing neutral faces as negative) and the neural structures and circuitry with which these errors are correlated (i.e., amygdala/ventromedial prefrontal cortex circuits; Rich et al., 2006). Results of such studies may suggest that certain forms of psychosocial intervention (e.g., psychoeducation, cognitive-behavioral therapy, or interpersonal therapy) can effectively teach emotion labeling skills but only among children who have shown an ability to mentalize or infer emotional states in others. In turn, demonstrating that such interventions influence aberrant neural pathways and result in symptom improvement, in part mediated by improved emotion recognition, would inform our understanding of developmental pathways in the onset of depression.

Whereas much of the work on depression and other types of psychopathology is, of necessity, naturalistic and correlational in nature, given ethical constraints on randomly assigning a developing person to key environmental or psychobiological conditions, the gold standard for clinical intervention and prevention research is the randomized clinical trial. The experimental nature of such investigations provides an unprecedented opportunity to make causal inferences in the field. The types of independent variables manipulated in clinical or prevention trials may be several steps removed from crucial, underlying etiological factors, given that such trials are primarily concerned with the practical, clinical goals of alleviating suffering and promoting competence rather than isolating primary causal variables. Nonetheless, careful research design and assiduous measurement of ancillary, psychological, and biological process variables through which intervention effects may occur can shed light on theory-driven mechanisms underlying healthy and pathological development (Cicchetti & Gunnar, 2008; Hinshaw, 2002; Howe et al., 2002).

Rather than waiting for a full-blown disorder to emerge, risk markers that portend a possible illness could be identified. Early identification and possible prevention could minimize the magnitude of the disease process and possible impairment. Prevention strategies become particularly relevant to the increasing diagnosis of the disorder in early childhood. As we progress with the ability to detect genetic and neurobiological markers of disease, prevention again emerges as an important future avenue to pursue. Such prevention strategies could also minimize the likelihood of the brain circuitry for depression becoming hard wired and increasingly recalcitrant to potential neuroplastic changes.

CONCLUSION AND FUTURE DIRECTIONS

Although it is evident that research on depression has engendered greater clarity with respect to clinical description, etiology, pathogenesis, psychosocial and drug treatment, and development, there remains much to examine in the future.

A developmental systems approach necessitates that research on depression must incorporate a multiple-levels-of-analysis perspective (Cicchetti & Natsuaki, 2014; Frodl et al., 2010, 2014; Hankin, 2012). A developmental psychopathology perspective underscores the importance of conducting ongoing prospective multiwave longitudinal studies that are properly designed and methodologically rigorous and that can provide an accurate portrayal of the life course trajectories of those afflicted with the varying subtypes of depression. Moreover, there is a strong need to be able to investigate depression before it emerges. What populations should be targeted to enhance the likelihood of observing depression at greater than population prevalence rates? What are the earlier precursors to depression across multiple levels of analysis? How can prodromal abnormal signs be identified within the framework of developmental psychopathology? A developmental perspective would also help to articulate and understand those factors that may contribute to the maintenance of depression over the life course, quite separate from those that might contribute to its etiology. In particular, a fuller comprehension of the role played by child physical, sexual, and emotional abuse and child neglect in the development of depression is needed (Cicchetti, Rogosch, Gunnar, & Toth, 2010; Heim, Newport, Mletzko, Miller, & Nemeroff, 2008).

Research in the area of endophenotypes should also be conducted. The endophenotype is a measurable component, unseen by the unaided eye, along the pathway between distal genotype and disease (Gottesman & Gould, 2003; Gottesman & Shields, 1972). Endophenotypes may be neurophysiological, endocrinological, neuroanatomical, cognitive, or neuropsychological in nature. Furthermore, the endophenotype is thought to represent a simpler clue to genetic underpinnings than the disease syndrome itself. The incorporation of endophenotypes will be extremely useful in advancing genomic, neuroimaging, neurobiological, and psychological investigations of depression.

Investigators and practitioners with a developmental perspective are interested not only in the differences between individuals with and without mental disorders but also in their similarities (Cicchetti, 1993; Zigler & Glick, 1986). Indeed, there are striking similarities between persons with depressive illness and their well counterparts. For example, children and adults with depression, just as with persons who are not disordered, experience a range of feelings, possess a need for connectedness with others, seek a sense of order in their worlds, strive for autonomy, and attempt to find meaning in their life experiences (Hinshaw & Cicchetti, 2000).

Individuals with depression typically shift from phases of normality to psychopathology and back. Almost all such individuals experience stages and phases of remission and relapse throughout the life span. Moreover, not only do persons with depression have periods of remission, but also an appreciable number manage to function in an adaptive and productive manner for prolonged periods of their lives. Accordingly, individuals with depression should not be reduced to their psychiatric diagnoses. Those persons with depression who have been successfully treated and those whose illnesses are in remission may be strikingly similar to persons who are without a mental disorder.

Research in developmental psychopathology has enhanced our understanding of risk, disorder, and resilience across the life course (see, e.g., Cicchetti, 2013, 2016). Advances in genomics, G×E interactions, and epigenetics; growth in the understanding of neurobiology and neural plasticity; and progress in the development of methodological and technological tools, including brain imaging, hormone assays, and statistical analysis of developmental change, pave the way for multiple-levels-of-analysis research programs aimed at elucidating the development and course of depression (Cicchetti & Curtis, 2006; Cicchetti & Natsuaki, 2014). Moreover, the information that is emanating from the field of developmental psychopathology can be integrated into the conceptual base and measurement armamentaria of scientists from diverse disciplines, even if they do not consider themselves to be developmental psychopathologists. These knowledge gains will not only benefit the scientific study of depression but also permit translation to informing development-based preventive strategies and interventions that will contribute to reducing the individual, familial, and societal burden of depressive illness.

ACKNOWLEDGMENTS

My work on this chapter was supported by grants from the National Institute of Mental Health (RO1MH45027 and RO1MH091070) and the Spunk Fund, Inc.

REFERENCES

Belsky, J., & Pluess, M. (2009). Beyond diathesis stress: differential susceptibility to environmental influences. *Psychological Bulletin, 135*, 885–908.

Bergman, L. R., & Magnusson, D. (1997). A person-oriented approach in research on developmental psychopathology. *Development and Psychopathology, 9*, 291–319.

von Bertalanffy, L. (1968). *General system theory: Foundations, development, applications.* New York: George Braziller.

Black, J., Jones, T. A., Nelson, C. A., & Greenough, W. T. (1998). Neuronal plasticity and the developing brain. In N. E. Alessi, J. T. Coyle, S. I. Harrison, & S. Eth (Eds.), *Handbook of child and adolescent psychiatry* (pp. 31–53). New York: Wiley.

Boyce, W. T., Frank, E., Jensen, P. S., Kessler, R. C., Nelson, C. A., & Steinberg, L. (1998). Social context in developmental psychopathology: recommendations for future research from the MacArthur Network on Psychopathology and Development. *Development and Psychopathology, 10*, 143–164.

Caspi, A., Hariri, A., Holmes, A., Uher, R., & Moffitt, T. E. (2010). Genetic sensitivity to the environment: the case of the serotonin transporter gene (5-HTT) and its implications for studying complex diseases and traits. *American Journal of Psychiatry, 167*, 509–527.

Caspi, A., McClay, J., Moffitt, T., Mill, J., Martin, J., & Craig, I. W. (2002). Role of genotype in the cycle of violence in maltreated children. *Science, 297*, 851–854.

Caspi, A., Sugden, K., Moffitt, T. E., Taylor, A., Craig, I. W., Harrington, H. L, ... Poulton, R. (2003). Influence of life stress on depression: moderation by a polymorphism in the 5-HTT gene. *Science, 301*, 386–389.

Causadias, J. M. (2013). A roadmap for the integration of culture into developmental psychopathology. *Development and Psychopathology, 25*(4 Pt 2), 1375–1398.

Charney, D. (2004). Psychobiological mechanisms of resilience and vulnerability: implications for successful adaptation to extreme stress. *American Journal of Psychiatry, 161*, 195–216.

Cicchetti, D. (1984). The emergence of developmental psychopathology. *Child Development, 55*, 1–7.

Cicchetti, D. (1990). A historical perspective on the discipline of developmental psychopathology. In J. Rolf, A. Masten, D. Cicchetti, K. Nuechterlein, & S. Weintraub (Eds.), *Risk and protective factors in the development of psychopathology* (pp. 2–28). New York: Cambridge University Press.

Cicchetti, D. (1993). Developmental psychopathology: reactions, reflections, projections. *Developmental Review, 13*, 471–502.

Cicchetti, D. (2002). The impact of social experience on neurobiological systems: illustration from a constructivist view of child maltreatment. *Cognitive Development, 17*, 1407–1428.

Cicchetti, D. (2006). Development and psychopathology. In D. Cicchetti, & D. J. Cohen (Eds.), *Developmental psychopathology* (Vol. 1, 2nd ed., pp. 1–23). New York: Wiley.

Cicchetti, D. (2010). Resilience under conditions of extreme stress: a multilevel perspective [Special article]. *World Psychiatry, 9*, 1–10.

Cicchetti, D. (2013). Annual research Review: resilient functioning in maltreated children—past, present, and future perspectives. *Journal of Child Psychology & Psychiatry, 54*(4), 402–422.

Cicchetti, D. (Ed.). (2016). *Developmental psychopathology* (3rd ed., Vols. 1–4). New York: Wiley.

Cicchetti, D., & Aber, J. L. (1998). Contextualism and developmental psychopathology. *Development and Psychopathology, 10*, 137–141.

Cicchetti, D., & Blender, J. A. (2006). A multiple-levels-of-analysis perspective on resilience: implications for the developing brain, neural plasticity, and preventive interventions. *Annals of the New York Academy of Sciences, 1094*, 248–258.

Cicchetti, D., & Cannon, T. D. (1999). Neurodevelopmental processes in the ontogenesis and epigenesis of psychopathology. *Development and Psychopathology, 11*, 375–393.

Cicchetti, D., & Cohen, D. J. (1995). Perspectives on developmental psychopathology. In D. Cicchetti, & D. J. Cohen (Eds.), *Developmental psychopathology: Theory method* (Vol. 1, pp. 3–20) New York: Wiley.

Cicchetti, D., & Curtis, W. J. (2006). The developing brain and neural plasticity: Implications for normality, psychopathology, and resilience. In D. Cicchetti & D. Cohen (Eds.), *Developmental Psychopathology (2nd ed.), Vol. 2: Developmental Neuroscience* (pp. 1–64). New York: Wiley.

Cicchetti, D., & Curtis, W. J. (2007). Multi-level perspectives on pathways to resilient functioning. *Development and Psychopathology, 19*(3), 627–629.

Cicchetti, D., & Garmezy, N. (1993). Prospects and promises in the study of resilience. *Development and Psychopathology, 5*, 497–502.

Cicchetti, D., & Gunnar, M. R. (2008). Integrating biological processes into the design and evaluation of preventive interventions. *Development and Psychopathology, 20*(3), 737–743.

Cicchetti, D., & Hinshaw, S. P. (2002). Prevention and intervention science: contributions to developmental theory [Special issue]. *Development and Psychopathology, 14*(4), 667–671.

Cicchetti, D., & Natsuaki, M. N. (Eds.). (2014). Multilevel developmental perspectives toward understanding internalizing disorders: current research and future directions [Special issue]. *Development and Psychopathology, 26*(4 pt 2), 1189–1190.

Cicchetti, D., & Pogge-Hesse, P. (1982). Possible contributions of the study of organically retarded persons to developmental theory. In E. Zigler, & D. Balla (Eds.), *Mental retardation: The developmental-difference controversy* (pp. 277–318). Hillsdale, NJ: Lawrence Erlbaum Associates.

Cicchetti, D., & Rogosch, F. A. (1996). Equifinality and multifinality in developmental psychopathology. *Development and Psychopathology, 8*, 597–600.

Cicchetti, D., & Rogosch, F. A. (2012). Gene by environment interaction and resilience: effects of child maltreatment and serotonin, corticotropin releasing hormone, dopamine, and oxytocin genes. *Development and Psychopathology, 24*, 411–427.

Cicchetti, D., & Rogosch, F. A. (1997). The role of self-organization in the promotion of resilience in maltreated children. *Development and Psychopathology, 9*, 797–815.

Cicchetti, D., Rogosch, F. A., Gunnar, M. R., & Toth, S. L. (2010). The differential impacts of early abuse on internalizing problems and diurnal cortisol activity in school-aged children. *Child Development, 25*, 252–269.

Cicchetti, D., Rogosch, F. A., & Toth, S. L. (1997). Ontogenesis, depressotypic organization, and the depressive spectrum. In S. S. Luthar, J. Burack, D. Cicchetti, & J. Weisz (Eds.), *Developmental psychopathology: Perspectives on adjustment, risk, and disorder* (pp. 273–313). New York: Cambridge University Press.

Cicchetti, D., & Schneider-Rosen, K. (1986). An organizational approach to childhood depression. In M. Rutter, C. Izard, & P. Read (Eds.), *Depression in young people, clinical and developmental perspectives* (pp. 71–134). New York: Guilford.

Cicchetti, D., & Sroufe, L. A. (2000). The past as prologue to the future: the times, they've been a changin'. *Development and Psychopathology, 12*, 255–264.

Cicchetti, D., & Toth, S. L. (1995). Developmental psychopathology and disorders of affect. In D. Cicchetti, & D. J. Cohen (Eds.), *Developmental psychopathology: Risk, disorder, and adaptation* (Vol. 2, pp. 369–420). New York: Wiley.

Cicchetti, D., & Toth, S. L. (1998). Perspectives on research and practice in developmental psychopathology. In W. Damon (Ed.), *Handbook of child psychology* (5th ed., Vol. 4, pp. 479–583). New York: Wiley.

Cicchetti, D., & Toth, S. L. (Eds.). (2000). Social policy implications of research in developmental psychopathology [Special issue]. *Development and Psychopathology, 12*(4), 551–554.

Cicchetti, D., & Toth, S. L. (Eds.). (2006). Translational research and developmental psychopathology [Special issue]. *Development and Psychopathology, 18*, 619–622.

Cicchetti, D., & Toth, S. L. (2009). The past achievements and future promises of developmental psychopathology: the coming of age of a discipline. *Journal of Child Psychology and Psychiatry, 50*, 16–25.

Cicchetti, D., & Tucker, D. (1994). Development and self-regulatory structures of the mind. *Development and Psychopathology, 6*, 533–549.

Colman, I., Murray, J., Abbott, R. A., Maughan, B., Kuh, D., & Croudace, T. J. (2009). Outcomes of conduct problems in adolescence: 40 year follow-up of national cohort. *British Medical Journal, 338*, a2981.

Cowan, W. M., Harter, D. H., & Kandel, E. R. (2000). The emergence of modern neuroscience: some implications for neurology and psychiatry. *Annual Review of Neuroscience, 23*, 343–391.

Curtis, W. J., & Cicchetti, D. (2003). Moving research on resilience into the 21st century: theoretical and methodological considerations in examining the biological contributors to resilience. *Development and Psychopathology, 15*, 773–810.

Cutuli, J. J., Raby, K. L., Cicchetti, D., Englund, M. M., & Egeland, B. (2012). Contributions of maltreatment and serotonin transporter genotype to depression in childhood, adolescence, and early adulthood. *Journal of Affective Disorders, 149*(1), 30–37.

DeBellis, M. D. (2001). Developmental traumatology: the psychobiological development of maltreated children and its implications for research, treatment, and policy. *Development and Psychopathology, 13*, 539–564.

Eisenberg, L. (1977). Development as a unifying concept in psychiatry. *British Journal of Psychiatry, 131*, 225–237.

Eisenberg, L. (1995). The social construction of the human brain. *American Journal of Psychiatry, 152*, 1563–1575.

von Eye, A., & Bergman, L. R. (2003). Research strategies in developmental psychopathology: dimensional identity and the person-oriented approach. *Development and Psychopathology, 15*, 553–580.

Frodl, T., Carballedo, A., Frey, E.-M., O'Keane, V., Skokauskas, N., Morris, D., ... Connor, T. (2014). Expression of glucocorticoid inducible genes is associated with reductions in cornu ammonis and dentate gyrus volumes in patients with major depressive disorder. *Development and Psychopathology, 26*(Special issue 4 Pt 2), 1209–1217.

Frodl, T., & O'Keane, V. (2013). How does the brain with cumulative stress? A review with focus developmental stress, HPA axis function and hippocampal structure in humans. *Neurobiology of Disease, 52*, 24–37.

Frodl, T., Reinhold, E., Koutsouleris, N., Donohoe, G., Bondy, B., Reiser, M., ... Meisenzahl, E. M. (2010). Childhood stress, serotonin transporter gene and brain structures in major depression. *Neuropsychopharmacology, 35*, 1383–1390.

Frodl, T., Szyf, M., Carballedo, A., Ly, V., Vaisheva, F., Morris, D., ... Booij, L. (2015). DNA methylation of the serotonin transporter gene (SLC6A4) is associated with brain function involved in processing emotional stimuli. *Journal of Psychiatry and Neuroscience, 40*(2), 140180.

Garcia-Coll, C., Akerman, A., & Cicchetti, D. (2000). Cultural influences on developmental processes and outcomes: Implications for the study of development and psychopathology. *Development and Psychopathology, 12*, 333–356.

Gibb, B. E., Beevers, C. G., & McGeary, J. E. (2013). Toward an integration of cognitive and genetic models of risk for depression. *Cognition and Emotion, 27*(2), 193–216.

Gottesman, I. I., & Gould, T. D. (2003). The endophenotype concept in psychiatry: etymology and strategic intentions. *American Journal of Psychiatry, 160*, 636–645.

Gottesman, I., & Shields, J. (1972). *Schizophrenia and genetics: A twin study vantage point*. New York: Academic Press.

Gottlieb, G. (1992). *Individual development and evolution: The genesis of novel behavior*. NewYork: Oxford University Press.

Gottlieb, G., & Halpern, C. T. (2002). A relational view of causality in normal and abnormal development. *Development and Psychopathology, 14*(3), 421–436.

Grillner, S., Kozlov, A., & Kotaleski, J. (2005). Integrative neuroscience: linking levels of analyses. *Current Opinion in Neurobiology, 15*, 614–621.

Hankin, B. L. (2012). Future directions in vulnerability to depression among youth: integrating risk factors and processes across multiple levels of analysis. *Journal of Clinical Child & Adolescent Psychology, 41*(5), 695–718.

Heim, C., & Binder, E. B. (2012). Current research trends in early life stress and depression: review of human studies on sensitive periods, gene-environment interactions, and epigenetics. *Experimental Neurology, 233*, 102–111.

Heim, C., Newport, J. D., Mletzko, T., Miller, A. H., & Nemeroff, C. B. (2008). The link between childhood trauma and depression: insights from HPA axis studies in humans. *Psychoneuroendocrinology, 33*, 693–710.

Hesse, P., & Cicchetti, D. (1982). Perspectives on an integrative theory of emotional development. *New Directions for Child Development, 16*, 3–48.

Hinshaw, S. P., & Cicchetti, D. (2000). Stigma and mental disorder: conceptions of illness, public attitudes, personal disclosure, and social policy. *Development and Psychopathology, 12*, 555–598.

Hinshaw, S. P. (2002). Intervention research, theoretical mechanisms, and causal processes related to externalizing behavior problems. *Development and Psychopathology, 14*(4), 789–818.

Howe, G. W., Reiss, D., & Yuh, J. (2002). Can prevention trials test theories of etiology? *Development and Psychopathology, 14*, 673–694.

Ialongo, N., Rogosch, F. A., Cicchetti, D., Toth, S. L., Buckley, J., Petras, H., & Neiderhiser, J. (2006). A developmental psychopathology approach to the prevention of mental health disorders. In D. Cicchetti, & D. J. Cohen (Eds.), *Developmental psychopathology* (2nd ed., Vol. 1, pp. 968–1018). New York: Wiley.

Kagan, J. (1971). *Change and continuity in infancy*. New York: Wiley.

Kaplan, R. M., Spittel, M. L., & Spotts, E. L. (2013). Advancing scientific inquiry by blurring research boundaries. *American Journal of Public Health, 103*(Suppl. 1), S4.

Karg, K., Burmeister, M., Shedden, K., & Sen, S. (2011). The serotonin transporter promoter variant (5-HTTLPR), stress, and depression meta-analysis revisited: evidence of genetic moderation. *Archives of General Psychiatry, 68*, 444–454.

Karmiloff-Smith, A. (2007). Atypical epigenesis. *Developmental Science, 10*(1), 84–88.

Kellam, S. G., & Rebok, G. W. (1992). Building developmental and etiological theory through epidemiologically based preventive intervention trials. In J. McCord, & R. E. Tremblay (Eds.), *Preventing antisocial behavior: Interventions from birth through adolescence* (pp. 162–195). New York: Guilford Press.

Luthar, S. S., & Cicchetti, D. (2000). The construct of resilience: implications for intervention and social policy. *Development and Psychopathology, 12*, 857–885.

Luthar, S. S., Cicchetti, D., & Becker, B. (2000). The construct of resilience: a critical evaluation and guidelines for future work. *Child Development, 71*, 543–562.

Luthar, S. S. (2006). Resilience in development: A synthesis of research across five decades. In D. Cicchetti & D. Cohen (Eds.), *Developmental Psychopathology (2nd ed.), Vol. 3. Risk, Disorder, and Adaptation* (pp. 739–795). Hoboken, NJ: Wiley.

Masten, A. S., Best, K. M., & Garmezy, N. (1990). Resilience and development: contributions from the study of children who overcome adversity. *Development and Psychopathology, 2*, 425–444.

Masten, A. S. (2001). Ordinary magic: resilience processes in development. *American Psychologist, 56*, 227–238.

Masten, A. S. (2014). *Ordinary magic: Resilience in development*. New York: Guilford.

Nelson, C. A., & Bloom, F. E. (1997). Child development and neuroscience. *Child Development, 68*, 970–987.

Noam, G. (1992). Development as the aim of clinical intervention. *Development and Psychopathology, 4*, 679–696.

Rende, R., & Waldman, I. (2006). Behavioral and molecular genetics and developmental psychopathology. In D. Cicchetti & D. Cohen (Eds.), *Developmental Psychopathology (2nd ed.), Vol. 2. Developmental Neuroscience* (pp. 427–464). New York: Wiley.

Rich, B. A., Vinton, D. T., Roberson-Nay, R., Hommer, R. E., Berghorst, L. H., McClure, E. B, ... Leibenluft, E. (2006). Limbic hyperactivation during processing of neutral facial expressions in children with bipolar disorder. *Proceedings of the National Academy of Sciences of the United States of America, 103*(23), 8900–8905.

Richters, J. E. (1997). The Hubble hypothesis and the developmentalist's dilemma. *Development and Psychopathology, 9*, 193–229.

Rutter, M. (1996). Transitions and turning points in developmental psychopathology: as applied to the age span between childhood and mid-adulthood. *International Journal of Behavioral Development, 19*(3), 603–626.

Rutter, M., & Garmezy, N. (1983). Developmental psychopathology. In E. M. Hetherington (Ed.), *Handbook of child psychology* (4th ed., Vol. 4, pp. 774–911). New York: Wiley.

Rutter, M. (1986). Child psychiatry: the interface between clinical and developmental research. *Psychological Medicine, 16*, 151–160.

Sameroff, A. J., & Chandler, M. J. (1975). Reproductive risk and the continuum of caretaking casualty. In F. D. Horowitz (Ed.), *Review of child development research* (Vol. 4, pp. 187–244).

Scarr, S., & McCartney, K. (1983). How people make their own environments: a theory of genotype-environment effects. *Child Development, 54*, 424–435.

Schildkraut, J. J., & Kety, S. S. (1967). Biogenic amines and emotion. *Science, 156*(3771), 21–37.

Sroufe, L. A. (1990). An organizational perspective on the self. In D. Cicchetti, & M. Beeghly (Eds.), *The self in transition: Infancy to childhood* (pp. 281–307). Chicago, IL: University of Chicago Press.

Sroufe, L. A., Egeland, B., & Kreutzer, T. (1990). The fate of early experience following developmental change: longitudinal approaches to individual adaptation in childhood. *Child Development, 61*, 1363–1373.

Sroufe, L. A., & Rutter, M. (1984). The domain of developmental psychopathology. *Child Development, 55*, 17–29. http://dx.doi.org/10.1111/1467-8624.ep7405165.

Sroufe, L. A. (2013). The promise of developmental psychopathology: past and present. *Development and Psychopathology, 25*, 1215–1224.

Thelen, E., & Smith, L. B. (1998). Dynamic systems theories. In W. Damon, & R. Lerner (Eds.), *Theoretical, models of human development: Vol. 1. Handbook of child psychology* (pp. 563–634). New York: John Wiley & Sons, Inc.

Toth, S. L., & Cicchetti, D. (1999). Developmental psychopathology and child psychotherapy. In S. Russ, & T. Ollendick (Eds.), *Handbook of psychotherapies with children and families* (pp. 15–44). New York: Plenum Press.

Toth, S. L., Rogosch, F. A., Manly, J. T., & Cicchetti, D. (2006). The efficacy of toddler-parent psychotherapy to reorganize attachment in the young offspring of mothers with major depressive disorder. *Journal of Consulting & Clinical Psychology, 74*(6), 1006–1016.

Wachs, T. D., & Plomin, R. (Eds.). (1991). *Conceptualization and measurement of organism-environment interaction*. West Lafayette, IN: Purdue University Press.

Werner, H., & Kaplan, B. (1963). *Symbol formation*. New York: Wiley.

Wilden, A. (1980). *System and structure*. London: Tavistock.

Zigler, E., & Glick, M. (1986). *A developmental approach to adult psychopathology*. New York: Wiley.

Chapter 2

Early Life Adversity and Risk for Depression: Alterations in Cortisol and Brain Structure and Function as Mediating Mechanisms

David Pagliaccio[1], Deanna M. Barch[1,2,3,4]

[1]*Program in Neuroscience, Washington University, St. Louis, MO, USA;* [2]*Department of Psychiatry, Washington University, St. Louis, MO, USA;* [3]*Department of Psychology, Washington University, St. Louis, MO, USA;* [4]*Department of Radiology, Washington University, St. Louis, MO, USA*

Depression is among the most common and disabling mental health conditions. Epidemiological results estimate that the lifetime morbid risk for a depressive episode (proportion of people who will eventually develop a disorder during their life) is 29.9% (Kessler, Petukhova, Sampson, Zaslavsky, & Wittchen, 2012). Furthermore, depression leads to a significant decrement in quality of life, loss of work productivity, increases in health care costs (Simon, 2003), and an average loss of over 20 years of life (Colton & Manderscheid, 2006). As of 2007, the national expenditures on care for mood and anxiety disorders were over $36 billion (Soni, 2010). In addition, depression has a relatively early onset, typically prior to or during one's twenties. The lifetime prevalence for a depressive episode among 13- to 17-year-olds is 12.6% (Kessler et al., 2012), but depression can set in even earlier among children/adolescents (see Costello, Mustillo, Erkanli, Keeler, & Angold, 2003) and has been characterized in children as young as preschool age (Luby, Heffelfinger, et al., 2009; Luby, Si, Belden, Tandon, & Spitznagel, 2009). Yet, the efficacy of common treatments for adult major depressive disorder (MDD) is generally quite low. Particularly, remission rates with antidepressant treatment are generally $<30\%$, while dropout rates from such trials are high (Pigott, Leventhal, Alter, & Boren, 2010). Similarly, low remission rates are observed

Systems Neuroscience in Depression. http://dx.doi.org/10.1016/B978-0-12-802456-0.00002-9

for cognitive behavioral therapy and psychodynamic therapies, with 40% of patients seeking additional treatment afterward (Driessen et al., 2013).

The experience of stressful life events (SLEs), symptom heterogeneity within depression (or different subtypes), comorbidity, genetic risk (e.g., family history), and low social support can all contribute to ineffective treatment and treatment resistance ($\sim$20–30% of patients fail to respond to at least one standard course of antidepressants) (Fava & Davidson, 2005). As such, understanding the role of factors such as stress in the etiological mechanisms underlying depression may be key to improved treatment and diagnostics. This chapter starts by briefly reviewing prior work linking stressors, particularly in early life, to depression as well as potential genetic moderators of this effect. We then review work on potential mediators of the relationships between stress and depression, namely cortisol function and brain structure/function. The hypothesized potential relationships among these factors are outlined in Figure 1, which provides a schematic representation of the mechanisms of interest in this chapter. Particularly, we will discuss a mechanism implicit in

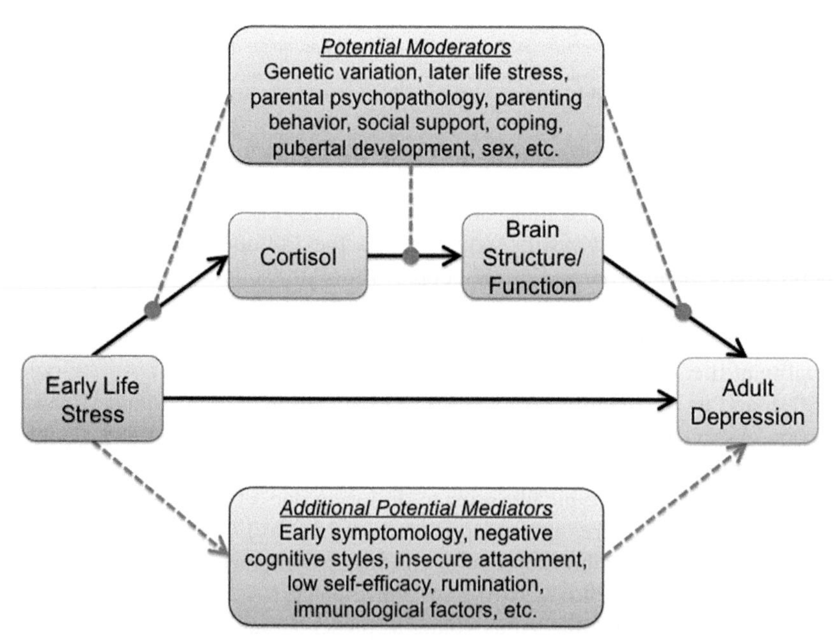

FIGURE 1 Potential mediators and moderators of the relationship between ELS and depression. This figure displays the proposed mechanism of interest by which alterations in cortisol and brain structure/function mediate the effects of ELS on depression. While this mechanism is the main focus of this chapter, we present a variety of other factors that have been suggested to mediate the effects of stress on the risk for depression. Similarly, while we briefly discuss genetic moderators, we also suggest other potential moderating factors in this schematic diagram.

much of the literature, namely that childhood stress leads to alterations in cortisol function, which in turn leads to alterations in the brain that contribute to the risk for MDD. Finally, we discuss some common limitations and future directions to this domain of work.

LINKING STRESS EXPOSURE AND DEPRESSION

Genetic versus environmental contributions to depression: A meta-analysis of genetic epidemiology studies found that the heritability of liability to MDD is ~37% (95% confidence interval [CI], 31–42%), which is on the lower end of heritability for psychiatric disorders. Further, although there is a minimal contribution of shared environmental factors (95% CI, 0–5%), there is a large effect of individual-specific environment of ~63% (95% CI, 58–67%) (Sullivan, Neale, & Kendler, 2000). This suggests that environmental factors are major contributors to the risk for depression. The experience of stressful and traumatic life events, particularly early life stress (ELS), has been cited as one of the foremost individual environmental factors contributing to MDD risk (Green et al., 2010; Kendler, Karkowski-Shuman, & Prescott, 1999; Kendler et al., 1995; Kendler, Kuhn, & Prescott, 2004; Kessler & Magee, 2009).

Proximal Stress: While this review focuses on the effects of ELS, there is also a clear and direct effect of more proximal stressors on the onset of MDD. Specifically, a large proportion of first onsets of MDD are preceded by a stressful event (e.g., Stroud, Davila, & Moyer, 2008). Some studies have aimed to clarify this effect based on characteristics of the provoking stressors. For example, the risk for depression was particularly elevated following threatening or interpersonal stressors, like assault, serious housing, marital, or financial problems, etc. (Kendler et al., 1999). In adults, events characterized by loss (decreased sense of connectedness) have been particularly related to depression onset, whereas events characterized by danger are more often associated with the onset of anxiety (Finlay-Jones & Brown, 1981; Kendler, Hettema, Butera, Gardner, & Prescott, 2003). A similar distinction has also been observed in a pediatric sample of twins (Eley & Stevenson, 2000). Other work has examined the impact of specific adult stressors on the risk for MDD. For instance, it has been suggested that nearly half of female victims of intimate partner violence will experience MDD (average odds ratio (OR), 3.8), with a dose-dependent relationship with the severity of violence (Golding, 1999). Low socioeconomic status (SES) has also been examined as a risk factor for depression; a meta-analysis indicated that low SES (index by income or education level) predicted increased odds for depression (OR, 1.81), particularly for persistent MDD (OR, 2.06) (Lorant et al., 2003). Low SES likely has more chronic and diffuse effects than acute stressors or may serve as a proxy for particular stressors, like financial stress, job insecurity, etc.

Early life stress: In addition to the effect of proximal stressors on the onset of MDD, ELS has been implicated in increasing the risk for MDD.

Various types of ELS have been studied in relation to MDD, including physical and sexual abuse, neglect, family discord/disruption, parental divorce, death of friends/family, natural disasters, low SES, parental psychopathology, etc. Epidemiological work by Kessler, Davis, and Kendler (1997) suggested that a large percentage of individuals with MDD have experienced at least some form of significant ELS (74.4% reporting at least one childhood adversity, including loss, parental psychopathology, interpersonal trauma/abuse, accidents/disasters, etc.). Furthermore, the experience of ELS predicted increased liability to all forms of adult psychopathology, though the most specific effects were observed for mood disorders. Particularly, the experience of parental divorce, physical or sexual abuse, accidents, and disasters all predicted increased odds of first onset of MDD (all OR $>$ 1.5, highest was 3.01 for repeated rape). Generally, these effects held when controlling for comorbid disorders, indicating some degree of specificity of ELS on MDD risk. Finally, this work also noted a potential time decay in the risk for pathology after an experience of certain types of ELS (parental divorce, physical attack, or sexual abuse) where the odds of MDD onset decreased for 10 years after the adversity (until reaching an OR of 1). Thus, the risk for onset of mood pathology was highest early after the stress, but "scars" persisted for up to ~10 years, though there is also a suggestion that individuals who did not experience pathology by that point might potentially be more resilient to developing later pathology (Kessler et al., 1997). Thus, there is considerable evidence that a variety of stressors experienced early in life can serve as potent risk factors for the development of MDD. However, work has further examined the specific effects of particular types of ELS, such as childhood sexual abuse, low SES, and even parental pathology, which may confer risk via both genetic and environmental mechanisms.

Childhood sexual abuse (CSA): A growing body of work has focused on the effects of particular early stressors or traumas, for example, CSA. CSA has been consistently associated with adult-onset MDD in both men and women, though reports of CSA are much more prevalent among women (for review, see Weiss, Longhurst, & Mazure, 1999). Particularly, results from the National Comorbidity Survey indicated that ~13.5% of women report a history of CSA versus only ~2.5% of men; these reports where split between isolated and repeated instances, and most were perpetrated by a relative or acquaintance. Importantly, CSA was associated with an increased risk for mood disorders among women (OR, 1.8 for MDD) where nearly 40% of women with MDD had a history of CSA. On the other hand, although numerically elevated, the risk was not significantly increased among men for MDD after CSA, though the risks for PTSD and substance use were increased in both men and women who experienced CSA. Interestingly, the effect of CSA was particularly salient among women with no other reported childhood adversities (OR, 3.8 for MDD), which the authors suggest might indicate particularly damaging effects of this unexpected betrayal of trust in an otherwise healthy environment

(Molnar, Buka, & Kessler, 2001). This presents an important contrast to other work showing overadditive effects of multiple types of ELS (e.g., Kessler et al., 1997). Importantly, this research also suggested some degree of temporal decay of the effects of abuse, as was also suggested more generally by Kessler et al. (1997). Again, the odds of depression were higher earlier after abuse (minor MDD in adolescents, OR, 15.5) than in adulthood (OR, 2.19 for MDD, 7.80 for recurrent MDD) (Collishaw et al., 2007).

Socioeconomic status: Low SES during one's childhood (generally index by household income, parental education attainment, or parental employment) has been cited as a risk factor for the development of MDD (hazard ratio[1] = 1.57) as well as for more recurrence/relapse (rate ratio = 1.61) (Gilman, Kawachi, Fitzmaurice, & Buka, 1999). This effect has been suggested to be robust, even controlling for a family history of MDD and adult SES, indicating a key role of low SES in childhood (Gilman, 2002). Poverty exposure before age 5 has also been shown to longitudinally predict increased anxious-depressed symptoms at age 14 (Spence, Najman, Bor, O'Callaghan, & Williams, 2002). There was also some evidence that the effects of low SES may be more salient among females (Gilman, 2002; Spence et al., 2002). Interestingly, other work has suggested that while children from poor families exhibited both more internalizing and more externalizing disorder symptomology, moving out of poverty only alleviated externalizing symptomology (though this was over a relatively short time scale of ~4 years during childhood) (Costello, Compton, Keeler, & Angold, 2003).

It is important to note that low SES is generally coincident with a variety of other specific chronic and acute stressors and thus may serve as a marker of generally elevated stress exposure or of particular stressors of relevance to the risk for MDD. Doucet and Doucet (2003) reviewed a variety of the means by which poverty can influence the risk for depression in women. Though they did not focus on childhood experiences of low SES, they highlighted many factors that can also affect children in low SES households either directly or indirectly via stress on their parents, including a greater experience of stressful events, greater rates of abuse, chronic financial stress/instability, discrimination, loss of financial and medical resources, parental stress, loss of social support, and social stress. Finally, it should also be noted that some work has suggested that low SES may be more related to anxiety than depressive disorders (Kessler et al., 1994; Miech, Caspi, Moffitt, Wright, & Silva, 1999).

Parental psychopathology: A parental history of psychopathology, particularly MDD, has often been cited as one of the biggest risk factors for developing depression oneself. Multiple prospective, longitudinal studies have shown that the offspring (children and adolescents) of depressed parents are at

1. In a Cox regression, a hazard ratio of 2 means that the event will occur twice as often at each time point given a one-unit increase in the predictor, i.e., this represents an instantaneous probability rather than a cumulative probability, as from a relative risk ratio.

an approximately three times greater risk for developing MDD as well as anxiety and substance use disorders (Lieb, Isensee, Höfler, Pfister, & Wittchen, 2002; Weissman et al., 2006; Williamson, Birmaher, & Axelson, 2004). Parental MDD is also associated with earlier MDD onset and a more severe course (Lieb et al., 2002). Meta-analyses have shown that both maternal MDD (Goodman et al., 2011) and paternal MDD (Kane & Garber, 2004) robustly predict elevated internalizing symptomology in offspring.

It is important to note that it is often difficult to parse the mechanisms by which parental MDD increases the risk among offspring from these results. As indicated by the heritability results outlined above, a portion of this risk is likely due to inherited genetic and epigenetic factors. But, importantly, parental depression has also been highlighted as an environmental risk factor, i.e., mediated through exposure to parental depression or coincident stressors. This has been robustly demonstrated across a variety of study designs/populations, including modeling of data from a children of twins design (Silberg, Maes, & Eaves, 2010) and work showing increased depression among genetically unrelated offspring of depressed parents by adoption (Tully, Iacono, & McGue, 2008) or by in vitro fertilization (Harold et al., 2011). Thus, increased early stress exposure is likely a salient mediator of the effect of parental depression. Particularly, parental psychopathology has been shown to cluster with other types of childhood adversity, including parental substance misuse and criminality, family violence, abuse, and neglect, where there was little evidence for specific effects of any particular type of stressor on pathology risk (Green et al., 2010). Further, several studies have implicated SLEs among offspring (Hammen, Hazel, Brennan, & Najman, 2012) and poor parenting/family environment (Burt et al., 2005; Garber & Cole, 2010) as important environmental mediators in the intergenerational transmission of depression. Parenting practices may also be a significant factor, i.e., mothers with a history of depression and particularly current depression show less positive and more negative parenting behaviors (Lovejoy, Graczyk, O'Hare, & Neuman, 2000). Importantly, a parental history of MDD may also serve as a moderator of the effects of stress on the risk for MDD. For example, work has found that SLEs only predicted MDD among girls with a history of MDD, potentially suggesting a gene × environment interaction or an interaction with other coincident factors, like differences in parenting, etc. (Silberg, Rutter, Neale, & Eaves, 2001).

COMPLEXITIES IN THE RELATIONSHIP OF STRESS TO DEPRESSION

Stress generation and gene–environment correlations: The stress generation hypothesis posits that depressed individuals are likely to behave in ways contributing to the occurrence of negative events in their lives (Hammen, 1991). Thus, it often behooves studies linking stress to depression risk to try to differentiate dependent events, those at least partially influenced/created by the individual, from independent events, those out of the individual's control. Particularly in retrospective studies, group differences in stress exposure between depressed and nondepressed individuals may capture generated stress rather than causal effects of stress on clinical outcomes. While stress generation appears to be a large factor at work in depression (for a review, see Liu & Alloy, 2010), independent stressors still have been found to predict MDD onset. For example, independent stressful events were associated with depression onset in a sample of female twins, though the probability of depression onset after dependent events was ~80% greater than following independent events, even when controlling for the threat level of stressors (Kendler et al., 1999). Stress generation also has important implications for parental depression as a risk factor for depression; particularly, it has been suggested that children of depressed mothers are experiencing more stress, in part generated by their mothers, and the behavioral challenges of these at-risk children are also contributing to increased stress experiences (Adrian & Hammen, 1993). This reciprocal relationship likely has depressogenic effects on both the mother and child, and these interpersonal stressors may mediate the effects of maternal depression on children's risk for MDD. Similarly, reciprocal longitudinal relationships have been suggested between the development of children's irritability symptomology and maternal depression (Wiggins, Mitchell, Stringaris, & Leibenluft, 2014), potentially relating to increases in parenting stress.

Though generally studied/conceptualized separately, stress generation may be of relevance to work examining gene–environment correlation (for a review see Jaffee & Price, 2007) or examining stress as a mediator of the genetic risk for depression. Particularly, it has been suggested that while parental depression predicts less positive parental engagement and greater parent-imposed SLEs, these factors only mediated the effects of parental depression on child depression in genetically related dyads versus genetically unrelated dyads conceived through in vitro fertilization (Rice, Lewis, Harold, & Thapar, 2013). This type of gene–environment correlation effect has also been examined by estimating the heritability of SLEs and examining SLEs as a mediator of genetic risk. For example, in a sample of older adult twins, the heritability of controllable life events was estimated at 43% while the heritability of uncontrollable events was estimated at 18% (Plomin, Lichtenstein, Pedersen, McClearn, & Nesselroade, 2004), suggesting that the tendency to select into or generate (controllable) stress may mediate some of the genetic influences on MDD. Particularly, it has been suggested that ~10–15% of the genetic liability for MDD is mediated through SLE exposure (Kendler & Karkowski-Shuman, 1997). Also, of relevance to ELS influences, associations between measures of parenting and depression may be due in part to shared genetic influences (Pike, McGuire, Hetherington, Reiss, & Plomin, 1996).

Early × later stress interactions: Several theories have been proposed by which the experience of ELS may moderate the effects of later life stress on the risk for depression (and other outcomes). These fall under the umbrella of stress–diathesis interactions (Monroe & Simons, 1991), where later stress interacts with the individual risk for depression, here determined by childhood adversity exposure, but are hypothesized to take several possible forms. The three commonly hypothesized interactions are presented in Figure 2. First, the stress-amplification theory (Figure 2(a)) posits an overadditive interaction between early and later stress where a history of childhood stress/adversity increases the risk for depression among those experiencing high levels of later stress. Some evidence for this hypothesis has been found; for example, in prepubertal girls, an experience of moderate early adversity amplified the effects of high levels of stress on depression risk (Rudolph & Flynn, 2007). Alternatively, the stress-inoculation theory (Figure 2(b)) suggests that ELS (though potentially only less severe stress) may buffer against the effects of later life stress on the risk for depression, particularly more severe later stress. This type of "steeling" effect (Chorpita & Barlow, 1998) potentially suggests that ELS may prompt individuals to develop adaptive coping mechanisms earlier, reducing the risk after later stress. There is some scattered evidence in the literature that, for example, chronic stress may reduce the impact of acute stressors on the risk for depression (McGonagle & Kessler, 1990) or that the emotional effects of rape are lower among those experiencing minor (but not large or no) recent life changes (Ruch, Chandler, & Harter, 1980). However, there is little direct evidence that childhood adversity buffers against later stress-induced risk for depression.

Finally, the stress-sensitization theory (Figure 2(c)) posits an under-additive interaction, where childhood stress increases the risk for depression following more mild stress (but not differentiating effects of severe stress), i.e., childhood stress makes individuals more sensitive to the effects of even mild stress (Harkness, Bruce, & Lumley, 2006). Evidence has also been found for this model; for example, women with a history of childhood adversity show an increased risk for depression at lower levels of stress, whereas childhood adversity does not differentiate risk under high stress conditions (Hammen, Henry, & Daley, 2000). This also may relate to the "kindling" hypothesis (Post, 1992), where the effect of stressors on the onset of MDD changes with each subsequent depressive episode (Kendler, Thornton, & Gardner, 2000), with later episodes requiring less and less severe provoking triggers (for a conceptual review, see Monroe & Harkness, 2005; for meta-analysis, see Stroud et al., 2008). Further, there is evidence that individuals at risk for depression (based on their twin's history of MDD) may be "prekindled" and show this altered relationship between stress and depression even at their first episode (Kendler, Thornton, & Gardner, 2001).

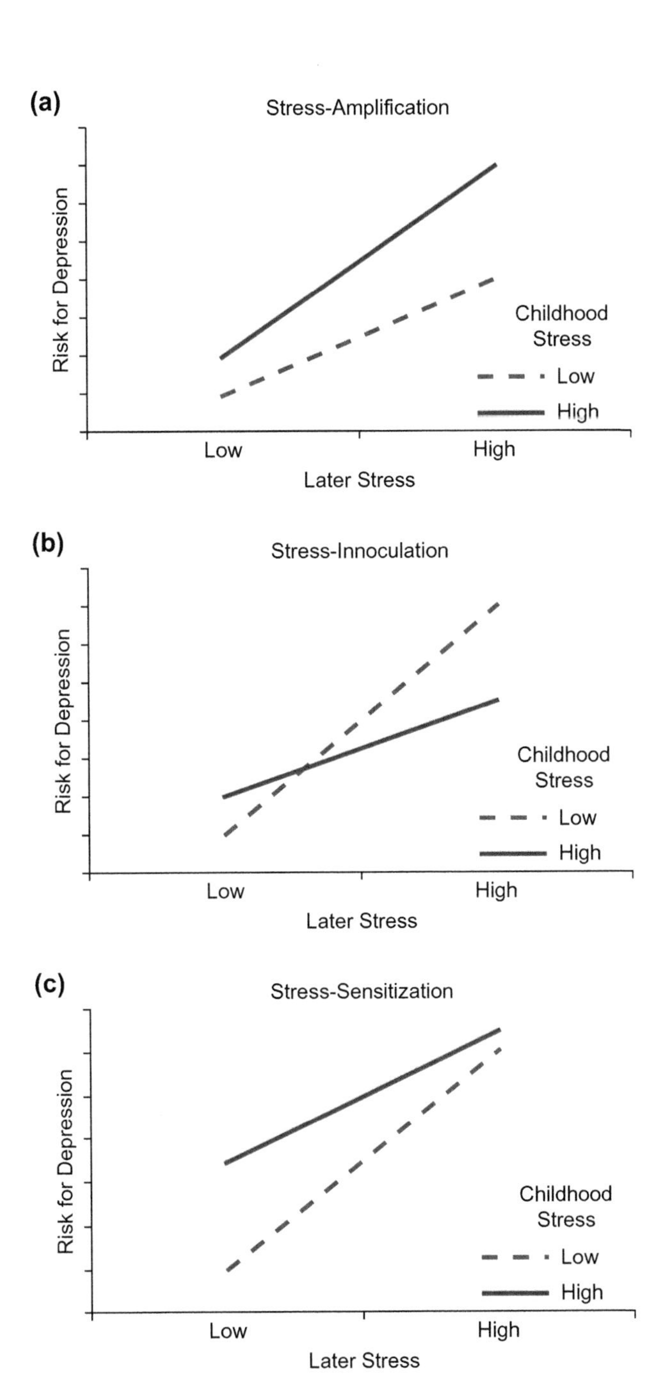

FIGURE 2 **Schematics of early × later life stress interaction theories.** These graphs display hypothetical patterns to schematize the three main theories regarding how early life and later life stress exposure interact to predict the risk for depression. The stress-amplification theory (a) predicts an overadditive interaction where greater early and later life stress both predict an increased risk for depression and experiencing ELS further amplifies the effects of later life stress. The stress-inoculation theory (b) predicts that the experience of ELS will buffer or steel an individual against the deleterious effects of later life stress. The stress-sensitization theory (c) predicts that experiencing ELS will make an individual more sensitive to the effects of later life stress, thus increasing the risk for depression even at low levels of stress but not differentiating the risk for depression in high stress conditions in later life.

Further, individuals developing depression may exhibit different interactions between early and later stress, e.g., stress sensitization versus stress amplification, than those who do not develop depression. One potential further connection here is that the prekindling effect could be mediated by early onset pathology. In other words, if the actual first onset of symptomology in those with a family history is early and is not being identified in these studies, then they may be kindled at their later adult "onsets" because they may have already had a nondiagnosed episode. There is some evidence that earlier MDD onset (and continued stress exposure) may mediate the effects of early adversity on later MDD (Turner & Butler, 2003). This, and the added genetic and environmental risk from a family history of MDD, could help to explain prekindling effects and should be examined further in future work.

Genetic moderators of the relationship between stress and mood pathology: Work has also begun to examine how normally occurring genetic variants may moderate the effects of stress on the risk for depression and other psychiatric and neural outcomes. Seminal work by Caspi et al. (2003) largely piqued interest in understanding the role of gene × environment interaction in depression by suggesting that carriers of the short allele of the serotonin transporter linked polymorphic region (5-HTTLPR) had an increased risk for depression following stress than long allele carriers (Caspi et al., 2003). While replications of this effect have been somewhat mixed, with one meta-analysis finding no significant gene × environment interaction (Risch et al., 2009), a larger meta-analysis did find significant moderation effects (Karg, Burmeister, Shedden, & Sen, 2011). This meta-analysis specifically noted a significant interaction between 5-HTTLPR allele and childhood maltreatment predicting the risk for depression. Further, it has been suggested that while the short allele may sensitize individuals to the depressogenic effect of stress, it may also lead to a decreased risk for MDD in those with positive early environments as compared to long allele carriers who showed little difference in MDD symptomology based on stress (Taylor et al., 2006). This has been conceptualized as biological sensitivity to context (Boyce & Ellis, 2005) or differential susceptibility (Belsky & Pluess, 2009), i.e., that 5-HTTLPR and other polymorphisms confer sensitivity to one's environment rather than only conferring risk for psychopathology. Rather, the pattern of interaction is confounded by the type of environmental factors studied, i.e., focusing only on negative factors limits our ability to see potentially beneficial effects of genetic variants in those with positive environments. Belsky and Pluess (2009) review similar evidence in the literature regarding other genetic variants of interest, e.g., MAOA, DRD4, etc. Additionally, genes relevant to the function of the *hypothalamic–pituitary–adrenal* (HPA) axis, e.g., FKBP5, have been implicated as moderators of the effects of stress on the risk for depression (Zimmermann et al., 2011) and other outcomes, like PTSD (Klengel et al., 2012).

MEDIATORS OF THE RELATIONSHIP BETWEEN STRESS AND MOOD PATHOLOGY

As reviewed above, there is robust literature linking stressor exposure and risk for depression. However, despite the strength of these links, it is not yet clear how the experience of stressors contributes to the risk for depression. Understanding the mechanisms that mediate the influence of stress on later pathology is critical if we are to develop effective preventative or early intervention efforts designed to stave off the deleterious effects of stress. As outlined in Figure 1, a number of important factors have been suggested as potentially critical mediators, including psychological factors, cortisol function, and changes in brain functions and structure. Below we briefly review the literature on each of these potential mediators.

Psychological mediators: A variety of psychological factors have been suggested to potentially mediate the effects of ELS on the risk for depression. The hypothesis in this domain is that ELS may alter or confer certain cognitive and emotional factors that in turn increase the risk for MDD. Decreases in self-efficacy (the personal belief about one's ability to control one's environment and circumstances) (Bandura, 1977) have been suggested to mediate ~40% of the effect of dependent life events on later MDD (Maciejewski, Prigerson, & Mazure, 2000). Insecure attachment style, negative cognitive style, and later life events have also been suggested as mediators of the effect of childhood maltreatment on later MDD (Hankin, 2006). Similarly, the development of cognitive schemas with themes of loss and worthlessness may also mediate relationships between childhood adversity and anhedonic symptomology in depressed adolescents (Lumley & Harkness, 2007). Ruminative tendencies have also been suggested as potential mediates between SLEs and depressive symptomology in adults (Michl, McLaughlin, Shepherd, & Nolen-Hoeksema, 2013). Finally, early-onset MDD symptoms may mediate the effects of childhood trauma on later outcomes as well (Hovens et al., 2012). While there are likely other psychosocial factors that may mediate the effects of ELS on the risk for depression, it is clear that negative cognitive styles/tendencies and subclinical symptomology are important constructs to consider in understanding the effects of ELS.

Cortisol Function as a Mediator between Stress and Depression

Cortisol and Depression

One of the major hypotheses as to a biological mechanism by which stress may contribute to the risk for MDD is via impacts on the HPA axis, which mediates the body's response to stress. Particularly, it has been suggested that stress exposure may lead to alterations in the reactivity/regulation of the HPA axis and thus cortisol output, contributing to the risk for MDD. MDD and ELS have been robustly linked to alterations in cortisol function, though findings vary based on the outcome of interest, i.e., basal cortisol, response to acute psychological stressors, pharmacological challenge, diurnal rhythms, etc. Nonetheless, the importance of cortisol function in depression has been noted in many studies, for example, suggesting that up to 90% of MDD patients may be characterized by an increased cortisol response to the administration of dexamethasone (DEX) and corticotropin-releasing hormone (CRH), where HPA axis dysregulation is most likely due to impaired feedback inhibition (Heuser, Yassouridis, & Holsboer, 1994). Dysregulation has also been observed in people with a high familial risk for MDD, indicating that HPA dysregulation or stress sensitivity may relate to familial vulnerability to depression or correlated increases in ELS exposure (Modell et al., 1998). Additionally, MDD patients who respond to treatment show a normalization of HPA axis abnormalities, i.e., a decrease in hyper-responsivity to stress (McKay & Zakzanis, 2010). This has lead to the idea that HPA normalization may be a common pathway of antidepressant effects (Holsboer & Barden, 1996). Relatedly, an older meta-analysis examined the utility of the DEX suppression test[2] (DST) for predicting treatment response. This study found that while DST suppression did not predict a short-term response to antidepressants, persistent nonsuppression after treatment strongly predicted early relapse and poor outcomes (Ribeiro, Tandon, Grunhaus, & Greden, 1993), again supporting this idea that HPA axis normalization is key to treatment outcomes.

Recent meta-analyses have noted more nuanced relationships between cortisol function and depression. Particularly, when examining cortisol response to psychosocial stressors, adults with MDD tended to exhibit similar baseline and stress-induced cortisol levels as compared to healthy controls, but MDD patients had much higher cortisol levels during recovery after stress ($d = 1.39$) (Burke, Davis, Otte, & Mohr, 2005). This is similar to the DST results above, implicating impaired cortisol negative feedback regulation in MDD. However, this effect may have been confounded by time of day, where higher baseline cortisol was noted when examined in the afternoon ($d = 0.83$). Interestingly, there was also a trend toward a more pronounced blunted reactivity-impaired recovery pattern among studies with older and more severely depressed patients (Burke et al., 2005). In a meta-analysis of 20 studies on basal salivary cortisol levels, MDD patients showed increased cortisol levels at morning acquisitions as compared to controls (2.58 nmol/L; $Z = 3.10$) and a smaller increase during the evening (0.27 nmol/L; $Z = 2.18$), though the distributions were highly overlapping across groups (Knorr, Vinberg, Kessing, & Wetterslev, 2010). Finally, an elevated DST response (Hedge's $G = 0.57$) and basal cortisol levels (Hedge's $G = 0.20$) have also been observed among depressed children and adolescents, with no time-of-day

2. The DST measures the degree to which ACTH secretion by the pituitary can be suppressed. Nonsuppression is indicative of stress-system dysfunction.

effects noted on differences in basal cortisol (Lopez-Duran, Kovacs, & George, 2009). Further, depressed children and adolescents showed normative responses to CRH infusion but elevated responses to psychological stressors, which has been suggested to implicate impaired negative feedback but intact activation, as seen in adults (Lopez-Duran et al., 2009). Kaufman, Martin, King, and Charney (2001) review similarities in cortisol (and other) dysfunction across childhood, adolescent, and adult MDD.

ELS and Cortisol

Given the links between ELS and depression described above, one potential pathway to altered cortisol function in depression is via the influence of ELS on cortisol function. While ELS has been related to cortisol function in many studies, to our knowledge, there has not been a comprehensive meta-analysis examining this relationship. We will review the general findings of this large literature but will not provide exhaustive detail on all prior work. Particularly, we will highlight the mixed findings relating ELS to cortisol as a function of age and ELS exposure across a variety of cortisol outcome measures.

ELS and Cortisol in Children/Adolescents

Studies examining cortisol function in children have typically found blunted levels after exposure to ELS, though some work has noted elevated levels. Blunted cortisol has been observed across a variety of outcome measures; for example, decreased morning and blunted production across the day have been observed in maltreated children (Gunnar & Vazquez, 2001) and in maltreated children with anxiety/depressive symptomology in other samples (Cicchetti, Rogosch, Gunnar, & Toth, 2010; van der Vegt, van der Ende, Huizink, Verhulst, & Tiemeier, 2010). Several studies have related blunted cortisol responses to laboratory stressors to ELS exposure in children (Pagliaccio et al., 2014) or to a history of abuse in adolescents (MacMillan et al., 2009). It has also been found that sexually abused girls may show blunted adrenocorticotrophic hormone (ACTH) but normal cortisol response to CRH administration (De Bellis et al., 1994). Gunnar and Quevedo (2008) have reviewed associations between ELS and cortisol function in abused, neglected, and deprived children in more depth, with an eye toward mechanisms underlying trauma vulnerability and PTSD. On the other hand, several studies have found opposing results. For example, Cicchetti and Rogosch (2001) did not identify differences in cortisol based on maltreatment history in children, though a subset of maltreated children did exhibit elevated cortisol predicted by a history of severe or repeated abuse. Increased morning cortisol has also been noted among maltreated children with internalizing symptomology (Cicchetti, 2009). Kaufman et al. (1997) noted elevated ACTH responses to CRH among depressed, maltreated children who were still living under conditions of intense family disruption and emotional abuse.

Institutional/Foster Care and Cortisol

A variety of studies have also examined the relationships between specific types of early adversities and cortisol function, for example, neglect during early institutional care. Generally, low cortisol levels have been found in children experiencing early institutional or foster care as assayed in toddlers (Bruce, Kroupina, Parker, & Gunnar, 2000), infants, and preschoolers (Dozier, Peloso, Lewis, Laurenceau, & Levine, 2008; Fisher, Burraston, & Pears, 2005). In other work, children in foster care showed lower waking cortisol than low-income community controls where cortisol levels were negatively correlated with severity of physical neglect (but positively related to severity of emotional maltreatment; Bruce, Fisher, Pears, & Levine, 2009). There has also been some suggestion of elevated morning cortisol in postinstitutionalized children exposed to deprivation (e.g., Gunnar, Morison, Chisholm, & Schuder, 2001). Importantly, work has also suggested that alterations in cortisol function may remit/improve after children are adopted into supportive families. Particularly, low/blunted cortisol began to remit in orphan-reared children 8 months after adoption (Bruce et al., 2000).

Childhood SES and Cortisol

Evidence for relationships between childhood SES and cortisol levels has been relatively sparse and mixed. Chen, Cohen, and Miller (2010) noted that lower neighborhood and lower family SES predicted lower mean salivary cortisol levels in adolescents. On the other hand, Lupien, King, Meaney, and McEwen (2000, 2001) found that low SES related to higher cortisol levels in 6- to 11-year-old children but did not find a significant relationship in 12- to 16-year-olds. Further, Goodman, McEwen, Huang, Dolan, and Adler (2005) found no relationship between parental SES and serum cortisol in adolescents. Li, Power, Kelly, Kirschbaum, and Hertzman (2007) found that childhood SES was associated with elevated early cortisol, but after simultaneous adjustment, only adult SES had an effect. Thus, these results have been quite mixed and may vary by age and sample. For more discussion and a general review of SES and cortisol, see Dowd, Simanek, and Aiello (2009).

Parental Depression, Stress, and Cortisol

Elevated morning cortisol and cortisol variability have been observed in 13-year-old children exposed to postnatal maternal depression, where this increase was not mediated by parental conflict or life events (Halligan, Herbert, Goodyer, & Murray, 2004). Similarly, elevated cortisol has been observed even earlier, in children exposed to postnatal depression (Essex, Klein, Cho, & Kalin, 2002). While these studies typically found specific effects of a family history of depression or exposure to maternal depression, they have also suggested that elevated parenting stress predicts elevated cortisol among infants across the day (Saridjan et al., 2010), suggesting some role for stress exposure as well. Additionally, adults with a family history of depression have been suggested to exhibit elevated cortisol levels as compared to those without a family history, and this difference was again not accounted for by parental attachment, life events, personality, or current mental state (Mannie, Harmer, & Cowen, 2007).

ELS and Cortisol in Adulthood

There is again mixed evidence for elevated versus blunted cortisol functioning in adults with a history of childhood adversity across a variety of outcome measures. In support of elevated cortical function, women with a history of maltreatment but no current psychiatric diagnoses showed elevated ACTH but not cortisol in response to a CRH challenge and a psychological stressor (Heim & Nemeroff, 2001; Heim, Newport, Heit, & Graham, 2000). This effect of maltreatment has similarly been shown in women with borderline personality (Rinne et al., 2002). A history of maltreatment also predicted an elevated ACTH and cortisol response to psychological stressors among depressed women (Heim & Nemeroff, 2001). Interestingly, early adversity and chronic stress exposure may show an overadditive interaction (similar to stress amplification), predicting peak cortisol to a psychological stressor in adults (Rao, Hammen, Ortiz, Chen, & Poland, 2008). Men with childhood trauma exposure exhibit increased ACTH and cortisol response after DEX/CRF administration, associated with severity, duration, and onset of abuse (Heim, Mletzko, Purselle, Musselman, & Nemeroff, 2008).

In support of blunted cortisol function in adulthood, other work has suggested that childhood maltreatment was associated with significantly lower cortisol and ACTH response to stressors, where neglect and physical abuse strongly predicted lower peak cortisol response (Carpenter et al., 2007). Blunted responses to psychosocial stressors have also been observed in adult women with a history of adversity (Carpenter, Shattuck, Tyrka, Geracioti, & Price, 2011). Similarly, enhanced cortisol suppression was seen during a DST in women with a history of CSA (Stein, Yehuda, Koverola, & Hanna, 1997) as well as diminished cortisol response to DEX/CRH challenge in those with self-reported childhood emotional abuse (Carpenter et al., 2009). A landmark study by Heim, Newport, Mletzko, Miller, and Nemeroff (2008) provided an in-depth characterization of relationships between ELS and stress-system functioning, suggesting some similarities and differences with alterations seen in MDD (and other pathology, like PTSD). The study suggested that ELS may account for much of the underlying alterations seen in MDD.

ELS and Cortisol: Summary

Overall, this work has tended to suggest that ELS is more consistently associated with blunted cortisol output in children/adolescents but provides mixed evidences for blunted versus elevated cortisol in adults with a history of ELS.

While blunted stress responsivity may actually be adaptive in the face of repeated stress, individuals with a familial history of MDD generally show elevated cortisol output, similar to what is generally seen in individuals with MDD themselves. More work is needed to fully parse the effects of ELS on the developmental trajectories of normal cortisol function, to examine whether certain types of stressors are differentially associated with elevated/blunted cortisol, and to relate stress-related blunting in cortisol to other outcomes, like brain developmental and psychopathology. Additionally, it will be important to examine elevated cortisol as a mediator of a family history of MDD on the risk for MDD and/or as a moderator of effects of ELS on the brain and psychopathology.

CHANGES IN BRAIN STRUCTURE AND FUNCTION AS A MEDIATOR BETWEEN STRESS AND DEPRESSION

As described above and highlighted in Figure 1, another component of the pathway linking ELS and depression may be that ELS contributes to alterations in brain structure and function, potentially via its impact on dysregulated HPA axis function and cortisol levels. To review the evidence for this, we first summarize, in broad strokes, the evidence for altered brain structure, function, and connectivity in depression. We then review the evidence for relationships between ELS and alternations in brain structure, function, and connectivity. Finally, we review the much smaller literature directly testing cortisol function as a mediator of the relationships between ELS and altered brain structure and function in humans.

Depression and Brain Structure

Depression has been related to a variety of alterations in brain structure and function; we will very briefly review meta-analyses to summarize current findings on these structural and functional alterations in MDD. Particularly, structural studies have noted alterations in gray matter density or volume in a variety of cortical and subcortical regions in MDD, often highlighting the anterior cingulate cortex (ACC), amygdala, and hippocampus. Cortically, meta-analyses of voxel-based morphometric (VBM) studies have noted decreased gray matter in regions of the rostral ACC, dorsomedial prefrontal cortex (dmPFC), dorsolateral PFC, and orbitofrontal cortex (OFC) in adults with MDD (Bora, Fornito, Pantelis, & Yücel, 2012; Koolschijn, van Haren, Lensvelt-Mulders, Hulshoff Pol, & Kahn, 2009; Sacher et al., 2012).

Subcortically, these meta-analyses have noted volumetric reductions in the hippocampus ($d = -0.41$), the putamen ($d = -0.48$), and the caudate ($d = -0.31$) (Koolschijn et al., 2009). Given the robust relationships between MDD and hippocampal volumes, two meta-analyses have specifically examined this and confirmed reductions in bilateral hippocampal volumes among depressed patients across measurement techniques (Campbell, Marriott, Nahmias, & MacQueen, 2004; Videbech & Ravnkilde, 2004). Further, the number of depressive episodes experienced negatively correlated with right hippocampal volumes (Videbech & Ravnkilde, 2004) where it has also been suggested that reductions are only observed in individuals experiencing persistent MDD (>2 years); volumes were also reduced in children with MDD (McKinnon, Yucel, Nazarov, & MacQueen, 2009). Studies have thus implicated stress-related mechanisms in the development of these volumetric alterations, though further longitudinal study is warranted. For example, work has found that depressed patients, particularly with chronic or relapsing MDD, show greater decline in hippocampal (and ACC, left amygdala, and right dmPFC) volumes over a 3-year period than healthy individuals (Frodl et al., 2008). Further, smaller hippocampal volumes predict a worse course of illness among patients with MDD (Frodl et al., 2004). Results regarding amygdala volumes in MDD, on the other hand, have been very mixed, even at the meta-analytic level. Studies have suggested that amygdala volumes are decreased in VBM studies of MDD (Sacher et al., 2012) or only decreased in patients with comorbid anxiety disorders or first-episode/drug-free samples (Bora et al., 2012), have found no difference by diagnostic group (Koolschijn et al., 2009), or have suggested larger left amygdala volumes among inpatients (but not outpatients) as compared to controls (Hajek et al., 2009). These mixed results point to key differences in sample populations that may account for these differences but that are not being fully captured here. Particularly, a greater characterization of severity, chronicity, treatment, and stress exposure might help to uncover the true causal factors at play linking amygdala volumes to MDD.

Depression and White Matter

Work has begun to examine alterations in white matter structure/integrity in depression using diffusion tensor imaging (DTI). A meta-analysis by Murphy and Frodl (2011) of seven studies examining DTI in depressed adults indicated decreased fractional anisotropy (FA) in the superior longitudinal fasciculus and increased FA in the fronto-occipital fasciculus, though the latter effect was largely driven by one study (Blood et al., 2010). A meta-analysis found particular decreases in FA in the genu of the corpus callosum in MDD (Wise et al., 2015). Decreases in FA indicate less restricted water flow along a track and have been suggested to indicate reduced white matter integrity or myelination. For a discussion of methodological and interpretation concerns, see Alexander, Lee, Lazar, and Field (2007).

Depression and Task-Related Brain Activity

Many studies have also examined alterations in neural response to emotional stimuli in depression using functional magnetic resonance imaging (see also Chapter 8 by I. Tendolkar on functional imaging in depression).

Meta-analyses have noted a decreased response to positive stimuli in the pregenual ACC, posterior cingulate, putamen, and several other regions, while responses were increased in the subgenual, rectus, inferior, middle, and superior frontal gyrus and other regions (Diener et al., 2012; Fitzgerald, Laird, Maller, & Daskalakis, 2008). Meta-analysis has also identified decreased activity in the striatum, thalamus, cerebellum, insula, and ACC but increased activity in visual and frontal regions to monetary rewards (Zhang, Chang, Guo, Zhang, & Wang, 2013). In response to negative stimuli, depressed patients often showed reduced activity in the regions of the PFC, subgenual, dorsal, and posterior cingulate as well as the insula but showed increased activity in the amygdala, putamen, insula, and other regions (Diener et al., 2012; Fitzgerald et al., 2008; Hamilton et al., 2012).

Depression and Functional Connectivity

Several studies have attempted to meta-analyze functional connectivity studies in depression, though this work is complicated by the diverse analyses methods used to study resting state functional connectivity (rsFC) (Biswal, Zerrin Yetkin, Haughton, & Hyde, 1995). Many studies examine voxel-wise correlations in connectivity with a given seed, seed-to-seed correlations, or more network-based approaches, while others focus on regional homogeneity (ReHo), a data-driven approach examining local correlations (Zang, Jiang, Lu, He, & Tian, 2004). Decreased connectivity in MDD has been noted in superior and middle temporal gyrus, the insula, precuneus, superior frontal gyrus, putamen, and thalamus while increased connectivity was noted in the middle and medial frontal gyrus, precuneus, inferior parietal lobule, ACC, hippocampus, posterior cingulate, and cerebellum (Sundermann, 2014). While these results point to important alterations in functional connectivity in MDD, it is difficult to meta-analyze the current literature, which varies greatly in analysis methods, i.e., examining ReHo versus seed/network-based approaches, and thus it is currently unclear at a meta-analytic level whether or not certain functional networks are particularly altered in MDD. A meta-analysis focused specifically on Chinese studies reporting alterations in ReHo in depression found increases in the medial PFC (mPFC)/dorsal anterior cingulate cortex ReHo and decreased ReHo in the cerebellum, postcentral gyrus, rolandic operculum, cuneus, and inferior parietal gyrus (Iwabuchi et al., 2015).

Summary of Neural Changes in Depression

Across modalities, these studies tend to indicate some common alterations in brain structure and function in individuals with depression, particularly highlighting dysfunction in the amygdala, hippocampus, regions of the cingulate (mainly dorsal, pregenual, and subgenual), and the PFC (mainly medial). However, more work is needed to characterize these relationships, as many of these findings tended to vary by severity, medication status, or by the task used to assay neural reactivity. Particularly, while amygdala reactivity is quite consistently elevated in MDD, findings on amygdala volumes are quite mixed. Finally, while work examining rsFC in depression has proliferated, inconsistent methodologies have largely limited attempts to provide meta-analyses of this growing body of work.

Cortisol and the Brain

As described above, stress dysregulation has been proposed as a major candidate hypothesis underpinning neural alterations in depression, particularly structural changes, though work has been examining relationships with function and connectivity as well. As a case example, we will describe literature linking cortisol to hippocampal volume. Importantly, while the hippocampus is a major source of inhibition on stress circuitry in the brain, the hippocampus is also particularly vulnerable to the neurotoxic effects of cortisol (Jacobson & Sapolsky, 1991). Prolonged, elevated cortisol levels can cause damage to the hippocampus, which may be observed as volumetric loss in humans. Consistent with this hypothesis, animal studies have shown chronic stress and corticosteroid administration to reduce dendritic branching and length of hippocampal CA3 pyramidal neurons (e.g., Conrad, LeDoux, Magariños, & McEwen, 1999; Magariños, McEwen, Flügge, & Fuchs, 1996; McKittrick et al., 2000; Watanabe, Gould, & McEwen, 1992) as well as to potentially cause cell death in the CA3 subfield and to impair neurogenesis in the dentate gyrus with prolonged stress exposure (Reagan & McEwen, 1997; Sapolsky, 2000). Particularly, it has been suggested that these cortisol effects may damage the hippocampus via a glutamate-mediated mechanism of excitotoxicity (Magariños et al., 1996; McEwen, 1997; Reagan & McEwen, 1997). Relatedly, stress and cortisol have also been suggested to decrease glutamate regulation by astrocytes or to lead to glial loss (e.g., Campbell & MacQueen, 2004; Cotter, Pariante, & Everall, 2001; Sapolsky, 2000). However, one study importantly showed that stress-induced loss of CA1 dendritic spines and decreases in spine length accounted for gray matter volume loss observed by magnetic resonance imaging (MRI) in mice, rather than neuronal or glial loss (Kassem et al., 2012).

A stress-/cortisol-mediated mechanism of change in hippocampal structure is also supported by the human literature. For example, in humans, high endogenous levels of cortisol across several years are associated with smaller hippocampal volumes in older adults (Lupien et al., 2005). Additionally, patients with Cushing's syndrome who exhibit hypercortisolemia tend to show reductions in hippocampal volume (Andela et al., 2015), where 24-h mean cortisol levels were negatively correlated with hippocampal volumes (Starkman, Gebarski, Berent, & Schteingart, 1992). Importantly, treatment of Cushing's syndrome can lead to enlargement of the hippocampus, associated with reductions in cortisol levels (Starkman et al., 1999). Given the role of the hippocampus in the regulation of the HPA axis, it is also important to note that these stress/cortisol effects on the hippocampus may also impact responsivity to future stressors. For example, there is some evidence from human studies that smaller hippocampal volumes may be a risk factor for developing PTSD after trauma exposure (Gilbertson et al., 2002). Understanding the role of stress and cortisol in determining brain structure may help explain, at least in part, the underlying neural alterations observed in depression. While we focused on the hippocampus here as a case example, cortisol has been linked to the structure and function of other regions as well. For example, Cushing's syndrome (Andela et al., 2015) and cortisol administration have been related to smaller amygdala volumes (Brown, Woolston, & Frol, 2008), and Cushing's syndrome has been linked to increased threat-related activity in the amygdala and hippocampus (Andela et al., 2015).

ELS and Brain Structure

Tables 1 and 2 review studies linking ELS/adversity to brain structure in child/adolescent and adult populations, respectively. ELS was broadly defined here and includes studies on childhood abuse, neglect, early institutional/orphanage care, poverty (or low SES), and general adversity/trauma. Studies were only reviewed if they examined brain volumes or gray matter density, i.e., using automated segmentations or hand-tracing to ascertain regional volumes or using VBM. Study inclusion was not restricted by age at scan, but studies were excluded if ELS was confounded entirely with diagnostic status, e.g., comparing individuals with PTSD secondary to childhood abuse to nonabused individuals. The hippocampus was the most examined region among these studies with 15 finding evidence for reduced volumes among individuals with ELS exposure, though 11 other studies reported no significant associations between ELS and hippocampal volume. Importantly, reductions in hippocampal volumes (and null results) were observed both in adults and younger populations and did not appear specific to any particular type of ELS. Also interestingly, studies finding significant effects tended to note either bilateral or left-specific reductions in hippocampal volume. While the mechanisms for this are unclear, there is evidence for differential heritability of the left and right hippocampus with the left being much less heritable (22.3% versus 77.7%), potentially allowing for a larger environmental effect on volume (Pagliaccio et al., 2015).

The amygdala has also received much attention in the literature, though as with studies in depressed samples, these results are more mixed. Two studies noted larger amygdala volumes in children/adolescents adopted out of institutions/orphanages (i.e., experiencing some degree of early neglect) as compared to controls not experiencing early institutional care. While one study found larger volumes in children who were adopted later versus early, i.e., experiencing more stress/neglect (Tottenham et al., 2010), the other noted

TABLE 1 Studies Examining Children and/or Adolescents

Study	Total *N* (Exposed *N*)	Stressor	Analysis	WBV	HC	Amygdala	Other	Notes
Teicher et al. (2004)	166 (28)	Abuse or neglect	ROI	–		–	↓CC	Also includes psychiatric controls
Mehta et al. (2009)	25 (14)	Institutional care	ROI	↓GMV, ↓WMV	NS	↑ by group, ↓with time institutionalized	NS CC	–
Rao et al. (2010)	87	Early life adversity	ROI	–	↓	–	–	Includes MDD and at risk
Tottenham et al. (2010)	78 (38)	Institutional care	ROI	NS	NS	↑	NS caudate	–
Hanson et al. (2010)	72 (31)	Physical abuse	Voxel-wise, tensor based	↓WBV		–	↓ various frontal, parietal, and temporal regions	–
Hanson, Chandra, Wolfe, and Pollak (2011)	317	SES	VBM	NS	↓	NS	–	–
Brain Development Cooperative Group (2011)	325	SES	ROI	NS	–	–	NS for various frontal, temporal, parietal, and occipital lobe, subcortical, cerebellum, and brainstem regions	Examined changes in longitudinal growth
Sheridan, Fox, Zeanah, McLaughlin, and Nelson (2012)	74 (54)	Institutional care	ROI	↓GMV	NS	NS	↓CC, NS basal ganglia	–

Continued

TABLE 1 Studies Examining Children and/or Adolescents—cont'd

Study	Total *N* (Exposed *N*)	Stressor	Analysis	WBV	HC	Amygdala	Other	Notes
Jednoróg et al. (2012)	23	SES	VBM/SBM	NS	↓	–	↓parahippocampal gyri, middle temporal gyri, insula, left fusiform gyrus, right inferior occipito-temporal region, and left superior/middle frontal gyrus	–
Noble, Houston, Kan, and Sowll (2012)	60	SES	ROI	NS	↓	↓	↓ ITG, NS L STG, L MTG, L IFG, left fusiform, ACC	–
Marcus Jenkins, Woollwy, Hooper, and De Bellis (2013)	102	SES	ROI	↓WBV, ↓GMV	–	–	–	–
Luby et al. (2013)	145	SES	ROI	↓GMV, ↓WMV	↓left, NS right	↓left, NS right	–	Mediated by parenting and SLE
Ganella et al. (2015)	91	CTQ	ROI	–	–	–	↑pituitary (females)	Examined longitudinal growth[a]
Hanson et al. (2015)	128	Physical abuse, orphanage/abandoned neglect, SES	ROI	–	↓	↓	–	–

Arrows indicate increased or decreased volume in the group exposed to ELS or with increasing amounts of stress exposure. NS, nonsignificant effects of ELS. WBV, whole brain volume; HC, hippocampus; GMV, gray matter volume; WMV, white matter volume; CC, corpus callosum; ITG, inferior temporal gyrus; STG, superior temporal gyrus; MTG, middle temporal gyrus; IFG, inferior frontal gyrus; ACC, anterior cingulate cortex; hemisphere noted by L=left, R=right; CTQ, Childhood Trauma Questionnaire; SES, socioeconomic status; MDD, major depressive disorder; ROI, region of interest-based analysis; VBM, voxel-based morphometry analysis.

[a]*Did not control for WBV.*

TABLE 2 Studies Examining Adults

Study	Total *N* (Exposed *N*)	Stressor	Analysis	WBV	HC	Amygdala	Other	Notes
Stein et al. (1997)	42 (21)	CSA	ROI	–	↓left, NS right	–	–	Women only
Driessen et al. (2000)	42	CTQ	ROI	–	↓	NS	–	Women only, includes BPD
Vythilingam et al. (2002)	46 (21)	Early trauma inventory (and MDD)	ROI	NS	↓left in MDD + abuse versus MDD and control, NS right	–	–	Women only, includes MDD
Pederson et al. (2004)	51 (34)	Abuse	ROI	–	NS	–	–	Women only, includes PTSD[a]
Golier et al. (2005)	47 (27)	Holocaust survivors	ROI	NS	NS	–	↑STG, ↑lateral temporal lobe	Older adults
Cohen et al. (2006)	265	Adverse childhood experiences, ELSQ	ROI	↓GMV	NS	NS	↓ACC and caudate, NS when controlling for total brain volume	–
Andersen et al. (2008)	42 (26)	CSA	ROI	–	↓	NS	↓CC, ↓frontal cortex	Women only
Lenze, Xiong, and Sheline (2008)	55	Childhood adversity (CECA)	ROI	–	NS	–	–	Women only, includes remitted MDD[a]

Continued

TABLE 2 Studies Examining Adults—cont'd

Study	Total *N* (Exposed *N*)	Stressor	Analysis	WBV	HC	Amygdala	Other	Notes
Weniger, Lange, Sachsse, and Irle (2008)	48 (23)	Traumatic Antecedent Questionnaire (TAQ)	ROI	NS	NS	NS	–	Women only, includes PTSD and DID
Tomoda, Navalta, Polcari, Sadato, and Teicher (2009)	37 (23)	CSA	VBM	–	–	–	↓visual cortex	Women only
Frodl, Reinhold, Koutsouleris, Reiser, and Meisenzahl (2010)	87	CTQ	ROI/VBM	–	↓	–	↓ prefrontal volumes in healthy controls with more physical abuse	MDD, MDD × ELS interaction effects
Van Harmelen et al. (2010)	181 (84)	CEM	VBM	NS	NS	NS	↓medial prefrontal gyrus	Includes unmedicated patients with depression and/or anxiety disorders
Baker et al. (2012)	173 (97)	Adverse childhood experiences, ELSQ	ROI	–	NS	NS	↓ACC, ↓insula; NS caudate	Varied by time of ELS exposure[a]
Butterworth, Cherbuin, Sachdev, and Anstey (2012)	431	SES, childhood poverty	ROI	NS	NS	NS	–	Middle-aged adults
Carballedo et al. (2012)	40 (20)	CTQ (and MDD risk)	ROI/VBM	NS	↓		↓ mPFC, ACC, dlPFC with family history + early abuse	Risk × abuse interaction

Staff et al. (2012)	249	SES	ROI	↓	↓	–	–	Older adults[b]
Teicher et al. (2012)	193	Adverse childhood experiences, CTQ	ROI	↓	–	–	–	–
Cavanagh et al. (2013)	42	SES	ROI	–	–	–	↓cerebellar gray matter	Males only
Chaney et al. (2014)	83 (30)	CTQ	ROI/ VBM	–	↓	–	↑left OFC, right dmPFC, NS dlPFC, ACC	Includes MDD
Gorka et al. (2014)	818	CTQ	ROI/ VBM	–	↓left, NS right	NS	↓mPFC	–
Opel et al. (2014)	170	CTQ	ROI	–	↓	–	–	Includes MDD
Holz et al. (2015)	167 (33)	SES	ROI/ VBM	NS	–	–	↓OFC ROI and inferior temporal gyrus, superior orbitofrontal gyrus, middle temporal gyrus, precentral gyrus, medial frontal gyrus, and the bilateral insula	–

Arrows indicate increased or decreased volume in the group exposed to ELS or with increasing amounts of stress exposure. NS, nonsignificant effects of ELS. WBV, whole brain volume; HC, hippocampus; GMV, gray matter volume; WMV, white matter volume; CC, corpus callosum; STG, superior temporal gyrus; ACC, anterior cingulate cortex; OFC, orbitofrontal cortex; mPFC, medial prefrontal cortex; dlPFC, dorsolateral prefrontal cortex; dmPFC, dorsomedial prefrontal cortex; hemisphere noted by L=left, R=right; CTQ, Childhood Trauma Questionnaire; SES, socioeconomic status; MDD, major depressive disorder; ROI, region of interest based analysis; VBM, voxel-based morphometry analysis; BPD, borderline personality disorder; DID, dissociative identity disorder.

[a]*Did not control for WBV.*

[b]*Did not control for WBV in the paper, but the results hold when doing so (author communication).*

larger volumes in previously institutionalized children but found a negative relationship between amygdala volumes and time spent in the institution (Mehta et al., 2009). Three other studies have noted reduced amygdala volumes in children/adolescents with early life adversity, specifically low SES or physical abuse. This might suggest potentially significant differentiations between neglect and other types of stressors/adversity. Interestingly, studies relating ELS to amygdala volumes measured in adulthood have typically found nonsignificant results. This may also suggest the potential for recovery over development, particularly as the amygdala continues to develop through adolescence (Giedd et al., 1996). Further, it has been suggested that the timing of stress along the developmental trajectory can greatly alter its influence (Tottenham, 2009).

Studies have begun to examine the effects of ELS across the rest of the brain, raising further regions of potential interest. Several studies have identified volumetric reductions in the ACC of adults associated with adverse childhood experiences. Volumetric reductions in mPFC/ACC and OFC have also been linked to ELS in children and adults. As noted above, volumetric changes in these regions have also been seen in depression. Several other regions have been less consistently associated with ELS or have only been reported in a single study. For example, one study has suggested smaller caudate volume while two others have found nonsignificant relationships between ELS and the caudate. Similarly, evidence for reductions in whole brain or corpus callosum volume have been mixed in the few studies reporting on these outcomes.

Diagnostic Status Confounds

As noted, we excluded several studies from the table if their examination of ELS was entirely confounded by diagnostic status. Much of this work has focused on individuals with PTSD secondary to some form of childhood trauma. A meta-analysis of these studies has suggested that hippocampal volumes are reduced bilaterally in adults with childhood maltreatment-related PTSD, whereas hippocampal (and amygdala) volumes did not differ in children with PTSD (Woon & Hedges, 2008). Thus, it was suggested that hippocampal volume reductions after childhood adversity may not be apparent until adulthood, as has been noted in some animal work as well (e.g., Isgor, Kabbaj, Akil, & Watson, 2004). There is also work showing reduced volumes in women with CSA and PTSD as compared to women with CSA but not PTSD or unexposed women (Bremner, 2003); this might speak to hippocampal volume as a marker of resilience/vulnerability to PTSD. This hypothesis has been tested more explicitly in a study of twins showing that smaller hippocampal volumes may be a risk factor for developing PTSD after trauma exposure (Gilbertson et al., 2002).

ELS and White Matter

Importantly, ELS has been linked to alterations in a variety of other neural outcomes as well. For example, studies have begun to examine the impact of ELS on white matter structure, though this literature is quite mixed. There is some evidence for white matter deficits in the corpus callosum, though the localization of this effect has varied—reduced FA in the genu in adult females (Paul et al., 2008) versus medial and posterior deficits in children with PTSD (Jackowski et al., 2008). Several other studies have pinpointed white matter deficits in individuals with ELS exposure to the uncinate fasciculus (Eluvathingal, 2006) and the arcuate fasciculus (Choi, Jeong, Rohan, Polcari, & Teicher, 2009). The risk for MDD based on a parental history has also been linked to lower FA in the corpus callosum, cingulum, superior longitudinal fasciculi, uncinate, and inferior fronto-occipital fasciculi (Huang, Fan, Williamson, & Rao, 2011). Further, ELS may interact with a familial risk for MDD relating to increased FA in the corpus callosum, fornix, inferior fronto-occipital fasciculi, and superior longitudinal fasciculus in those at elevated risk but decreased FA in those without a family history of MDD (Frodl et al., 2012).

ELS and Brain Function

Finally, ELS has been related to alterations in neural responsivity and connectivity. This work again has largely focused on neural responsivity to emotional stimuli. These studies have consistently found that ELS predicted elevated responsivity to negative emotional stimuli, particularly in the amygdala, across a variety of types of ELS (e.g., Bogdan, Williamson, & Hariri, 2012; Dannlowski et al., 2011; Ganzel, Kim, Gilmore, Tottenham, & Temple, 2013; Pagliaccio et al., 2015a; Suzuki, Luby, Botteron, & Dietrich, 2014; Tottenham et al., 2011). As described above, increased amygdala reactivity to negative stimuli has also been seen consistently in depression. It has also been suggested that ELS relates to hypo-activity in the nucleus accumbens and to greater depressive symptomology (though activity did not mediate the effects of ELS on MDD) (Goff et al., 2012). For a review, see Goff and Tottenham (2014). ELS has also been suggested to impair the regulation of emotion where the experience of childhood poverty predicted reduced vlPFC and dlPFC activity during emotion regulation in adulthood, and this effect was mediated by chronic stress exposure across childhood/adolescence (Kim et al., 2013). Finally, work has begun to link ELS to alterations in functional connectivity as well. This work has generally noted alterations in connectivity between the amygdala and mPFC/ACC in children/adolescents, though the form of this relationship differed somewhat by sample (Burghy et al., 2012; Gee et al., 2013; Pagliaccio et al., 2015b).

Summary of ELS and Brain Structure/Function

This review of the literature on the relationship between ELS and brain structure/function points to several key findings. There is evidence for reduced hippocampal volume associated with ELS but more mixed evidence in regard to the direction of amygdala effects, with some studies finding increased volume and some deceased. Both of these findings are consistent with the results from the depression literature. As of yet, there is mixed evidence as to white matter changes associated with ELS (and MDD), though alterations in the corpus callosum have been noted with both ELS and MDD. However, there is consistent evidence for increased amygdala reactivity to negative stimuli associated with ELS, similar to what has been observed in MDD patients, where studies have begun to suggest particularly that ELS may account for these MDD effects. Common structural, functional, and connectivity alterations in the ACC/mPFC have also been noted in studies of ELS and MDD. Figure 3 highlights some of these main regions of interest showing alterations relating to ELS and MDD, including the amygdala, hippocampus, mPFC, subgenual ACC, and pregenual and rostral ACC.

DIRECT TESTS OF THE ELS-CORTISOL-BRAIN-DEPRESSION MECHANISM

Many studies have implicated a mechanism by which ELS increases the risk for depression mediated by cortisol and neural alterations, but few human studies have explicitly tested this. Ideally, such tests would involve prospective longitudinal data examining cortisol, neural, and clinical outcomes preceding ELS and following through later development to examine the

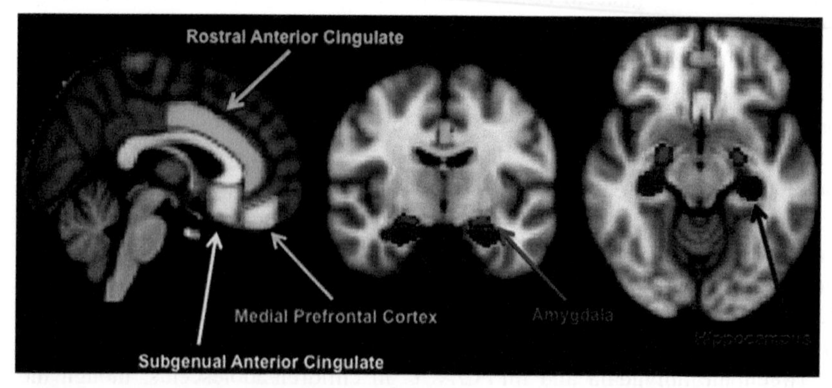

FIGURE 3 Major brain areas showing alterations with ELS and depression. This figure displays several of the key regions of interest that have shown structural/functional alterations associated with ELS and depression, including the hippocampus (blue), amygdala (red), mPFC (pink), subgenual cingulate (yellow), and rostral anterior cingulate (green). These regions were extracted from the Harvard–Oxford Cortical and Subcortical Atlases.

causal relationships between these factors. As this is quite difficult and costly, most studies have examined retrospective relationships with longitudinal or cross-sectional data. Further insight can also be gained from intervention studies aimed at treating/alleviating ELS or cortisol effects (Slopen, McLaughlin, & Shonkoff, 2014), e.g., studying poverty alleviation interventions (Fernald & Gunnar, 2009), moving children from orphanages to foster care (Zeanah et al., 2003), or pharmacological treatments impacting cortisol function (Zobel et al., 2000). Here, we will review work that has explicitly tested some step of this proposed ELS-cortisol-brain-depression mechanism. Additionally, where prior studies have not tested mediation but have provided the necessary statistics, we have calculated Sobel tests of mediation posthoc (http://quantpsy.org/sobel/sobel.htm). This was only possible when sufficient information was provided to calculate the t-statistics for the effect of the independent variable on the proposed mediator (path a) and for the effect of the mediator on the dependent variable, controlling for the independent variable (path b). While this is not the ideal method for examining mediation (Hayes, 2009, 2012, 2013), this did allow us to further synthesize prior literature that can inform this issue and will hopefully prompt others to explicitly test statistical mediation in their future work relating ELS, cortisol, neural, and/or clinical outcomes.

ELS, Depression, and Brain Changes

First, it has been suggested that ELS exposure might account for observed relationships between MDD and neural alterations. These studies tended not to test a mediation model similar to what is presented in Figure 1. However, given the temporal ordering of events, these results can likely be conceptualized as showing that ELS is the causal agent predicting neural change as associated MDD outcomes. Particularly, while hippocampal volume was reduced in a sample of MDD patients, childhood trauma exposure (CTQ scores) was suggested to account for this relationship, i.e., MDD patients experience more childhood adversity, and this predicted volumes (Opel et al., 2014). Our posthoc mediation calculations indicated that CTQ scores did in fact mediate the relationship between MDD and volumes of both the left ($t_a = 5.68$, $t_b = -3.06$, Sobel $Z = 2.69$, $p = 0.007$) and right hippocampus ($t_a = 5.68$, $t_b = -3.54$, Sobel $Z = 3.00$, $p = 0.002$). Conversely, other work has found relationships between ELS and smaller hippocampal volumes that were not mediated by MDD or PTSD experience (Teicher, Anderson, & Polcari, 2012). A similar relationship has been shown where ELS may also account for relationships between MDD and amygdala reactivity ($t_a = 3.96$, $t_b = 3.55$, Sobel $Z = 2.64$, $p = 0.008$) (Grant, Cannistraci, Hollon, Gore, & Shelton, 2011). Unfortunately, insufficient information was provided in these studies to test whether or not neural changes mediated relationships between ELS and MDD, though the directionality of any potential effects would not be distinguishable without longitudinal data. Thus, these mediation effects should be thought of as showing, for example, that CTQ scores account for effects of MDD, rather than addressing more mechanistic mediation hypotheses.

With that said, other work has explicitly tested this type of mediation relationship in a longitudinal study and found that smaller hippocampal volumes partially mediate the effect of early life adversity on MDD at longitudinal follow-up (Rao et al., 2010). Work has also suggested that reduced gray matter in the left hippocampus and mPFC mediates the effects of ELS on trait anxiety levels (Gorka, Hanson, Radtke, & Hariri, 2014). These studies are more in line with the proposed mechanism in Figure 1.

Cortisol as a Mediator of ELS Effects on Brain Structure/Function

Other work has linked ELS to neural outcomes via alteration in cortisol function. Work from our lab has suggested that stress cortisol levels at preschool age mediate relationships between ELS experience and both left hippocampal and amygdala volumes, where higher cortisol levels longitudinally predicted smaller limbic volumes at school age (Pagliaccio et al., 2014). Interestingly, a composite risk index of several physiological markers, including cortisol (and blood glucose, body mass index, blood pressure, triglycerides, etc.), was shown to mediate the effects of early SES on cerebellar volumes (Cavanagh et al., 2013). It has also been suggested that cortisol may mediate relationships between ELS and amygdala-mPFC resting state functional connectivity with ELS in females (Burghy et al., 2012) and after early institutional care (Gee et al., 2013). Further, it was suggested that more negative amygdala-mPFC connectivity mediated relationships between childhood cortisol levels and adolescent depression severity (Burghy et al., 2012). Work from our lab has suggested that greater ELS exposure predicts weaker amygdala-ACC connectivity, where weaker connectivity and elevated concurrent anxiety acted as serial mediators of the effect of ELS on worsening anxiety at a longitudinal follow-up (Pagliaccio et al., 2015b). Finally, other work has suggested that childhood poverty predicts reduced default mode network connectivity, which predicts higher anticipatory cortisol, but unfortunately insufficient information was provided to test mediated effects (Sripada, Swain, Evans, Welsh, & Liberzon, 2015).

Summary of Tests of Mediated Effects

While much of the work reviewed here helps to inform our understandings of the mechanistic relationships between ELS, cortisol, brain alterations, and the risk for depression, few studies have really begun to explicitly test these hypothesized mediation effects. Here, we focused on studies examining neural alterations, but there may be other studies in the literature that inform the degree to which cortisol alterations mediate relationships between ELS and psychopathology. However, a review of these studies is beyond the scope of this chapter. Despite the power of this mediation approach and the added knowledge to be gained by testing mediation, we should note several caveats. Particularly, the amount of variance accounted for in these relationships is sometimes quite small. Further, much of the interpretation of these mediation effects relies on temporal precedence. That is to say that while cortisol may serve as a statistical mediator of relationships between stress experienced earlier in life and brain volumes assessed later in life, this does not assure that the actual underlying mechanism functions this way. For example, brain volumes instead could mediate relationships between ELS and cortisol. Alternatively, in this example, cortisol could serve as a marker of another factor that is the actual causal contributor, for example neuroinflammation, i.e., ELS could cause correlated increases in cortisol and inflammation, but inflammation rather than cortisol is what acts on brain structure. Additionally, it is important to note that there are likely many moderators of any potential mechanisms linking ELS to the MDD. For example, as we noted previously, genetic variation may moderate the ELS-related risk for MDD; this may be due to a moderation of effects of ELS on the brain. Several studies have begun to examine this where our work suggests that the additive effect of several single nucleotide polymorphisms in HPA axis-related genes moderates the effects of ELS on hippocampal and amygdala volume (Pagliaccio et al., 2014) and amygdala functional connectivity (Pagliaccio et al., 2015b). These results suggest that the intrinsic functioning of the HPA axis may be a particularly salient moderator of the effects of stress. For a more in-depth review of neuroimaging studies examining gene × environmental interactions, see Bogdan, Pagliaccio, Baranger, and Hariri (2015).

Limitations

There are a variety of common limitations to studying the relationships between ELS and depression, particularly to elucidating the mechanisms underlying this. First, most studies are retrospective in their assessment of childhood adversity, leading to issues with recall bias. This includes both the potential for inaccurate recall/characterization of childhood adversity by adult reporters and, more importantly, differentially inaccurate recall as a function of current pathology, e.g., if depressed individuals are more likely to over-report childhood stress (Brewin, Andrews, & Gotlib, 1993; Hardt & Rutter, 2004). This may potentially be aided by using more in-depth or interview-based assessments of life stress (Cohen, Kessler, & Gordon, 1997), by acquiring information from multiple reporters, or by examining state reporting of abuse rather than questionnaire/checklist methods. While these may alleviate some issues with recall bias, they also have their own limitations, e.g., collection time and difficulty, dealing with discordance between reporters, biases from other reporters, underreporting to state agencies, etc.

Thus, prospective longitudinal data may be the ideal method for addressing this type of question, which clearly comes with its own challenges, e.g., attrition, low base rate of severe ELS, ethical imperatives to intervene/report when observing ongoing trauma/abuse, etc. These costs and benefits must be weighted in future work to best characterize ELS and the outcomes of interest.

Additional complications/limitations that were mentioned previously include the clustering of early stressors and the independence/dependence of events. While both independent and dependent events have been linked to depression, more care must be given to characterizing the type and impact of ELS. Further, different stressors are often coincident, making it difficult to ascertain specific effects of any given type of stressor. Similarly, it is often difficult to ascertain the chronicity and timing of stressors. The confounding of ELS with later pathology has also been an issue in determining the specific effects of stress on outcomes, like cortisol function and the brain. For example, there is a body of literature examining the comparison of brain structure across individuals with PTSD secondary to childhood trauma and individuals unexposed to trauma. While this can be informative about the neural substrates of PTSD, the confounding of PTSD and ELS limits our ability to make specific claims about the effects of ELS.

SUMMARY AND POINTERS TO THE FUTURE

Here, we reviewed the role of ELS in establishing its link to the risk for depression and to examining evidence for alterations to cortisol function and brain structure/function as potential mediating mechanisms. Particularly, we described epidemiological evidence for ELS in general increasing the risk for depression as well as work relating specific types of stressors to MDD, like poverty and childhood abuse. Further, we discussed parental psychopathology as a form of ELS, particularly in the context of the stress generation hypothesis.

We then reviewed the literature linking both MDD and ELS to alterations in cortisol function and brain structure/function. Here, we saw relatively consistent relationships between MDD and elevated cortisol but mixed association with ELS, though studies generally pointed toward blunted cortisol after ELS. The exception to this was parental psychopathology, which has been more commonly linked to elevated cortisol. Such alterations in cortisol may provide a mechanism by which parental depression increases the risk for depression in offspring and should also be examined as a mechanism by which parental depression could moderate the effects of ELS. In other words, it may be that elevated cortisol makes individuals with a family history of MDD more susceptible to the deleterious effects of stress. On the other hand, findings of blunted cortisol function after ELS do complicate the support for the proposed mechanism. It will be important to further examine the relationship between ELS and cortisol to fully understand this, particularly probing a variety of outcomes, like diurnal rhythms and responsivity to acute stressors, and to examine the change in cortisol function over development. Further, cumulative cortisol exposure could be a further important outcome to examine, which is also elevated in MDD, as assessed by hair cortisol (Staufenbiel, Penninx, Spijker, Elzinga, & van Rossum, 2013). This may be particularly true if individuals exposed to ELS continue to experience increased levels of environmental stressors; cumulative cortisol exposure could increase despite a tendency toward blunted responsivity.

While the literature examining alterations in brain structure and function is still growing, strong parallels are being found between relationships with ELS and MDD. Particularly, the literature highlights decreased hippocampal volumes, finds mixed evidence related to amygdala volumes, and provides strong evidence for elevated amygdala reactivity. While these are the most consistent findings, the literature is not unanimous, likely in part due to the small and heterogeneous samples often examined that limit power. Thus, it will be important for future work to assure that sufficiently large samples are being ascertained to clearly establish these relationships as well as to better characterize both ELS exposure and MDD to determine what is accounting for these mixed results. Furthermore, while many studies have focused on the amygdala and hippocampus a priori, more whole brain examinations have been conducted. This is important both for hypothesis generation and to create a more holistic understanding of neural alterations associated with ELS and MDD.

Finally, we addressed data that have explicitly examined some facets of an ELS-cortisol-brain-MDD mechanism. There is a small but growing body of work that states that ELS may account for depression-related differences in the brain or that more directly examines cortisol as a mediator between ELS and the brain and/or the brain as a mediator between ELS and pathology. While much of this work relies on temporal precedence to establish directionality, it does begin to provide evidence for aspects of the mechanisms outlined in Figure 1. Furthermore, it appears that a variety of studies have the data available but have not examined mediation and moderation effects that can help build our understandings of the causal relationships between ELS, cortisol, brain structure/function, and depression. Given the ease and accessibility of testing a variety of models (e.g., Hayes, 2012, 2013), this should be pursued more frequently, particularly when longitudinal data is available, in order to establish stronger mechanistic understandings linking ELS to depression.

ACKNOWLEDGMENTS

Work by D.P. was supported by a grant from the National Institute of General Medical Sciences (5T32GM081739). Sections of this chapter have been adapted from D.P.'s dissertation work Pagliaccio, D. (2015). The Effects of HPA Axis Genetic Variation and Early Life Stress on Cortisol Levels in Preschool Age Children and on Amygdala and Hippocampus Volumes, Reactivity, and Connectivity at School Age (Doctoral dissertation).

REFERENCES

Adrian, C., & Hammen, C. (1993). Stress exposure and stress generation in children of depressed mothers. *Journal of Consulting and Clinical Psychology, 61*(2), 354–359. http://dx.doi.org/10.1037/0022-006X.61.2.354.

Alexander, A. L., Lee, J. E., Lazar, M., & Field, A. S. (2007). Diffusion tensor imaging of the brain. *Neurotherapeutics, 4*(3), 316–329. http://dx.doi.org/10.1016/j.nurt.2007.05.011.

Andela, C. D., Van Haalen, F., Ragnarsson, O., Papakokkinou, E., Johannsson, G., Santos, A., ... Pereira, A. M. (2015). Mechanisms in endocrinology: Cushing's syndrome causes irreversible effects on the human brain: a systematic review of structural and functional MRI studies. *European Journal of Endocrinology, 173*(1), R1–R14. http://dx.doi.org/10.1530/EJE-14-1101.

Andersen, S. L., Tomada, A., Vincow, E. S., Valente, E., Polcari, A., & Teicher, M. H. (2008). Preliminary evidence for sensitive periods in the effect of childhood sexual abuse on regional brain development. *Journal of Neuropsychiatry and Clinical Neurosciences, 20*(3), 292–301.

Baker, L. M., Williams, L. M., Korgaonkar, M. S., Cohen, R. A., Heaps, J. M., & Paul, R. H. (2012). Impact of early vs. late childhood early life stress on brain morphometrics. *Brain Imaging and Behavior, 7*(2), 196–203.

Bandura, A. (1977). Self-efficacy: toward a unifying theory of behavioral change. *Psychological Review, 84*(2), 191–215. http://dx.doi.org/10.1037/0033-295X.84.2.191.

Belsky, J., & Pluess, M. (2009). Beyond diathesis stress: differential susceptibility to environmental influences. *Psychological Bulletin, 135*(6), 885–908. http://dx.doi.org/10.1037/a0017376.

Biswal, B., Zerrin Yetkin, F., Haughton, V. M., & Hyde, J. S. (1995). Functional connectivity in the motor cortex of resting human brain using echo-planar MRI. *Magnetic Resonance in Medicine, 34*(4), 537–541. http://dx.doi.org/10.1002/mrm.1910340409.

Blood, A. J., Iosifescu, D. V., Makris, N., Perlis, R. H., Kennedy, D. N., Dougherty, D. D., ... Breiter, H. C. (2010). Microstructural abnormalities in subcortical reward circuitry of subjects with major depressive disorder. *PLoS One, 5*(11), e13945. http://dx.doi.org/10.1371/journal.pone.0013945.

Bogdan, R., Pagliaccio, D., Baranger, D. A., & Hariri, A. R. (2015). Genetic moderation of stress effects on corticolimbic circuitry. *Neuropsychopharmacology.*

Bogdan, R., Williamson, D. E., & Hariri, A. R. (2012). Mineralocorticoid receptor iso/val (rs5522) genotype moderates the association between previous childhood emotional neglect and amygdala reactivity. *American Journal of Psychiatry, 169*(5), 515–522.

Bora, E., Fornito, A., Pantelis, C., & Yücel, M. (2012). Gray matter abnormalities in Major Depressive Disorder: a meta-analysis of voxel based morphometry studies. *Journal of Affective Disorders, 138*(1–2), 9–18. http://dx.doi.org/10.1016/j.jad.2011.03.049.

Boyce, W. T., & Ellis, B. J. (2005). Biological sensitivity to context: I. An evolutionary-developmental theory of the origins and functions of stress reactivity. *Development and Psychopathology, 17*(2), 271–301.

Brain Development Cooperative Group. (2011). Total and regional brain volumes in a population-based normative sample from 4 to 18 years: the NIH MRI study of normal brain development. *Cerebral Cortex, 22*(1), 1–12.

Bremner, J. D. (2003). MRI and PET study of deficits in hippocampal structure and function in women with childhood sexual abuse and posttraumatic stress disorder. *American Journal of Psychiatry, 160*(5), 924–932. http://dx.doi.org/10.1176/appi.ajp.160.5.924.

Brewin, C. R., Andrews, B., & Gotlib, I. H. (1993). Psychopathology and early experience: a reappraisal of retrospective reports. *Psychological Bulletin, 113*(1), 82–98.

Brown, E. S., Woolston, D. J., & Frol, A. B. (2008). Amygdala volume in patients receiving chronic corticosteroid therapy. *Biological Psychiatry, 63*(7), 705–709. http://dx.doi.org/10.1016/j.biopsych.2007.09.014.

Bruce, J., Kroupina, M., Parker, S., & Gunnar, M. (2000). The relationships between cortisol patterns, growth retardation, and developmental delays in post-institutionalized children. http://doi.org/10.1002/dev.20333/full.

Bruce, J., Fisher, P. A., Pears, K. C., & Levine, S. (2009). Morning cortisol levels in preschool-aged foster children: differential effects of maltreatment type. *Developmental Psychobiology, 51*(1), 14–23. http://dx.doi.org/10.1002/dev.20333.

Burghy, C. A., Stodola, D. E., Ruttle, P. L., Molloy, E. K., Armstrong, J. M., Oler, J. A., … Birn, R. M. (2012). Developmental pathways to amygdala-prefrontal function and internalizing symptoms in adolescence. *Nature Neuroscience, 15*(12), 1736–1741. http://dx.doi.org/10.1038/nn.3257.

Burke, H. M., Davis, M. C., Otte, C., & Mohr, D. C. (2005). Depression and cortisol responses to psychological stress: a meta-analysis. *Psychoneuroendocrinology, 30*(9), 846–856. http://dx.doi.org/10.1016/j.psyneuen.2005.02.010.

Burt, K. B., van Dulmen, M. H. M., Carlivati, J., Egeland, B., Alan Sroufe, L., Forman, D. R., … Carlson, E. A. (2005). Mediating links between maternal depression and offspring psychopathology: the importance of independent data. *Journal of Child Psychology and Psychiatry, 46*(5), 490–499. http://dx.doi.org/10.1111/j.1469-7610.2004.00367.x.

Butterworth, P., Cherbuin, N., Sachdev, P., & Anstey, K. J. (2012). The association between financial hardship and amygdala and hippocampal volumes: results from the PATH through life project. *Social Cognitive and Affective Neuroscience, 7*(5), 548–556.

Campbell, S., & MacQueen, G. (2004). The role of the hippocampus in the pathophysiology of major depression. *Journal of Psychiatry and Neuroscience, 29*(6), 417–426.

Campbell, S., Marriott, M., Nahmias, C., & MacQueen, G. M. (2004). Lower hippocampal volume in patients suffering from depression: a meta-analysis. *American Journal of Psychiatry, 161*(4), 598–607. http://dx.doi.org/10.1176/appi.ajp.161.4.598.

Carballedo, A., Lisiecka, D., Fagan, A., Saleh, K., Ferguson, Y., Connolly, G., … Frodl, T. (2012). Early life adversity is associated with brain changes in subjects at family risk for depression. *World Journal of Biological Psychiatry, 13*(8), 569–578.

Carpenter, L. L., Carvalho, J. P., Tyrka, A. R., Wier, L. M., Mello, A. F., Mello, M. F., … Price, L. H. (2007). Decreased adrenocorticotropic hormone and cortisol responses to stress in healthy adults reporting significant childhood maltreatment. *Biological Psychiatry, 62*(10), 1080–1087. http://dx.doi.org/10.1016/j.biopsych.2007.05.002.

Carpenter, L. L., Shattuck, T. T., Tyrka, A. R., Geracioti, T. D., & Price, L. H. (2011). Effect of childhood physical abuse on cortisol stress response. *Psychopharmacology, 214*(1), 367–375. http://dx.doi.org/10.1007/s00213-010-2007-4.

Carpenter, L. L., Tyrka, A. R., Ross, N. S., Khoury, L., Anderson, G. M., & Price, L. H. (2009). Effect of childhood emotional abuse and age on cortisol responsivity in adulthood. *Biological Psychiatry, 66*(1), 69–75. http://dx.doi.org/10.1016/j.biopsych.2009.02.030.

Caspi, A., Sugden, K., Moffitt, T. E., Taylor, A., Craig, I. W., Harrington, H., … Poulton, R. (2003). Influence of life stress on depression: moderation by a polymorphism in the 5-HTT gene. *Science, 301*(5631), 386–389. http://dx.doi.org/10.1126/science.1083968.

Cavanagh, J., Krishnadas, R., Batty, G. D., Burns, H., Deans, K. A., Ford, I., … McLean, J. (2013). Socioeconomic status and the cerebellar grey matter volume. Data from a well-characterised population sample. *Cerebellum, 12*(6), 882–891. http://dx.doi.org/10.1007/s12311-013-0497-4.

Chaney, A., Carballedo, A., Amico, F., Fagan, A., Skokauskas, N., Meaney, J., & Frodl, T. (2014). Effect of childhood maltreatment on brain structure in adult patients with major depressive disorder and healthy participants. *Journal of Psychiatry and Neuroscience, 39*(1), 50.

Chen, E., Cohen, S., & Miller, G. E. (2010). How low socioeconomic status affects 2-year hormonal trajectories in children. *Psychological Science, 21*(1), 31–37. http://dx.doi.org/10.1177/0956797609355566.

Choi, J., Jeong, B., Rohan, M. L., Polcari, A. M., & Teicher, M. H. (2009). Preliminary evidence for white matter tract abnormalities in young adults exposed to parental verbal abuse. *Biological Psychiatry, 65*(3), 227–234. http://dx.doi.org/10.1016/j.biopsych.2008.06.022.

Chorpita, B. F., & Barlow, D. H. (1998). The development of anxiety: the role of control in the early environment. *Psychological Bulletin, 124*(1), 3–21.

Cicchetti, D. (2009). Neuroendocrine functioning in maltreated children. In D. Cicchetti, & E. F. Walker (Eds.), *Neurodevelopmental mechanisms in psychopathology* (pp. 345–365). Cambridge: Cambridge University Press. http://dx.doi.org/10.1017/CBO9780511546365.016.

Cicchetti, D., & Rogosch, F. A. (2001). The impact of child maltreatment and psychopathology on neuroendocrine functioning. *Development and Psychopathology, 13*(4), 783–804.

Cicchetti, D., Rogosch, F. A., Gunnar, M. R., & Toth, S. L. (2010). The differential impacts of early physical and sexual abuse and internalizing problems on daytime cortisol rhythm in school-aged children. *Child Development, 81*(1), 252–269. http://dx.doi.org/10.1111/j.1467-8624.2009.01393.x.

Cohen, R. A., Grieve, S., Hoth, K. F., Paul, R. H., Sweet, L., Tate, D., … Williams, L. M. (2006). Early life stress and morphometry of the adult anterior cingulate cortex and caudate nuclei. *Biological Psychiatry, 59*(10), 975–982.

Cohen, S., Kessler, R. C., & Gordon, L. U. (1997). *Measuring stress: A guide for health and social scientists*. Oxford University Press.

Collishaw, S., Pickles, A., Messer, J., Rutter, M., Shearer, C., & Maughan, B. (2007). Resilience to adult psychopathology following childhood maltreatment: evidence from a community sample. *Child Abuse and Neglect, 31*(3), 211–229. http://dx.doi.org/10.1016/j.chiabu.2007.02.004.

Colton, C. W., & Manderscheid, R. W. (2006). Congruencies in increased mortality rates, years of potential life lost, and causes of death among public mental health clients in eight states. *Preventing Chronic Disease, 3*(2), A42.

Conrad, C. D. C., LeDoux, J. E. J., Magariños, A. M. A., & McEwen, B. S. B. (1999). Repeated restraint stress facilitates fear conditioning independently of causing hippocampal CA3 dendritic atrophy. *Behavioral Neuroscience, 113*(5), 902–913.

Costello, E. J., Compton, S. N., Keeler, G., & Angold, A. (2003). Relationships between poverty and psychopathology. *Journal of the American Medical Association, 290*(15), 2023. http://dx.doi.org/10.1001/jama.290.15.2023.

Costello, E. J., Mustillo, S., Erkanli, A., Keeler, G., & Angold, A. (2003). Prevalence and development of psychiatric disorders in childhood and adolescence. *Archives of General Psychiatry, 60*(8), 837–844. http://dx.doi.org/10.1001/archpsyc.60.8.837.

Cotter, D. R. D., Pariante, C. M. C., & Everall, I. P. I. (2001). Glial cell abnormalities in major psychiatric disorders: the evidence and implications. *Brain Research Bulletin, 55*(5), 585–595.

Dannlowski, U., Stuhrmann, A., Beutelmann, V., Zwanzger, P., Lenzen, T., Grotegerd, D., … Kugel, H. (2011). Limbic scars: long-term consequences of childhood maltreatment revealed by functional and structural magnetic resonance imaging. *Biological Psychiatry*, 1–8. http://dx.doi.org/10.1016/j.biopsych.2011.10.021.

De Bellis, M. D., Chrousos, G. P., Dorn, L. D., Burke, L., Helmers, K., Kling, M. A., … Putnam, F. W. (1994). Hypothalamic-pituitary-adrenal axis dysregulation in sexually abused girls. *Journal of Clinical Endocrinology and Metabolism, 78*(2), 249–255. http://dx.doi.org/10.1210/jcem.78.2.8106608.

Diener, C., Kuehner, C., Brusniak, W., Ubl, B., Wessa, M., & Flor, H. (2012). A meta-analysis of neurofunctional imaging studies of emotion and cognition in major depression. *NeuroImage*, 1–40. http://dx.doi.org/10.1016/j.neuroimage.2012.04.005.

Doucet, D., & Doucet, J. (2003). Poverty, inequality, and discrimination as sources of depression among US women. *Psychology of Women Quarterly, 27*(2), 101–113. http://dx.doi.org/10.1111/1471-6402.00090.

Dowd, J. B., Simanek, A. M., & Aiello, A. E. (2009). Socio-economic status, cortisol and allostatic load: a review of the literature. *International Journal of Epidemiology, 38*(5), 1297–1309. http://dx.doi.org/10.1093/ije/dyp277.

Dozier, M., Peloso, E., Lewis, E., Laurenceau, J.-P., & Levine, S. (2008). Effects of an attachment-based intervention on the cortisol production of infants and toddlers in foster care. *Development and Psychopathology, 20*(3), 845–859. http://dx.doi.org/10.1017/S0954579408000400.

Driessen, M., Herrmann, J., Stahl, K., Zwaan, M., Meier, S., Hill, A., … Petersen, D. (2000). Magnetic resonance imaging volumes of the hippocampus and the amygdala in women with borderline personality disorder and early traumatization. *Archives of General Psychiatry, 57*(12), 1115–1122.

Driessen, E., Van, H. L., Don, F. J., Peen, J., Kool, S., Westra, D., … Dekker, J. J. M. (2013). The efficacy of cognitive-behavioral therapy and psychodynamic therapy in the outpatient treatment of major depression: a randomized clinical trial. *American Journal of Psychiatry, 170*(9), 1041–1050. http://dx.doi.org/10.1176/appi.ajp.2013.12070899.

Eley, T. C., & Stevenson, J. (2000). Specific life events and chronic experiences differentially associated with depression and anxiety in young twins. *Journal of Abnormal Child Psychology, 28*(4), 383–394. http://dx.doi.org/10.1097/00004703-200102000-00016.

Eluvathingal, T. J. (2006). Abnormal brain connectivity in children after early severe socioemotional deprivation: a diffusion tensor imaging study. *Pediatrics, 117*(6), 2093–2100. http://dx.doi.org/10.1542/peds.2005-1727.

Essex, M. J., Klein, M. H., Cho, E., & Kalin, N. H. (2002). Maternal stress beginning in infancy may sensitize children to later stress exposure: effects on cortisol and behavior. *Biological Psychiatry, 52*(8), 776–784.

Fava, M., & Davidson, K. G. (2005). Definition and epidemiology of treatment-resistant depression. *Psychiatric Clinics of North America, 19*(2), 1–22. http://dx.doi.org/10.1016/s0193-953x(05)70283-5.

Fernald, L. C. H., & Gunnar, M. R. (2009). Poverty-alleviation program participation and salivary cortisol in very low-income children. *Social Science and Medicine, 68*(12), 2180–2189. http://dx.doi.org/10.1016/j.socscimed.2009.03.032.

Finlay-Jones, R., & Brown, G. W. (1981). Types of stressful life event and the onset of anxiety and depressive disorders. *Psychological Medicine, 11*(4), 803–815.

Fisher, P. A., Burraston, B., & Pears, K. (2005). The early intervention foster care program: permanent placement outcomes from a randomized trial. *Child Maltreatment, 10*(1), 61–71. http://dx.doi.org/10.1177/1077559504271561.

Fitzgerald, P. B., Laird, A. R., Maller, J., & Daskalakis, Z. J. (2008). A meta-analytic study of changes in brain activation in depression. *Human Brain Mapping, 29*(6), 683–695. http://dx.doi.org/10.1002/hbm.20426.

Frodl, T., Carballedo, A., Fagan, A. J., Lisiecka, D., Ferguson, Y., & Meaney, J. F. (2012). Effects of early-life adversity on white matter diffusivity changes in patients at risk for major depression. *Journal of Psychiatry and Neuroscience, 37*(1), 37–45. http://dx.doi.org/10.1503/jpn.110028.

Frodl, T. S., Koutsouleris, N., Bottlender, R., Born, C., Jäger, M., Scupin, I., ... Meisenzahl, E. M. (2008). Depression-related variation in brain morphology over 3 years: effects of stress? *Archives of General Psychiatry, 65*(10), 1156–1165.

Frodl, T., Meisenzahl, E. M., Zetzsche, T., Höhne, T., Banac, S., Schorr, C., ... Möller, H. J. (2004). Hippocampal and amygdala changes in patients with major depressive disorder and healthy controls during a 1-year follow-up. *Journal of Clinical Psychiatry, 65*(4), 492–499.

Frodl, T., Reinhold, E., Koutsouleris, N., Reiser, M., & Meisenzahl, E. M. (2010). Interaction of childhood stress with hippocampus and prefrontal cortex volume reduction in major depression. *Journal of Psychiatric Research, 44*(13), 799–807.

Ganella, D. E., Allen, N. B., Simmons, J. G., Schwartz, O., Kim, J. H., Sheeber, L., & Whittle, S. (2015). Early life stress alters pituitary growth during adolescence-A longitudinal study. *Psychoneuroendocrinology, 53*, 185–194.

Ganzel, B. L., Kim, P., Gilmore, H., Tottenham, N., & Temple, E. (2013). Stress and the healthy adolescent brain: evidence for the neural embedding of life events. *Development and Psychopathology, 25*(4 pt 1), 879–889. http://dx.doi.org/10.1017/S0954579413000242.

Garber, J., & Cole, D. A. (2010). Intergenerational transmission of depression: a launch and grow model of change across adolescence. *Development and Psychopathology, 22*(4), 819–830. http://dx.doi.org/10.1017/S0954579410000489.

Gee, D. G., Gabard-Durnam, L. J., Flannery, J., Goff, B., Humphreys, K. L., Telzer, E. H., ... Tottenham, N. (2013). Early developmental emergence of human amygdala–prefrontal connectivity after maternal deprivation. *Proceedings of the National Academy of Sciences of the United States of America, 100*, 15638–15643.

Giedd, J. N., Vaituzis, A. C., Hamburger, S. D., Lange, N., Rajapakse, J. C., Kaysen, D., ... Rapoport, J. L. (1996). Quantitative MRI of the temporal lobe, amygdala, and hippocampus in normal human development: ages 4–18 years. *Journal of Comparative Neurology, 366*(2), 223–230.

Gilbertson, M. W., Shenton, M. E., Ciszewski, A., Kasai, K., Lasko, N. B., Orr, S. P., & Pitman, R. K. (2002). Smaller hippocampal volume predicts pathologic vulnerability to psychological trauma. *Nature Neuroscience, 5*(11), 1242–1247. http://dx.doi.org/10.1038/nn958.

Gilman, S. E. (2002). Socioeconomic status in childhood and the lifetime risk of major depression. *International Journal of Epidemiology, 31*(2), 359–367. http://dx.doi.org/10.1093/ije/31.2.359.

Gilman, S. E., Kawachi, I., Fitzmaurice, G. M., & Buka, S. (1999). Socio-economic status, family disruption and residential stability in childhood: relation to onset, recurrence and remission of major depression. *Psychological Medicine, 33*(8), 1341–1355. http://dx.doi.org/10.1017/S0033291703008377.

Goff, B., & Tottenham, N. (2014). Early-life adversity and adolescent depression: mechanisms involving the ventral striatum. *CNS Spectrums*, 1–9. http://dx.doi.org/10.1017/S1092852914000674.

Goff, B., Gee, D. G., Telzer, E. H., Humphreys, K. L., Gabard-Durnam, L., Flannery, J., & Tottenham, N. (2012). Reduced nucleus accumbens reactivity and adolescent depression following early-life stress. *Neuroscience*, 1–35. http://dx.doi.org/10.1016/j.neuroscience.2012.12.010.

Golding, J. M. (1999). Intimate partner violence as a risk factor for mental disorders: a meta-analysis. *Journal of Family Violence, 14*(2), 99–132. http://dx.doi.org/10.1023/A:1022079418229.

Golier, J. A., Yehuda, R., De Santi, S., Segal, S., Dolan, S., & de Leon, M. J. (2005). Absence of hippocampal volume differences in survivors of the Nazi Holocaust with and without post-traumatic stress disorder. *Psychiatry Research, 139*(1), 53–64.

Goodman, E., McEwen, B. S., Huang, B., Dolan, L. M., & Adler, N. E. (2005). Social inequalities in biomarkers of cardiovascular risk in adolescence. *Psychosomatic Medicine, 67*(1), 9.

Goodman, S. H., Rouse, M. H., Connell, A. M., Broth, M. R., Hall, C. M., & Heyward, D. (2011). Maternal depression and child psychopathology: a meta-analytic review. *Clinical Child and Family Psychology Review, 14*(1), 1–27. http://dx.doi.org/10.1007/s10567-010-0080-1.

Gorka, A. X., Hanson, J. L., Radtke, S. R., & Hariri, A. R. (2014). Reduced hippocampal and medial prefrontal gray matter mediate the association between reported childhood maltreatment and trait anxiety in adulthood and predict sensitivity to future life stress. *Biology of Mood and Anxiety Disorders, 4*(1), 12. http://dx.doi.org/10.1186/2045-5380-4-12.

Grant, M. M., Cannistraci, C., Hollon, S. D., Gore, J., & Shelton, R. (2011). Childhood trauma history differentiates amygdala response to sad faces within MDD. *Journal of Psychiatric Research, 45*(7), 886–895. http://dx.doi.org/10.1016/j.jpsychires.2010.12.004.

Green, J. G., McLaughlin, K. A., Berglund, P. A., Gruber, M. J., Sampson, N. A., Zaslavsky, A. M., & Kessler, R. C. (2010). Childhood adversities and adult psychiatric disorders in the national comorbidity survey replication I: associations with first onset of DSM-IV disorders. *Archives of General Psychiatry, 67*(2), 113–123. http://dx.doi.org/10.1001/archgenpsychiatry.2009.186.

Gunnar, M. R., Morison, S. J., Chisholm, K., & Schuder, M. (2001). Salivary cortisol levels in children adopted from Romanian orphanages. *Development and Psychopathology, 13*(3), 611–628.

Gunnar, M. R., & Quevedo, K. M. (2008). Early care experiences and HPA axis regulation in children: a mechanism for later trauma vulnerability. *Progress in Brain Research, 167*, 137–149. http://dx.doi.org/10.1016/S0079-6123(07)67010-1.

Gunnar, M. R., & Vazquez, D. M. (2001). Low cortisol and a flattening of expected daytime rhythm: potential indices of risk in human development. *Development and Psychopathology, 13*(3), 515–538.

Hajek, T., Kopecek, M., Kozeny, J., Gunde, E., Alda, M., & Höschl, C. (2009). Amygdala volumes in mood disorders — meta-analysis of magnetic resonance volumetry studies. *Journal of Affective Disorders, 115*(3), 395–410. http://dx.doi.org/10.1016/j.jad.2008.10.007.

Halligan, S. L., Herbert, J., Goodyer, I. M., & Murray, L. (2004). Exposure to postnatal depression predicts elevated cortisol in adolescent offspring. *Biological Psychiatry, 55*(4), 376–381. http://dx.doi.org/10.1016/j.biopsych.2003.09.013.

Hamilton, J. P., Etkin, A., Furman, D. J., Lemus, M. G., Johnson, R. F., & Gotlib, I. H. (2012). Functional neuroimaging of major depressive disorder: a meta-analysis and new integration of baseline activation and neural response data. *American Journal of Psychiatry, 169*(7), 693–703.

Hammen, C. (1991). Generation of stress in the course of unipolar depression. *Journal of Abnormal Psychology, 100*(4), 555–561.

Hammen, C., Hazel, N. A., Brennan, P. A., & Najman, J. (2012). Intergenerational transmission and continuity of stress and depression: depressed women and their offspring in 20 years of follow-up. *Psychological Medicine, 42*(5), 931–942. http://dx.doi.org/10.1017/S0033291711001978.

Hammen, C., Henry, R., & Daley, S. E. (2000). Depression and sensitization to stressors among young women as a function of childhood adversity. *Journal of Consulting and Clinical Psychology, 68*(5), 782–787. http://dx.doi.org/10.1037//0022-006X.68.5.782.

Hankin, B. L. (2006). Childhood maltreatment and psychopathology: prospective tests of attachment, cognitive vulnerability, and stress as mediating processes. *Cognitive Therapy and Research, 29*(6), 645–671. http://dx.doi.org/10.1007/s10608-005-9631-z.

Hanson, J. L., Chandra, A., Wolfe, B. L., & Pollak, S. D. (2011). Association between income and the hippocampus. *PLoS One, 6*(5), e18712.

Hanson, J. L., Chung, M. K., Avants, B. B., Shirtcliff, E. A., Gee, J. C., Davidson, R. J., & Pollak, S. D. (2010). Early stress is associated with alterations in the orbitofrontal cortex: a tensor-based morphometry investigation of brain structure and behavioral risk. *Journal of Neuroscience, 30*(22), 7466–7472.

Hanson, J. L., Nacewicz, B. M., Sutterer, M. J., Cayo, A. A., Schaefer, S. M., Rudolph, K. D., … Davidson, R. J. (2015). Behavioral problems after early life stress: contributions of the hippocampus and amygdala. *Biological Psychiatry, 77*(4), 314–323.

Hardt, J., & Rutter, M. (2004). Validity of adult retrospective reports of adverse childhood experiences: review of the evidence. *Journal of Child Psychology and Psychiatry, and Allied Disciplines, 45*(2), 260–273. http://dx.doi.org/10.1111/j.1469-7610.2004.00218.x.

Harkness, K. L., Bruce, A. E., & Lumley, M. N. (2006). The role of childhood abuse and neglect in the sensitization to stressful life events in adolescent depression. *Journal of Abnormal Psychology, 115*(4), 730–741. http://dx.doi.org/10.1037/0021-843X.115.4.730.

Harold, G. T., Rice, F., Hay, D. F., Boivin, J., van den Bree, M., & Thapar, A. (2011). Familial transmission of depression and antisocial behavior symptoms: disentangling the contribution of inherited and environmental factors and testing the mediating role of parenting. *Psychological Medicine, 41*(6), 1175–1185. http://dx.doi.org/10.1017/S0033291710001753.

Hayes, A. F. (2009). Beyond Baron and Kenny: statistical mediation analysis in the new millennium. *Communication Monographs, 76*(4), 408–420. http://dx.doi.org/10.1080/03637750903310360.

Hayes, A.F. (2012). PROCESS: a versatile computational tool for observed variable mediation, moderation, and conditional process modeling. [White paper]. Retrieved from http://www.afhayes.com/public/process2012.pdf.

Hayes, A. F. (2013). *Introduction to mediation, moderation, and conditional process analysis*. Guilford Press.

Heim, C., Mletzko, T., Purselle, D., Musselman, D. L., & Nemeroff, C. B. (2008). The dexamethasone/corticotropin-releasing factor test in men with major depression: role of childhood trauma. *Biological Psychiatry, 63*(4), 398–405. http://dx.doi.org/10.1016/j.biopsych.2007.07.002.

Heim, C., & Nemeroff, C. B. (2001). The role of childhood trauma in the neurobiology of mood and anxiety disorders: preclinical and clinical studies. *Biological Psychiatry, 49*(12), 1023–1039. http://dx.doi.org/10.1016/S0006-3223(01)01157-X.

Heim, C., Newport, D. J., Heit, S., & Graham, Y. P. (2000). Pituitary-adrenal and autonomic responses to stress in women after sexual and physical abuse in childhood. *Journal of the American Medical Association, 284*(5), 592–597.

Heim, C., Newport, D. J., Mletzko, T., Miller, A. H., & Nemeroff, C. B. (2008). The link between childhood trauma and depression: insights from HPA axis studies in humans. *Psychoneuroendocrinology, 33*(6), 693–710. http://dx.doi.org/10.1016/j.psyneuen.2008.03.008.

Heuser, I., Yassouridis, A., & Holsboer, F. (1994). The combined dexamethasone/CRH test: a refined laboratory test for psychiatric disorders. *Journal of Psychiatric Research, 28*(4), 341–356. http://dx.doi.org/10.1016/0022-3956(94)90017-5.

Holsboer, F., & Barden, N. (1996). Antidepressants and hypothalamic-pituitary-adrenocortical regulation. *Endocrine Reviews, 17*(2), 187–205.

Holz, N. E., Boecker, R., Hohm, E., Zohsel, K., Buchmann, A. F., Blomeyer, D., … Laucht, M. (2015). The long-term impact of early life poverty on orbitofrontal cortex volume in adulthood: results from a prospective study over 25 years. *Neuropsychopharmacology, 40*(4), 1–9.

Hovens, J. G. F. M., Giltay, E. J., Wiersma, J. E., Spinhoven, P., Penninx, B. W. J. H., & Zitman, F. G. (2012). Impact of childhood life events and trauma on the course of depressive and anxiety disorders. *Acta Psychiatrica Scandinavica, 126*(3), 198–207. http://dx.doi.org/10.1111/j.1600-0447.2011.01828.x.

Huang, H., Fan, X., Williamson, D. E., & Rao, U. (2011). White matter changes in healthy adolescents at familial risk for unipolar depression: a diffusion tensor imaging study. *Neuropsychopharmacology, 36*(3), 684–691. http://dx.doi.org/10.1038/npp.2010.199.

Isgor, C., Kabbaj, M., Akil, H., & Watson, S. J. (2004). Delayed effects of chronic variable stress during peripubertal-juvenile period on hippocampal morphology and on cognitive and stress axis functions in rats. *Hippocampus, 14*(5), 636–648. http://dx.doi.org/10.1002/hipo.10207.

Iwabuchi, S. J., Krishnadas, R., Li, C., Auer, D., Radua, J., & Palaniyappan, L. (2015). Localized connectivity in depression: a meta-analysis of resting state functional imaging studies. *Neuroscience and Biobehavioral Reviews*, 1–29. http://dx.doi.org/10.1016/j.neubiorev.2015.01.006.

Jackowski, A. P., Douglas-Palumberi, H., Jackowski, M., Win, L., Schultz, R. T., Staib, L. W., … Kaufman, J. (2008). Corpus callosum in maltreated children with posttraumatic stress disorder: a diffusion tensor imaging study. *Psychiatry Research, 162*(3), 256–261. http://dx.doi.org/10.1016/j.pscychresns.2007.08.006.

Jacobson, L., & Sapolsky, R. (1991). The role of the hippocampus in feedback regulation of the hypothalamic-pituitary-adrenocortical axis. *Endocrine Reviews, 12*(2), 118–134.

Jaffee, S. R., & Price, T. S. (2007). Gene-environment correlations: a review of the evidence and implications for prevention of mental illness. *Molecular Psychiatry, 12*(5), 432–442. http://dx.doi.org/10.1038/sj.mp.4001950.

Jednoróg, K., Altarelli, I., Monzalvo, K., Fluss, J., Dubois, J., Billard, C., … Ramus, F. (2012). The influence of socioeconomic status on children's brain structure. *PLoS One, 7*(8), e42486.

Kane, P., & Garber, J. (2004). The relations among depression in fathers, children's psychopathology, and father-child conflict: a meta-analysis. *Clinical Psychology Review, 24*(3), 339–360. http://dx.doi.org/10.1016/j.cpr.2004.03.004.

Karg, K., Burmeister, M., Shedden, K., & Sen, S. (2011). The serotonin transporter promoter variant (5-HTTLPR), stress, and depression meta-analysis revisited: evidence of genetic moderation. *Archives of General Psychiatry, 68*(5), 444–454. http://dx.doi.org/10.1001/archgenpsychiatry.2010.189.

Kassem, M. S., Lagopoulos, J., Stait-Gardner, T., Price, W. S., Chohan, T. W., Arnold, J. C., … Bennett, M. R. (2012). Stress-induced grey matter loss determined by MRI is primarily due to loss of dendrites and their synapses. *Molecular Neurobiology, 47*(2), 645–661. http://dx.doi.org/10.1007/s12035-012-8365-7.

Kaufman, J., Birmaher, B., Perel, J., Dahl, R. E., Moreci, P., Nelson, B., … Ryan, N. D. (1997). The corticotropin-releasing hormone challenge in depressed abused, depressed nonabused, and normal control children. *Biological Psychiatry, 42*(8), 669–679. http://dx.doi.org/10.1016/s0006-3223(96)00470-2.

Kaufman, J., Martin, A., King, R., & Charney, D. (2001). Are child-, adolescent-, and adult-onset depression one and the same disorder? *Biological Psychiatry, 49*(12), 980–1001.

Kendler, K. S., Hettema, J. M., Butera, F., Gardner, C. O., & Prescott, C. A. (2003). Life event dimensions of loss, humiliation, entrapment, and danger in the prediction of onsets of major depression and generalized anxiety. *Archives of General Psychiatry, 60*(8), 1–8. http://dx.doi.org/10.1001/archpsyc.60.8.789.

Kendler, K. S., & Karkowski-Shuman, L. (1997). Stressful life events and genetic liability to major depression: genetic control of exposure to the environment ? *Psychological Medicine, 27*(3), 1–9. http://dx.doi.org/10.1017/s0033291797004716.

Kendler, K. S., Karkowski-Shuman, L., & Prescott, C. A. (1999). Causal relationship between stressful life events and the onset of major depression. *American Journal of Psychiatry, 156*(6), 837.

Kendler, K. S., Kessler, R. C., Walters, E. E., MacLean, C., Neale, M. C., Heath, A. C., & Eaves, L. J. (1995). Stressful life events, genetic liability, and onset of an episode of major depression in women. *American Journal of Psychiatry, 152*(6), 833–842.

Kendler, K. S., Kuhn, J. W., & Prescott, C. A. (2004). Childhood sexual abuse, stressful life events and risk for major depression in women. *Psychological Medicine, 34*(08), 1475. http://dx.doi.org/10.1017/S003329170400265X.

Kendler, K. S., Thornton, L. M., & Gardner, C. O. (2000). Stressful life events and previous episodes in the etiology of major depression in women: an evaluation of the "kindling" hypothesis. *American Journal of Psychiatry, 157*(8), 1243–1251

Kendler, K. S., Thornton, L. M., & Gardner, C. O. (2001). Genetic risk, number of previous depressive episodes, and stressful life events in predicting onset of major depression. *American Journal of Psychiatry, 158*(4), 582–586. http://dx.doi.org/10.1176/appi.ajp.158.4.582.

Kessler, R. C., & Magee, W. J. (2009). Childhood adversities and adult depression: basic patterns of association in a US national survey. *Psychological Medicine, 23*(3), 679. http://dx.doi.org/10.1017/S0033291700025460.

Kessler, R. C., McGonagle, K. A., Zhao, S., Nelson, C. B., Hughes, M., Eshleman, S., … Kendler, K. S. (1994). Lifetime and 12-month prevalence of DSM-III-R psychiatric disorders in the United States: results from the national comorbidity survey. *Archives of General Psychiatry, 51*(1), 8–19. http://dx.doi.org/10.1001/archpsyc.1994.03950010008002.

Kessler, R. C., Davis, C. G., & Kendler, K. S. (1997). Childhood adversity and adult psychiatric disorder in the US National Comorbidity Survey. *Psychological Medicine, 27*(5), 1–19. http://dx.doi.org/10.1017/s0033291797005588.

Kessler, R. C., Petukhova, M., Sampson, N. A., Zaslavsky, A. M., & Wittchen, H.-U. (2012). Twelve-month and lifetime prevalence and lifetime morbid risk of anxiety and mood disorders in the United States. *International Journal of Methods in Psychiatric Research, 21*(3), 169–184. http://dx.doi.org/10.1002/mpr.1359.

Kim, P., Evans, G. W., Angstadt, M., Ho, S. S., Sripada, C. S., Swain, J. E., … Phan, K. L. (2013). Effects of childhood poverty and chronic stress on emotion regulatory brain function in adulthood. *Proceedings of the National Academy of Sciences, 110*(46), 18442–18447. http://dx.doi.org/10.1073/pnas.1308240110.

Klengel, T., Mehta, D., Anacker, C., Rex-Haffner, M., Pruessner, J. C., Pariante, C. M., … Binder, E. B. (2012). Allele-specific FKBP5 DNA demethylation mediates gene–childhood trauma interactions. *Nature Neuroscience, 16*(1), 33–41. http://dx.doi.org/10.1038/nn.3275.

Knorr, U., Vinberg, M., Kessing, L. V., & Wetterslev, J. (2010). Salivary cortisol in depressed patients versus control persons: a systematic review and meta-analysis. *Psychoneuroendocrinology, 35*(9), 1275–1286. http://dx.doi.org/10.1016/j.psyneuen.2010.04.001.

Koolschijn, P. C. M. P., van Haren, N. E. M., Lensvelt-Mulders, G. J. L. M., Hulshoff Pol, H. E., & Kahn, R. S. (2009). Brain volume abnormalities in major depressive disorder: a meta-analysis of magnetic resonance imaging studies. *Human Brain Mapping, 30*(11), 3719–3735. http://dx.doi.org/10.1002/hbm.20801.

Lenze, S. N., Xiong, C., & Sheline, Y. I. (2008). Childhood adversity predicts earlier onset of major depression but not reduced hippocampal volume. *Psychiatry Research, 162*(1), 39–49.

Li, L., Power, C., Kelly, S., Kirschbaum, C., & Hertzman, C. (2007). Life-time socio-economic position and cortisol patterns in mid-life. *Psychoneuroendocrinology, 32*(7), 824–833. http://dx.doi.org/10.1016/j.psyneuen.2007.05.014.

Lieb, R., Isensee, B., Höfler, M., Pfister, H., & Wittchen, H.-U. (2002). Parental major depression and the risk of depression and other mental disorders in offspring: a prospective-longitudinal community study. *Archives of General Psychiatry, 59*(4), 365–374.

Liu, R. T., & Alloy, L. B. (2010). Stress generation in depression: a systematic review of the empirical literature and recommendations for future study. *Clinical Psychology Review, 30*(5), 582–593. http://dx.doi.org/10.1016/j.cpr.2010.04.010.

Lopez-Duran, N. L., Kovacs, M., & George, C. J. (2009). Hypothalamic–pituitary–adrenal axis dysregulation in depressed children and adolescents: a meta-analysis. *Psychoneuroendocrinology, 34*(9), 1272–1283. http://dx.doi.org/10.1016/j.psyneuen.2009.03.016.

Lorant, V., Deliège, D., Eaton, W., Robert, A., Philippot, P., & Ansseau, M. (2003). Socioeconomic inequalities in depression: a meta-analysis. *American Journal of Epidemiology, 157*(2), 98–112.

Lovejoy, M. C., Graczyk, P. A., O'Hare, E., & Neuman, G. (2000). Maternal depression and parenting behavior. *Clinical Psychology Review, 20*(5), 561–592. http://dx.doi.org/10.1016/S0272-7358(98)00100-7.

Luby, J. L., Belden, A., Botteron, K., Marrus, N., Harms, M. P., Babb, C., … Barch, D. (2013). The effects of poverty on childhood brain development. *Journal of American Medical Association Pediatrics, 167*(12), 1135.

Luby, J. L., Heffelfinger, A. K., Mrakotsky, C., Brown, K. M., Hessler, M. J., Wallis, J. M., & Spitznagel, E. L. (2009). The clinical picture of depression in preschool children. *Journal of the American Academy of Child and Adolescent Psychiatry, 42*(3), 340–348. http://dx.doi.org/10.1097/00004583-200303000-00015.

Luby, J. L., Si, X., Belden, A., Tandon, M., & Spitznagel, E. (2009). Preschool depression: homotypic continuity and course over 24 months. *Archives of General Psychiatry, 66*(8), 897.

Lumley, M. N., & Harkness, K. L. (2007). Specificity in the relations among childhood adversity, early maladaptive schemas, and symptom profiles in adolescent depression. *Cognitive Therapy and Research, 31*(5), 639–657. http://dx.doi.org/10.1007/s10608-006-9100-3.

Lupien, S. J., Fiocco, A., Wan, N., Maheu, F., Lord, C., Schramek, T., & Tu, M. T. (2005). Stress hormones and human memory function across the lifespan. *Psychoneuroendocrinology, 30*(3), 225–242. http://dx.doi.org/10.1016/j.psyneuen.2004.08.003.

Lupien, S. J., King, S., Meaney, M. J., & McEwen, B. S. (2000). Child's stress hormone levels correlate with mother's socioeconomic status and depressive state. *Biological Psychiatry, 48*(10), 976–980. http://dx.doi.org/10.1016/s0006-3223(00)00965-3.

Lupien, S. J., King, S., Meaney, M. J., & McEwen, B. S. (2001). Can poverty get under your skin? basal cortisol levels and cognitive function in children from low and high socioeconomic status. *Development and Psychopathology, 13*(3), 653–676.

Maciejewski, P. K., Prigerson, H. G., & Mazure, C. M. (2000). Self-efficacy as a mediator between stressful life events and depressive symptoms. Differences based on history of prior depression. *British Journal of Psychiatry, 176*, 373–378.

MacMillan, H. L., Georgiades, K., Duku, E. K., Shea, A., Steiner, M., Niec, A., … Schmidt, L. A. (2009). Cortisol response to stress in female youths exposed to childhood maltreatment: results of the youth mood project. *Biological Psychiatry, 66*(1), 62–68. http://dx.doi.org/10.1016/j.biopsych.2008.12.014.

Magariños, A. M., McEwen, B. S., Flügge, G., & Fuchs, E. (1996). Chronic psychosocial stress causes apical dendritic atrophy of hippocampal CA3 pyramidal neurons in subordinate tree shrews. *Journal of Neuroscience, 16*(10), 3534–3540.

Mannie, Z. N., Harmer, C. J., & Cowen, P. J. (2007). Increased waking salivary cortisol levels in young people at familial risk of depression. *American Journal of Psychiatry, 164*(4), 617–621. http://dx.doi.org/10.1176/appi.ajp.164.4.617.

Marcus Jenkins, J. V., Woolley, D. P., Hooper, S. R., & De Bellis, M. D. (2013). Direct and indirect effects of brain volume, socioeconomic status and family stress on child IQ. *Journal of Child and Adolescent Behavior, 1*(2).

McEwen, B. S. (1997). Possible mechanisms for atrophy of the human hippocampus. *Molecular Psychiatry, 2*(3), 255–262.

McGonagle, K. A., & Kessler, R. C. (1990). Chronic stress, acute stress, and depressive symptoms. *American Journal of Community Psychology, 18*(5), 681–706.

McKay, M. S., & Zakzanis, K. K. (2010). The impact of treatment on HPA axis activity in unipolar major depression. *Journal of Psychiatric Research, 44*(3), 183–192. http://dx.doi.org/10.1016/j.jpsychires.2009.07.012.

McKinnon, M. C., Yucel, K., Nazarov, A., & MacQueen, G. M. (2009). A meta-analysis examining clinical predictors of hippocampal volume in patients with major depressive disorder. *Journal of Psychiatry and Neuroscience, 34*(1), 41–54.

McKittrick, C. R., Magariños, A. M., Blanchard, D. C., Blanchard, R. J., McEwen, B. S., & Sakai, R. R. (2000). Chronic social stress reduces dendritic arbors in CA3 of hippocampus and decreases binding to serotonin transporter sites. *Synapse, 36*(2), 85–94.

Mehta, M. A., Golembo, N. I., Nosarti, C., Colvert, E., Mota, A., Williams, S. C. R., … Sonuga-Barke, E. J. (2009). Amygdala, hippocampal and corpus callosum size following severe early institutional deprivation: the English and Romanian adoptees study pilot. *Journal of Child Psychology and Psychiatry, 50*(8), 943–951. http://dx.doi.org/10.1111/j.1469-7610.2009.02084.x.

Michl, L. C., McLaughlin, K. A., Shepherd, K., & Nolen-Hoeksema, S. (2013). Rumination as a mechanism linking stressful life events to symptoms of depression and anxiety: longitudinal evidence in early adolescents and adults. *Journal of Abnormal Psychology, 122*(2), 339–352. http://dx.doi.org/10.1037/a0031994.

Miech, R. A., Caspi, A., Moffitt, T. E., Wright, B. R. E., & Silva, P. A. (1999). Low socioeconomic status and mental disorders: a longitudinal study of selection and causation during young adulthood. *American Journal of Sociology, 104*(4), 1096–1131. http://dx.doi.org/10.1086/210137.

Modell, S., Lauer, C. J., Schreiber, W., Huber, J., Krieg, J. C., & Holsboer, F. (1998). Hormonal response pattern in the combined DEX-CRH test is stable over time in subjects at high familial risk for affective disorders. *Neuropsychopharmacology, 18*(4), 253–262. http://dx.doi.org/10.1016/S0893-133X(97)00144-9.

Molnar, B. E., Buka, S., & Kessler, R. C. (2001). Child sexual abuse and subsequent psychopathology: results from the National Comorbidity Survey. *American Journal of Public Health, 91*(5), 753–760. http://dx.doi.org/10.2105/AJPH.91.5.753.

Monroe, S. M., & Harkness, K. L. (2005). Life stress, the "Kindling" hypothesis, and the recurrence of depression: considerations from a life stress perspective. *Psychological Review, 112*(2), 417–445. http://dx.doi.org/10.1037/0033-295X.112.2.417.

Monroe, S. M., & Simons, A. D. (1991). Diathesis-stress theories in the context of life stress research: implications for the depressive disorders. *Psychological Bulletin, 110*(3), 406–425. http://dx.doi.org/10.1037/0033-2909.110.3.406.

Murphy, M. L., & Frodl, T. (2011). Meta-analysis of diffusion tensor imaging studiesshows altered fractional anisotropy occurring indistinct brain areas in association with depression. *Biology of Mood and Anxiety Disorders, 1*(1), 3. http://dx.doi.org/10.1186/2045-5380-1-3.

Noble, K. G., Houston, S. M., Kan, E., & Sowell, E. R. (2012). Neural correlates of socioeconomic status in the developing human brain. *Developmental Science, 15*(4), 516–527.

Opel, N., Redlich, R., Zwanzger, P., Grotegerd, D., Arolt, V., Heindel, W., ... Dannlowski, U. (2014). Hippocampal atrophy in major depression: a function of childhood maltreatment rather than diagnosis? *Neuropsychopharmacology, 39*(12), 2723–2731. http://dx.doi.org/10.1038/npp.2014.145.

Pagliaccio, D., Barch, D. M., Bogdan, R., Wood, P. K., Lynskey, M. T., Heath, A. C., & Agrawal, A. (2015). Shared predisposition in the association between cannabis use and subcortical brain structure. *JAMA Psychiatry, 72*(10), 994–1001.

Pagliaccio, D., Luby, J. L., Bogdan, R., Agrawal, A., Gaffrey, M. S., Belden, A. C., ... Barch, D. M. (2014). Stress-system genes and life stress predict cortisol levels and amygdala and hippocampal volumes in children. *Neuropsychopharmacology, 39*(5), 1245–1253. http://dx.doi.org/10.1038/npp.2013.327.

Pagliaccio, D., Luby, J. L., Bogdan, R., Agrawal, A., Gaffrey, M. S., Belden, A. C., ... Barch, D. M. (2015a). HPA axis genetic variation, pubertal status, and sex interact to predict amygdala and hippocampus responses to negative emotional faces in school-age children. *NeuroImage, 109*, 1–11. http://dx.doi.org/10.1016/j.neuroimage.2015.01.017.

Pagliaccio, D., Luby, J. L., Bogdan, R., Agrawal, A., Gaffrey, M. S., Belden, A. C., ... Barch, D. M. (2015b). Amygdala functional connectivity, HPA axis genetic variation, and life stress in children and relations to anxiety and emotion regulation. *Journal of Abnormal Psychology, 124*(4), 817.

Paul, R., Henry, L., Grieve, S. M., Guilmette, T. J., Niaura, R., Bryant, R., ... Gordon, E. (2008). The relationship between early life stress and microstructural integrity of the corpus callosum in a non-clinical population. *Neuropsychiatric Disease and Treatment, 4*(1), 193–201.

Pederson, C. L., Maurer, S. H., Kaminski, P. L., Zander, K. A., Peters, C. M., Stokes Crowe, L. A., & Osborn, R. E. (2004). Hippocampal volume and memory performance in a community-based sample of women with posttraumatic stress disorder secondary to child abuse. *Journal of Traumatic Stress, 17*(1), 37–40.

Pigott, H. E., Leventhal, A. M., Alter, G. S., & Boren, J. J. (2010). Efficacy and effectiveness of antidepressants: current status of research. *Psychotherapy and Psychosomatics, 79*(5), 267–279. http://dx.doi.org/10.1159/000318293.

Pike, A., McGuire, S., Hetherington, E. M., Reiss, D., & Plomin, R. (1996). Family environment and adolescent depressive symptoms and antisocial behavior: a multivariate genetic analysis. *American Psychological Association, 32*(4), 590.

Plomin, R., Lichtenstein, P., Pedersen, N. L., McClearn, G. E., & Nesselroade, J. R. (2004). Genetic influence on life events during the last half of the life span. *Psychology and Aging, 5*(1), 1–6. http://dx.doi.org/10.1037//0882-7974.5.1.25.

Post, R. M. (1992). Transduction of psychosocial stress into the neurobiology of recurrent affective disorder. *American Journal of Psychiatry, 149*(8), 999–1010.

Rao, U., Chen, L.-A., Bidesi, A. S., Shad, M. U., Thomas, M. A., & Hammen, C. L. (2010). Hippocampal changes associated with early-life adversity and vulnerability to depression. *Biological Psychiatry, 67*(4), 357–364. http://dx.doi.org/10.1016/j.biopsych.2009.10.017.

Rao, U., Hammen, C., Ortiz, L. R., Chen, L.-A., & Poland, R. E. (2008). Effects of early and recent adverse experiences on adrenal response to psychosocial stress in depressed adolescents. *Biological Psychiatry, 64*(6), 521–526. http://dx.doi.org/10.1016/j.biopsych.2008.05.012.

Reagan, L. P., & McEwen, B. S. (1997). Controversies surrounding glucocorticoid-mediated cell death in the hippocampus. *Journal of Chemical Neuroanatomy, 13*(3), 149–167.

Ribeiro, S. C., Tandon, R., Grunhaus, L., & Greden, J. F. (1993). The DST as a predictor of outcome in depression: a meta-analysis. *American Journal of Psychiatry, 150*(11), 1618–1629. http://dx.doi.org/10.1176/ajp.150.11.1618.

Rice, F., Lewis, G., Harold, G. T., & Thapar, A. (2013). Examining the role of passive gene-environment correlation in childhood depression using a novel genetically sensitive design. *Development and Psychopathology, 25*(1), 37–50. http://dx.doi.org/10.1017/S0954579412000880.

Rinne, T., de Kloet, E. R., Wouters, L., Goekoop, J. G., DeRijk, R. H., & van den Brink, W. (2002). Hyperresponsiveness of hypothalamic-pituitary-adrenal axis to combined dexamethasone/corticotropin-releasing hormone challenge in female borderline personality disorder subjects with a history of sustained childhood abuse. *Biological Psychiatry, 52*(11), 1102–1112.

Risch, N., Herrell, R., Lehner, T., Liang, K.-Y., Eaves, L., Hoh, J., ... Merikangas, K. R. (2009). Interaction between the serotonin transporter gene (5-HTTLPR), stressful life events, and risk of depression. *Journal of the American Medical Association, 301*(23), 2462–2471.

Ruch, L. O., Chandler, S. M., & Harter, R. A. (1980). Life change and rape impact. *Journal of Health and Social Behavior, 21*(3), 248–260.

Rudolph, K. D., & Flynn, M. (2007). Childhood adversity and youth depression: Influence of gender and pubertal status. *Development and Psychopathology, 19*(2). http://dx.doi.org/10.1017/S0954579407070241.

Sacher, J., Neumann, J., Fünfstück, T., Soliman, A., Villringer, A., & Schroeter, M. L. (2012). Mapping the depressed brain: a meta-analysis of structural and functional alterations in major depressive disorder. *Journal of Affective Disorders, 140*(2), 142–148. http://dx.doi.org/10.1016/j.jad.2011.08.001.

Sapolsky, R. (2000). Glucocorticoids and hippocampal atrophy in neuropsychiatric disorders. *Archives of General Psychiatry, 57*(10), 925–935.

Saridjan, N. S., Huizink, A. C., Koetsier, J. A., Jaddoe, V. W., Mackenbach, J. P., Hofman, A., ... Tiemeier, H. (2010). Do social disadvantage and early family adversity affect the diurnal cortisol rhythm in infants? the Generation R Study. *Hormones and Behavior, 57*(2), 247–254. http://dx.doi.org/10.1016/j.yhbeh.2009.12.001.

Sheridan, M. A., Fox, N. A., Zeanah, C. H., McLaughlin, K. A., & Nelson, C. A. (2012). Variation in neural development as a result of exposure to institutionalization early in childhood. *Proceedings of the National Academy of Sciences, 109*(32), 12927–12932.

Silberg, J. L., Maes, H., & Eaves, L. J. (2010). Genetic and environmental influences on the transmission of parental depression to children's depression and conduct disturbance: an extended Children of Twins study. *Journal of Child Psychology and Psychiatry, 51*(6), 734–744. http://dx.doi.org/10.1111/j.1469-7610.2010.02205.x.

Silberg, J., Rutter, M., Neale, M., & Eaves, L. (2001). Genetic moderation of environmental risk for depression and anxiety in adolescent girls. *British Journal of Psychiatry, 179*(2), 116–121. http://dx.doi.org/10.1192/bjp.179.2.116.

Simon, G. (2003). Social and economic burden of mood disorders. *Biological Psychiatry, 54*(3), 208–215. http://dx.doi.org/10.1016/S0006-3223(03)00420-7.

Slopen, N., McLaughlin, K. A., & Shonkoff, J. P. (2014). Interventions to improve cortisol regulation in children: a systematic review. *Pediatrics, 133*(2), 312–326. http://dx.doi.org/10.1542/peds.2013-1632.

Soni, A. (2010). *Statistical brief #303: Anxiety and mood disorders: Use and expenditures for adults 18 and older, U.S. civilian noninstitutionalized population, 2007* (pp. 1–5).

Spence, S. H., Najman, J. M., Bor, W., O'Callaghan, M. J., & Williams, G. M. (2002). Maternal anxiety and depression, poverty and marital relationship factors during early childhood as predictors of anxiety and depressive symptoms in adolescence. *Journal of Child Psychology and Psychiatry, and Allied Disciplines, 43*(4), 457–469. http://dx.doi.org/10.1111/1469-7610.00037.

Sripada, R. K., Swain, J. E., Evans, G. W., Welsh, R. C., & Liberzon, I. (2015). Childhood poverty and stress reactivity are associated with aberrant functional connectivity in default mode network. *Neuropsychopharmacology, 39*(9), 2244–2251. http://dx.doi.org/10.1038/npp.2014.75.

Staff, R. T., Murray, A. D., Ahearn, T. S., Mustafa, N., Fox, H. C., & Whalley, L. J. (2012). Childhood socioeconomic status and adult brain size: childhood socioeconomic status influences adult hippocampal size. *Annals of Neurology, 71*(5), 653–660.

Starkman, M. N., Gebarski, S. S., Berent, S., & Schteingart, D. E. (1992). Hippocampal formation volume, memory dysfunction, and cortisol levels in patients with Cushing's syndrome. *Biological Psychiatry, 32*(9), 756–765.

Starkman, M. N., Giordani, B., Gebarski, S. S., Berent, S., Schork, M. A., & Schteingart, D. E. (1999). Decrease in cortisol reverses human hippocampal atrophy following treatment of Cushing's disease. *Biological Psychiatry, 46*(12), 1595–1602.

Staufenbiel, S. M., Penninx, B. W. J. H., Spijker, A. T., Elzinga, B. M., & van Rossum, E. F. C. (2013). Hair cortisol, stress exposure, and mental health in humans: a systematic review. *Psychoneuroendocrinology, 38*(8), 1220–1235. http://dx.doi.org/10.1016/j.psyneuen.2012.11.015.

Stein, M. B., Yehuda, R., Koverola, C., & Hanna, C. (1997). Enhanced dexamethasone suppression of plasma cortisol in adult women traumatized by childhood sexual abuse. *Biological Psychiatry, 42*(8), 680–686. http://dx.doi.org/10.1016/S0006-3223(96)00489-1.

Stroud, C. B., Davila, J., & Moyer, A. (2008). The relationship between stress and depression in first onsets versus recurrences: a meta-analytic review. *Journal of Abnormal Psychology, 117*(1), 206–213. http://dx.doi.org/10.1037/0021-843X.117.1.206.

Sullivan, P. F., Neale, M. C., & Kendler, K. S. (2000). Genetic epidemiology of major depression: review and meta-analysis. *American Journal of Psychiatry, 157*(10), 1552–1562.

Sundermann, B. (2014). Toward literature-based feature selection for diagnostic classification: a meta-analysis of resting-state fMRI in depression. *Frontiers in Human Neuroscience, 60*, 1–12. http://dx.doi.org/10.3389/fnhum.2014.00692/abstract.

Suzuki, H., Luby, J. L., Botteron, K. N., & Dietrich, R. (2014). Early life stress and trauma and enhanced limbic activation to emotionally valenced faces in depressed and healthy children. *Journal of the American Academy of Child and Adolescent Psychiatry, 53*(7), 800–813.

Taylor, S. E., Way, B. M., Welch, W. T., Hilmert, C. J., Lehman, B. J., & Eisenberger, N. I. (2006). Early family environment, current adversity, the serotonin transporter promoter polymorphism, and depressive symptomatology. *Biological Psychiatry, 60*(7), 671–676. http://dx.doi.org/10.1016/j.biopsych.2006.04.019.

Teicher, M. H., Anderson, C. M., & Polcari, A. (2012). Childhood maltreatment is associated with reduced volume in the hippocampal subfields CA3, dentate gyrus, and subiculum. *Proceedings of the National Academy of Sciences, 109*(9), E563–E572.

Teicher, M. H., Dumont, N. L., Ito, Y., Vaituzis, C., Giedd, J. N., & Andersen, S. L. (2004). Childhood neglect is associated with reduced corpus callosum area. *Biological Psychiatry, 56*(2), 80–85.

Tomoda, A., Navalta, C. P., Polcari, A., Sadato, N., & Teicher, M. H. (2009). Childhood sexual abuse is associated with reduced gray matter volume in visual cortex of young women. *Biological Psychiatry, 66*(7), 642–648.

Tottenham, N., Hare, T. A., Quinn, B. T., McCarry, T. W., Nurse, M., Gilhooly, T., ... Casey, B. J. (2010). Prolonged institutional rearing is associated with atypically large amygdala volume and difficulties in emotion regulation. *Developmental Science, 13*(1), 46–61. http://dx.doi.org/10.1111/j.1467-7687.2009.00852.x.

Tottenham, N., Hare, T. A., Millner, A., Gilhooly, T., Zevin, J. D., & Casey, B. J. (2011). Elevated amygdala response to faces following early deprivation. *Developmental Science, 14*(2), 190–204. http://dx.doi.org/10.1111/j.1467-7687.2010.00971.x.

Tottenham, N., & Sheridan, M. A. (2009). A review of adversity, the amygdala and the hippocampus: a consideration of developmental timing. *Frontiers in Human Neuroscience, 3*, 68. http://dx.doi.org/10.3389/neuro.09.068.2009.

Tully, E. C., Iacono, W. G., & McGue, M. (2008). An adoption study of parental depression as an environmental liability for adolescent depression and childhood disruptive disorders. *American Journal of Psychiatry, 165*(9), 1148–1154. http://dx.doi.org/10.1176/appi.ajp.2008.07091438.

Turner, H. A., & Butler, M. J. (2003). Direct and indirect effects of childhood adversity on depressive symptoms in young adults. *Journal of Youth and Adolescence, 32*(2), 89–103. http://dx.doi.org/10.1023/A:1021853600645.

van der Vegt, E. J. M., van der Ende, J., Huizink, A. C., Verhulst, F. C., & Tiemeier, H. (2010). Childhood adversity modifies the relationship between anxiety disorders and cortisol secretion. *Biological Psychiatry, 68*(11), 1048–1054. http://dx.doi.org/10.1016/j.biopsych.2010.07.027.

Van Harmelen, A.-L., van Tol, M.-J., van der Wee, N. J. A., Veltman, D. J., Aleman, A., Spinhoven, P., ... Elzinga, B. M. (2010). Reduced medial prefrontal cortex volume in adults reporting childhood emotional maltreatment. *Biological Psychiatry, 68*(9), 832–838.

Videbech, P., & Ravnkilde, B. (2004). Hippocampal volume and depression: a meta-analysis of MRI studies. *American Journal of Psychiatry, 161*(11), 1957–1966. http://dx.doi.org/10.1176/appi.ajp.161.11.1957.

Vythilingam, M., Heim, C., Newport, J., Miller, A. H., Anderson, E., Bronen, R., ... Bremner, J. D. (2002). Childhood trauma associated with smaller hippocampal volume in women with major depression. *American Journal of Psychiatry, 159*(12), 2072–2080.

Watanabe, Y., Gould, E., & McEwen, B. S. (1992). Stress induces atrophy of apical dendrites of hippocampal CA3 pyramidal neurons. *Brain Research, 588*(2), 341–345. http://dx.doi.org/10.1016/0006-8993(92)91597-8.

Weiss, E. L., Longhurst, J. G., & Mazure, C. M. (1999). Childhood sexual abuse as a risk factor for depression in women: psychosocial and neurobiological correlates. *American Journal of Psychiatry, 156*(6), 816–828. http://dx.doi.org/10.1130/spe253-p13.

Weissman, M. M., Wickramaratne, P., Nomura, Y., Warner, V., Pilowsky, D., & Verdeli, H. (2006). Offspring of depressed parents: 20 years later. *American Journal of Psychiatry, 163*(6), 1001–1008. http://dx.doi.org/10.1176/appi.ajp.163.6.1001.

Weniger, G., Lange, C., Sachsse, U., & Irle, E. (2008). Amygdala and hippocampal volumes and cognition in adult survivors of childhood abuse with dissociative disorders. *Acta Psychiatrica Scandinavica, 118*(4), 281–290.

Wiggins, J. L., Mitchell, C., Stringaris, A., & Leibenluft, E. (2014). Developmental trajectories of irritability and bidirectional associations with maternal depression. *Journal of the American Academy of Child and Adolescent Psychiatry, 53*(11), 1191–1205. http://dx.doi.org/10.1016/j.jaac.2014.08.005, e1–4.

Williamson, D. E., Birmaher, B., & Axelson, D. A. (2004). First episode of depression in children at low and high familial risk for depression. *Journal of the American Academy of Child and Adolescent Psychiatry, 43*(3), 291–297. http://dx.doi.org/10.1097/00004583-200403000-00010.

Wise, T., Radua, J., Nortje, G., Cleare, A. J., Young, A. H., & Arnone, D. (2015). Voxel-based meta-analytical evidence of structural disconnectivity in major depression and bipolar disorder. *Biological Psychiatry*. http://dx.doi.org/10.1016/j.biopsych.2015.03.004.

Woon, F. L., & Hedges, D. W. (2008). Hippocampal and amygdala volumes in children and adults with childhood maltreatment-related posttraumatic stress disorder: a meta-analysis. *Hippocampus, 18*(8), 729–736. http://dx.doi.org/10.1002/hipo.20437.

Zang, Y., Jiang, T., Lu, Y., He, Y., & Tian, L. (2004). Regional homogeneity approach to fMRI data analysis. *NeuroImage, 22*(1), 394–400. http://dx.doi.org/10.1016/j.neuroimage.2003.12.030.

Zeanah, C. H., Nelson, C. A., Fox, N. A., Smyke, A. T., Marshall, P., Parker, S. W., & Koga, S. (2003). Designing research to study the effects of institutionalization on brain and behavioral development: the Bucharest Early Intervention Project. *Development and Psychopathology, 15*(4), 885–907.

Zhang, W.-N., Chang, S.-H., Guo, L.-Y., Zhang, K.-L., & Wang, J. (2013). The neural correlates of reward-related processing in major depressive disorder: a meta-analysis of functional magnetic resonance imaging studies. *Journal of Affective Disorders, 151*(2), 531–539. http://dx.doi.org/10.1016/j.jad.2013.06.039.

Zimmermann, P., Brückl, T., Nocon, A., Pfister, H., Binder, E. B., Uhr, M., ... Holsboer, F. (2011). Interaction of FKBP5 gene variants and adverse life events in predicting depression onset: results from a 10-year prospective community study. *American Journal of Psychiatry, 168*(10), 1107–1116. http://dx.doi.org/10.1176/appi.ajp.2011.10111577.

Zobel, A. W., Nickel, T., Künzel, H. E., Ackl, N., Sonntag, A., Ising, M., & Holsboer, F. (2000). Effects of the high-affinity corticotropin-releasing hormone receptor 1 antagonist R121919 in major depression: the first 20 patients treated. *Journal of Psychiatric Research, 34*(3), 171–181.

Part II

Experimental Systems Neuroscience

Chapter 3

Novel Targets in the Glutamate and Nitric Oxide Neurotransmitter Systems for the Treatment of Depression

Eileen O'Toole, Marika V. Doucet, Eoin Sherwin, Andrew Harkin
Neuropsychopharmacology Research Group, School of Pharmacy and Pharmaceutical Sciences, Trinity College Institute of Neuroscience, Trinity College Dublin, Dublin, Ireland

INTRODUCTION

The *N*-methyl-D-aspartate receptor (NMDA-R) antagonist ketamine has repeatedly shown antidepressant effects with a rapid onset of action and sustained therapeutic effects following a single dose in a number of clinical studies (Murrough, 2012). Such a profile of clinical efficacy has gone some way in addressing the largely unmet clinical need associated with treating depression and the limitations of antidepressants, which are typically associated with a slow onset of action and a substantial incidence of treatment resistance. However, a wider use of ketamine in the clinic is limited by its poor bioavailability following oral administration; potential for abuse; side effects including psychotomimetic properties (Wolff & Winstock, 2006); and a requirement for repeated administrations and/or add-on therapy to maintain its effects (aan het Rot et al., 2010; Ibrahim et al., 2012; Mathew et al., 2010; Segmiller et al., 2013). Investigations are therefore aimed at developing "cleaner" ketamine compounds that could demonstrate a fast onset of action and sustained efficacy without the adverse effects associated with ketamine. In this regard, targets located downstream of the NMDA-R, such as nitric oxide synthase (NOS), may represent novel avenues for treating neuropsychiatric disorders. Of particular interest, NOS inhibitors have repeatedly demonstrated antidepressant and anxiolytic properties in preclinical studies, but the development of highly selective inhibitors for the neuronal isoform (nNOS) with favorable pharmacokinetic properties has been difficult to bring to fruition. Since nNOS couples to the NMDA-R via the postsynaptic protein PSD-95, strategies that target the

Systems Neuroscience in Depression. http://dx.doi.org/10.1016/B978-0-12-802456-0.00003-0

PSD-95/nNOS interaction may also represent an interesting locus for the treatment of neuropsychiatric disorders (Doucet, Harkin, & Dev, 2012). Here, we review the rationale for the development of glutamate and nitric oxide (NO)-modulating drugs for treating depression and highlight findings in the field which provide a basis for ongoing drug development.

OVERVIEW OF GLUTAMATE NEUROTRANSMISSION IN THE BRAIN

Glutamate is the main mediator of excitatory neurotransmission in the mammalian brain and has a relatively uniform distribution in the central nervous system (CNS). Glutamate neurotransmission is of particular importance in modulating synaptic plasticity and for the mechanisms underlying higher order processes such as learning and memory, but under pathological conditions it can become excitotoxic and has the potential to trigger neurotoxicity associated with several brain injuries. The role of glutamate in health and disease has been covered by numerous reviews (Mattson, 2008; Meldrum, 2000; Sanacora, Zarate, Krystal, & Manji, 2008; Scannevin & Huganir, 2000; Willard & Koochekpour, 2013), and key points regarding glutamate neurotransmission are summarized in Box 1.

BOX 1 Glutamate Production and Transmission

1. Glutamate is synthesized *de novo* from glucose via the Krebs cycle and the transamination of α-oxoglutarate or recycled through the glutamate/glutamine cycle. Glutamate is then transported into and stored at high concentrations in vesicles by glutamate transporters (VGlut) and released by exocytosis in the synaptic cleft in a Ca^{2+}-dependent manner.
2. Glutamate is rapidly removed from the synapse through reuptake into the presynaptic terminal via class II and III metabotropic glutamate receptors (mGluRs) and also through high-affinity excitatory amino acid transporters (EAATs) located on the surface of astrocytes. Astrocytes possess the enzyme glutamine synthetase, which converts glutamate to glutamine, which can then be transported back to neurons where a glutaminase enzyme converts it to glutamate, thereby allowing glutamate to be recycled and completing the glutamate/glutamine cycle.
3. The family of glutamate receptors is divided into two distinct classes based on sequence homology, pharmacology, and electrophysiological properties (Scannevin & Huganir, 2000). They include mGluRs and ionotropic glutamate receptors (iGluRs).
4. mGluRs are composed of mGluR1-8, all of which are G protein-coupled receptors (GPCRs), which mediate their effects through G proteins and intracellular signaling cascades, including activation of the phospholipase C/inositol 1,4,5-triphosphate/diacylglycerol (PLC/IP3/DAG) pathway or inhibition of adenylate cyclase (AC).

Box 1 Glutamate Production and Transmission—cont'd

5. iGluRs are ligand-gated receptors that allow cation influx upon glutamate binding and include α-amino-3-hydroxy-5-methyl-4-isoxazole propionate (AMPA), kainate, and NMDA-Rs. Kainate receptors and AMPA receptors (AMPA-Rs) mediate fast electrophysiological responses to glutamate, whereas NMDA-Rs mediate slower responses.

OVERVIEW OF THE FUNCTIONAL ROLE AND REGULATION OF nNOS IN THE BRAIN

NO is a free radical gaseous messenger and is regarded as an atypical neurotransmitter in the brain. It is implicated in modulating a diverse range of functions in the CNS, including roles in synaptic plasticity, retrograde signaling, neurodevelopment, and excitotoxicity. As such, NO has been the subject of a number of comprehensive reviews, which may be referred to for a more detailed coverage of its role as a chemical messenger in the body (Akyol, Zoroglu, Armutcu, Sahin, & Gurel, 2004; Alderton, Cooper, & Knowles, 2001; Bredt & Snyder, 1992; Guix, Uribesalgo, Coma, & Munoz, 2005; Mungrue, Bredt, Stewart, & Husain, 2003; Vallance & Leiper, 2002; Zhou & Zhu, 2009). A number of key points are summarized in Box 2.

BOX 2 Nitric Oxide (NO) and Nitric Oxide Synthase (NOS)

1. NO is synthesized by a family of enzymes, NOS enzymes, which have an almost ubiquitous distribution in the body. There are three subtypes of NOS enzymes with a sequence homology of 50–60%, named according to the cell type and/or conditions in which they were first described: endothelial NOS (eNOS or NOS3), neuronal NOS (nNOS or NOS1), and inducible NOS (iNOS or NOS2). Mitochondria are also known to contain a NOS enzyme, termed mitochondrial (mt) NOS "mtNOS," which is thought to be a nNOS isoform with two posttranslational modifications (Elfering, Sarkela, & Giulivi, 2002; Guix et al., 2005).
2. All NOS isoforms catalyze the oxidation of L-arginine to citrulline and NO in the presence of nicotinamide adenine dinucleotide phosphate (NADPH) and O_2. Other functional prosthetic groups to enzymatic activity include tetrahydrobiopterin (BH_4), iron protoporphyrin IX (heme), flavin adenine dinucleotide (FAD), and flavin adenine mononucleotide (FMN) (Figure 1). nNOS and eNOS are constitutively expressed in mammalian cells and are both Ca^{2+}-calmodulin-dependent enzymes that generate small quantities of NO, lasting only minutes. Constitutive NOS exists as an inactive monomer, requiring dimerization for activation. By contrast, iNOS, a Ca^{2+}-calmodulin-independent enzyme, requires *de novo* synthesis and is expressed following inflammatory or immunological stimulation in microglia, astrocytes, macrophages, and other cells, generating significant quantities of NO that can persist for hours or days.

Continued

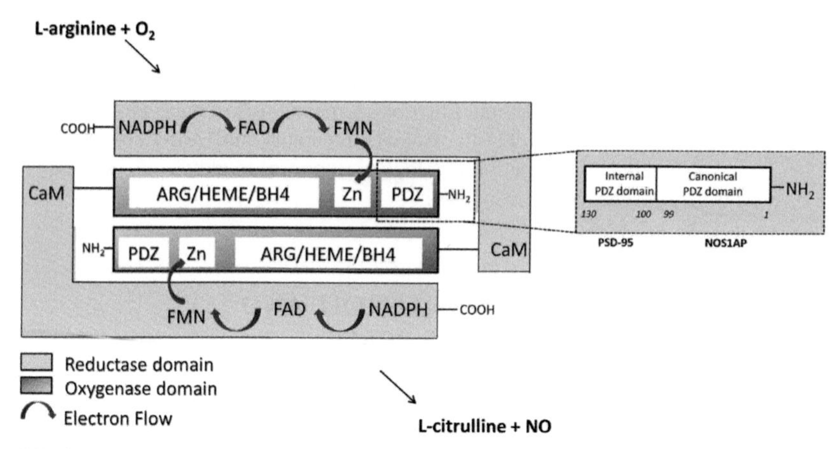

FIGURE 1 **Neuronal nitric oxide synthase (nNOS) dimer.** nNOS is active in dimer form and consists of two identical monomer subunits. Each monomer has a reductase domain (C-terminal) and an oxygenase domain (N-terminal), which are separated by a calmodulin-binding motif. The reductase domain contains FAD and FMN binding sites, whereas the oxygenase domain contains the binding sites for the substrate L-arginine (ARG) as well as binding sites for heme, tetrahydrobiopterin (BH_4), and zinc (Zn), which facilitates dimerization. Electrons donated by NADPH transfer from the reductase domain of one monomer to the oxygenase domain of the second monomer via FAD, FMN, and calmodulin. *Inset:* The PDZ domain of nNOS is located at the N-terminus (NH_2) of nNOS and is composed of a canonical PDZ domain at amino acid residues 1–99, which binds to several proteins, including NOS1AP, and also contains an internal PDZ domain at residues 100–130, which binds to PSD-95.

Box 2 Nitric Oxide (NO) and Nitric Oxide Synthase (NOS)—cont'd

3. NO is a gas with a half-life of about 30 s to 1 min, which diffuses freely across cell membranes with a paracrine mode of signaling, allowing both anterograde and retrograde diffusion.
4. NO is a physiological mediator in the gastrointestinal, cardiovascular, respiratory, and nervous systems. With regard to the CNS, NO is described as a neurotransmitter with atypical characteristics: it is not stored in synaptic vesicles, is not released by exocytosis, and NO does not act upon receptor proteins on the postsynaptic membrane.
5. In the CNS, NO activates a soluble guanylate cyclase (sGC) to generate cyclic guanosine monophosphate (cGMP), which acts as a second messenger and subsequently activates protein kinases (e.g., PKG I and II) and phosphodiesterases.
6. Overproduction of NO through the overactivation of the NMDA-R has been implicated in major depression and anxiety, while hypofunction of the NMDA-R has been linked to schizophrenia. These changes appear to be region dependent (Bernstein, Bogerts, & Keilhoff, 2005; Dhir & Kulkarni, 2011).

nNOS REGULATION BY PDZ-BINDING PROTEINS

nNOS activity can be regulated through physical interactions with several adaptor proteins, the binding of which occurs mainly via the PDZ (PSD-95/discs large/ZO-1 homologous) domain located at the N-terminal of nNOS. PDZ domains refer to a region of ~90 residues that act as modules for protein–protein interactions (Scannevin & Huganir, 2000). These regions have structurally well-defined pockets that can accommodate PDZ motifs that are generally found at the extreme carboxyl terminal of interacting proteins (Dev, 2004). PDZ domains can also self- or heterodimerize, giving rise to large protein complexes. The PDZ domain of nNOS is composed of the core, or canonical, PDZ domain located at residues 1–99 and also consists of a flanking region located at residues 100–130, referred to as the internal PDZ domain. The PDZ domain of nNOS is known to interact with various proteins including α_1-syntrophin, vac14, phosphofructokinase isoform M P13-kinase (PFK-M), nNOS-interacting DHHC domain-containing protein (NIDD), the serotonin (5-HT) transporter (SERT), and $5HT_{2B}$. For a review, see Doucet et al. (2012). Here, we focus on nNOS PDZ domain interactions with nitric oxide synthase 1 adaptor protein (NOS1AP) and PSD-95, as disruption of these interactions has been linked with the production of antidepressant and anxiolytic activity in animal models.

Of particular interest, the internal PDZ motif of nNOS binds to PSD-95, a scaffolding protein with important roles in postsynaptic signal transduction and localizing receptors, kinases, phosphatases, and other signaling molecules at the synapse. Importantly, nNOS activation via NMDA-R requires an interaction with PSD-95. PSD-95 contains three PDZ domains (PDZ1-3), and the first PDZ domain of PSD-95 interacts with the extreme carboxy terminal (ct) of the NR2A/B subunit of NMDA-R. This leaves the second PDZ domain of PSD-95 (PDZ2) free to interact with nNOS, which uses its internal PDZ domain and a β-finger sequence to form a ternary complex composed of NMDA-R/PSD-95 (PDZ1) and PSD-95 (PDZ2)/nNOS interactions (Christopherson, Hillier, Lim, & Bredt, 1999). Taken together, this ternary complex created by PDZ domain interactions between NMDA-R, PSD-95, and nNOS acts to scaffold nNOS to the NMDA receptor, and NMDA-induced Ca^{2+} influx is therefore efficiently coupled to the activation of nNOS and the resulting production of NO (Doucet et al., 2012; Guix et al., 2005).

The canonical PDZ domain of nNOS binds to the nNOS adaptor protein NOS1AP, also known as CAPON (carboxy-terminal PSD-95-Dlg-ZO1 [PDZ] ligand of nNOS). NOS1AP was originally regarded as an inhibitor of PSD-95/nNOS interaction based on findings that overexpression of NOS1AP resulted in a loss of PSD-95/nNOS complexes in transfected cells (Jaffrey, Snowman, Eliasson, Cohen, & Snyder, 1998). However, this view has been challenged with functional studies, suggesting that NOS1AP may act as a mediator rather than an inhibitor of NMDA-R/nNOS signaling, and several conceptual models have been proposed that may better describe the nature of the interaction between PSD-95 and NOS1AP with nNOS (Courtney, Li, & Lai, 2014). These conceptual models suggest that NOS1AP may act to modify rather than inhibit nNOS signaling and focus on research that shows that NOS1AP binds to the canonical PDZ domain of nNOS, while PSD-95 binds to the internal PDZ motif—meaning that the binding of PSD-95 does not necessarily exclude NOS1AP from also binding in addition to PSD-95. NOS1AP has been implicated in mediating excitotoxicity, with data demonstrating that NOS1AP interacts with p38MAPK-activating kinase (MKK3), leading to downstream cell death cascades (Li et al., 2013). NOS1AP's best-known effector protein is dexamethasone-induced Ras protein 1 (Dexras1; Fang et al., 2000), a member of the Ras family of small G proteins induced by dexamethasone. NOS1AP therefore acts to bring nNOS into proximity to Dexras1, enabling NO to activate Dexras1 by *S*-nitrosation. Dexras1 has been shown to negatively regulate the phosphorylation of extracellular signal regulated kinase (ERK), a kinase involved in various signaling pathways mediating responses such as cell growth and synaptogenesis (Zhu et al., 2014).

NO SIGNALING: cGMP-DEPENDENT AND cGMP-INDEPENDENT PATHWAYS

sGC is regarded as the most sensitive receptor for NO. NO activates sGC by binding to the heme site on the protein and inducing a conformational change. sGC activation triggers the formation of cGMP from guanosine triphosphate (GTP). cGMP acts as a second messenger, and its production results in activation of protein kinases including PKG I and II, phosphodiesterases (PDE III), and cGMP-dependent kinase II (cGKII). NO can also mediate effects independent of sGC activation and cGMP production. For example, NO can interact directly or indirectly with metals, reduced thiols, molecular oxygen, and other reactive oxygen species, an example being the association of NO with superoxide ion (O_2^-) that leads to the formation of peroxynitrite ($ONOO^-$), an oxidant that can modify proteins and lipids by direct oxidative reactions or via indirect, radical-mediated mechanisms (Pacher, Beckman, & Liaudet, 2007; Vallance & Leiper, 2002). Furthermore, NO can interact with cyclooxygenase and heme oxygenase I, proteins involved in inflammatory and stress response systems, as well as regulating the Akt kinase signaling pathway and the transcription factor cyclic-AMP-responsive-element-binding protein (CREB; Calabrese et al., 2007). NO can also directly modify proteins through *S*-nitrosation (also called *S*-nitrosylation), a process which is increasingly regarded as an important mechanism by which NO can exert effects on processes such as synaptic plasticity, synaptogenesis, and neurogenesis (Hardingham, Dachtler, & Fox, 2013; Santos, Martínez-Ruiz, & Araújo, 2015). *S*-nitrosation refers to a posttranslational modification of thiol groups (mainly cysteines) of various target proteins, leading to functional modifications. For example, NO mediates *S*-nitrosation of the GluR1 AMPA-R subunits, which has been reported to increase AMPA-R membrane insertion, thereby facilitating long-term potentiation and synaptic plasticity (Selvakumar et al., 2013). In addition, Dexras1 is known to be S-nitrosated by NO, a process which is facilitated by the scaffolding properties of NOS1AP (Fang et al., 2000) (Figure 2).

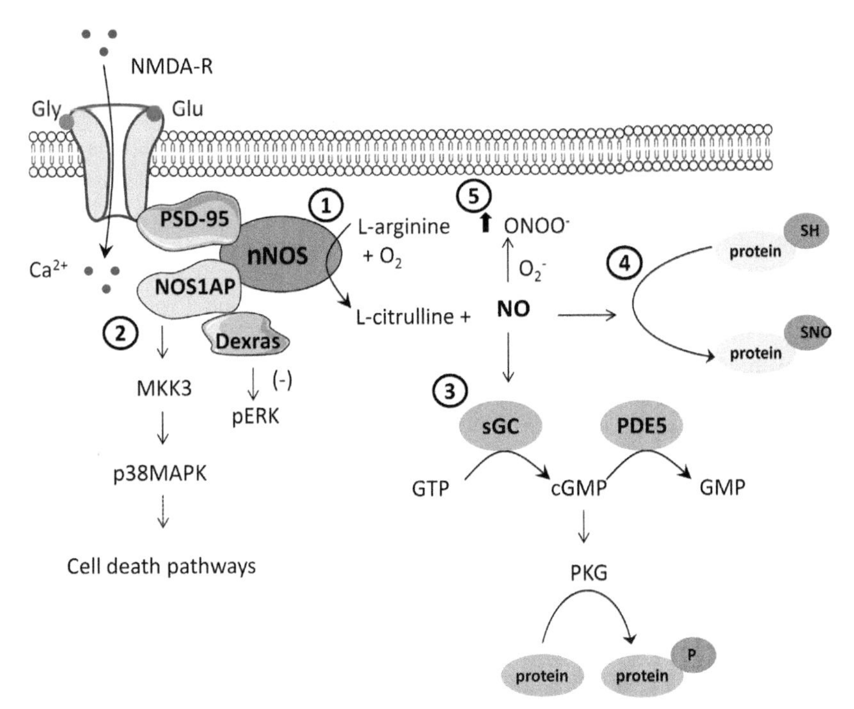

FIGURE 2 **Nitric oxide signaling.** (1) The NMDA-R is coupled to nNOS via PDZ domain protein interactions with the PSD-95 scaffolding protein, thereby allowing NMDA-R-mediated Ca^{2+} influx to activate nNOS, which triggers the production of NO from L-arginine and O_2. (2) nNOS also couples to NOS1AP via a PDZ domain, an interaction that has been reported to be linked to the activation of p38MAPK pathways and cell death. NOS1AP also links to the G protein, Dexras, which has been shown to have a negative influence on the phosphorylation of ERK. (3) The main physiological receptor for NO is sGC, which catalyzes the conversion of GTP to cGMP, a signaling molecule with many targets, including the activation of protein kinase G (PKG), which subsequently mediates the phosphorylation of various proteins. (4) NO can also act independent of sGC and cGMP production and in large quantities covalently bonds with protein thiol groups (protein-SH) to form S-nitroso-proteins (protein-SNO). (5) Large amounts of NO can also combine with the superoxide radical (O_2^-) to form the neurotoxic reactive oxygen species peroxynitrite when combined with O_2. Abbreviations: Ca^{2+}, calcium ion; Dexras, dexamethasone-induced Ras protein; GMP, guanosine monophosphate; cGMP, cyclic GMP; GTP, guanosine triphosphate; Glu, glutamate; Gly, glycine; MKK3, p38MAPK-activating kinase; nNOS, neuronal nitric oxide synthase; NMDA-R, *N*-methyl-D-aspartate receptor; NO, nitric oxide; NOS1AP, NOS1 interacting protein; $ONOO^-$, peroxynitrite; O_2^-, superoxide anion; p38MAPK, p38 mitogen-activated protein kinase; pERK, phosphorylated extracellular signal-related kinase; PDE5, phosphodiesterase5; PKG, protein kinase G; PSD-95, postsynaptic density-95; sGC, soluble guanylate cyclase.

BIOLOGICAL MECHANISMS UNDERLYING THE PATHOPHYSIOLOGY OF DEPRESSION

Despite intensive research, the neurobiology of depression remains poorly understood. The treatment of depression has been predominantly based on the monoamine hypothesis, which posits that a deficiency of serotonin (5-hydroxytryptamine or 5-HT), noradrenaline, and/or dopamine manifest in depressive symptoms. Traditional antidepressants either prevent neuronal reuptake (for example, serotonin reuptake inhibitors [SSRIs]) or prevent the degradation (for example, monoamine oxidase inhibitors [MAOIs]) of serotonin, noradrenaline, and/or dopamine. Although these antidepressants improve mood in some patients, they also suffer from serious limitations, with studies such as the Sequenced Treatment Alternatives to Relieve Depression (STAR*D) trial showing that only one-third of patients receiving antidepressants achieved remission with initial treatment. Of the nonresponders, another 30% approximately achieved remission when the antidepressant was augmented with sustained-release bupropion or busperone (Trivedi et al., 2006). This leaves a large proportion of patients still failing to show an improvement in mood. It is also noteworthy from this STAR*D trial that patients who did not respond to the initial treatment were more likely to fail successive treatments with other antidepressant drugs (Nelson, 2006; Rush et al., 2006).

Another consideration with standard antidepressants is the time lag between drug administration and the onset of therapeutic effects. Monoamine-based antidepressants increase the concentration of monoamines at the synapse within minutes or hours, yet weeks of treatment are required before their mood-enhancing properties become evident. This temporal disparity between the immediate increase of synaptic monoamines and the onset of therapeutic response suggests that other alterations in the brain are responsible for the antidepressant properties of these agents and that classical antidepressants instead may act on substrates considerably upstream of targets that are ultimately responsible for their effects. These observations have prompted a paradigm shift from a central monoamine hypothesis of depression to focusing on other observations that have been made in patients suffering from depression. In this regard, extensive research has led to the development of numerous alternative hypotheses of depression, which can be reviewed in more detail and include the neuroendocrine (Holsboer, 2000), brain-derived neurotrophic factor (BDNF) and neurotrophin (Groves, 2007), inflammatory (Dantzer, O'Connor, Freund, Johnson, & Kelley, 2008), neuroplasticity (Pittenger & Duman, 2008), and glutamate (Sanacora, Treccani, & Popoli, 2012) hypotheses of depression.

EVIDENCE LINKING GLUTAMATE TO DEPRESSION

There is growing evidence to suggest that the glutamatergic system plays a role in the pathophysiology of depression. For a review, see Sanacora et al. (2008). First, altered glutamate levels have been reported in the plasma, serum, and cerebrospinal fluid of depressed individuals. In addition, postmortem studies have documented aberrant glutamate levels in brain areas such as the frontal cortex (Hashimoto, Sawa, & Iyo, 2007), and magnetic resonance spectroscopy studies have also reported glutamate alterations in individuals suffering from mood disorders (Yildiz-Yesiloglu & Ankerst, 2006). Furthermore, expression levels and binding properties of glutamate receptor subtypes have been reported to be altered in patients suffering from depression (see Doucet et al., 2012).

Further to findings of altered glutamate in depression is research indicating prominent reductions in astrocyte number and function in limbic areas in the brain of patients suffering from depression. For a review, see Cotter, Pariante, and Everall (2001); Niciu, Henter, Sanacora, and Zarate Jr (2014); and Rajkowska and Miguel-Hidalgo (2007). Since astrocytes regulate glutamate levels, it is possible that the astrocytic pathology in depressed patients may have a functional relevance to the alterations in glutamate levels also observed in the illness. In support of this, blocking astrocytic glutamate uptake by microinjection of the GLT1 inhibitor, dihydrokainic acid, in the prefrontal cortex (PFC) of rats induces anhedonia, a core symptom of depression (John et al., 2012). This suggests that increased extracellular glutamate on foot of impaired astrocytic glutamate uptake may play a role in inducing anhedonia-like behaviors. In addition, pharmacological astrocytic ablation through the use of the selective astrocyte toxin, L-alpha aminoadipic acid (L-AAA), is sufficient to induce depression-like symptoms in rats (Banasr & Duman, 2008). Another study showed that L-AAA-induced glial loss was accompanied by progressive neuronal loss and dendritic atrophy in the lesion site of the mPFC, leading the authors to suggest that astrocytic dysfunction may create knock-on effects for neuronal integrity, which may result in observed behavioral impairments in cognition (Lima et al., 2014).

NMDA Receptor Antagonists as Antidepressants

In further support for the role of glutamate in depression is the consistently found observation that rapid antidepressant effects are achieved with NMDA receptor antagonists, both clinically and in animal models of depression. The noncompetitive NMDA receptor antagonist, ketamine, has received particular attention in this regard. Initial studies showed that a low dose of intravenously administered ketamine (0.5 mg/kg) produces an antidepressant response within hours, even in patients who are otherwise treatment-resistant to two or more typical antidepressants (Berman et al., 2000; Zarate et al., 2006). Further studies demonstrated that ketamine displayed rapid and robust antidepressant action in cohorts of treatment-resistant depressed patients, and that suicidal ideation was significantly reduced after ketamine infusion (DiazGranados et al., 2010).

While these rapid and robust antidepressant effects are promising, a broader use of ketamine in the clinic is limited due to its adverse effects. Ketamine possesses sedative and psychotomimetic side effects, and repeated exposure to ketamine increases the risk of severe psychosis, dissociative episodes, and euphoria (Machado-Vieira, Salvadorc, Diazgranados, & Zarate, 2009). In addition, repeated administrations or/and add-on therapy are needed in order to maintain the effects of ketamine (aan het Rot et al., 2010; Ibrahim et al., 2012; Mathew et al., 2010; Segmiller et al., 2013). Due to these side effects, ketamine is instead mainly regarded as a research tool, and elucidation of the molecular mechanisms underpinning its therapeutic profile may provide clues as to novel targets for the development of rapidly acting antidepressants without the side effects associated with ketamine. In this regard, the cellular effects underlying ketamine's antidepressant action have been studied by a number of research groups using animal models of depression. Subanesthetic doses of ketamine have been shown to result in acute and sustained (2 weeks) antidepressant-like effects in rodent models of depression. Another NMDA-R antagonist, MK-801, and the selective NR2B antagonist, Ro25-6981, also produce similar antidepressant effects, but these effects are not sustained for as long as those observed with ketamine (Maeng et al., 2008).

Compelling evidence suggests that ketamine may elicit its antidepressant effects through increasing synaptogenesis. For example, ketamine administration has been shown to rapidly activate the mammalian target of rapamycin (mTOR) signaling pathway in the PFC of rats, accompanied by increased levels of synaptic signaling proteins, leading to an increased number and function of new spine synapses in the PFC. Blocking the mTOR pathway prevented ketamine-induced synaptogenesis and antidepressant-like effects (Li et al., 2010). These authors also report that ketamine is sufficient to reverse chronic unpredictable stress (CUS)-induced anhedonic and anxiogenic behaviors in rats. CUS was also found to decrease the expression levels of synaptic proteins and spine number, and the frequency/amplitude of synaptic currents in layer V pyramidal neurons in the PFC—effects that could be reversed by ketamine administration (Li et al., 2011).

Another study found that the effects of ketamine on synaptogenesis and antidepressant-like behavior depend on inhibitory phosphorylation of glycogen synthase kinase-3 (GSK-3), which prevents its usual inhibitory influence on mTOR, by showing that the effects of a subeffective dose of ketamine were potentiated when given with a single dose of lithium chloride (a nonselective GSK-3 inhibitor) or a preferential GSK-3b inhibitor (Liu et al., 2013).

Ketamine's antidepressant-like effects are also reported to be dependent on the synthesis of BDNF, based on evidence showing that ketamine deactivates eukaryotic elongation factor 2 (eEF2) kinase, which leads to a reduction in eEF2 phosphorylation and desuppression of translation of BDNF (Autry et al., 2011; Monteggia, Gideons, & Kavalali, 2013). An important role for AMPA-Rs has been implicated in mediating the effects observed with ketamine administration, as the AMPA-R antagonist, NBQX, was capable of attenuating ketamine's antidepressant-like effects (Koike, Iijima, & Chaki, 2011; Maeng et al., 2008). Interestingly, ketamine has been shown to lead to an increase in presynaptic glutamate release, and it has been suggested that ketamine may act to increase AMPA-R to NMDA-R throughput in critical neuronal circuits, since the latter receptor is blocked by ketamine (Maeng & Zarate, 2007). Ketamine administration has also been shown to regulate phosphorylation of the AMPA-R subunit GluR1 (Maeng et al., 2008). NMDA-R-dependent regulation of this subunit has been implicated in AMPA-R trafficking and membrane insertion and has been shown to occur via NO-mediated S-nitrosylation of GluR1 (Selvakumar et al., 2013) and also by phosphorylation of GluR1 by downstream activation of cGMP-dependent kinase cGKII (Serulle et al., 2007), lending support to the suggestion that ketamine may act by altering AMPA-R properties (Figure 3).

EVIDENCE LINKING NO TO DEPRESSION

In line with evidence of a potential hyperactive glutamatergic system in major depressive disorder (MDD), one could suggest that the nitrergic system, downstream of glutamate signaling, may also be contributing toward depression pathophysiology. In support of this, increased nNOS immunoreactivity has been documented in the CA1 and subiculum of the hippocampus of depressed and bipolar individuals (Oliveira, Guimaraes, & Deakin, 2008). With increasing interest in developing biomarkers for psychiatric disorders, attempts have been made to detect NO-related biomarkers for depression. The detection of such biomarkers is problematic, however, as NO is rapidly metabolized. Moreover, it is debatable as to whether or not changes in NO levels in plasma reflect changes of the gaseous substance in the CNS. A potential caveat of circulating NO as a biomarker for depression is that the changes reported in depressed individuals most likely represent a chronic low-grade inflammation observed in MDD through iNOS activation rather than a glutamate-nNOS mechanism (Miller, Maletic, & Raison, 2009). Nonetheless, several studies have reported increased circulating plasma levels of nitrate metabolites in depressed patients (Akpinar, Yaman, Demirdas, & Onal, 2013; Suzuki, Yagi, Nakaki, Kanba, & Asai, 2001; Talarowska et al., 2012). One clinical study reported no significant difference in plasma nitrite levels between healthy controls and depressed individuals but found that treatment with an antidepressant (SSRIs) over an 8-week period significantly reduced

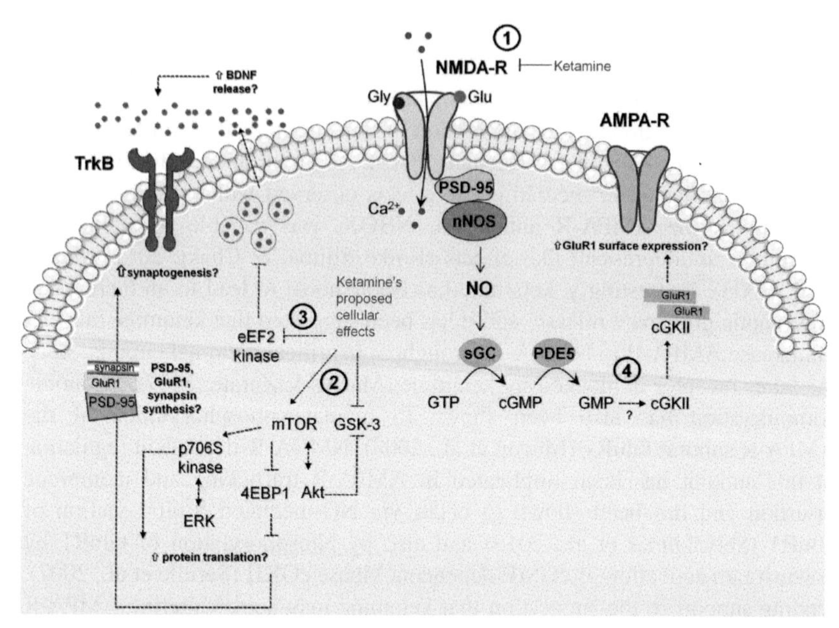

FIGURE 3 Molecular mechanisms thought to underlie the antidepressant effects of the NMDA-R antagonist ketamine. (1) Ketamine is a noncompetitive NMDA-R antagonist and the cellular effects underlying its antidepressant properties are reported to include: (2) mTOR pathway activation, including phosphorylation and activation of eukaryotic initiation factor 4E-binding protein 1 (4EBP1), p70S6 kinase and activated ERK, including ERK1 and ERK2, and Akt. These effects occur in line with increased production of proteins associated with synapse formation such as PSD-95, GluR1, and synapsin. Ketamine also inhibits the phosphorylation of glycogen synthase kinase-3 (GSK-3), which prevents its usual inhibitory influence on mTOR. (3) Ketamine has been shown to inhibit eEF2 kinase leading to dephosphorylation of eEF2 and augmentation of BDNF synthesis, a growth factor with well-known roles in neuronal growth and synapse formation. (4) The antidepressant properties of ketamine are prevented by AMPA-R antagonism, indicating a role for AMPA-R in mediating ketamine's effects. NMDA-R-dependent regulation of AMPA-R trafficking has been shown to occur via NO-mediated S-nitrosylation of GluR1 and also by phosphorylation of GluR1 by downstream activation of cGMP-dependent kinase cGKII. Abbreviations: BDNF, brain-derived neurotrophic factor; Ca^{2+}, calcium ion; Dexras, dexamethasone-induced Ras protein; ERK, extracellular signal-related kinase; GMP, guanosine monophosphate; cGMP, cyclic GMP; GSK-3, glycogen synthase kinase-3; eEf2 kinase, eukaryotic elongation factor-2 kinase; GTP, guanosine triphosphate; Glu, glutamate; Gly, glycine; MKK3, p38MAPK-activating kinase; mTOR, mammalian target of rapamycin; nNOS, neuronal nitric oxide ynthase; NMDA-R, *N*-methyl-D-aspartate receptor; NO, nitric oxide; PDE5, phosphodiesterase5; PKG, protein kinase G; PSD-95, postsynaptic density-95; sGC, soluble guanylate cyclase; TrKB, tropomyosin receptor kinase B; 4EBP1, eukaryotic translation initiation factor 4E-binding protein 1.

plasma nitrite levels in depressed patients (Herken et al., 2007). In addition to biomarker studies, genetic analyses have led to the identification of risk alleles of candidate genes that increase one's susceptibility to developing an affective disorder. While a stronger association exists between NOS1 gene polymorphisms and schizophrenia, several studies suggest a potential link between genetic variations of this gene and depression. For a review, see Freudenberg, Alttoa, and Reif (2015). Short allele carriers of the NOS1 exIf-VNTR are found to display higher levels of neuroticism and anxiety. Exposure to environmental adversity only served to increase the degree of neuroticism and anxiety in these risk allele carriers (Kurrikoff et al., 2012). A study by Sarginson et al. (2014) revealed that NOS1 genotypes interact with environmental adversity (financial hardship) to give rise to depression, suggesting that polymorphisms of NOS1 give rise to affective disorders when they are combined with environmental adversity (see Table 1 for NOS1 gene polymorphisms associated with neuropsychiatric disorders and personality traits).

Preclinically, evidence suggests potential mechanisms through which excessive central nNOS activity may produce a depressive-like phenotype. Both acute and chronic restraint stress have been shown to induce an increase in nNOS expression in limbic areas of the rat brain such as the hippocampal formation, hypothalamus, and the amygdala (de Oliveira et al., 2000; Echeverry, Guimaraes, & Del Bel, 2004). Exposure of Flinders Sensitive Line (FSL; an animal model of depression) rat to escapable/inescapable stress leads to changes in the expression of proteins engaged along the NMDA-R/nNOS pathway that may contribute to its well-acknowledged depressive-like phenotype. Escapable/inescapable stress led to an increase in the expression of gene transcripts for the NR1 subunit of the NMDA-R, PSD-95, and nNOS in the hippocampus of FSL rats, along with an increase in the nNOS protein and NOS activity (Wegener et al., 2010).

NO and the Hypothalamic-Pituitary-Adrenal (HPA) Axis

Further evidence to suggest a role of NO in mediating the physiological stress response is observed at the level of the hypothalamus and the endocrine system. Hyperactivity of the HPA axis is a hallmark of stress-related illnesses and is associated with a diminished feedback inhibition by endogenous glucocorticoids. For a review, see Pariante and Lightman (2008). It has been suggested that the stress-related glucocorticoid hormones are capable of increasing glutamate-mediated neurotransmission and, as a result, NOS activity in the CNS. Deletion of the nNOS gene was shown to mitigate the transcriptional activity of vasopressin in the hypothalamus of mice (Orlando, Langnaese, Schulz, Wolf, & Engelmann, 2008). Moreover, nNOS knockout mice display a reduced ability to produce adrenaline from the adrenal glands following an acute stressor, which suggests that nNOS regulates the sympathoadrenal system (Orlando et al., 2008).

Chronic corticosterone supplementation in rats, which has been shown to produce anxiogenic and depression-like effects in animals (O'Donovan, Dalton, Harkin, & McLoughlin, 2014), resulted in an increase in nNOS immunoreactivity in the amygdala, paraventricular nucleus (PVN) of the

TABLE 1 NOS-1 Gene Polymorphisms Associated with Neuropsychiatric Disorders and Personality Traits

	NOS-1 Gene Polymorphisms Associated with Disease	Notes	References
Schizophrenia	rs3782219, rs3782221, rs3782206, rs41279104, rs2682826	Not confirmed in other studies	Cui et al. (2010), Fallin et al. (2005), Okumura, Okochi et al. (2009), Reif et al. (2006), Riley et al. (2010), Shinkai, Ohmori, Hori, and Nakamura (2002) and Tang et al. (2008)
Bipolar disorder		No association found	Buttenschøn et al. (2004), Okumura, Kishi et al. (2009) and Reif et al. (2006)
Major depression		No association found	Gałecki et al. (2011), Okumura, Kishi et al. (2009) and Yu et al. (2002)
Cognition	rs6490121, rs41279104, VNTR exon 1f		Donohoe et al. (2009), Kopf et al. (2011), Reif et al. (2011) and Rose et al. (2012)
Suicide	rs2682826, rs1353939, rs693534	Not confirmed in other studies	Giegling et al. (2011) and Rujescu et al. (2008)
Personality traits (impulsivity, neuroticism, extraversion, anxiety, depressiveness)	VNTR exon 1f, rs2293054, rs3741475, rs10774909, rs7298903	rs7298903 not confirmed in other studies	Hoogman et al. (2011), Kopf et al. (2011), Kurrikoff et al. (2012), Laas et al. (2010), Luciano et al. (2010, 2012), Reif et al. (2009, 2011) and Retz, Reif, Freitag, Retz-Junginger, and Rösler (2010)

hypothalamus, and in several subfields of the hippocampus associated with a facilitation in step-down inhibitory avoidance indicative of an increased fear/anxiety state (Santos, Cespedes, & Viana, 2014). This increase in nNOS is not surprising, as Zhou and colleagues have previously shown that corticosterone increases nNOS expression in the hippocampus of mice, which is facilitated through the mineralocorticoid receptor, coinciding with a depression-like behavior in the forced swim test (FST) and tail suspension test (TST) (Zhou et al., 2011). These two tests are routinely used preclinical models of depression-like behavior and are based on the development of immobility in rodents, which is thought to represent a state of "behavioral despair" in response to an inescapable stressor. Moreover, this depression-like behavior induced by chronic corticosterone supplementation was reversed by chronic (21-day) treatment with the selective nNOS inhibitor, 7-nitroindazole (7-NI) (Zhou et al., 2011). The authors suggest that hippocampal NMDA-derived nNOS is an upstream repressor of glucocorticoid receptor function through cGMP-dependent and cGMP-independent (e.g., free radical generation, nitrosylation) mechanisms that, ultimately, give rise to depressive behaviors.

NO and Monoamine Interactions

Further corroborating a potential role for the nitrergic system in depression pathophysiology are the interactions reported between NO and monoamine transmission. Several reports suggest that increasing NO concentration using L-arginine or NO donors alters dopamine, noradrenaline, and serotonin release in *in vitro* and *in vivo* experiments (Dhir & Kulkarni, 2011; Kiss, 2000). The nature of these effects are contradictory, however, with studies reporting that NO can exert both positive and inhibitory effects on monoamine release (Prast & Philippu, 2001). Of particular interest, a closely linked connection between NO and serotonergic neurotransmission has been reported by several studies. For example, a physical association has been reported between nNOS and the serotonin transporter, SERT, through a PDZ interaction (Chanrion et al., 2007). Chanrion and colleagues demonstrated that in cells transfected with SERT and nNOS, the interaction of these two proteins culminated to have a negative influence upon the activity of SERT, which is believed to be a consequence of reduced trafficking of SERT to the plasma membrane. Moreover, in cells transfected with SERT and nNOS, serotonin reuptake via the transporter enhances the activity of nNOS. The same study revealed that mice with a genetic deletion of nNOS display enhanced cortical reuptake of 5-HT. They conclude that this alteration in *in vivo* 5-HT reuptake is most likely a consequence of a loss of the interaction of SERT and nNOS. Such results suggest a reciprocal functional relationship between these two proteins.

Evidence also suggests that the neuronal NOS isoform shares a connection with the 5-HT_{1A} receptor in terms of the receptor's downstream signaling. The 5-HT_{1A} receptor exists as a presynaptic autoreceptor, regulating 5-HT release from serotonergic neurons, and also acts as a heteroreceptor in limbic brain regions. In the hippocampus, it is found on the postsynaptic neuron (Ögren et al., 2008), and it is within this brain region where the receptor is believed to influence behaviors such as anxiety. Zhang and colleagues demonstrated that serotonin, acting through the 5-HT$_{1A}$ receptor, suppresses hippocampal nNOS expression. Blockade of the 5-HT$_{1A}$ receptor resulted in an increase in hippocampal nNOS expression and anxiety-like behavior in mice (Zhang et al., 2010). This 5-HT$_{1A}$—nNOS interaction that regulates behavior is believed to involve the transcription factor CREB. Inhibition of nNOS using 7-NI resulted in an increase in phosphorylated CREB in the hippocampus, which was accompanied by an anxiolytic-like response in the elevated plus maze (EPM; Zhu et al., 2006). Moreover, stimulation of 5-HT$_{1A}$ receptors promotes CREB phosphorylation, whereas receptor antagonism mitigates *p*-CREB production (Zhang et al., 2010). Such results suggest that stimulation of 5-HT$_{1A}$ receptors facilitates an anxiolytic-like response through suppression of nNOS and a concomitant increase in CREB activation. Another study revealed that microinjection of the 5HT$_{1A}$ receptor antagonist, WAY100635, into the dorsal hippocampus counteracted the antidepressant-like responses of both fluoxetine and the nNOS inhibitor, N^{ω}-propyl-L-arginine, in the rat FST, further suggesting that 5HT$_{1A}$ receptor activation and serotonergic signaling are involved in the antidepressant-like effects of nNOS inhibition (Hiroaki-Sato, Sales, Biojone, & Joca, 2014).

NOS Inhibitors Exhibit Antidepressant-Like Properties

The inhibition of the NOS enzyme can be achieved mainly through two pharmacological interventions. The enzyme can be inhibited by a competitive (nonselective) antagonism of the L-arginine substrate site, effectively blocking the metabolism of the amino acid into citrulline and the concomitant production of the NO by-product. A consequence of using such NOS inhibitors is that all isoforms of the NO-producing enzyme are inhibited, which can be unfavorable if the objective is to inhibit the neuronal isoform selectively. Nonetheless, nonselective NOS inhibitors such as L-nitroarginine (L-NA), L-nitroarginine methylester (L-NAME), and L-NG-monomethyl arginine citrate (L-NMMA) have been reported to produce antidepressant and anxiolytic-like effects in preclinical behavioral paradigms (Gigliucci, Buckley, Nunan, O'Shea, & Harkin, 2010; Harkin, Connor, Walsh, St John, & Kelly, 2003; Joung et al., 2012). While nonselective NOS inhibitors produce antidepressant-like effects, they also facilitate changes in the vasculature as a consequence of eNOS inhibition, which is an undesirable off-target effect, were such drugs to be used clinically (Figure 4).

Specific isoforms can be selectively inhibited, which is more advantageous given that selective inhibition of the neuronal isoform spares inhibition of eNOS and iNOS. The nNOS inhibitor, 1-(2-Trifluoromethylphenyl)-imidazole

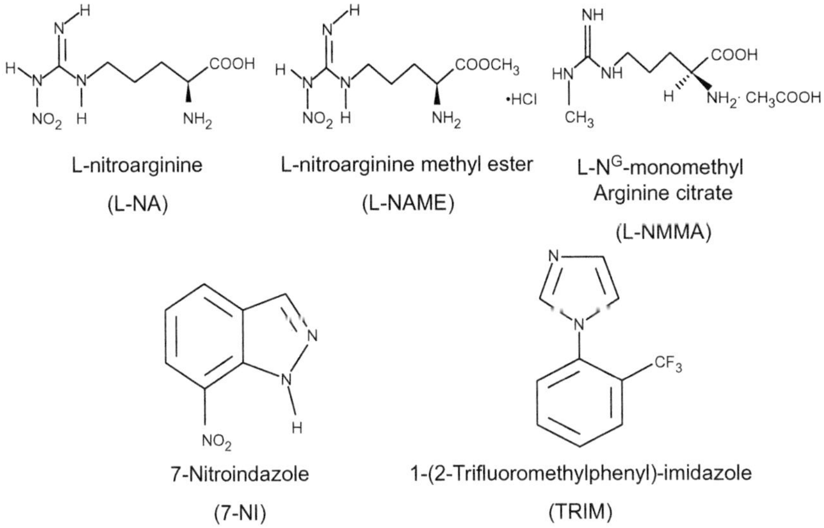

L-nitroarginine (L-NA)

L-nitroarginine methyl ester (L-NAME)

L-N^G-monomethyl Arginine citrate (L-NMMA)

7-Nitroindazole (7-NI)

1-(2-Trifluoromethylphenyl)-imidazole (TRIM)

FIGURE 4 Structure of frequently used NOS inhibitors.

(TRIM), was shown to have antidepressant-like effects in the FST and TST (Doucet, Levine, Dev, & Harkin, 2013; Ulak et al., 2010) and has also shown similar efficacy as the standard antidepressant fluoxetine in attenuating behavioral changes associated with unpredictable chronic mild stress (CMS) in mice (Mutlu, Ulak, Laugeray, & Belzung, 2009), a commonly used murine model of depression. TRIM displayed antidepressant-like efficacy in the FST even in the face of serotonin depletion following *p*-chlorophenylalanine pretreatment (Ulak et al., 2010). This is somewhat equivocal given the reported serotonin dependency of NOS inhibitors previously reported (Harkin et al., 2003) but may relate to the extent of serotonin depletion obtained in such tests. However, the antidepressant-like effects of selective nNOS inhibitors may also be facilitated through modulating glutamatergic neurotransmission due to the association of the nitrergic system with the NMDA receptor (see above). An additional nNOS inhibitor, 7-NI, was also shown to display antidepressant-like efficacy in the FST while also curbing stress-induced neuronal activation (Silva, Aguiar, Diniz, Guimaraes, & Joca, 2012). Joung and colleagues reported that 7-NI also possesses anxiolytic-like properties in rats exposed to the EPM (Joung et al., 2012). Although both 7-NI and TRIM are efficacious in the FST and EPM, reports suggest that 7-NI has a negative effect upon memory processing, impairing spatial working memory in the Morris water maze, and object recognition and passive avoidance at doses ranging from 30 to 45 mg/kg (Mutlu, Ulak, & Belzung, 2011). TRIM was not found to have any effect on these cognitive parameters at any dose tested, which makes it a drug of choice when assessing the antidepressant and anxiolytic properties of neuronal NOS inhibitors. An interesting observation regarding the effects of NOS inhibitors is the differential actions of selective and nonselective inhibitors on stress-related physiological processes. Joung and colleagues showed that although both L-NAME and 7-NI produced an anxiolytic-like effect in the EPM, L-NAME was found to attenuate stress-induced corticosterone production, whereas 7-NI did not. Moreover, L-NAME reduced the level of the stress-induced NO metabolites nitrate and nitrite (collectively referred to as NOx) in the plasma, an effect that was not observed with 7-NI. However, treatment with the nNOS inhibitor led to a reduction in central NOx metabolite levels while also reducing the number of NADPH-positive neurons in the PVN of the hypothalamus and in the lateral dorsal tegmental nucleus (Joung et al., 2012).

INHIBITION OF NMDA-R/PSD-95/nNOS PROTEIN INTERACTIONS

Despite promising preclinical studies, the use of nNOS inhibitors as potential antidepressants is problematic. nNOS is expressed in several tissues other than the brain and regulates important physiological functions, for example, in skeletal muscle contraction and relaxation of the sphincter muscle in the gut. Therefore, not surprisingly, nNOS inhibitors are reported to cause considerable gastrointestinal and genitourinary side effects (Vallance & Leiper, 2002). Moreover, drugs such as TRIM and 7-NI, which display apparent preferential inhibition of nNOS, are reported to display efficacy in inhibiting other NOS isoforms (Vallance & Leiper, 2002).

One strategy for overcoming issues with directly targeting the NMDA-R or nNOS is to instead target specific elements of the NMDA-R/PSD-95/nNOS interaction. This may represent a promising strategy for providing potential antidepressant effects without the adverse effects associated with direct NMDA-R or nNOS inhibition (Aarts et al., 2002; Cao et al., 2005; Doucet et al., 2012, 2013; Florio et al., 2009; Li Zhou et al., 2010). Disruption of the interaction between NMDA-R and PSD-95 has been achieved successfully by Aarts et al. (2002) who transduced neurons with a peptide that competitively disrupted this interaction while maintaining NMDA-R's ability to participate in synaptic activity and calcium influx. This peptide comprised the last nine ct residues of NR2B fused to the cell membrane transduction domain of the human immunodeficiency virus-type 1 (HIV-1) Tat protein to create the cell membrane-permeable Tat-NR2B9c. The Tat-NR2B9c peptide showed neuroprotective properties against excitotoxic insults *in vitro* and reduced focal ischemic brain damage as well as improving neurological scores in a middle cerebral artery occlusion (MCAO) rat model of stroke *in vivo*. Specific inhibition of the PSD-95/nNOS interface has been achieved using cell-permeable fusion proteins based on the first 300 residues of nNOS. First, studies by Cao et al. (2005) developed an $nNOS_{1\text{-}300}$ protein fragment, which contains the PDZ domain and β-finger of nNOS and can therefore act as a competitive inhibitor of the interaction between the PDZ motif of nNOS and the PDZ2 of

PSD-95. This protein fragment was shown to prevent the induction of stress-activated protein kinase p38 and glutamate-induced neuronal death. In addition, a protein called Tat-nNOS was developed, which consists of residues 1–299 of PSD-95 (Florio et al., 2009). Tat-nNOS effectively disrupts the PSD-95/nNOS interaction *in vitro* with an IC_{50} of 300 nM and shows efficacy in inhibiting acute thermal hyperalgesia and chronic mechanical allodynia in a rodent model of neuropathic pain (Florio et al., 2009). Furthermore, lentiviral-mediated overexpression of $nNOS_{1-300}$ in the right cerebral cortex of MCAO mice was capable of significantly reducing the infarct area and improved neurological scores (Zhou et al., 2010).

Small Molecule Inhibition of PSD-95/nNOS Interaction: ZL006, IC87201

Small molecular weight compounds have also been developed that inhibit the interaction of PSD-95 with nNOS. Such compounds represent the optimal method of specific protein disruption because they may circumvent problems associated with the delivery of biologically active peptides, which are subject to rapid degradation and poor pharmacokinetic properties (Doucet et al., 2012). First, IC87201 (2-((1H-benzo [d] [1,2,3] triazol-5-ylamino) methyl)-4,6-dichlorophenol) was identified in a high-throughput screen and found to disrupt the PSD-95/nNOS interaction with an IC_{50} of 31 μM without inhibiting nNOS catalytic activity (Florio et al., 2009). Intrathecal administration of IC87201 in a mouse model of NMDA-induced thermal hyperalgesia resulted in antinociceptive effects at a dose of 1 pmol as analyzed by a reduction in warm water tail flick latency. In addition, IC87201 was also effective after intraperitoneal injection with an EC_{50} of 0.1 mg/kg. IC87201 has also shown efficacy in abolishing mechanical allodynia after intrathecal (at 50 and 100 nmol dose) and intraperitoneal (2 mg/kg) administration in a rat model of sciatic nerve constriction. These results suggest that IC87201 can cross the blood–brain barrier (BBB). A structurally related molecule was later identified, called ZL006 (4-(3,5-dichloro-2-hydroxy-benzylamino)-2-hydroxybenzoic acid). This compound has been shown to inhibit NMDA-R-dependent NO synthesis in cortical neurons with an IC_{50} of 82 nM and is also able to cross the BBB after systemic administration. ZL006 is believed to bind to the β-finger of nNOS by forming an ionic bond between its carboxyl group and the amino group in Arg-121 of the nNOS PDZ domain and a hydrophobic bond between the hydrophobic ring of ZL006 and Leu-107 or Phe-111 of nNOS, thereby interrupting the conformational change of nNOS PDZ domain (Zhou et al., 2010). ZL006 has shown preclinical efficacy in an MCAO model of stroke, whereby mice treated intravenously with ZL006 at 1.5 mg/kg at 1 h after reperfusion showed improvements in neurological scores and a reduction in infarct size as compared to nontreated controls. Importantly, ZL006 did not inhibit nNOS catalytic activity or NMDA-R function, nor did it affect behavioral measures associated with NMDA-R antagonism, such as spatial memory or aggressive behavior, thus indicating that this molecule disrupts PSD-95/nNOS without causing adverse effects and lends support to its potential for drug development.

Data from our laboratory indicates that both IC87201 (0.01–2 mg/kg) and ZL006 (10 mg/kg) display antidepressant-like effects by reducing immobility time in the FST and TST following a single administration in mice (Doucet et al., 2013). These effects are novel in their mode of action in that they become apparent 24 and 72 h following drug administration in a modified version of the FST and TST, which is dependent on prior exposure to the behavioral tests. These effects are in contrast to the acute behavioral effects observed with the classical antidepressant imipramine, which are observed 1 h following drug administration. Similar delayed and sustained effects were observed with a single administration of TRIM (50 mg/kg) and ketamine (30 mg/kg) in the TST following a pretest exposure. The novel delayed and sustained antidepressant effects observed with inhibitors of the NMDA-R/PSD-95/nNOS interaction may be indicative of adaptive changes to neural plasticity. IC87201 did not display anxiolytic effects as assessed by subjecting mice to the light dark box and EPM. The lack of anxiolytic activity implies that the compound produces a selective antidepressant-related action in keeping with the antidepressant actions of ketamine, which has not been reported to produce anxiolytic activity either in experimental animals or in the clinic.

Small Molecule Inhibition of nNOS/NOS1AP Interaction: ZLc-002

Interestingly, disruption of the interaction between nNOS and NOS1AP (aka CAPON) has been shown to provide anxiolytic-like effects in animal models, indicating that inhibition of this protein interaction may also serve as a promising target for the development of rapid acting agents for stress-related illnesses (Zhu et al., 2014). Overexpression of NOS1AP in the hippocampus of mice resulted in anxiogenic-like behaviors, while disruption of nNOS/NOS1AP by overexpressing peptides designed from C-terminal amino acids of NOS1AP was capable of producing anxiolytic-like effects. These effects were also observed in response to delivery of Tat-CAPON-12C (a peptide composed of Tat and the 12 C-terminal amino acids of NOS1AP) to the hippocampus of mice as well as through administration of a small molecule inhibitor of the nNOS/NOS1AP interface, ZLc-002. The authors show that mice exposed to CMS exhibit increased nNOS/NOS1AP coupling in the hippocampus concurrent with displaying anxiogenic-like behaviors, and that disruption of nNOS/NOS1AP can ameliorate these CMS-induced anxiogenic-like behaviors. Dexras1 ERK signaling was found to be

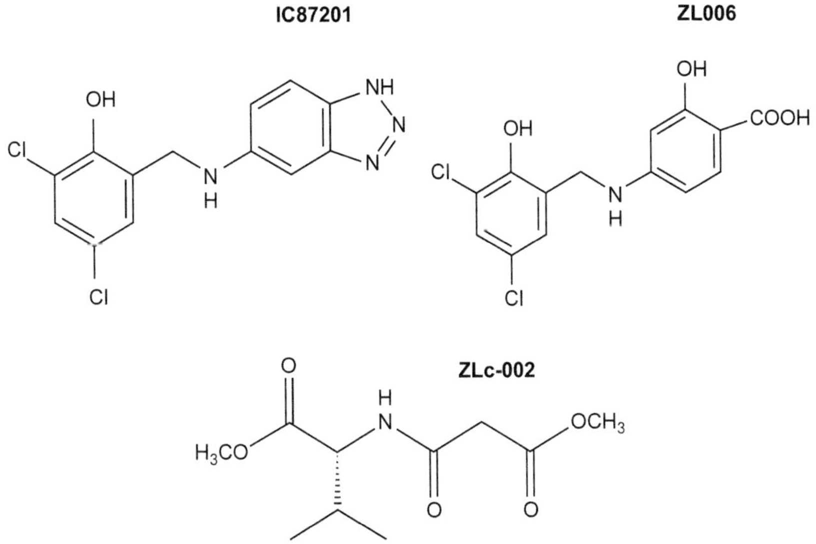

FIGURE 5 Structure of small molecular weight inhibitors of the PSD-95/nNOS (IC87201; ZL006), and nNOS/NOS1AP (ZLc-002) interactions.

involved in mediating the anxiolytic-like effects of these nNOS-NOS1AP blockers (Figure 5).

DOWNSTREAM OF NO: A ROLE FOR sGC, cGMP, AND PDE IN DEPRESSION

Consistent with the efficacy of NMDA-R antagonists and NOS inhibitors in possessing antidepressant properties are the findings that inhibition of elements downstream of NO production are also capable of producing antidepressant-like effects. As previously mentioned, NO activates sGC (Hall & Attwell, 2008), which converts GTP to the important intracellular messenger cGMP (Hall & Attwell, 2008; Hobbs, Higgs, & Moncada, 1999). cGMP is subsequently degraded by PDE to GMP, and the inhibition of PDE by agents such as sildenafil leads to an increase in cGMP levels. Notably, methylene blue, an inhibitor of sGC, is known to have prophylactic effects on patients with manic depressive psychosis when orally administered in addition to lithium (Naylor, Martin, Hopwood, & Watson, 1986). Preclinically, inhibition of sGC by the selective sGC inhibitor 1H-(1,2,4)oxadiazolo(4,3-a) quinoxalin-1-one (ODQ) at 10 and 20 mg/kg decreases immobility time in the FST, comparable to the effects produced by the NOS inhibitor 7-NI and the classical antidepressant imipramine (Heiberg, Wegener, & Rosenberg, 2002), thus implicating a role for this NO signaling pathway in mediating the antidepressant-like effects of NOS inhibition.

It has also been reported that the antidepressant-like actions of the SSRI escitalopram in the mouse FST are dependent not only on NMDA and NO but also on cGMP production, by showing that pretreatment with NMDA, the NO substrate L-arginine, and the PDE5 inhibitor sildenafil prevented the observed antidepressant-like effects of escitalopram. Inhibition of NO with 7-NI and sGC with methylene blue and ODQ in combination with subeffective doses of escitalopram was also able to reduce the immobility time in the FST, mirroring results obtained from the inhibition of nNOS with 7-NI (Zomkowski, Engel, Gabilan, & Rodrigues, 2010). In addition, the NOS inhibitor 3-Br-7-NI and ODQ administration prevented chronic unpredictable mild stress (CUMS)-induced depression-like behavioral changes, including reduced sucrose preference, body weight, and locomotor activity as well as increased immobility time in the FST (Yazir, Utkan, & Aricioglu, 2012). These inhibitors were also capable of attenuating against CUMS-induced reductions in BDNF protein levels in the hippocampus, indicating that the NO-sGC-cGMP pathway may play a role in the reduction of BDNF levels observed during stress. This pathway has also been implicated in mediating the antidepressant-like effects that are observed with various agents that have shown antidepressant efficacy, including the centrally acting analgesic tramadol (Jesse, Bortolatto, Savegnago, Rocha, & Nogueira, 2008) as well as zinc (Rosa, Lin, Calixto, Santos, & Rodrigues, 2003), adenosine (Kaster, Rosa, Santos, & Rodrigues, 2005), and ascorbic acid (Moretti et al., 2011). Data suggests that discrete pathways are involved in the antidepressant-like responses of NMDA-R antagonism and NO signaling inhibition in the ventral medial PFC in the rat FST by showing that the AMPA-R antagonist NBQX blocked the effects of NMDA-R antagonism but not nNOS or sGC inhibition (Pereira, Romano, Wegener, & Joca, 2015).

NMDA-R/NO PATHWAY IN NEURONAL PLASTICITY: A MECHANISM FOR PRODUCING ANTIDEPRESSANT EFFECTS

Exposure to chronic stress is known to cause neuronal atrophy and volumetric reductions in several limbic brain regions associated with mood, and chronic, but not acute, treatment with conventional antidepressant treatment has been reported to attenuate these deficits (Malberg, Eisch, Nestler, & Duman, 2000). Interestingly, reports are also linking the antidepressant effects observed with acute ketamine treatment to a rapid increase in neurotrophic factor production (particularly BDNF) and an increase in neurogenesis, dendritic length, and spine density in the hippocampus and PFC (Duman & Li, 2012). These reports suggest that the therapeutic effects of antidepressants may involve structural and functional neuroplastic changes in the brain. For a review, see Pittenger and Duman (2008). Since the NMDA-R is functionally coupled to nNOS, it is a possibility that NOS inhibitors may also mediate their antidepressant-like behavioral effects through increasing synaptic markers and through producing neuroplastic changes. It has been shown that the effects of ketamine (Zhang et al., 2013) and MK-801 (Dhir & Kulkarni, 2008) in decreasing immobility time in the FST in rodents can be prevented by pretreatment with L-arginine, suggesting that ketamine's mode of action in this test may involve the L-arginine/NO pathway. In addition, treatment with subactive doses of both ketamine along with the NOS inhibitor, L-NAME, can reduce immobility time as compared to either drug alone (Zhang et al., 2013), and pretreatment with the nNOS inhibitor 7-NI potentiated the action of a subeffective dose of MK-801 (Dhir & Kulkarni, 2008). These reports link the antidepressant effects of NMDA-R antagonism to the NO pathway and highlight the possibility that the antidepressant effects observed with NOS inhibition, or specific disruption of NMDA-R/NO signaling using small molecule inhibitors, may involve similar molecular mechanisms and may also increase neurotrophic production and neuroplastic changes in brain regions relevant to depression. In support of this, an increasing number of reports indicate that NO plays a major role in the regulation of synaptic signaling events, neuronal plasticity, and neurogenesis.

Adult neurogenesis occurs in two main regions in the brain: the dentate gyrus of the hippocampal formation and the subventricular zone (SVZ) and its projection through the rostral migratory stream to the olfactory bulb. Several reports indicate that nNOS inhibition promotes adult neurogenesis in these regions in rodents (Matarredona, Murillo-Carretero, Moreno-López, & Estrada, 2004; Moreno-López, Noval, González-Bonet, & Estrada, 2000; Packer et al., 2003; Zhu et al., 2006) and that NO inhibits adult neurogenesis in the SVZ (Romero-Grimaldi, Moreno-López, & Estrada, 2008). It has been reported that nNOS contributes to chronic stress-induced depression by suppressing hippocampal neurogenesis in mice (Zhou et al., 2007). These authors report hippocampal nNOS overexpression after 4 days of exposure to chronic stress (which persists up to 56 days) concomitant with behavioral despair indicative of a depression-like phenotype in mice and a reduction in hippocampal neurogenesis. Such behavioral effects and impairments in neurogenesis were prevented and reversed in null mutant mice lacking the nNOS gene (nNOS −/− mice) and in mice treated with 7-NI. In addition, the standard antidepressant fluoxetine has been shown to inhibit the atrophy of hippocampal neurons in the CMS rodent depression model by decreasing NOS expression, thereby linking NO to neuronal atrophy (Luo & Tan, 2000).

CONCLUSIONS

Despite intensive research, little progress has been made in developing novel medications for mood disorders that can offer advantages over the classical monoaminergic-based antidepressants. A large body of evidence points to glutamate and NO abnormalities in patients suffering from depression, and compelling research suggests that targeting elements of glutamatergic and NO signaling produces rapid and robust antidepressant effects. Most notably, the NMDA-R antagonist ketamine has displayed rapid antidepressant efficacy within hours of administration, even in otherwise treatment-resistant patients. Likewise, NOS inhibition is known to produce antidepressant-like effects in several animal models of depression. Selective inhibition of protein interactions involved in the NMDA-R/PSD-95/nNOS pathway has shown antidepressant and anxiolytic efficacy. Our laboratory has shown that selectively uncoupling the PSD-95 from nNOS interaction using small molecule compounds, namely, IC87201 and ZL006, can produce rapid and sustained antidepressant-like efficacy in a mouse model of depression. These results mirror those obtained using ketamine and the nNOS inhibitor, TRIM, suggesting that selectively targeting the NMDA/PSD-95/nNOS interface may harness the antidepressant efficacy obtained with ketamine without incurring its adverse side effects. An increase in the expression of synaptic proteins coupled with increased synaptogenesis and spine density has been associated with ketamine treatment concomitant with its antidepressant behavioral effects, which is of interest considering that chronic stress has been shown to induce neuronal atrophy and structural alterations in limbic regions associated with mood disorders. Such a mechanism needs to be confirmed for agents that target the NDMA-R/nNOS/cGMP or NOS1AP signaling pathways. In summary, continued research into the antidepressant-like effects observed with glutamatergic and NO modulating compounds holds considerable promise for the development of a new class of medication that provides patients suffering from depression with rapid and effective symptomatic relief.

REFERENCES

Aarts, M., Liu, Y., Liu, L., Besshoh, S., Arundine, M., Gurd, J. W., ... Tymianski, M. (2002). Treatment of ischemic brain damage by perturbing NMDA receptor-PSD-95 protein interactions. *Science (New York, N.Y.), 298*, 846–850.

Akpinar, A., Yaman, G. B., Demirdas, A., & Onal, S. (2013). Possible role of adrenomedullin and nitric oxide in major depression. *Progress in Neuro-Psychopharmacology and Biological Psychiatry, 46*, 120–125.

Akyol, O., Zoroglu, S. S., Armutcu, F., Sahin, S., & Gurel, A. (2004). Nitric oxide as a physiopathological factor in neuropsychiatric disorders. *In Vivo, 18*, 377–390.

Alderton, W. K., Cooper, C. E., & Knowles, R. G. (2001). Nitric oxide synthases: structure, function and inhibition. *Biochemistry Journal, 357*, 593–615.

Autry, A. E., Adachi, M., Nosyreva, E., Na, E. S., Los, M. F., Cheng, P-f., ... Monteggia, L. M. (2011). NMDA receptor blockade at rest triggers rapid behavioural antidepressant responses. *Nature, 475*, 91–95.

Banasr, M., & Duman, R. S. (2008). Glial loss in the prefrontal cortex is sufficient to induce depressive-like behaviors. *Biological Psychiatry, 64*, 863–870.

Berman, R. M., Cappiello, A., Anand, A., Oren, D. A., Heninger, G. R., Charney, D. S., & Krystal, J. H. (2000). Antidepressant effects of ketamine in depressed patients. *Biological Psychiatry, 47*, 351–354.

Bernstein, H.-G., Bogerts, B., & Keilhoff, G. (2005). The many faces of nitric oxide in schizophrenia. A review. *Schizophrenia Research, 78*, 69–86.

Bredt, D. S., & Snyder, S. H. (1992). Nitric oxide, a novel neuronal messenger. *Neuron, 8*, 3–11.

Buttenschøn, H. N., Mors, O., Ewald, H., McQuillin, A., Kalsi, G., Lawrence, J., & Kruse, T. (2004). No association between a neuronal nitric oxide synthase (NOS1) gene polymorphism on chromosome 12q24 and bipolar disorder. *American Journal of Medical Genetics Part B: Neuropsychiatric Genetics, 124*, 73–75.

Calabrese, V., Mancuso, C., Calvani, M., Rizzarelli, E., Butterfield, D. A., & Stella, A. M. G. (2007). Nitric oxide in the central nervous system: neuroprotection versus neurotoxicity. *Nature Reviews Neuroscience, 8*, 766–775.

Cao, J., Viholainen, J. I., Dart, C., Warwick, H. K., Leyland, M. L., & Courtney, M. J. (2005). The PSD95-nNOS interface: a target for inhibition of excitotoxic p38 stress-activated protein kinase activation and cell death. *Journal of Cell Biology, 168*, 117–126.

Chanrion, B., la Cour, C. M., Bertaso, F., Lerner-Natoli, M., Freissmuth, M., Millan, M., … Marin, P. (2007). Physical interaction between the serotonin transporter and neuronal nitric oxide synthase underlies reciprocal modulation of their activity. *Proceedings of the National Academy of Sciences, 104*, 8119–8124.

Christopherson, K. S., Hillier, B. J., Lim, W. A., & Bredt, D. S. (1999). PSD-95 assembles a ternary complex with the *N*-methyl-D-aspartic acid receptor and a bivalent neuronal NO synthase PDZ domain. *Journal of Biological Chemistry, 274*, 27467–27473.

Cotter, D. R., Pariante, C. M., & Everall, I. P. (2001). Glial cell abnormalities in major psychiatric disorders: the evidence and implications. *Brain Research Bulletin, 55*, 585–595.

Courtney, M. J., Li, L.-L., & Lai, Y. Y. (2014). Mechanisms of NOS1AP action on NMDA receptor-nNOS signaling. *Frontiers in Cellular Neuroscience, 8*.

Cui, H., Nishiguchi, N., Yanagi, M., Fukutake, M., Mouri, K., Kitamura, N., … Hishimoto, A. (2010). A putative cis-acting polymorphism in the NOS1 gene is associated with schizophrenia and NOS1 immunoreactivity in the postmortem brain. *Schizophrenia Research, 121*, 172–178.

Dantzer, R., O'Connor, J. C., Freund, G. G., Johnson, R. W., & Kelley, K. W. (2008). From inflammation to sickness and depression: when the immune system subjugates the brain. *Nature Reviews Neuroscience, 9*, 46–56.

Dev, K. K. (2004). Making protein interactions druggable: targeting PDZ domains. *Nature Reviews Drug Discovery, 3*, 1047–1056.

Dhir, A., & Kulkarni, S. (2008). Possible involvement of nitric oxide (NO) signaling pathway in the anti-depressant-like effect of MK-801 (dizocilpine), a NMDA receptor antagonist in mouse forced swim test. *Indian Journal of Experimental Biology, 46*, 164.

Dhir, A., & Kulkarni, S. K. (2011). Nitric oxide and major depression. *Nitric Oxide: Biology and Chemistry: Official Journal of the Nitric Oxide Society, 24*, 125–131.

DiazGranados, N., Ibrahim, L., Brutsche, N., Ameli, R., Henter, I. D., Luckenbaugh, D. A., … Zarate, C. A., Jr. (2010). Rapid resolution of suicidal ideation after a single infusion of an NMDA antagonist in patients with treatment-resistant major depressive disorder. *Journal of Clinical Psychiatry, 71*, 1605.

Donohoe, G., Walters, J., Morris, D. W., Quinn, E. M., Judge, R., Norton, N., … Muglia, P. (2009). Influence of NOS1 on verbal intelligence and working memory in both patients with schizophrenia and healthy control subjects. *Archives of General Psychiatry, 66*, 1045–1054.

Doucet, M. V., Harkin, A., & Dev, K. K. (2012). The PSD-95/nNOS complex: new drugs for depression? *Pharmacology and Therapeutics, 133*, 218–229.

Doucet, M. V., Levine, H., Dev, K. K., & Harkin, A. (2013). Small molecule inhibitors at the PSD-95/nNOS interface have antidepressant-like properties in mice. *Neuropsychopharmacology: Official Publication of the American College of Neuropsychopharmacology, 8*, 1575–1584.

Duman, R. S., & Li, N. (2012). A neurotrophic hypothesis of depression: role of synaptogenesis in the actions of NMDA receptor antagonists. *Philosophical Transactions of the Royal Society B: Biological Sciences, 367*, 2475–2484.

Echeverry, M., Guimaraes, F., & Del Bel, E. (2004). Acute and delayed restraint stress-induced changes in nitric oxide producing neurons in limbic regions. *Neuroscience, 125*, 981–993.

Elfering, S. L., Sarkela, T. M., & Giulivi, C. (2002). Biochemistry of mitochondrial nitric-oxide synthase. *Journal of Biological Chemistry, 277*, 38079–38086.

Fallin, M. D., Lasseter, V. K., Avramopoulos, D., Nicodemus, K. K., Wolyniec, P. S., McGrath, J. A., ... Huganir, R. L. (2005). Bipolar I disorder and schizophrenia: a 440–single-nucleotide polymorphism screen of 64 candidate genes among Ashkenazi Jewish case-parent trios. *American Journal of Human Genetics, 77*, 918–936.

Fang, M., Jaffrey, S. R., Sawa, A., Ye, K., Luo, X., & Snyder, S. H. (2000). Dexras1: a G protein specifically coupled to neuronal nitric oxide synthase via CAPON. *Neuron, 28*, 183–193.

Florio, S. K., Loh, C., Huang, S. M., Iwamaye, A. E., Kitto, K. F., Fowler, K. W., ... Lai, Y. (2009). Disruption of nNOS-PSD95 protein-protein interaction inhibits acute thermal hyperalgesia and chronic mechanical allodynia in rodents. *British Journal of Pharmacology, 158*, 494–506.

Freudenberg, F., Alttoa, A., & Reif, A. (2015). Neuronal nitric oxide synthase (NOS1) and its adaptor, NOS1AP, as a genetic risk factors for psychiatric disorders. *Genes, Brain and Behavior, 14*(1), 46–63.

Gałecki, P., Maes, M., Florkowski, A., Lewiński, A., Gałecka, E., Bieńkiewicz, M., & Szemraj, J. (2011). Association between inducible and neuronal nitric oxide synthase polymorphisms and recurrent depressive disorder. *Journal of Affective Disorders, 129*, 175–182.

Giegling, I., Calati, R., Porcelli, S., Hartmann, A. M., Möller, H.-J., De Ronchi, D., ... Serretti, A. (2011). NCAM1, TACR1 and NOS genes and temperament: a study on suicide attempters and controls. *Neuropsychobiology, 64*, 32–37.

Gigliucci, V., Buckley, K. N., Nunan, J., O'Shea, K., & Harkin, A. (2010). A role for serotonin in the antidepressant activity of NG-nitro-L-arginine, in the rat forced swimming test. *Pharmacology Biochemistry and Behavior, 94*, 524–533.

Groves, J. (2007). Is it time to reassess the BDNF hypothesis of depression? *Molecular Psychiatry, 12*, 1079–1088.

Guix, F. X., Uribesalgo, I., Coma, M., & Munoz, F. J. (2005). The physiology and pathophysiology of nitric oxide in the brain. *Progress in Neurobiology, 76*, 126–152.

Hall, C. N., & Attwell, D. (2008). Assessing the physiological concentration and targets of nitric oxide in brain tissue. *Journal of Physiology, 586*, 3597–3615.

Hardingham, N., Dachtler, J., & Fox, K. (2013). The role of nitric oxide in pre-synaptic plasticity and homeostasis. *Frontiers in Cellular Neuroscience, 7*.

Harkin, A., Connor, T. J., Walsh, M., St John, N., & Kelly, J. P. (2003). Serotonergic mediation of the antidepressant-like effects of nitric oxide synthase inhibitors. *Neuropharmacology, 44*, 616–623.

Hashimoto, K., Sawa, A., & Iyo, M. (2007). Increased levels of glutamate in brains from patients with mood disorders. *Biological Psychiatry, 62*, 1310–1316.

Heiberg, I. L., Wegener, G., & Rosenberg, R. (2002). Reduction of cGMP and nitric oxide has antidepressant-like effects in the forced swimming test in rats. *Behavioural Brain Research, 134*, 479–484.

Herken, H., Gurel, A., Selek, S., Armutcu, F., Ozen, M. E., Bulut, M., ... Akyol, O. (2007). Adenosine deaminase, nitric oxide, superoxide dismutase, and xanthine oxidase in patients with major depression: impact of antidepressant treatment. *Archives of Medical Research, 38*, 247–252.

Hiroaki-Sato, V. A., Sales, A. J., Biojone, C., & Joca, S. R. (2014). Hippocampal nNOS inhibition induces an antidepressant-like effect: involvement of 5HT1A receptors. *Behavioural Pharmacology, 25*, 187–196.

Hobbs, A. J., Higgs, A., & Moncada, S. (1999). Inhibition of nitric oxide synthase as a potential therapeutic target. *Annual Reviews Pharmacology and Toxicology, 39*, 191–220.

Holsboer, F. (2000). The corticosteroid receptor hypothesis of depression. *Neuropsychopharmacology, 23*, 477–501.

Hoogman, M., Aarts, E., Zwiers, M., Slaats-Willemse, D., Naber, M., Onnink, M., ... Franke, B. (2011). Nitric oxide synthase genotype modulation of impulsivity and ventral striatal activity in adult ADHD patients and healthy comparison subjects. *American Journal of Psychiatry, 168*, 1099–1106.

Ibrahim, L., Diazgranados, N., Franco-Chaves, J., Brutsche, N., Henter, I. D., Kronstein, P., ... Zarate, C. A., Jr. (2012). Course of improvement in depressive symptoms to a single intravenous infusion of ketamine vs add-on riluzole: results from a 4-week, double-blind, placebo-controlled study. *Neuropsychopharmacology, 37*, 1526–1533.

Jaffrey, S. R., Snowman, A. M., Eliasson, M. J., Cohen, N. A., & Snyder, S. H. (1998). CAPON: a protein associated with neuronal nitric oxide synthase that regulates its interactions with PSD95. *Neuron, 20*, 115–124.

Jesse, C. R., Bortolatto, C. F., Savegnago, L., Rocha, J. B., & Nogueira, C. W. (2008). Involvement of L-arginine–nitric oxide–cyclic guanosine monophosphate pathway in the antidepressant-like effect of tramadol in the rat forced swimming test. *Progress in Neuro-Psychopharmacology and Biological Psychiatry, 32*, 1838–1843.

John, C. S., Smith, K. L., Van't Veer, A., Gompf, H. S., Carlezon, W. A., Cohen, B. M., ... Bechtholt-Gompf, A. J. (2012). Blockade of astrocytic glutamate uptake in the prefrontal cortex induces anhedonia. *Neuropsychopharmacology: Official Publication of the American College of Neuropsychopharmacology, 37*, 2467–2475.

Joung, H.-Y., Jung, E.-Y., Kim, K., Lee, M.-S., Her, S., & Shim, I. (2012). The differential role of NOS inhibitors on stress-induced anxiety and neuroendocrine alterations in the rat. *Behavioural Brain Research, 235*, 176–181.

Kaster, M. P., Rosa, A. O., Santos, A. R., & Rodrigues, A. L. S. (2005). Involvement of nitric oxide–cGMP pathway in the antidepressant-like effects of adenosine in the forced swimming test. *International Journal of Neuropsychopharmacology, 8*, 601–606.

Kiss, J. P. (2000). Role of nitric oxide in the regulation of monoaminergic neurotransmission. *Brain Research Bulletin, 52*, 459–466.

Koike, H., Iijima, M., & Chaki, S. (2011). Involvement of AMPA receptor in both the rapid and sustained antidepressant-like effects of ketamine in animal models of depression. *Behavioural Brain Research, 224*, 107–111.

Kopf, J., Schecklmann, M., Hahn, T., Dresler, T., Dieler, A. C., Herrmann, M. J., ... Reif, A. (2011). NOS1 ex1f-VNTR polymorphism influences prefrontal brain oxygenation during a working memory task. *NeuroImage, 57*, 1617–1623.

Kurrikoff, T., Lesch, K.-P., Kiive, E., Konstabel, K., Herterich, S., Veidebaum, T., ... Harro, J. (2012). Association of a functional variant of the nitric oxide synthase 1 gene with personality, anxiety, and depressiveness. *Development and Psychopathology, 24*, 1225–1235.

Laas, K., Reif, A., Herterich, S., Eensoo, D., Lesch, K.-P., & Harro, J. (2010). The effect of a functional NOS1 promoter polymorphism on impulsivity is moderated by platelet MAO activity. *Psychopharmacology, 209*, 255–261.

Li, L.-L., Ginet, V., Liu, X., Vergun, O., Tuittila, M., Mathieu, M., … Courtney, M. J. (2013). The nNOS-p38MAPK pathway is mediated by NOS1AP during neuronal death. *Journal of Neuroscience, 33*, 8185–8201.

Li, N., Lee, B., Liu, R.-J., Banasr, M., Dwyer, J. M., Iwata, M., … Duman, R. S. (2010). mTOR-dependent synapse formation underlies the rapid antidepressant effects of NMDA antagonists. *Science (New York, N.Y.), 329*, 959–964.

Li, N., Liu, R.-J., Dwyer, J. M., Banasr, M., Lee, B., Son, H., … Duman, R. S. (2011). Glutamate *N*-methyl-D-aspartate receptor antagonists rapidly reverse behavioral and synaptic deficits caused by chronic stress exposure. *Biological Psychiatry, 69*, 754–761.

Lima, A., Sardinha, V. M., Oliveira, A. F., Reis, M., Mota, C., Silva, M. A., … Oliveira, J. F. (2014). Astrocyte pathology in the prefrontal cortex impairs the cognitive function of rats. *Molecular Psychiatry*, 1–8.

Liu, R.-J., Fuchikami, M., Dwyer, J. M., Lepack, A. E., Duman, R. S., & Aghajanian, G. K. (2013). GSK-3 inhibition potentiates the synaptogenic and antidepressant-like effects of subthreshold doses of ketamine. *Neuropsychopharmacology: Official Publication of the American College of Neuropsychopharmacology, 38*, 2268–2277.

Luciano, M., Houlihan, L. M., Harris, S. E., Gow, A. J., Hayward, C., Starr, J. M., & Deary, I. J. (2010). Association of existing and new candidate genes for anxiety, depression and personality traits in older people. *Behavior Genetics, 40*, 518–532.

Luciano, M., Huffman, J. E., Arias-Vásquez, A., Vinkhuyzen, A. A., Middeldorp, C. M., Giegling, I., … Janzing, J. (2012). Genome-wide association uncovers shared genetic effects among personality traits and mood states. *American Journal of Medical Genetics Part B: Neuropsychiatric Genetics, 159*, 684–695.

Luo, L., & Tan, R. (2000). Fluoxetine inhibits dendrite atrophy of hippocampal neurons by decreasing nitric oxide synthase expression in rat depression model. *Acta Pharmacologica Sinica, 22*, 865–870.

Machado-Vieira, R., Salvadore, G., Diazgranados, N., & Zarate, C. A. (2009). Ketamine and the next generation of antidepressants with a rapid onset of action. *Pharmacology & Therapeutics, 123*, 143–150.

Maeng, S., & Zarate, C. A. (2007). The role of glutamate in mood disorders: results from the ketamine in major depression study and the presumed cellular mechanism underlying its antidepressant effects. *Current Psychiatry Reports, 9*, 467–474.

Maeng, S., Zarate, C. A., Du, J., Schloesser, R. J., McCammon, J., Chen, G., & Manji, H. K. (2008). Cellular mechanisms underlying the antidepressant effects of ketamine: role of alpha-amino-3-hydroxy-5-methylisoxazole-4-propionic acid receptors. *Biological Psychiatry, 63*, 349–352.

Malberg, J. E., Eisch, A. J., Nestler, E. J., & Duman, R. S. (2000). Chronic antidepressant treatment increases neurogenesis in adult rat hippocampus. *Journal of Neuroscience, 20*, 9104–9110.

Matarredona, E. R., Murillo-Carretero, M., Moreno-López, B., & Estrada, C. (2004). Nitric oxide synthesis inhibition increases proliferation of neural precursors isolated from the postnatal mouse subventricular zone. *Brain Research, 995*, 274–284.

Mathew, S. J., Murrough, J. W., aan het Rot, M., Collins, K. A., Reich, D. L., & Charney, D. S. (2010). Riluzole for relapse prevention following intravenous ketamine in treatment-resistant depression: a pilot randomized, placebo-controlled continuation trial. *International Journal of Neuropsychopharmacology, 13*, 71–82.

Mattson, M. P. (2008). Glutamate and neurotrophic factors in neuronal plasticity and disease. *Annals of the New York Academy of Sciences, 1144*, 97–112.

Meldrum, B. S. (2000). Glutamate as a neurotransmitter in the brain: review of physiology and pathology. *The Journal of Nutrition, 130*, 1007S–1015S.

Miller, A. H., Maletic, V., & Raison, C. L. (2009). Inflammation and its discontents: the role of cytokines in the pathophysiology of major depression. *Biological Psychiatry, 65*, 732–741.

Monteggia, L. M., Gideons, E., & Kavalali, E. T. (2013). The role of eukaryotic elongation factor 2 kinase in rapid antidepressant action of ketamine. *Biological Psychiatry, 73*, 1199–1203.

Moreno-López, B., Noval, J. A., González-Bonet, L. G., & Estrada, C. (2000). Morphological bases for a role of nitric oxide in adult neurogenesis. *Brain Research, 869*, 244–250.

Moretti, M., de Freitas, A. E., Budni, J., Fernandes, S. C. P., de Oliveira Balen, G., & Rodrigues, A. L. S. (2011). Involvement of nitric oxide–cGMP pathway in the antidepressant-like effect of ascorbic acid in the tail suspension test. *Behavioural Brain Research, 225*, 328–333.

Mungrue, I. N., Bredt, D. S., Stewart, D. J., & Husain, M. (2003). From molecules to mammals: what's NOS got to do with it? *Acta Physiologica Scandinavica, 179*, 123–135.

Murrough, J. W. (2012). Ketamine as a novel antidepressant: from synapse to behavior. *Clinical Pharmacology and Therapeutics, 91*, 303–309.

Mutlu, O., Ulak, G., & Belzung, C. (2011). Effects of nitric oxide synthase inhibitors 1-(2-trifluoromethylphenyl)–imidazole (TRIM) and 7-nitroindazole (7-NI) on learning and memory in mice. *Fundamental and Clinical Pharmacology, 25*, 368–377.

Mutlu, O., Ulak, G., Laugeray, A., & Belzung, C. (2009). Effects of neuronal and inducible NOS inhibitor 1-[2-(trifluoromethyl) phenyl] imidazole (TRIM) in unpredictable chronic mild stress procedure in mice. *Pharmacology Biochemistry and Behavior, 92*, 82–87.

Naylor, G. J., Martin, B., Hopwood, S., & Watson, Y. (1986). A two-year double-blind crossover trial of the prophylactic effect of methylene blue in manic depressive psychosis. *Biological Psychiatry, 21*, 915–920.

Nelson, J. C. M. (2006). The STAR*D study: a four-course meal that leaves us wanting more. *The American Journal of Psychiatry, 163*, 1864–1866.

Niciu, M. J., Henter, I. D., Sanacora, G., & Zarate, C. A., Jr. (2014). Glial abnormalities in substance use disorders and depression: does shared glutamatergic dysfunction contribute to comorbidity? *World Journal of Biological Psychiatry, 15*, 2–16.

O'Donovan, S., Dalton, V., Harkin, A., & McLoughlin, D. M. (2014). Effects of brief pulse and ultrabrief pulse electroconvulsive stimulation on rodent brain and behaviour in the corticosterone model of depression. *International Journal of Neuropsychopharmacology, 17*, 1477–1486.

Ögren, S. O., Eriksson, T. M., Elvander-Tottie, E., D'Addario, C., Ekström, J. C., Svenningsson, P., ... Stiedl, O. (2008). The role of 5-HT 1A receptors in learning and memory. *Behavioural Brain Research, 195*, 54–77.

Okumura, T., Kishi, T., Okochi, T., Ikeda, M., Kitajima, T., Yamanouchi, Y., ... Inada, T. (2009). Genetic association analysis of functional polymorphisms in neuronal nitric oxide synthase 1 gene (NOS1) and mood disorders and fluvoxamine response in major depressive disorder in the Japanese population. *Neuropsychobiology, 61*, 57–63.

Okumura, T., Okochi, T., Kishi, T., Ikeda, M., Kitajima, T., Yamanouchi, Y., ... Ujike, H. (2009). No association between polymorphisms of neuronal oxide synthase 1 gene (NOS1) and schizophrenia in a Japanese population. *Neuromolecular Medicine, 11*, 123–127.

de Oliveira, R. M., Aparecida Del Bel, E., Mamede-Rosa, M. L., Padovan, C. M., Deakin, J. F., & Guimarães, F. S. (2000). Expression of neuronal nitric oxide synthase mRNA in stress-related brain areas after restraint in rats. *Neuroscience Letters, 289*, 123–126.

Oliveira, R., Guimaraes, F., & Deakin, J. (2008). Expression of neuronal nitric oxide synthase in the hippocampal formation in affective disorders. *Brazilian Journal of Medical and Biological Research, 41*, 333–341.

Orlando, G., Langnaese, K., Schulz, C., Wolf, G., & Engelmann, M. (2008). Neuronal nitric oxide synthase gene inactivation reduces the expression of vasopressin in the hypothalamic paraventricular nucleus and of catecholamine biosynthetic enzymes in the adrenal gland of the mouse: research report. *Stress: The International Journal on the Biology of Stress, 11*, 42–51.

Pacher, P., Beckman, J. S., & Liaudet, L. (2007). Nitric oxide and peroxynitrite in health and disease. *Physiological Reviews, 87*, 315–424.

Packer, M. A., Stasiv, Y., Benraiss, A., Chmielnicki, E., Grinberg, A., Westphal, H., ... Enikolopov, G. (2003). Nitric oxide negatively regulates mammalian adult neurogenesis. *Proceedings of the National Academy of Sciences, 100*, 9566–9571.

Pariante, C. M., & Lightman, S. L. (2008). The HPA axis in major depression: classical theories and new developments. *Trends in Neurosciences, 31*, 464–468.

Pereira, V. S., Romano, A., Wegener, G., & Joca, S. R. (2015). Antidepressant-like effects induced by NMDA receptor blockade and NO synthesis inhibition in the ventral medial prefrontal cortex of rats exposed to the forced swim test. *Psychopharmacology*, 1–11.

Pittenger, C., & Duman, R. S. (2008). Stress, depression, and neuroplasticity: a convergence of mechanisms. *Neuropsychopharmacology, 33*, 88–109.

Prast, H., & Philippu, A. (2001). Nitric oxide as modulator of neuronal function. *Progress in Neurobiology, 64*, 51–68.

Rajkowska, G., & Miguel-Hidalgo, J. J. (2007). Gliogenesis and glial pathology in depression. *CNS and Neurological Disorders Drug Targets, 6*, 219–233.

Reif, A., Herterich, S., Strobel, A., Ehlis, A., Saur, D., Jacob, C., ... Walter, U. (2006). A neuronal nitric oxide synthase (NOS-I) haplotype associated with schizophrenia modifies prefrontal cortex function. *Molecular Psychiatry, 11*, 286–300.

Reif, A., Jacob, C. P., Rujescu, D., Herterich, S., Lang, S., Gutknecht, L., ... Giegling, I. (2009). Influence of functional variant of neuronal nitric oxide synthase on impulsive behaviors in humans. *Archives of General Psychiatry, 66*, 41–50.

Reif, A., Kiive, E., Kurrikoff, T., Paaver, M., Herterich, S., Konstabel, K., ... Harro, J. (2011). A functional NOS1 promoter polymorphism interacts with adverse environment on functional and dysfunctional impulsivity. *Psychopharmacology, 214*, 239–248.

Retz, W., Reif, A., Freitag, C. M., Retz-Junginger, P., & Rösler, M. (2010). Association of a functional variant of neuronal nitric oxide synthase gene with self-reported impulsiveness, venturesomeness and empathy in male offenders. *Journal of Neural Transmission, 117*, 321–324.

Riley, B., Thiselton, D., Maher, B. S., Bigdeli, T., Wormley, B., McMichael, G. O., ... Walsh, D. (2010). Replication of association between schizophrenia and ZNF804A in the Irish case–control study of schizophrenia sample. *Molecular psychiatry, 15*, 29–37.

Romero-Grimaldi, C., Moreno-López, B., & Estrada, C. (2008). Age-dependent effect of nitric oxide on subventricular zone and olfactory bulb neural precursor proliferation. *Journal of Comparative Neurology, 506*, 339–346.

Rosa, A. O., Lin, J., Calixto, J. B., Santos, A. R. S., & Rodrigues, A. L. S. (2003). Involvement of NMDA receptors and L-arginine-nitric oxide pathway in the antidepressant-like effects of zinc in mice. *Behavioural Brain Research, 144*, 87–93.

Rose, E. J., Greene, C., Kelly, S., Morris, D. W., Robertson, I. H., Fahey, C., ... McGrath, J. (2012). The NOS1 variant rs6490121 is associated with variation in prefrontal function and grey matter density in healthy individuals. *NeuroImage, 60*, 614–622.

aan het Rot, M., Collins, K. A., Murrough, J. W., Perez, A. M., Reich, D. L., Charney, D. S., & Mathew, S. J. (2010). Safety and efficacy of repeated-dose intravenous ketamine for treatment-resistant depression. *Biological Psychiatry, 67*, 139–145.

Rujescu, D., Giegling, I., Mandelli, L., Schneider, B., Hartmann, A. M., Schnabel, A., ... Serretti, A. (2008). NOS-I and -III gene variants are differentially associated with facets of suicidal behavior and aggression-related traits. *American Journal of Medical Genetics Part B: Neuropsychiatric Genetics, 147*, 42–48.

Rush, A. J. M., Trivedi, M. H., Wisniewski, S. R., Nierenberg, A. A., Stewart, J. W., Warden, D., ... Lebowitz, B. D. (2006). Acute and longer-term outcomes in depressed outpatients requiring one or several treatment steps: a STAR*D report. *The American Journal of Psychiatry, 163*, 1905–1917.

Sanacora, G., Treccani, G., & Popoli, M. (2012). Towards a glutamate hypothesis of depression: an emerging frontier of neuropsychopharmacology for mood disorders. *Neuropharmacology, 62*, 63–77.

Sanacora, G., Zarate, C. A., Krystal, J. H., & Manji, H. K. (2008). Targeting the glutamatergic system to develop novel, improved therapeutics for mood disorders. *Nature Reviews. Drug Discovery, 7*, 426–437.

Santos, T. B., Cespedes, I. C., & Viana, M. B. (2014). Chronic corticosterone administration facilitates aversive memory retrieval and increases GR/NOS immunoreactivity. *Behavioural Brain Research, 267*, 46–54.

Santos, A., Martínez-Ruiz, A., & Araújo, I. (2015). S-nitrosation and neuronal plasticity. *British Journal of Pharmacology, 172*(6), 1468–1478.

Sarginson, J. E., Deakin, J. F., Anderson, I. M., Downey, D., Thomas, E., Elliott, R., & Juhasz, G. (2014). Neuronal nitric oxide synthase (NOS1) polymorphisms interact with financial hardship to affect depression risk. *Neuropsychopharmacology, 39*, 2857–2866.

Scannevin, R. H., & Huganir, R. L. (2000). Postsynaptic organization and regulation of excitatory synapses. *Nature Reviews. Neuroscience, 1*, 133–141.

Segmiller, F., Ruther, T., Linhardt, A., Padberg, F., Berger, M., Pogarell, O., ... Schule, C. (2013). Repeated S-ketamine infusions in therapy resistant depression: a case series. *Journal of Clinical Pharmacology, 53*, 996–998.

Selvakumar, B., Jenkins, M. A., Hussain, N. K., Huganir, R. L., Traynelis, S. F., & Snyder, S. H. (2013). S-nitrosylation of AMPA receptor GluA1 regulates phosphorylation, single-channel conductance, and endocytosis. *Proceedings of the National Academy of Sciences of the United States of America, 110*, 1077–1082.

Serulle, Y., Zhang, S., Ninan, I., Puzzo, D., McCarthy, M., Khatri, L., ... Ziff, E. B. (2007). A GluR1-cGKII interaction regulates AMPA receptor trafficking. *Neuron, 56*, 670–688.

Shinkai, T., Ohmori, O., Hori, H., & Nakamura, J. (2002). Allelic association of the neuronal nitric oxide synthase (NOS1) gene with schizophrenia. *Molecular Psychiatry, 7*, 560–563.

Silva, M., Aguiar, D. C., Diniz, C. R., Guimaraes, F. S., & Joca, S. R. (2012). Neuronal NOS inhibitor and conventional antidepressant drugs attenuate stress-induced fos expression in overlapping brain regions. *Cellular and Molecular Neurobiology, 32*, 443–453.

Suzuki, E., Yagi, G., Nakaki, T., Kanba, S., & Asai, M. (2001). Elevated plasma nitrate levels in depressive states. *Journal of Affective Disorders, 63*, 221–224.

Talarowska, M., Gałecki, P., Maes, M., Orzechowska, A., Chamielec, M., Bartosz, G., & Kowalczyk, E. (2012). Nitric oxide plasma concentration associated with cognitive impairment in patients with recurrent depressive disorder. *Neuroscience Letters, 510*, 127–131.

Tang, W., Huang, K., Tang, R., Zhou, G., Fang, C., Zhang, J., ... Shi, Y. (2008). Evidence for association between the 5′ flank of the NOS1 gene and schizophrenia in the Chinese population. *International Journal of Neuropsychopharmacology, 11*, 1063–1071.

Trivedi, M. H., Fava, M., Wisniewski, S. R., Thase, M. E., Quitkin, F., Warden, D., ... Biggs, M. M. (2006). Medication augmentation after the failure of SSRIs for depression. *New England Journal of Medicine, 354*, 1243–1252.

Ulak, G., Mutlu, O., Tanyeri, P., Komsuoglu, F. I., Akar, F. Y., & Erden, B. F. (2010). Involvement of serotonin receptor subtypes in the antidepressant-like effect of TRIM in the rat forced swimming test. *Pharmacology Biochemistry and Behavior, 95*, 308–314.

Vallance, P., & Leiper, J. (2002). Blocking NO synthesis: how, where and why? *Nature Reviews Drug Discovery, 1*, 939–950.

Wegener, G., Harvey, B. H., Bonefeld, B., Müller, H. K., Volke, V., Overstreet, D. H., & Elfving, B. (2010). Increased stress-evoked nitric oxide signalling in the Flinders sensitive line (FSL) rat: a genetic animal model of depression. *International Journal of Neuropsychopharmacology: Official Scientific Journal of the Collegium Internationale Neuropsychopharmacologicum (CINP), 13*, 461–473.

Willard, S. S., & Koochekpour, S. (2013). Glutamate, glutamate receptors, and downstream signaling pathways. *International Journal of Biological Sciences, 9*, 948–959.

Wolff, K., & Winstock, A. R. (2006). Ketamine: from medicine to misuse. *CNS Drugs, 20*, 199–218.

Yazir, Y., Utkan, T., & Aricioglu, F. (2012). Inhibition of neuronal nitric oxide synthase and soluble guanylate cyclase prevents depression-like behaviour in rats exposed to chronic unpredictable mild stress. *Basic Clinical Pharmacology and Toxicology, 111*, 154–160.

Yildiz-Yesiloglu, A., & Ankerst, D. P. (2006). Review of 1H magnetic resonance spectroscopy findings in major depressive disorder: a meta-analysis. *Psychiatry Research: Neuroimaging, 147*, 1–25.

Yu, Y., Chen, T.-J., Wang, Y.-C., Liou, Y.-J., Hong, C.-J., & Tsai, S.-J. (2002). Association analysis for neuronal nitric oxide synthase gene polymorphism with major depression and fluoxetine response. *Neuropsychobiology, 47*, 137–140.

Zarate, C. A., Singh, J. B., Carlson, P. J., Brutsche, N. E., Ameli, R., Luckenbaugh, D. A., ... Manji, H. K. (2006). A randomized trial of an *N*-methyl-D-aspartate antagonist in treatment-resistant major depression. *Archives of General Psychiatry, 63*, 856–864.

Zhang, G.-F., Wang, N., Shi, J.-Y., Xu, S.-X., Li, X.-M., Ji, M.-H., ... Yang, J.-J. (2013). Inhibition of the L-arginine–nitric oxide pathway mediates the antidepressant effects of ketamine in rats in the forced swimming test. *Pharmacology Biochemistry and Behavior, 110*, 8–12.

Zhang, J., Huang, X., Ye, M., Luo, C., Wu, H., Hu, Y., ... Zhu, D. (2010). Neuronal nitric oxide synthase alteration accounts for the role of 5-HT_{1A} receptor in modulating anxiety-related behaviors. *Journal of Neuroscience, 30*, 2433–2441.

Zhou, Q. G., Hu, Y., Hua, Y., Hu, M., Luo, C. X., Han, X., ... Zhu, D. Y. (2007). Neuronal nitric oxide synthase contributes to chronic stress-induced depression by suppressing hippocampal neurogenesis. *Journal of Neurochemistry, 103*, 1843–1854.

Zhou, L., Li, F., Xu, H.-B., Luo, C.-X., Wu, H.-Y., Zhu, M.-M., ... Zhu, D.-Y. (2010). Treatment of cerebral ischemia by disrupting ischemia-induced interaction of nNOS with PSD-95. *Nature Medicine, 16*, 1439–1443.

Zhou, L., & Zhu, D. Y. (2009). Neuronal nitric oxide synthase: structure, subcellular localization, regulation, and clinical implications. *Nitric Oxide, 20*, 223–230.

Zhou, Q. G., Zhu, L. J., Chen, C., Wu, H. Y., Luo, C. X., Chang, L., & Zhu, D. Y. (2011). Hippocampal neuronal nitric oxide synthase mediates the stress-related depressive behaviors of glucocorticoids by downregulating glucocorticoid receptor. *J Neurosci, 31*, 7579–7590.

Zhu, X., Hua, Y., Jiang, J., Zhou, Q., Luo, C., Han, X., ... Zhu, D. (2006). Neuronal nitric oxide synthase-derived nitric oxide inhibits neurogenesis in the adult dentate gyrus by down-regulating cyclic AMP response element binding protein phosphorylation. *Neuroscience, 141*, 827–836.

Zhu, L.-J., Li, T.-Y., Luo, C.-X., Jiang, N., Chang, L., Lin, Y.-H., ... Lu, W. (2014). CAPON-nNOS coupling can serve as a target for developing new anxiolytics. *Nature medicine, 20*, 1050–1054.

Zomkowski, A. D., Engel, D., Gabilan, N. H., & Rodrigues, A. L. S. (2010). Involvement of NMDA receptors and l-arginine-nitric oxide-cyclic guanosine monophosphate pathway in the antidepressant-like effects of escitalopram in the forced swimming test. *European Neuropsychopharmacology, 20*, 793–801.

Chapter 4

Anti-Inflammatory and Immune-Modulatory Therapeutic Approaches in Major Depression

Norbert Müller
Department of Psychiatry and Psychotherapy, Ludwig-Maximilian University of Munich, Munich, Germany

INTRODUCTION

Activation of the inflammatory response system in major depression (MD) is well documented (Maes, 1994; Maes et al., 1992; Müller, Hofschuster, Ackenheil, Mempel, & Eckstein, 1993; Myint, Leonard, Steinbusch, & Kim, 2005; Rothermundt, Arolt, Peters, et al., 2001). Two meta-analyses clearly showed elevated interleukin-6 (IL-6) levels in patients with MD (Dowlati et al., 2010; Howren, Lamkin, & Suls, 2009). However, the findings of the two meta-analyses differed regarding levels of the inflammatory markers C-reactive protein (CRP), IL-1, IL-1RA, and TNF-α. In general, the inflammatory response system appears to be activated, but the levels of the different markers vary across studies. MD is a disorder often triggered by stress. It has been shown—often based on genetic disposition—that early life stress or separation stress are associated with an increase of proinflammatory cytokines, leading to an activation of the immune system and proinflammatory prostaglandins. Prostaglandin E2 (PGE_2) is an important mediator of inflammation (Song et al., 1998). Increased PGE_2 in the saliva, serum, and cerebrospinal fluid (CSF) of depressed patients has been described previously (Calabrese et al., 1986; Linnoila et al., 1983; Nishino, Ueno, Ohishi, Sakai, & Hayaishi, 1989; Ohishi, Ueno, Nishino, Sakai, & Hayaishi, 1988). The enzyme cyclooxygenase-2 (COX-2) is involved in the function of PGE_2 in the inflammatory pathway. In the brain, the activation of microglia cells and astrocytes is crucial, because proinflammatory molecules are produced and released in the brain. The interactions between the immune system and neurotransmitters, the tryptophan-kynurenine system, and the glutatmatergic

Systems Neuroscience in Depression. http://dx.doi.org/10.1016/B978-0-12-802456-0.00004-2

neurotransmission are further links between stress, depression, and the immune system. Accordingly, anti-inflammatory therapy, e.g., with the COX-2 inhibitor celecoxib, is effective in depression.

The roles of microglial cells and astrocytes as mediators of inflammation in the brain are discussed as well as the influence of a disturbed blood–brain barrier (BBB) facilitating the invasion of inflammatory molecules and immune cells, including monocytes as well as T and B cells.

THE CELLULAR BASIS OF INFLAMMATION IN THE CENTRAL NERVOUS SYSTEM

Microglia cell and astrocytes have been shown to play thc central role in regulating neuroinflammation (Brambilla et al., 2005; Farina, Aloisi, & Meinl, 2007). Astrocytes comprise approximately 80% of the brain cells and are the most abundant type of glial cells in the central nervous system (CNS), while microglial cells comprise about 15% of the brain cells. Astrocytes have a strategic location: they are in close contact with CNS resident cells (neurons, microglia, oligodendrocytes, and other astrocytes) and are part of the immune and inflammatory systems in the CNS. Among the cytokines, TNF-α, INF-γ, IL-1, and IL-6 are the main astrocytic activators (Farina et al., 2007). Astrocytes are strongly involved in increased BBB permeability, endothelial cell activation, monocyte and microglia activation, and B-cell survival and differentiation. Astrocytes produce and release a series of chemokines and are thus involved in the recruitment of monocytes and macrophages, dendritic cells, and T and B cells in the CNS.

Astrocytes, however, also store and release neurotransmitters (Gao & Hong, 2008). For example, an important function of astrocytes is the regulation and storage of glutamate, which is actively transported into astrocytes (Aronica et al., 2003). Therefore, astrocytes play a role not only as immune regulators and hosts of kynurenine metabolism but also in the bioavailability of neurotransmitters, including glutamate, the most abundant neurotransmitter, which is involved in psychiatric disorders such as MD.

BBB DISTURBANCE IN MAJOR DEPRESSION

By far most studies of biological parameters, including immune parameters in MD, have examined components of blood cells and serum, although the CSF reflects disturbances in the CNS much better. During acute inflammation immune cells invade the CNS parenchyma through disturbed BBB, i.e., through the endothelium of the small vessels and the tight junctions of astrocytes around the vessels. This invasion is mediated by cytokines, chemokines, adhesion molecules, and other mediators of inflammation. Signs of inflammation in the CNS are a disturbance of the BBB, increased immunoglobulins, and, especially in acute inflammatory states, an increase in the cell number in the CSF.

Analysis of the CSF is the gold standard for diagnosing CNS inflammatory disorders (Schwartz & Shechter, 2010; Wildemann, Oschmann, & Reiber, 2010). The rough method of "routine" analysis of CSF parameters, however, only discovers gross changes in the CSF. More subtle changes in the CSF of MD patients may occur in a higher percentage of patients. Pathological changes in the CSF were observed in a subgroup of around 25–30% of patients suffering from MD (Hampel et al., 1995; Hampel, Kotter, Moller, 1997; Hampel, Kotter, Padberg, Korschenhausen, Moller, 1999). These changes included an increased production of immunoglobulin G (IgG) and an increased BBB permeability, as is found in inflammatory states. Studies with the advanced CSF methodology confirmed and extended these previous studies in that about 15% of cases with therapy-resistant depression provided evidence of low-grade classical neuroinflammation and about 25% blood–CSF barrier dysfunction (Bechter et al., 2010). With some overlap to these pathological findings, about 30% of patients demonstrated inflammatory activation patterns on CSF cells (Maxeiner et al., 2009). In addition, more than 30% showed, with little overlap to the other findings, considerably increased CSF-neopterin, likely indicating some immune-inflammatory condition (Kuehne, Reiber, Bechter, Hagberg, & Fuchs, 2013).

THE MODEL OF "SICKNESS BEHAVIOR" FOR DEPRESSION

An animal model for MD is "sickness behavior," i.e., the behavioral, vegetative, cognitive, and emotional reaction of an organism to infection and inflammation (Dantzer, 2001; Dantzer, O'Connor, Freund, Johnson, & Kelley, 2008). However, increasing evidence from animal models and clinical studies indicates that this "sickness behavior" represents a highly differentiated adjustment reaction with the aim to specifically fight an infection and promote survival (Dantzer et al., 2007; Hart, 1988). The proinflammatory cytokines IL-1 beta (IL-1β), IL-6, and TNF-α, which are released by macrophages during the early innate immune response, play a central role in sickness behavior. They are involved in the communication between the peripheral immune response and the CNS (Dantzer et al., 2008; Raison, Capuron, & Miller, 2006). A peripheral immune response, e.g., induced by macrophages, results in increased production and release of cytokines in the CNS. Cytokines can convey signals to the CNS via different routes, including activation of afferent vagal fibers that "project" to the core of the "solitary tract" and higher viscerosensory centers via cytokine-specific transport molecules that express CNS endothelial tissue. Moreover, the circumventricular organs lack a BBB (Dantzer et al., 2008; Raison et al., 2006). As soon as they reach the CNS, signals from cytokines can be amplified through the central cytokines network, which has important effects on neurotransmitter metabolism, neuroendocrine functions, synaptic plasticity, and behavior (Raison et al., 2006).

In humans, the involvement of cytokines in the regulation of sickness behavior has been studied by administering the bacterial endotoxin Lipopolysaccharide (LPS) to healthy volunteers (Reichenberg et al., 2001). The levels of anxiety, depression, and cognitive impairment were found to be related to the levels of circulating cytokines (Reichenberg et al., 2001, 2002).

THE PROINFLAMMATORY IMMUNE STATE IN MAJOR DEPRESSION

A high blood level of CRP is a common marker for an inflammatory process. Higher than normal CRP levels have been repeatedly observed in depression, for example, in severely depressed inpatients (Lanquillon, Krieg, Bening-Abu-Shach, & Vedder, 2000), and high CRP levels have been found to be associated with the severity of depression (Häfner et al., 2008). Higher CRP levels were also observed in remitted patients after a depressive state, in both men (Danner, Kasl, Abramson, & Vaccarino, 2003; Ford & Erlinger, 2004) and women (Cizza et al., 2009; Kling et al., 2007). In a sample of older healthy persons, CRP levels (and IL-6 levels) were predictive of cognitive symptoms of depression 12 years later (Gimeno, Marmot, & Singh-Manoux, 2008).

Characteristics of immune activation in MD include increased numbers of circulating lymphocytes and phagocytic cells; upregulated serum levels of markers of immune activation (neopterin, soluble IL-2 receptors); higher serum concentrations of positive acute phase proteins (APPs), coupled with reduced levels of negative APPs; and increased release of proinflammatory cytokines, such as IL-1β, IL-2, TNF-α, and IL-6 through activated macrophages and IFN-γ through activated T cells (Irwin, 1999; Maes, Meltzer, Bosmans, et al., 1995; Maes, Meltzer, Buckley, Bosmans, 1995; Mikova, Yakimova, Bosmans, Kenis, & Maes, 2001; Müller et al., 1993; Müller & Schwarz, 2002; Nunes et al., 2002) (Table 1).

TABLE 1 Candidates for Immune Markers Related to Major Depression

Disease-Related Markers	Markers for Antidepressant Response	Markers for Response to Immune-Related Therapy
IL-6	IL-6	IL-6
TNF-α	Quinolinic acid	CRP
CRP	TNF-α	TNF-α
Neopterin		TNFR1
		TNFR2
		Kynurenine/tryptophan

Increased numbers of peripheral mononuclear cells in MD have been described by different research groups (Herbert & Cohen, 1993; Rothermundt, Arolt, Fenker, et al., 2001; Seidel et al., 1996). In accordance with the findings of increased monocytes and macrophages, an increased level of neopterin has also been described (Bonaccorso et al., 1998; Duch, Woolf, Nichol, Davidson, & Garbutt, 1984; Dunbar, Hill, Neale, & Mellsop, 1992; Maes et al., 1994). The role of cellular immunity, cytokines, and the innate and adaptive immune systems in depression has been reviewed (Maes, 2011; Müller, Myint, & Schwarz, 2011).

Accordingly, it has been well known for many years that efficient antidepressant treatment is associated with a decrease of proinflammatory cytokines, in particular IL-1 and IL-6 and partly TNF-α (Dowlati et al., 2010; Hannestad, DellaGioia, & Bloch, 2011; Liu, Ho, & Mak, 2012).

THE ROLE OF CYTOKINES AND THE HORMONES OF THE HYPOTHALAMUS-PITUITARY-ADRENAL AXIS

As a product of activated monocytes and macrophages, IL-6 is a frequently investigated immune parameter in patients suffering from MD (Berk, Wadee, Kuschke, & O'Neill-Kerr, 1997; Frommberger et al., 1997; Maes, Meltzer, Bosmans, et al., 1995; Maes et al., 1997; Sluzewska et al., 1996; Song et al., 1998). Most of the publications report a marked increase of in vitro IL-6 production (Maes et al., 1993) or serum IL-6 levels in depressed patients (Berk et al., 1997; Frommberger et al., 1997; Maes, Meltzer, Bosmans, et al., 1995; Maes et al., 1997; Sluzewska et al., 1996; Song et al., 1998), although there are a few contradictory results, indicating reduced (Katila, Appelberg, Hurme, & Rimon, 1994), or unaltered serum IL-6 levels (Brambilla & Maggioni, 1998). An age-related increase of IL-6 serum values was reported in patients with MD (Ershler et al., 1993). The potential influence of possibly interfering variables, however, such as smoking, gender, recent infections, and prior medication to IL-6 release and concentration must be considered and may contribute to differing results (Haack et al., 1999).

IL-6 contributes to indoleamine 2,3-dioxygenase (IDO) activation by its stimulatory effect on PGE_2, which acts as cofactor in the activation of IDO. This fits with a report on the correlation of increased IL-6 production in vitro with decreased tryptophan levels in depressed patients that emphasizes the influence of IL-6 on the serotonin metabolism in depressed patients (Maes et al., 1993).

There is no doubt that IL-6 is involved in the modulation of the HPA axis, and increased availability of IL-6 in the hypothalamus is associated with increased HPA activity (Plata-Salaman, 1991). Activation of the HPA axis is one of the best-documented changes in MD (Roy et al., 1987). Stress acts as a predisposing factor for MD, and an increased susceptibility to stress has repeatedly been described in patients with MD, even prior to their first exacerbation of the disorder. Psychosocial stressors frequently precede the onset of MD. Additionally, an altered HPA axis physiology and dysfunctions of the extrahypothalamic corticotropin releasing hormone (CRH) system have been consistently found in subjects with MD (Hasler, Drevets, Manji, & Charney, 2004). Several studies demonstrate that MD patients exhibit higher baseline cortisol levels or at least much higher cortisol levels during the recovery period after psychological stress (Burke, Davis, Otte, & Mohr, 2005).

The effect of chronic stress on the peripheral immune system and its relevance for MD has been extensively discussed (O'brien, Scott, & Dinan, 2004). In vivo evidence suggests that stress-induced elevation of glucocorticoids also enhances immune function within the CNS through microglia activation and proliferation. Animal studies show that stress induces an enhanced expression of proinflammatory factors like IL-1β (Nguyen et al., 1998; Pugh et al., 1999), macrophage migration inhibitory factor (Bacher et al., 1998; Niino, Ogata, Kikuchi, Tashiro, & Nishihira, 2000; Suzuki et al., 2000) and COX-2 (Madrigal et al., 2003) in the brain.

On the other hand, proinflammatory cytokines such as IL-1 and IL-6 are known to stimulate the HPA axis via hypothalamic neurons. For example, the release of the CRH and growth hormone releasing hormone is stimulated by IL-1 (Berkenbosch, van Oers, del Rey, Tilders, & Besedovsky, 1987; Besedovsky, del Rey, Sorkin, & Dinarello, 1986) and the central IL-1 upregulation leads to stimulation of CRH, the HPA axis, and the sympathetic nervous system (Sundar, Cierpial, Kilts, Ritchie, & Weiss, 1990; Weiss, Quan, & Sundar, 1994). Therefore, a vicious circle may be induced if the stress response is not limited, as it is discussed in MD.

Elevation of these proinflammatory factors is accompanied by dendritic atrophy and neuronal death within the hippocampus (Sapolsky, 1985; Woolley, Gould, & McEwen, 1990), which are also found in brains of subjects with MD (Campbell & Macqueen, 2004). These detrimental effects of glucocorticoids in the CNS are mediated by a rise in extracellular glutamate (Moghaddam, Bolinao, Stein-Behrens, & Sapolsky, 1994; Stein-Behrens, Lin, & Sapolsky, 1994) and subsequent overstimulation of the NMDA receptor. Such an overstimulation of the NMDA receptor results in excitotoxic neuronal damage (Takahashi et al., 2002). Nair and Bonneau (2006) demonstrated that restraint-induced psychological stress stimulates proliferation of microglia, which was prevented by a blockade of corticosterone synthesis, the glucocorticoid receptor, or the NMDA receptor. These data show that stress-induced microglia proliferation is mediated by corticosterone-induced, NMDA receptor-mediated activation within the CNS. Moreover, NMDA receptor activation during stress leads again to increased expression of COX-2 and PGE_2. Both COX-2 and PGE_2 are able to stimulate microglia activation.

THE INFLUENCE OF STRESS ON THE IMMUNE SYSTEM

The influence of stress on the immune system should be seen to be as varied as the way stress is experienced and processed. Stress research differentiates between eustress and distress, acute and chronic stress, and somatic and mental stress. One and the same stressor can have quite different effects on the immune system, depending, for example, on personality factors, coping mechanisms, and current mental and somatic conditions. Acute stress causes a rapid upregulation of immune parameters, e.g., in the paradigm of the first parachute jump: the number and activity of natural killer cells in the blood—parts of the innate immune response that react very quickly as the first "barrier" of the immune system—increase sharply right before the jump but then fall again very quickly, i.e., within minutes (Schedlowski, Benschop, & Schmidt, 1995). The situation is similar in another stress paradigm, "Football trainer during a match": the concentration of immunoglobulin A (IgA) in saliva increases sharply shortly before the match but then decreases again quickly (Kugler, Reintjes, Tewes, & Schedlowski, 1996). Chronic stress is different; the example of dentistry students during exam preparations shows that IgA levels decrease over months and then increase again slowly after the exam period is over (Jemmott et al., 1983).

The association between chronic, mostly aversively experienced stress and downregulation of the immune system has been known for a long time. It often seems plausible from one's own experience and was systematically investigated as early as the 1980s. One of the findings was that both the exacerbation and course of viral infections is affected by stress (Glaser & Kiecolt-Glaser, 1986; Laudenslager et al., 1988). When stress levels are high, exacerbations are more common, and the course is less favorable. The antibody titers of neurotropic viruses are also affected by stress (Glaser & Kiecolt-Glaser, 1986). It has also been known for a long time that stress has (unfavorable) effects on the exacerbation and course of autoimmune diseases. Stress "weakens" the immune defense, whereby speaking of a "strong" or "weak" immune system does not do justice to the complex regulation mechanisms of the immune system, with upregulation of proinflammatory cytokines and downregulation of anti-inflammatory cytokines. The key element is the imbalance of the immune response, which is related to inadequate immune activation.

KINDLING AND SENSITIZATION OF THE IMMUNE RESPONSE: BASIS FOR THE STRESS-INDUCED INFLAMMATORY RESPONSE IN PSYCHIATRIC DISORDERS

The immune response and the release of cytokines can become more sensitized for activating stimuli by a kindling process: the initial immune response, i.e., the release of cytokines and other mediators of immune activation, is initiated as a result of exposure to a certain stimulus. Thereafter, re-exposure to the same stimulus, e.g., stress or infection, is associated with an increased release of cytokines, or a weaker stimulus is necessary for the same activation process. This "sensitization" or "kindling" may be due to the memory function of the acquired immune system (Furukawa, del Rey, Monge-Arditi, & Besedovsky,

1998; Sparkman & Johnson, 2008). Stress-associated release of IL-6 was shown to reactivate (prenatal) conditioned processes (Zhou, Kusnecov, Shurin, DePaoli, & Rabin, 1993). In healthy persons, a second stimulus (e.g., systemic inflammation, stress) led to immune activation, associated with cellular proliferation, and an increased production and release of proinflammatory cytokines (Frank, Baratta, Sprunger, Watkins, & Maier, 2007). This mechanism is a key mechanism for triggering an immune activation and inflammation, e.g., the stress-induced immune activation leading to psychopathological symptoms. A sensitization process in the immune system is in accordance with the view that after an infection during early childhood, reinfection or another stimulation of the immune system in later stages of life might be associated with a boosted release of sensitized cytokines, resulting in neurotransmitter disturbances.

Sensitization phenomena play a role in stress-related, cytokine-induced, neurotransmitter-mediated behavioral abnormalities, i.e., the cytokine response to a stimulus increases, whereas the intensity of the stimulus decreases (Sparkman & Johnson, 2008). In animal experiments, however, cytokines promote greater neurotransmitter responses when the animals are re-exposed to the cytokine (Anisman & Merali, 2003), for example, TNF-α (Hayley, Wall, & Anisman, 2002). In the CNS, the stress-induced activation and proliferation of microglia may mediate these cytokine effects (Nair & Bonneau, 2006).

THE VULNERABILITY-STRESS-INFLAMMATION MODEL OF DEPRESSION

The vulnerability-stress model of mental disorders, which was first postulated over 30 years ago for schizophrenic disorders (Zubin & Spring, 1977), focuses on the role of physical and mental stress as the trigger for a psychotic episode. It has also been known for a long time, however, that different types of stress can trigger a depressive episode and may even play a role in the pathogenesis of depressions. The concept states that in people with an increased vulnerability for depression, stress can trigger an exacerbation of depressive symptoms. This vulnerability could represent the result of a genetic disposition or could be acquired during early childhood development. The inflammation theory of depression fits this model well, as experiments in animal models and also data from studies in patients show. A study in an animal model that simulates the induction of an infection and the associated proinflammatory immune response in the mother in a late stage of pregnancy found an increased risk for the appearance of a depression in adulthood in the offspring (Meyer, Feldon, & Fatemi, 2009), while in schizophrenia—probably on the background of a genetic vulnerability—offspring showed an increased risk of illness if the proinflammatory immune response was induced in the mother during the second trimester of the pregnancy. However, an increased risk could also be shown in the offspring during later stages of CNS development.

A study in men with a depressive illness who had been exposed to higher stress levels in early childhood showed increased reactivity of inflammatory parameters to psychosocial stress, i.e., in this group the inflammatory parameters increased more after stress than in healthy controls (Pace et al., 2006). A similar result was obtained in a birth cohort of 1000 individuals with a history of mistreatments in childhood: people with a history of maltreatments more often had depression and increased inflammatory parameters (Danese et al., 2008). These data show that stress in early childhood is a vulnerability factor for the later occurrence of a depression associated with inflammation.

The mechanisms underlying the joint occurrence of stress and inflammation were studied in animal experiments, and stress was repeatedly shown to be associated with an increase in proinflammatory cytokines (Sparkman & Johnson, 2008).

INFECTIONS AND AUTOIMMUNE DISORDERS AS RISK FACTORS FOR MD

Results of a very interesting population-based Danish register study support the view that an infection or an autoimmune disease significantly increases the risk of getting a depressive disorder. This population-based prospective cohort study of 78 million people included 3.6 million people born between 1945 and 1996. The follow up was documented from 1977 to 2010. All persons with the diagnosis of an affective disorder according to ICD-8, ICD-9, and ICD-10 were included in case they had at least one hospital contact as an in- or outpatient due to affective disorder (including bipolar disorder). Every hospital contact due to autoimmune disorder or infection (excluding HIV/AIDS) prior to psychiatric diagnosis was recorded. More than 91,000 affective disorder cases were identified, ~30,000 diagnosed with infection, and >4000 were diagnosed with autoimmune disease.

Hospitalization for infection significantly increased the risk for later mood disorder by 62% (IRR 1.62), whereas hospitalization for autoimmune disease significantly increased the risk for later mood disorder by 45% (IRR 1,45). Both risk factors interacted and increased the risk to IRR 2.35. The risks were higher for hepatitis infection (IRR 2.82) as compared to sepsis or CNS infections. The risk for mood disorder increased with the proximity to the infection, with the highest risk within the first year (IRR 2.70; Benros et al., 2013).

INFLAMMATION INFLUENCES THE METABOLISM OF SEROTONIN AND NORADRENALIN IN DEPRESSION

Overwhelming evidence collected over the past 40 years suggests that disturbances in serotonergic and noradrenergic neurotransmission are crucial factors in the pathogenesis of MD (Coppen et al., 1988; Matussek, 1966). Although the pathogenesis of the disturbed serotonergic and noradrenergic mechanisms is still unclear, the involvement of the proinflammatory immune state might be crucial.

The proinflammatory cytokine IL-1β increases the metabolism of serotonin and noradrenalin within the hypothalamus, prefrontal cortex, hippocampus, and amygdala (Anisman & Merali, 1999; Day, Curran, Watson, & Akil, 1999; Linthorst, Flachskamm, Muller Preuss, Holsboer, & Reul, 1995; Merali, Lacosta, & Anisman, 1997; Shintani et al., 1995; Song, Merali, & Anisman, 1999). Similar but less pronounced effects on central monoamine activity after stimulation with LPS or poly:IC have been observed for several mediators such as IL-6 and TNF-α (Song et al., 1999; Zalcman et al., 1994).

Similarly, administration of the proinflammatory cytokine IFN-α is associated with reduced levels of serotonin in the prefrontal cortex (Asnis et al., 2003). Accordingly, symptoms of depression were observed in many patients treated with IFN-α (Raison et al., 2006; Schäfer et al., 2004).

Moreover, two indirect pathways of the tryptophan/kynurenine metabolism contribute to the induction of symptoms of depression by proinflammatory cytokines: (1) the increased metabolism of serotonin and (2) the increased production of NMDA agonists, i.e., glutamatergic products of kynurenine metabolism, after activation of the enzyme IDO by proinflammatory cytokines (Müller & Schwarz, 2007; Müller, Schwarz, & Riedel, 2005). Proinflammatory molecules such as PGE_2 or TNF-α, however, induce synergistically with IFN the increase of IDO activity (Braun, Longman, & Albert, 2005; Kwidzinski et al., 2005; Robinson, Hale, & Carlin, 2005).

Increased activity of the glutamatergic system in the peripheral blood of depressive patients was repeatedly shown (Altamura et al., 1993; Kim, Schmid-Burgk, Claus, & Kornhuber, 1982; Mauri et al., 1998), although this result could not be replicated by all groups (Maes, Verkerk, Vandoolaeghe, Lin, & Scharpe, 1998). The inconsistency of the findings, however, might be due to methodological problems (Kugaya & Sanacora, 2005). Support for increased glutamatergic activity in depression comes from magnetic resonance spectroscopy: elevated glutamate levels were found in the occipital cortex of unmedicated subjects with MD (Sanacora et al., 2004). Furthermore, NMDA antagonists such as MK-801 (Maj, Rogoz, Skuza, & Sowinska, 1992; Trullas & Skolnick, 1990), ketamine (Yilmaz, Schulz, Aksoy, & Canbeyli, 2002), memantine (Ossowska, Klenk-Majewska, & Szymczyk, 1997), amantadine (Huber, Dietrich, & Emrich, 1999; Stryjer et al., 2003), and others (Kugaya & Sanacora, 2005) have exhibited antidepressant effects in humans. The partial NMDA receptor agonist D-cycloserine demonstrated antidepressant effects at high doses (Crane, 1959).

Several mechanisms can cause depressive states:

1. a direct influence of proinflammatory cytokines on serotonin and noradrenalin metabolism (Besedovsky et al., 1983; Song & Leonard, 2000; Zalcman et al., 1994);
2. an imbalance of the type 1 and type 2 immune responses leading to increased tryptophan and serotonin metabolism by activation of IDO in the CNS (Myint & Kim, 2003; Schwarz, Chiang, Müller, & Ackenheil, 2001);
3. a decreased availability of tryptophan and serotonin (Maes et al., 1994); and

4. a disturbance of kynurenine metabolism, with an imbalance in favor of the production of the NMDA receptor agonist quinolinic acid (Myint & Kim, 2003; Myint, Kim, et al., 2007).

EFFECTS OF THE PROINFLAMMATORY IMMUNE ACTIVATION ON THE KYNURENINE METABOLISM IN DEPRESSION

The enzyme IDO metabolizes tryptophan to kynurenine. Kynurenine is then converted to quinolinic acid via the intermediate 3-HK by the enzyme kynurenine hydroxylase. Both IDO and kynurenine hydroxylase are induced by the type 1 cytokine IFN-γ. The activity of IDO is an important regulatory component in the control of lymphocyte proliferation, the activation of the type 1 immune response, and the regulation of the tryptophan metabolism (Mellor & Munn, 1999). It induces a halt in the lymphocyte cell cycle due to the catabolism of tryptophan (Munn et al., 1999). In contrast to the type 1 cytokines, the type 2 cytokines IL-4 and IL-10 inhibit the IFN-γ-induced IDO-mediated tryptophan catabolism (Weiss et al., 1999; see Figure 1). IDO is located in several cell types, including monocytes and microglial cells (Alberati, Ricciardi, Kohler, & Cesura, 1996). An IFN-γ-induced, IDO-mediated decrease of CNS tryptophan availability may lead to a serotonergic deficiency in the CNS, since tryptophan availability is one of the limitations of the serotonin synthesis. Other proinflammatory molecules such as PGE_2 or TNF-α, however, induce synergistically with IFN-γ the increase of IDO activity (Braun et al., 2005; Kwidzinski et al., 2005; Robinson et al., 2005). Accordingly, increased levels of PGE_2 and TNF-α were described in MD (e.g., Linnoila et al., 1983; Mikova et al., 2001; see Figure 2).

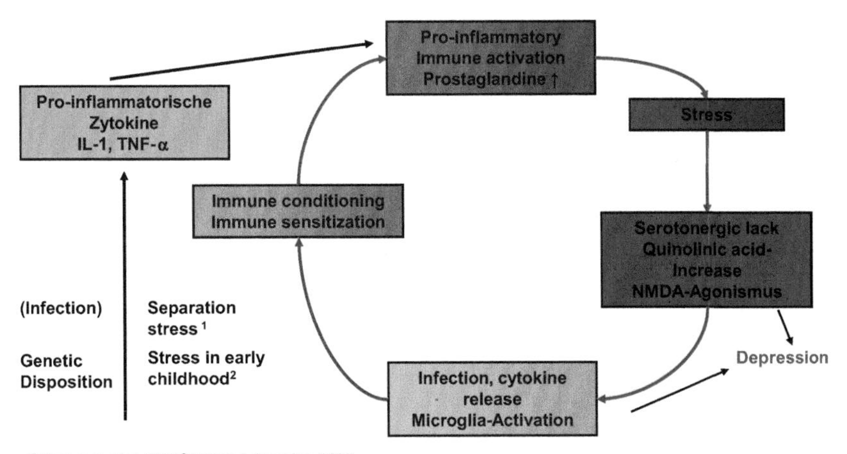

FIGURE 1 Vulnerability-stress-inflammation hypothesis of depression.

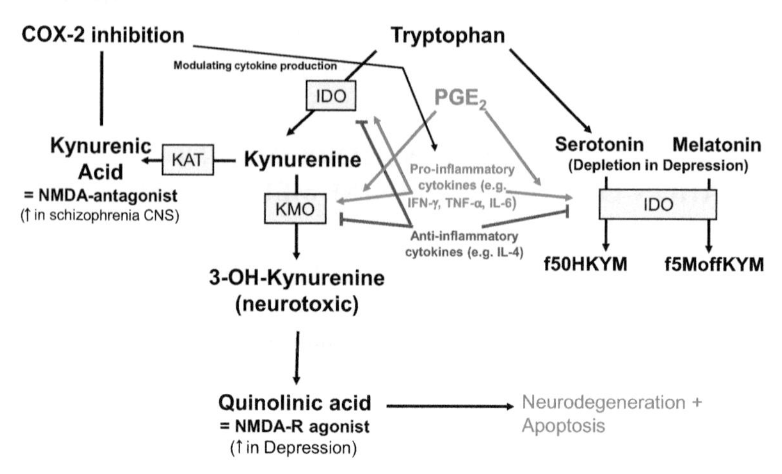

FIGURE 2 The tryptophan/kynurenine metabolism—possible implications for psychiatric disorders.

Low levels of 5-hydroxyindoleacetic acid (5-HIAA)—the metabolite of serotonin—in the CSF of suicidal persons have repeatedly been observed (Lidberg, Belfrage, Bertilsson, Evenden, & Asberg, 2000; Mann & Malone, 1997; Nordstrom et al., 1994). This gives additional evidence for a possible link between the type 1 cytokine IFN-γ and the IDO-related reduction of serotonin availability in the CNS of suicidal patients.

An interesting study showed that immunotherapy with IFN-α was followed by an increase of depressive symptoms and serum kynurenine concentrations on the one hand and reduced concentrations of tryptophan and serotonin on the other hand (Bonaccorso et al., 2002). The kynurenine/tryptophan ratio, which reflects the activity of IDO, increased significantly. Changes in depressive symptoms were significantly positively correlated with kynurenine and negatively correlated with serotonin concentrations (Bonaccorso et al., 2002). This study and others (Capuron et al., 2003) clearly show that the IDO activity is increased by IFN, leading to an increased kynurenine production and depletion of tryptophan and serotonin. The further metabolism of kynurenine, however, seems to play an additional crucial role for the psychopathological states, quinolinic acid in depression (Leonard & Myint, 2006; Steiner et al., 2011), and kynurenic acid in schizophrenia (Olsson et al., 2009).

In addition to the effects of the proinflammatory immune response on serotonin metabolism, other neurotransmitter systems, in particular the catecholaminergic system, are also involved in depression (Matussek, 1988). Although the relationship of immune activation and the lack of catecholaminergic neurotransmission has not been well studied, the increase of the monoamino-oxidase (MAO) activity, which leads to decreased noradrenergic neurotransmission, might be an indirect effect of the increased production of kynurenine and quinolinic acid (Schiepers, Wichers, & Maes, 2005).

CNS VOLUME LOSS IN NEUROIMAGING STUDIES: A CONSEQUENCE OF AN INFLAMMATORY PROCESS?

A loss of brain volume has been observed in MD. Male patients with a first episode of MD had significantly smaller hippocampal total and gray matter volumes than healthy male comparison subjects (Frodl et al., 2002). In a long-term study a significant higher decline of volume in several CNS regions as compared to healthy controls was observed (Frodl et al., 2008). The pathophysiology of this volume loss is unclear. Glial reductions have been consistently found in brain circuits known to be involved in mood disorders, such as the limbic and prefrontal cortex (Cotter, Pariante, & Rajkowska, 2002; Öngur, Drevets, & Price, 1998; Rajkowska, 2003; Rajkowska et al., 1999; Rajkowska, Halaris, Selemon, 2001). Studies showed that the number of astrocytes is reduced in patients suffering from MD (Johnston-Wilson et al., 2000; Miguel-Hidalgo et al., 2000; Si, Miguel-Hidalgo, O'Dwyer, Stockmeier, & Rajkowska, 2004), although the data are not fully consistent (Davis et al., 2002).

CNS INFLAMMATION IN MD: FINDINGS FROM POSITRON EMISSION TOMOGRAPHY SCANS

A study on the translocator protein density measured by positron emission tomography as marker for the activation of microglia, which was interpreted as a measure of neuroinflammation in MD patients, showed a very interesting outcome. The authors observed with this in vivo measurement an increase of activated microglia in 20 patients during an episode of MD as compared to 20 healthy controls. Patients with the MD episode were medication-free for at least 6 weeks. All participants were otherwise healthy and nonsmokers. The significant elevation of microglia activation was preferably found in the prefrontal cortex, the nucleus' accumbens (ACC), and in the insula, all being structures that are known to be involved in MD. The elevation was 26% in the prefrontal cortex in patients with MD, 32% in the ACC, and 33% in the insula. Moreover, a statistically significant correlation of the microglia activation with the severity of depression ($r = 0.63$; $p < 0.001$) in the ACC was described, i.e., the higher the translocator protein density, the more pronounced the severity of MD (Setiawan et al., 2015).

COX-2 INHIBITION AS AN EXAMPLE OF AN ANTI-INFLAMMATORY THERAPEUTIC APPROACH IN MD

COX-2 inhibitors influence the CNS serotonergic system, either directly or via CNS immune mechanisms. In a rat model, treatment with rofecoxib was followed by an increase of serotonin in the frontal and the temporoparietal cortex (Sandrini, Vitale, & Pini, 2002). Therefore COX-2 inhibitors would be expected to show a clinical antidepressant effect. In the depression animal model of the bulbectomized rat, a decrease in hypothalamic cytokine levels and a change in behavior have been observed after chronic celecoxib treatment

(Myint, Steinbusch, et al., 2007). In another animal model of depression, however, the mixed COX-1/COX-2 inhibitor acetylsalicylic acid showed an additional antidepressant effect by accelerating the antidepressant effect of fluoxetine (Brunello et al., 2006). A significant therapeutic effect of the COX-2 inhibitor celecoxib in MD was also found in a randomized, double-blind pilot add-on study of reboxetine and celecoxib versus reboxetine and placebo (Müller et al., 2006). Interestingly, the ratio of kynurenine to tryptophan, which represents the activity of the proinflammatory cytokine-driven enzyme IDO, predicted the antidepressant response to the celecoxib therapy. Patients with a high activity of IDO, i.e., a high proinflammatory activity, responded better to celecoxib. Another randomized, double-blind study in 50 depressed patients suffering from MD also showed a significantly better outcome of the COX-2 inhibitor celecoxib plus fluoxetine than with fluoxetine alone (Akhondzadeh et al., 2009). This finding was replicated using the combination of sertraline and celecoxib in 40 depressed patients (Abbasi, Hosseini, Modabbernia, Ashrafi, & Akhondzadeh, 2012). Interestingly, the blood levels of IL-6 predicted the antidepressant response both in the sertraline (plus placebo) and in the celecoxib (plus sertraline) groups.

A meta-analysis of the efficacy of adjunctive celecoxib treatment for patients with MD included 150 patients. The study concluded that adjunctive treatment with nonsteroidal anti-inflammatory drugs (NSAIDs), particularly celecoxib, can be a promising strategy for patients with depressive disorder. However, future studies with a larger sample size and longer study duration are needed to confirm the efficacy and tolerability of NSAIDs for depression (Na, Lee, Lee, Cho, & Jung, 2013).

ANTI-INFLAMMATORY COMPOUNDS OTHER THAN NSAIDs

The anti-TNF-α antibody infliximab, which blocks the interaction of TNF-α with cell surface receptors and was developed for the therapy of inflammatory joint disorders and psoriasis, showed a highly significant effect on symptoms of depression in psoriasis patients (Tyring et al., 2006). In a placebo-controlled add-on study using infliximab, however, an overall antidepressant effect could not be shown in a study of treatment-resistant depressed patients. Three infusions of infliximab or placebo were given in a 12-week trial ($n = 60$) in partly medication-free ($n = 23$) nonresponders to antidepressant therapy. No overall better outcome of infliximab versus placebo could be shown. There was, however, a significant interaction between treatment, time, and baseline CRP ($\leq$5 mg/L). Patients with a higher baseline CRP had a higher response rate to infliximab (62%) versus placebo (33%). Moreover, the baseline concentrations of TNF-α, sTNFR1 and sTNFR2 were significantly higher in infliximab responders ($p \leq 0.01$). Additionally, infliximab responders exhibited a significantly higher decrease of CRP ($p \leq 0.01$) than nonresponders (Raison et al., 2012).

Interestingly, there are also preliminary findings that angiotensin II AT1 receptor blockade has anti-inflammatory effects in the CNS and ameliorates stress, anxiety, and CNS inflammation (Benicky et al., 2011; Saavedra, Sanchez-Lemus, & Benicky, 2011).

A meta-analysis of 10 publications reporting on 14 trials (6262 participants) showed very interesting results of anti-inflammatory treatment in MD: 10 trials evaluated the use of NSAIDs ($n = 4258$) and 4 investigated cytokine inhibitors ($n = 2004$). The pooled effect estimate suggested that anti-inflammatory treatment reduced depressive symptoms (Standard Mean Differences (SMD), −0.34; 95% Confidence Interval (CI), −0.57 to −0.11; I2 = 90%) compared with placebo. This effect was observed in studies including patients with depression (SMD, −0.54; 95% CI, −1.08 to −0.01; I2 = 68%) and depressive symptoms (SMD, −0.27; 95% CI, −0.53 to −0.01; I2 = 68%). The heterogeneity of the studies was not explained by differences in inclusion of clinical depression versus depressive symptoms or the use of NSAIDs versus cytokine inhibitors. Subanalyses emphasized the antidepressant properties of the selective COX-2 inhibitor celecoxib (SMD, −0.29; 95% CI, −0.49 to −0.08; I2 = 73%) on remission (Odds Ratio (OR), 7.89; 95% CI, 2.94 to 21.17; I2 = 0%) and response (OR, 6.59; 95% CI, 2.24 to 19.42; I2 = 0%). Among the six studies reporting on adverse effects, no evidence was found of an increased number of gastrointestinal or cardiovascular events after 6 weeks or infections after 12 weeks of anti-inflammatory treatment as compared with placebo. All trials were associated with a high risk of bias owing to potentially compromised internal validity. The analysis suggests that anti-inflammatory treatment, in particular celecoxib, decreases depressive symptoms without increased risks of adverse effects. This study supports a proof of concept concerning the use of anti-inflammatory treatment in depression (Köhler et al., 2014).

INFLAMMATORY PATHOGENESIS IN MD AND SCHIZOPHRENIA: THE END OF THE KRAEPELINIAN DICHOTOMY?

The Kraepelinian dichotomy of schizophrenia and affective disorder has been discussed for many years. Several findings of modern biological psychiatry show an overlap of both disorders, and in particular, genetic researchers postulate new diagnostic classifications that have greater biological validity and possibly in the future will allow for selecting treatments based on underlying pathogenesis (Craddock & Owen, 2010). With respect to the mechanisms of inflammation, different patterns of type 1−type 2 immune activation seem associated with schizophrenia versus MD (Müller, 2014). The large number of immune and inflammatory genes involved makes it plausible that various types of immune balance may be involved. Differences in immune activation may have different impacts on the IDO activation and the tryptophan-kynurenine metabolism, which may result in an increased production of kynurenic acid in schizophrenia and increased quinolinic acid levels in depression. Such differences may associate with an imbalance in the glutamatergic neurotransmission, contributing to an overweight of NMDA activity in depression. In schizophenia, however, NMDA antagonism results in reduced glutamatergic neurotransmission. Moreover, differential activation of microglia cells and astrocytes in schizophrenia and depression may be an additional immune-related mechanism contributing to different outcome of the tryptophan/kynurenine metabolism.

Although there is strong evidence for the view that the interactions of the immune system, IDO, serotonergic system, and glutamatergic neurotransmission play a key role in schizophrenia and depression, several gaps, e.g., the roles of genetics, disease course, sex, different psychopathological states, etc., have to be bridged by intense further research. Moreover, COX-2 inhibition is only one example of possible therapeutic approaches acting on these mechanisms. Also, the effects of COX-2 inhibition in the CNS as well as the different components of the inflammatory system, the kynurenine-metabolism, and the glutamatergic neurotransmission need careful further scientific evaluation.

Moreover, similar pathological influences may lead to different psychopathological states due to different localizations of the pathological processes in the CNS. Neuropathological and neuroimaging studies show that—albeit with a broad range of overlap—different brain regions are involved in schizophrenia (e.g., the hippocampus; Dutt et al., 2011; Radonic et al., 2011) and depression (e.g., the habenula; Savitz et al., 2011; Winter, Vollmayr, Djodari-Irani, Klein, & Sartorius, 2011).

Different final pathways may lead, despite some common final pathway, to different syndromes of schizophrenia and depression. Inflammation, including low-grade inflammation, is a general pathway of the body representing a response to a lot of different noxae and pathogens. Despite some overlap, which is apparent between pathophysiological mechanisms identified in schizophrenia and affective disorders, the differential mechanisms are also of major interest. Similar phenomena, overlap, and some specificity are found in medicine in general, not only in psychiatry. To develop preventive strategies in high-risk individuals and in patients (Müller & Schwarz, 2008) such refined understanding is of outstanding interest.

ACKNOWLEDGMENTS

Parts of this chapter have been published before. The work was supported by the Foundation Immunität und Seele.

REFERENCES

Abbasi, S. H., Hosseini, F., Modabbernia, A., Ashrafi, M., & Akhondzadeh, S. (2012). Effect of celecoxib add-on treatment on symptoms and serum IL-6 concentrations in patients with major depressive disorder: randomized double-blind placebo-controlled study. *Journal of Affective Disorders, 141*(2–3), 308–314.

Akhondzadeh, S., Jafari, S., Raisi, F., Nasehi, A. A., Ghoreishi, A., Salehi, B., ... Kamalipour, A. (2009). Clinical trial of adjunctive celecoxib treatment in patients with major depression: a double blind and placebo controlled trial. *Depression and Anxiety, 26*(7), 607–611.

Alberati, G. D., Ricciardi, C. P., Kohler, C., & Cesura, A. M. (1996). Regulation of the kynurenine metabolic pathway by interferon-gamma in murine cloned macrophages and microglial cells. *Journal of Neurochemistry, 66*, 996–1004.

Altamura, C. A., Mauri, M. C., Ferrara, A., Moro, A. R., D'Andrea, G., & Zamberlan, F. (1993). Plasma and platelet excitatory amino acids in psychiatric disorders. *American Journal of Psychiatry, 150*, 1731–1733.

Anisman, H., & Merali, Z. (1999). Anhedonic and anxiogenic effects of cytokine exposure. *Advances in Experimental Medicine and Biology, 461*, 199–233.

Anisman, H., & Merali, Z. (2003). Cytokines, stress and depressive illness: brain-immune interactions. *Annals of Medicine, 35*, 2–11.

Aronica, E., Gorter, J. A., Ijlst-Keizers, H., Rozemuller, A. J., Yankaya, B., Leenstra, S., & Troost, D. (2003). Expression and functional role of mGluR3 and mGluR5 in human astrocytes and glioma cells: opposite regulation of glutamate transporter proteins. *European Journal of Neuroscience, 17*, 2106–2118.

Asnis, G. M., De La, G. R., Kohn, S. R., Reinus, J. F., Henderson, M., & Shah, J. (2003). IFN-induced depression: a role for NSAIDs. *Psychopharmacology Bulletin, 37*, 29–50.

Bacher, M., Meinhardt, A., Lan, H. Y., Dhabhar, F. S., Mu, W., Metz, C. N., ... Bucala, R. (1998). MIF expression in the rat brain: implications for neuronal function. *Molecular Medicine, 4*, 217–230.

Bechter, K., Reiber, H., Herzog, S., Fuchs, D., Tumani, H., & Maxeiner, H. G. (2010). Cerebrospinal fluid analysis in affective and schizophrenic spectrum disorders: identification of subgroups with immune responses and blood-CSF barrier dysfunction. *Journal of Psychiatric Research, 44*, 321–330.

Benicky, J., Sanchez-Lemus, E., Honda, M., Pang, T., Orecna, M., Wang, J., ... Saavedra, J. M. (2011). Angiotensin II AT(1) receptor blockade ameliorates brain inflammation. *Neuropsychopharmacology, 36*, 857–870.

Benros, M. E., Waltoft, B. L., Nordentoft, M., Ostergaard, S. D., Eaton, W. W., Krogh, J., & Mortensen, P. B. (2013). Autoimmune diseases and severe infections as risk factors for mood disorders: a nationwide study. *JAMA Psychiatry, 70*, 812–820.

Berkenbosch, F., van Oers, J., del Rey, A., Tilders, F., & Besedovsky, H. (1987). Corticotropin-releasing factor-producing neurons in the rat activated by interleukin-1. *Science, 238*, 524–526.

Berk, M., Wadee, A. A., Kuschke, R. H., & O'Neill-Kerr, A. (1997). Acute phase proteins in major depression. *Journal of Psychosomatic Research, 43*, 529–534.

Besedovsky, H., del Rey, A., Sorkin, E., & Dinarello, C. A. (1986). Immunoregulatory feedback between interleukin-1 and glucocorticoid hormones. *Science, 233*, 652–654.

Besedovsky, H., del Rey, A., Sorkin, E., Da, P. M., Burri, R., & Honegger, C. (1983). The immune response evokes changes in brain noradrenergic neurons. *Science, 221*, 564–566.

Bonaccorso, S., Lin, A. H., Verkerk, R., Van Hunsel, F., Libbrecht, I., Scharpe, S., ... Maes, M. (1998). Immune markers in fibromyalgia: comparison with major depressed patients and normal volunteers. *Journal of Affective Disorders, 48*, 75–82.

Bonaccorso, S., Marino, V., Puzella, A., Pasquini, M., Biondi, M., Artini, M., ... Maes, M. (2002). Increased depressive ratings in patients with hepatitis C receiving interferon-alpha-based immunotherapy are related to interferon-alpha-induced changes in the serotonergic system. *Journal of Clinical Psychopharmacology, 22*, 86–90.

Brambilla, R., Bracchi-Ricard, V., Hu, W. H., Frydel, B., Bramwell, A., Karmally, S., ... Bethea, J. R. (2005). Inhibition of astroglial nuclear factor kappaB reduces inflammation and improves functional recovery after spinal cord injury. *Journal of Experimental Medicine, 202*, 145–156.

Brambilla, F., & Maggioni, M. (1998). Blood levels of cytokines in elderly patients with major depressive disorder. *Acta Psychiatrica Scandinavica, 97*, 309–313.

Braun, D., Longman, R. S., & Albert, M. L. (2005). A two-step induction of indoleamine 2,3 dioxygenase (IDO) activity during dendritic-cell maturation. *Blood, 106*, 2375–2381.

Brunello, N., Alboni, S., Capone, G., Benatti, C., Blom, J. M., Tascedda, F., ... Mendlewicz, J. (2006). Acetylsalicylic acid accelerates the antidepressant effect of fluoxetine in the chronic escape deficit model of depression. *International Clinical Psychopharmacology, 21*, 219–225.

Burke, H. M., Davis, M. C., Otte, C., & Mohr, D. C. (2005). Depression and cortisol responses to psychological stress: a meta-analysis. *Psychoneuroendocrinology, 30*, 846–856.

Calabrese, J. R., Skwerer, R. G., Barna, B., Gulledge, A. D., Valenzuela, R., Butkus, A., ... Krupp, N. E. (1986). Depression, immunocompetence, and prostaglandins of the E series. *Psychiatry Research, 17*, 41–47.

Campbell, S., & Macqueen, G. (2004). The role of the hippocampus in the pathophysiology of major depression. *Journal of Psychiatry and Neuroscience, 29*, 417–426.

Capuron, L., Neurauter, G., Musselman, D. L., Lawson, D. H., Nemeroff, C. B., Fuchs, D., & Miller, A. H. (2003). Interferon-alpha-induced changes in tryptophan metabolism: relationship to depression and paroxetine treatment. *Biological Psychiatry, 54*, 906–914.

Cizza, G., Eskandari, F., Coyle, M., Krishnamurthy, P., Wright, E. C., Mistry, S., & Csako, G. (2009). Plasma CRP levels in premenopausal women with major depression: a 12-month controlled study. *Hormone and Metabolic Research, 41*(8), 641–648.

Coppen, A., & Swade, C. (1988). 5-HT and depression: the present position. In M. Briley, & G. Fillion (Eds.), *New concepts in depression. Pierre fabre monograph series* (pp. 120–136). London: MacMillan Press.

Cotter, D., Pariante, C., & Rajkowska, G. (2002). Glial pathology in major psychiatric disorders. In G. Agam, R. H. Belmaker, & I. Everall (Eds.), *The post-mortem brain in psychiatric research* (pp. 291–324). Boston: Kluwer Academic Publishers.

Craddock, N., & Owen, M. J. (2010). The Kraepelinian dichotomy - going, going... but still not gone. *British Journal of Psychiatry, 196*, 92–95.

Crane, G. (1959). Cyloserine as an antidepressant agent. *American Journal of Psychiatry, 115*, 1025–1026.

Danese, A., Moffitt, T. E., Pariante, C. M., Ambler, A., Poulton, R., & Caspi, A. (2008). Elevated inflammation levels in depressed adults with a history of childhood maltreatment. *Archives of General Psychiatry, 65*, 409–415.

Danner, M., Kasl, S. V., Abramson, J. L., & Vaccarino, V. (2003). Association between depression and elevated C-reactive protein. *Psychosomatic Medicine, 65*, 347–356.

Dantzer, R. (2001). Cytokine-induced sickness behavior: where do we stand? *Brain Behavior and Immunity, 15*, 7–24.

Dantzer, R., Bluthe, R. M., Castanon, N., Kelly, K. W., Konsman, J. P., Laye, S., ... Parnet, P. (2007). Cytokines, sickness behaviior, and depression. In R. Ader (Ed.), *Psychoneuroimmunology* (pp. 281–318). New York: Elsevier.

Dantzer, R., O'Connor, J. C., Freund, G. G., Johnson, R. W., & Kelley, K. W. (2008). From inflammation to sickness and depression: when the immune system subjugates the brain. *Nature Reviews Neuroscience, 9*, 46–56.

Davis, S., Thomas, A., Perry, R., Oakley, A., Kalaria, R. N., & O'Brien, J. T. (2002). Glial fibrillary acidic protein in late life major depressive disorder: an immunocytochemical study. *Journal of Neurology, Neurosurgery and Psychiatry, 73*, 556–560.

Day, H. E., Curran, E. J., Watson, S. J., Jr., & Akil, H. (1999). Distinct neurochemical populations in the rat central nucleus of the amygdala and bed nucleus of the stria terminalis: evidence for their selective activation by interleukin-1beta. *Journal of Comparative Neurology, 413*, 113–128.

Dowlati, Y., Herrmann, N., Swardfager, W., Liu, H., Sham, L., Reim, E. K., & Lanctot, K. L. (2010). A meta-analysis of cytokines in major depression. *Biological Psychiatry, 67*, 446–457.

Duch, D. S., Woolf, J. H., Nichol, C. A., Davidson, J. R., & Garbutt, J. C. (1984). Urinary excretion of biopterin and neopterin in psychiatric disorders. *Psychiatry Research, 11*, 83–89.

Dunbar, P. R., Hill, J., Neale, T. J., & Mellsop, G. W. (1992). Neopterin measurement provides evidence of altered cell-mediated immunity in patients with depression, but not with schizophrenia. *Psychologie Medicale, 22*, 1051–1057.

Dutt, A., Ganguly, T., Shaikh, M., Walshe, M., Schulze, K., Marshall, N., ... Bramon, E. (2011). Association between hippocampal volume and P300 event related potential in psychosis: support for the Kraepelinian divide. *Neuroimage, 59*(2), 997–1003.

Ershler, W. B., Sun, W. H., Binkley, N., Gravenstein, S., Volk, M. J., Kamoske, G., ... Weindruch, R. (1993). Interleukin-6 and aging: blood levels and mononuclear cell production increase with advancing age and in vitro production is modifiable by dietary restriction. *Lymphokine and Cytokine Research, 12*, 225–230.

Farina, C., Aloisi, F., & Meinl, E. (2007). Astrocytes are active players in cerebral innate immunity. *Trends in Immunology, 28*, 138–145.

Ford, D. E., & Erlinger, T. P. (2004). Depression and C-reactive protein in US adults: data from the Third National Health and Nutrition Examination Survey. *Archives of Internal Medicine, 164*, 1010–1014.

Frank, M. G., Baratta, M. V., Sprunger, D. B., Watkins, L. R., & Maier, S. F. (2007). Microglia serve as a neuroimmune substrate for stress-induced potentiation of CNS pro-inflammatory cytokine responses. *Brain Behavior and Immunity, 21*, 47–59.

Frodl, T. S., Koutsouleris, N., Bottlender, R., Born, C., Jager, M., Scupin, I., ... Meisenzahl, E. M. (2008). Depression-related variation in brain morphology over 3 years: effects of stress? *Archives of General Psychiatry, 65*, 1156–1165.

Frodl, T., Meisenzahl, E. M., Zetzsche, T., Born, C., Groll, C., Jager, M., ... Moller, H. J. (2002). Hippocampal changes in patients with a first episode of major depression. *American Journal of Psychiatry, 159*, 1112–1118.

Frommberger, U. H., Bauer, J., Haselbauer, P., Fraulin, A., Riemann, D., & Berger, M. (1997). Interleukin-6-(IL-6) plasma levels in depression and schizophrenia: comparison between the acute state and after remission. *European Archives of Psychiatry and Clinical Neuroscience, 247*, 228–233.

Furukawa, H., del Rey, A., Monge-Arditi, G., & Besedovsky, H. O. (1998). Interleukin-1, but not stress, stimulates glucocorticoid output during early postnatal life in mice. *Annals of the New York Academy of Sciences, 840*, 117–122.

Gao, H. M., & Hong, J. S. (2008). Why neurodegenerative diseases are progressive: uncontrolled inflammation drives disease progression. *Trends in Immunology, 29*, 357–365.

Gimeno, D., Marmot, M. G., & Singh-Manoux, A. (2008). Inflammatory markers and cognitive function in middle-aged adults: the Whitehall II study. *Psychoneuroendocrinology, 33*, 1322–1334.

Glaser, R., & Kiecolt-Glaser, J. K. (1986). Stress and immune function. *Clinical Neuropharmacology, 9*(Suppl. 4), 485–487.

Haack, M., Hinze-Selch, D., Fenzel, T., Kraus, T., Kuhn, M., Schuld, A., & Pollmacher, T. (1999). Plasma levels of cytokines and soluble cytokine receptors in psychiatric patients upon hospital admission: effects of confounding factors and diagnosis. *Journal of Psychiatric Research, 33*, 407–418.

Häfner, S., Baghai, T. C., Eser, D., Schüle, C., Rupprecht, R., Bondy, B., ... von Schacky, C. (2008). C-reactive protein is associated with polymorphisms of the angiotensin-converting enzyme gene in major depressed patients. *Journal of Psychiatric Research, 42*, 163–165.

Hampel, H., Kotter, H. U., & Moller, H. J. (1997). Blood-cerebrospinal fluid barrier dysfunction for high molecular weight proteins in Alzheimer disease and major depression: indication for disease subsets. *Alzheimer Disease and Associated Disorders, 11*, 78–87.

Hampel, H., Kotter, H. U., Padberg, F., Korschenhausen, D. A., & Moller, H. J. (1999). Oligoclonal bands and blood–cerebrospinal-fluid barrier dysfunction in a subset of patients with Alzheimer disease: comparison with vascular dementia, major depression, and multiple sclerosis. *Alzheimer Disease and Associated Disorders, 13*, 9–19.

Hampel, H., Muller, S. F., Berger, C., Haberl, A., Ackenheil, M., & Hock, C. (1995). Evidence of blood-cerebrospinal fluid-barrier impairment in a subgroup of patients with dementia of the Alzheimer type and major depression: a possible indicator for immunoactivation. *Dementia, 6*, 348–354.

Hannestad, J., DellaGioia, N., & Bloch, M. (2011). The effect of antidepressant medication treatment on serum levels of inflammatory cytokines: a meta-analysis. *Neuropsychopharmacology, 36*, 2452–2459.

Hart, B. L. (1988). Biological basis of the behavior of sick animals. *Neuroscience and Biobehavioral Reviews, 12*, 123–137.

Hasler, G., Drevets, W. C., Manji, H. K., & Charney, D. S. (2004). Discovering endophenotypes for major depression. *Neuropsychopharmacology, 29*, 1765–1781.

Hayley, S., Wall, P., & Anisman, H. (2002). Sensitization to the neuroendocrine, central monoamine and behavioural effects of murine tumor necrosis factor-alpha: peripheral and central mechanisms. *European Journal of Neuroscience, 15*, 1061–1076.

Herbert, T. B., & Cohen, S. (1993). Depression and immunity: a meta-analytic review. *Psychological Bulletin, 113*, 472–486.

Howren, M. B., Lamkin, D. M., & Suls, J. (2009). Associations of depression with C-reactive protein, IL-1, and IL-6: a meta-analysis. *Psychosomatic Medicine, 71*, 171–186.

Huber, T. J., Dietrich, D. E., & Emrich, H. M. (1999). Possible use of amantadine in depression. *Pharmacopsychiatry, 32*, 47–55.

Irwin, M. (1999). Immune correlates of depression. *Advances in Experimental Medicine and Biology, 461*, 1–24.

Jemmott, J. B., 3rd, Borysenko, J. Z., Borysenko, M., McClelland, D. C., Chapman, R., Meyer, D., & Benson, H. (1983). Academic stress, power motivation, and decrease in secretion rate of salivary secretory immunoglobulin A. *Lancet, 1*, 1400–1402.

Johnston-Wilson, N. L., Sims, C. D., Hofmann, J.-P., Anderson, L., Shore, A. D., Torrey, E. F., & Yolken, R. H. (2000). Disease-specific alterations in frontal cortex brain proteins in schizophrenia, bipolar disorder, and major depressive disorder. *Molecular Psychiatry, 5*, 142–149.

Katila, H., Appelberg, B., Hurme, M., & Rimon, R. (1994). Plasma levels of interleukin-1 beta and interleukin-6 in schizophrenia, other psychoses, and affective disorders. *Schizophrenia Research, 12*, 29–34.

Kim, J. S., Schmid-Burgk, W., Claus, D., & Kornhuber, H. H. (1982). Increased serum glutamate in depressed patients. *Archiv fuer Psychiatrie und Nervenkrankheiten, 232*, 299–304.

Kling, M. A., Alesci, S., Csako, G., Costello, R., Luckenbaugh, D. A., Bonne, O., ... Neumeister, A. (2007). Sustained low-grade pro-inflammatory state in unmedicated, remitted women with major depressive disorder as evidenced by elevated serum levels of the acute phase proteins C-reactive protein and serum amyloid A. *Biological Psychiatry, 62*, 309–313.

Köhler, O., Benros, M. E., Nordentoft, M., Farkouh, M. E., Iyengar, R. L., Mors, O., & Krogh, J. (2014). Effect of anti-inflammatory treatment on depression, depressive symptoms, and adverse effects: a systematic review and meta-analysis of randomized clinical trials. *JAMA Psychiatry, 71*, 1381–1391.

Kuehne, L. K., Reiber, H., Bechter, K., Hagberg, L., & Fuchs, D. (October 2013). Cerebrospinal fluid neopterin is brain-derived and not associated with blood CSF barrier dysfunction. *Acta Neurologica Scandinavica, 47*(10), 1417–1422.

Kugaya, A., & Sanacora, G. (2005). Beyond monoamines: glutamatergic function in mood disorders. *CNS Spectrums, 10*, 808–819.

Kugler, J., Reintjes, F., Tewes, V., & Schedlowski, M. (1996). Competition stress in soccer coaches increases salivary. Immunoglobulin A and salivary cortisol concentrations. *Journal of Sports Medicine and Physical Fitness, 36*, 117–120.

Kwidzinski, E., Bunse, J., Aktas, O., Richter, D., Mutlu, L., Zipp, F., ... Bechmann, I. (2005). Indolamine 2,3-dioxygenase is expressed in the CNS and down-regulates autoimmune inflammation. *FASEB Journal, 19*, 1347–1349.

Lanquillon, S., Krieg, J. C., Bening-Abu-Shach, U., & Vedder, H. (2000). Cytokine production and treatment response in major depressive disorder. *Neuropsychopharmacology, 22*, 370–379.

Laudenslager, M. L., Fleshner, M., Hofstadter, P., Held, P. E., Simons, L., & Maier, S. F. (1988). Suppression of specific antibody production by inescapable shock: stability under varying conditions. *Brain Behavior and Immunity, 2*, 92–101.

Leonard, B. E., & Myint, A. (2006). Inflammation and depression: is there a causal connection with dementia? *Neurotoxicity Research, 10*, 149–160.

Lidberg, L., Belfrage, H., Bertilsson, L., Evenden, M. M., & Asberg, M. (2000). Suicide attempts and impulse control disorder are related to low cerebrospinal fluid 5-HIAA in mentally disordered violent offenders. *Acta Psychiatrica Scandinavica, 101*, 395–402.

Linnoila, M., Whorton, A. R., Rubinow, D. R., Cowdry, R. W., Ninan, P. T., & Waters, R. N. (1983). CSF prostaglandin levels in depressed and schizophrenic patients. *Archives of General Psychiatry, 40*, 405–406.

Linthorst, A. C., Flachskamm, C., Muller Preuss, P., Holsboer, F., & Reul, J. M. (1995). Effect of bacterial endotoxin and interleukin-1 beta on hippocampal serotonergic neurotransmission, behavioral activity, and free corticosterone levels: an in vivo microdialysis study. *Journal of Neuroscience, 15*, 2920–2934.

Liu, Y., Ho, R. C., & Mak, A. (2012). Interleukin (IL)-6, tumour necrosis factor alpha (TNF-alpha) and soluble interleukin-2 receptors (sIL-2R) are elevated in patients with major depressive disorder: a meta-analysis and meta-regression. *Journal of Affective Disorders, 139*, 230–239.

Madrigal, J. L., Garcia-Bueno, B., Moro, M. A., Lizasoain, I., Lorenzo, P., & Leza, J. C. (2003). Relationship between cyclooxygenase-2 and nitric oxide synthase-2 in rat cortex after stress. *European Journal of Neuroscience, 18*, 1701–1705.

Maes, M. (1994). Cytokines in major depression [letter; comment]. *Biological Psychiatry, 36*, 498–499.

Maes, M. (April 29, 2011). Depression is an inflammatory disease, but cell-mediated immune activation is the key component of depression. *Progress in Neuropsychopharmacology and Biological Psychiatry, 35*(3), 664–675.

Maes, M., Bosmans, E., De Jongh, R., Kenis, G., Vandoolaeghe, E., & Neels, H. (1997). Increased serum IL-6 and IL-1 receptor antagonist concentrations in major depression and treatment resistant depression. *Cytokine, 9*, 853–858.

Maes, M., Meltzer, H. Y., Bosmans, E., Bergmans, R., Vandoolaeghe, E., Ranjan, R., & Desnyder, R. (1995). Increased plasma concentrations of interleukin-6, soluble interleukin-6, soluble interleukin-2 and transferrin receptor in major depression. *Journal of Affective Disorders, 34*, 301–309.

Maes, M., Meltzer, H. Y., Buckley, P., & Bosmans, E. (1995). Plasma-soluble interleukin-2 and transferrin receptor in schizophrenia and major depression. *European Archives of Psychiatry and Clinical Neuroscience, 244*, 325–329.

Maes, M., Scharpe, S., Meltzer, H. Y., Bosmans, E., Suy, E., Calabrese, J., & Cosyns, P. (1993). Relationships between interleukin-6 activity, acute phase proteins, and function of the hypothalamic-pituitary-adrenal axis in severe depression. *Psychiatry Research, 49*, 11–27.

Maes, M., Scharpe, S., Meltzer, H. Y., Okayli, G., Bosmans, E., D'Hondt, P., ... Cosyns, P. (1994). Increased neopterin and interferon-gamma secretion and lower availability of L-tryptophan in major depression: further evidence for an immune response. *Psychiatry Research, 54*, 143–160.

Maes, M., Stevens, W., DeClerck, L., Bridts, C., Peeters, D., Schotte, C., & Cosyns, P. (1992). Immune disorders in depression: higher T helper/T suppressor-cytotoxic cell ratio. *Acta Psychiatrica Scandinavica, 86*, 423–431.

Maes, M., Verkerk, R., Vandoolaeghe, E., Lin, A., & Scharpe, S. (1998). Serum levels of excitatory amino acids, serine, glycine, histidine, threonine, taurine, alanine and arginine in treatment-resistant depression: modulation by treatment with antidepressants and prediction of clinical responsivity. *Acta Psychiatrica Scandinavica, 97*, 302–308.

Maj, J., Rogoz, Z., Skuza, G., & Sowinska, H. (1992). Effects of MK-801 and antidepressant drugs in the forced swimming test in rats. *European Neuropsychopharmacology, 2*, 37–41.

Mann, J. J., & Malone, K. M. (1997). Cerebrospinal fluid amines and higher-lethality suicide attempts in depressed inpatients. *Biological Psychiatry, 41*, 162–171.

Matussek, N. (1966). Neurobiologie und depression. *Medlzinische Monatsschrift, 3*, 109–112.

Matussek, N. (1988). *Catecholamines and mood: Neuroendocrine aspects; Current topics in neuroendocrinology* (pp. 141–181). Heidelberg, New York.

Mauri, M. C., Ferrara, A., Boscati, L., Bravin, S., Zamberlan, F., Alecci, M., & Invernizzi, G. (1998). Plasma and platelet amino acid concentrations in patients affected by major depression and under fluvoxamine treatment. *Neuropsychobiology, 37*, 124–129.

Maxeiner, H. G., Rojewski, M. T., Schmitt, A., Tumani, H., Bechter, K., & Schmitt, M. (2009). Flow cytometric analysis of T cell subsets in paired samples of cerebrospinal fluid and peripheral blood from patients with neurological and psychiatric disorders. *Brain Behavior and Immunity, 23*, 134–142.

Mellor, A. L., & Munn, D. H. (1999). Tryptophan catabolism and T-cell tolerance: immunosuppression by starvation? *Immunology Today, 20*, 469–473.

Merali, Z., Lacosta, S., & Anisman, H. (1997). Effects of interleukin-1beta and mild stress on alterations of norepinephrine, dopamine and serotonin neurotransmission: a regional microdialysis study. *Brain Research, 761*, 225–235.

Meyer, U., Feldon, J., & Fatemi, S. H. (2009). In-vivo rodent models for the experimental investigation of prenatal immune activation effects in neurodevelopmental brain disorders. *Neuroscience and Biobehavioral Reviews, 33*, 1061–1079.

Miguel-Hidalgo, J. J., Baucom, C., Dilley, G., Overholser, J. C., Meltzer, H. Y., Stockmeier, C. A., & Rajkowska, G. (2000). Glial fibrillary acidic protein immunoreactivity in the prefrontal cortex distinguishes younger from older adults in major depressive disorder. *Biological Psychiatry, 48*, 861–873.

Mikova, O., Yakimova, R., Bosmans, E., Kenis, G., & Maes, M. (2001). Increased serum tumor necrosis factor alpha concentrations in major depression and multiple sclerosis. *European Neuropsychopharmacology, 11*, 203–208.

Moghaddam, B., Bolinao, M. L., Stein-Behrens, B., & Sapolsky, R. (1994). Glucocorticoids mediate the stress-induced extracellular accumulation of glutamate. *Brain Research, 655*, 251–254.

Müller, N. (2014). Immunology of schizophrenia. *Neuroimmunomodulation, 21*, 109–116.

Müller, N., Hofschuster, E., Ackenheil, M., Mempel, W., & Eckstein, R. (1993). Investigations of the cellular immunity during depression and the free interval: evidence for an immune activation in affective psychosis. *Progress in Neuropsychopharmacology and Biological Psychiatry, 17*, 713–730.

Müller, N., Myint, A. M., & Schwarz, M. J. (2011). Inflammatory biomarkers and depression. *Neurotoxicity Research, 19*, 308–318.

Müller, N., & Schwarz, M. J. (2002). Immunology in anxiety and depression. In S. Kasper, J. A. den Boer, & J. M. A. Sitsen (Eds.), *Handbook of depression and anxiety* (pp. 267–288). New York: Marcel Dekker.

Müller, N., & Schwarz, M. J. (2007). The immune-mediated alteration of serotonin and glutamate: towards an integrated view of depression. *Molecular Psychiatry*, 1–13.

Müller, N., & Schwarz, M. J. (2008). A psychoneuroimmunological perspective to Emil Kraepelins dichotomy: schizophrenia and major depression as inflammatory CNS disorders. *European Archives of Psychiatry and Clinical Neuroscience, 258*(Suppl. 2), 97–106.

Müller, N., Schwarz, M. J., Dehning, S., Douhet, A., Cerovecki, A., Goldstein-Müller, B., ... Riedel, M. (2006). The cyclooxygenase-2 inhibitor celecoxib has therapeutic effects in major depression: results of a double-blind, randomized, placebo controlled, add-on pilot study to reboxetine. *Molecular Psychiatry, 11*, 680–684.

Müller, N., Schwarz, M. J., & Riedel, M. (2005). COX-2 inhibition in schizophrenia: focus on clinical effects of celecoxib therapy and the role of TNF-alpha. In W. W. Eaton (Ed.), *Medical and psychiatric comorbidity over the course of life* (pp. 265–276). Washington DC: American Psychiatric Publishing.

Munn, D. H., Shafizadeh, E., Attwood, J. T., Bondarev, I., Pashine, A., & Mellor, A. L. (1999). Inhibition of T cell proliferation by macrophage tryptophan catabolism. *Journal of Experimental Medicine, 189*, 1363–1372.

Myint, A. M., & Kim, Y. K. (2003). Cytokine-serotonin interaction through IDO: a neurodegeneration hypothesis of depression. *Medical Hypotheses, 61*, 519–525.

Myint, A. M., Kim, Y. K., Verkerk, R., Scharpe, S., Steinbusch, H., & Leonard, B. (2007). Kynurenine pathway in major depression: evidence of impaired neuroprotection. *Journal of Affective Disorders, 98*, 143–151.

Myint, A. M., Leonard, B. E., Steinbusch, H. W., & Kim, Y. K. (2005). Th1, Th2, and Th3 cytokine alterations in major depression. *Journal of Affective Disorders, 88*, 167–173.

Myint, A. M., Steinbusch, H. W., Goeghegan, L., Luchtman, D., Kim, Y. K., & Leonard, B. E. (2007). Effect of the COX-2 inhibitor celecoxib on behavioural and immune changes in an olfactory bulbectomised rat model of depression. *Neuroimmunomodulation, 14*, 65–71.

Nair, A., & Bonneau, R. H. (2006). Stress-induced elevation of glucocorticoids increases microglia proliferation through NMDA receptor activation. *Journal of Neuroimmunology, 171*, 72–85.

Na, K. S., Lee, K. J., Lee, J. S., Cho, Y. S., & Jung, H. Y. (2013). Efficacy of adjunctive celecoxib treatment for patients with major depressive disorder: a meta-analysis. *Progress in Neuro-psychopharmacology and Biological Psychiatry, 48C*, 79–85. http://dx.doi.org/10.1016/j.pnpbp.2013.09.006.

Nguyen, K. T., Deak, T., Owens, S. M., Kohno, T., Fleshner, M., Watkins, L. R., & Maier, S. F. (1998). Exposure to acute stress induces brain interleukin-1beta protein in the rat. *Journal of Neuroscience, 18*, 2239–2246.

Niino, M., Ogata, A., Kikuchi, S., Tashiro, K., & Nishihira, J. (2000). Macrophage migration inhibitory factor in the cerebrospinal fluid of patients with conventional and optic-spinal forms of multiple sclerosis and neuro-Behcet's disease. *Journal of Neurological Sciences, 179*, 127–131.

Nishino, S., Ueno, R., Ohishi, K., Sakai, T., & Hayaishi, O. (1989). Salivary prostaglandin concentrations: possible state indicators for major depression. *American Journal of Psychiatry, 146*, 365–368.

Nordstrom, P., Samuelsson, M., Asberg, M., Traskman, B. L., Aberg, W. A., Nordin, C., & Bertilsson, L. (1994). CSF 5-HIAA predicts suicide risk after attempted suicide. *Suicide and Life-Threatening Behavior, 24*, 1–9.

Nunes, S. O. V., Reiche, E. M. V., Morimoto, H. K., Matsuo, T., Itano, E. N., Xavier, E. C. D., … Kaminami, M. S. (2002). Immune and hormonal activity in adults suffering from depression. *Brazilian Journal of Medical and Biological Research, 35*, 581–587.

O'Brien, S. M., Scott, L. V., & Dinan, T. G. (2004). Cytokines: abnormalities in major depression and implications for pharmacological treatment. *Human Psychopharmacology, 19*, 397–403.

Ohishi, K., Ueno, R., Nishino, S., Sakai, T., & Hayaishi, O. (1988). Increased level of salivary prostaglandins in patients with major depression. *Biological Psychiatry, 23*, 326–334.

Olsson, S. K., Andersson, A. S., Linderholm, K. R., Holtze, M., Nilsson-Todd, L. K., Schwieler, L., … Erhardt, S. (2009). Elevated levels of kynurenic acid change the dopaminergic response to amphetamine: implications for schizophrenia. *International Journal of Neuropsychopharmacology, 12*, 501–512.

Ongur, D., Drevets, W. C., & Price, J. L. (1998). Glial reduction in the subgenual prefrontal cortex in mood disorders. *Proceedings of the National Academy of Sciences of the United States of America, 95*, 13290–13295.

Ossowska, G., Klenk-Majewska, B., & Szymczyk, G. (1997). The effect of NMDA antagonists on footshock-induced fighting behavior in chronically stressed rats. *Journal of Physiology and Pharmacology, 48*, 127–135.

Pace, T. W., Mletzko, T. C., Alagbe, O., Musselman, D. L., Nemeroff, C. B., Miller, A. H., & Heim, C. M. (2006). Increased stress-induced inflammatory responses in male patients with major depression and increased early life stress. *American Journal of Psychiatry, 163*, 1630–1633.

Plata-Salaman, C. R. (1991). Immunoregulators in the nervous system. *Neuroscience and Biobehavioral Reviews, 15*, 185–215.

Pugh, C. R., Nguyen, K. T., Gonyea, J. L., Fleshner, M., Wakins, L. R., Maier, S. F., & Rudy, J. W. (1999). Role of interleukin-1 beta in impairment of contextual fear conditioning caused by social isolation. *Behavioural Brain Research, 106*, 109–118.

Radonic, E., Rados, M., Kalember, P., Bajs-Janovic, M., Folnegovic-Smalc, V., & Henigsberg, N. (2011). Comparison of hippocampal volumes in schizophrenia, schizoaffective and bipolar disorder. *Collegium Antropologicum, 35*(Suppl. 1), 249–252.

Raison, C. L., Capuron, L., & Miller, A. H. (2006). Cytokines sing the blues: inflammation and the pathogenesis of depression. *Trends in Immunology, 27*, 24–31.

Raison, C. L., Rutherford, R. E., Woolwine, B. J., Shuo, C., Schettler, P., Drake, D. F., ... Miller, A. H. (2012). A randomized controlled trial of the tumor necrosis factor antagonist infliximab for treatment-resistant depression: the role of baseline inflammatory biomarkers. *Archives of General Psychiatry*, 1–11.

Rajkowska, G. (2003). Depression: what we can learn from postmortem studies. *Neuroscientist, 9*, 273–284.

Rajkowska, G., Halaris, A., & Selemon, L. D. (2001). Reductions in neuronal and glial density characterize the dorsolateral prefrontal cortex in bipolar disorder. *Biological Psychiatry, 49*, 741–752.

Rajkowska, G., Miguel-Hidalgo, J. J., Wei, J., Dilley, G., Pittman, S. D., Meltzer, H. Y., ... Stockmeier, C. A. (1999). Morphometric evidence for neuronal and glial ;prefrontal cell pathology in major depression. *Biological Psychiatry, 45*, 1085–1098.

Reichenberg, A., Kraus, T., Haack, M., Schuld, A., Pollmacher, T., & Yirmiya, R. (2002). Endotoxin-induced changes in food consumption in healthy volunteers are associated with TNF-alpha and IL-6 secretion. *Psychoneuroendocrinology, 27*, 945–956.

Reichenberg, A., Yirmiya, R., Schuld, A., Kraus, T., Haack, M., Morag, A., & Pollmacher, T. (2001). Cytokine-associated emotional and cognitive disturbances in humans. *Archives of General Psychiatry, 58*, 445–452.

Robinson, C. M., Hale, P. T., & Carlin, J. M. (2005). The role of IFN-gamma and TNF-alpha-responsive regulatory elements in the synergistic induction of indoleamine dioxygenase. *Journal of Interferon and Cytokine Research, 25*, 20–30.

Rothermundt, M., Arolt, V., Fenker, J., Gutbrodt, H., Peters, M., & Kirchner, H. (2001). Different immune patterns in melancholic and non-melancholic major depression. *European Archives of Psychiatry and Clinical Neuroscience, 251*, 90–97.

Rothermundt, M., Arolt, V., Peters, M., Gutbrodt, H., Fenker, J., Kersting, A., & Kirchner, H. (2001). Inflammatory markers in major depression and melancholia. *Journal of Affective Disorders, 63*, 93–102.

Roy, A., Pickar, D., Paul, S., Doran, A., Chrousos, G. P., & Gold, P. W. (1987). CSF corticotropin-releasing hormone in depressed patients and normal control subjects. *American Journal of Psychiatry, 144*, 641–645.

Saavedra, J. M., Sanchez-Lemus, E., & Benicky, J. (2011). Blockade of brain angiotensin II AT1 receptors ameliorates stress, anxiety, brain inflammation and ischemia: therapeutic implications. *Psychoneuroendocrinology, 36*, 1–18.

Sanacora, G., Gueorguieva, R., Epperson, C. N., Wu, Y. T., Appel, M., Rothman, D. L., ... Mason, G. F. (2004). Subtype-specific alterations of gamma-aminobutyric acid and glutamate in patients with major depression. *Archives of General Psychiatry, 61*, 705–713.

Sandrini, M., Vitale, G., & Pini, L. A. (2002). Effect of rofecoxib on nociception and the serotonin system in the rat brain. *Inflammation Research, 51*, 154–159.

Sapolsky, R. M. (1985). A mechanism for glucocorticoid toxicity in the hippocampus: increased neuronal vulnerability to metabolic insults. *Journal of Neuroscience, 5*, 1228–1232.

Savitz, J. B., Nugent, A. C., Bogers, W., Roiser, J. P., Bain, E. E., Neumeister, A., ... Drevets, W. C. (2011). Habenula volume in bipolar disorder and major depressive disorder: a high-resolution magnetic resonance imaging study. *Biological Psychiatry, 69*, 336–343.

Schäfer, M., Horn, M., Schmidt, F., Schmid-Wendtner, M. H., Volkenandt, M., Ackenheil, M., ... Schwarz, M. J. (2004). Correlation between sICAM-1 and depressive symptoms during adjuvant treatment of melanoma with interferon-alpha. *Brain Behavior and Immunity, 18*, 555–562.

Schedlowski, M., Benschop, R. J., & Schmidt, R. E. (1995). Psychological aspects of stress immunology. *Immunology Today, 16*, 165.

Schiepers, O. J., Wichers, M. C., & Maes, M. (2005). Cytokines and major depression. *Progress in Neuropsychopharmacology and Biological Psychiatry, 29*, 201–217.

Schwartz, M., & Shechter, R. (2010). Protective autoimmunity functions by intracranial immunosurveillance to support the mind: the missing link between health and disease. *Molecular Psychiatry, 15*, 342–354.

Schwarz, M. J., Chiang, S., Müller, N., & Ackenheil, M. (2001). T-helper-1 and T-helper-2 responses in psychiatric disorders. *Brain Behavior and Immunity, 15*, 340–370.

Seidel, A., Arolt, V., Hunstiger, M., Rink, L., Behnisch, A., & Kirchner, H. (1996). Major depressive disorder is associated with elevated monocyte counts. *Acta Psychiatrica Scandinavica, 94*, 198–204.

Setiawan, E., Wilson, A. A., Mizrahi, R., Rusjan, P. M., Miler, L., Rajkowska, G., ... Meyer, J. H. (2015). Role of translocator protein density, a marker of neuroinflammation, in the brain during major depressive episodes. *JAMA Psychiatry, 72*, 268–275.

Shintani, F., Nakaki, T., Kanba, S., Sato, K., Yagi, G., Shiozawa, M., ... Asai, M. (1995). Involvement of interleukin-1 in immobilization stress-induced increase in plasma adrenocorticotropic hormone and in release of hypothalamic monoamines in the rat. *Journal of Neuroscience, 15*, 1961–1970.

Si, X., Miguel-Hidalgo, J. J., O'Dwyer, G., Stockmeier, C. A., & Rajkowska, G. (2004). Age-dependent reductions in the level of glial fibrillary acidic protein in the prefrontal cortex in major depression. *Neuropsychopharmacology, 29*, 2088–2096.

Sluzewska, A., Rybakowski, J., Bosmans, E., Sobieska, M., Berghmans, R., Maes, M., & Wiktorowicz, K. (1996). Indicators of immune activation in major depression. *Psychiatry Research, 64*, 161–167.

Song, C., & Leonard, B. E. (2000). *Fundamentals of psychoneuroimmunology*. Chichester, New York: J Wiley and Sons.

Song, C., Lin, A., Bonaccorso, S., Heide, C., Verkerk, R., Kenis, G., ... Maes, M. (1998). The inflammatory response system and the availability of plasma tryptophan in patients with primary sleep disorders and major depression. *Journal of Affective Disorders, 49*, 211–219.

Song, C., Merali, Z., & Anisman, H. (1999). Variations of nucleus accumbens dopamine and serotonin following systemic interleukin-1, interleukin-2 or interleukin-6 treatment. *Neuroscience, 88*, 823–836.

Sparkman, N. L., & Johnson, R. W. (2008). Neuroinflammation associated with aging sensitizes the brain to the effects of infection or stress. *Neuroimmunomodulation, 15*, 323–330.

Stein-Behrens, B. A., Lin, W. J., & Sapolsky, R. M. (1994). Physiological elevations of glucocorticoids potentiate glutamate accumulation in the hippocampus. *Journal of Neurochemistry, 63*, 596–602.

Steiner, J., Walter, M., Gos, T., Guillemain, G. J., Benstein, H. G., Samyai, Z., ... Myin, A. M. (August 10, 2011). Severe Depression is associated with increased microglial quinolinic acid in subregions of the anterior cinculate gyrus: evidence for an immune-modulated glutamatergic neurotransmission? *Journal of Neuroinflammation, 8*, 94.

Stryjer, R., Strous, R. D., Shaked, G., Bar, F., Feldman, B., Kotler, M., ... Weizman, A. (2003). Amantadine as augmentation therapy in the management of treatment-resistant depression. *International Clinical Psychopharmacology, 18*, 93–96.

Sundar, S. K., Cierpial, M. A., Kilts, C., Ritchie, J. C., & Weiss, J. M. (1990). Brain IL-1-induced immunosuppression occurs through activation of both pituitary-adrenal axis and sympathetic nervous system by corticotropin-releasing factor. *Journal of Neuroscience, 10*, 3701–3706.

Suzuki, T., Ogata, A., Tashiro, K., Nagashima, K., Tamura, M., Yasui, K., & Nishihira, J. (2000). Japanese encephalitis virus up-regulates expression of macrophage migration inhibitory factor (MIF) mRNA in the mouse brain. *Biochimica et Biophysica Acta, 1517*, 100–106.

Takahashi, T., Kimoto, T., Tanabe, N., Hattori, T. A., Yasumatsu, N., & Kawato, S. (2002). Corticosterone acutely prolonged *N*-methyl-D-aspartate receptor-mediated Ca2+ elevation in cultured rat hippocampal neurons. *Journal of Neurochemistry, 83*, 1441–1451.

Trullas, R., & Skolnick, P. (1990). Functional antagonists at the NMDA receptor complex exhibit antidepressant actions. *European Journal of Pharmacology, 185*, 1–10.

Tyring, S., Gottlieb, A., Papp, K., Gordon, K., Leonardi, C., Wang, A., ... Krishnan, R. (2006). Etanercept and clinical outcomes, fatigue, and depression in psoriasis: double-blind placebo-controlled randomised phase III trial. *Lancet, 367*, 29–35.

Weiss, G., Murr, C., Zoller, H., Haun, M., Widner, B., Ludescher, C., & Fuchs, D. (1999). Modulation of neopterin formation and tryptophan degradation by Th1- and Th2-derived cytokines in human monocytic cells. *Clinical and Experimental Immunology, 116*, 435–440.

Weiss, J. M., Quan, N., & Sundar, S. K. (1994). Immunological consequences of interleukin-1 in the brain. *Neuropsychopharmacology, 10*, 833.

Wildemann, B., Oschmann, P., & Reiber, H. (2010). *Laboratory diagnosis in neurology*. Stuttgart: Thieme.

Winter, C., Vollmayr, B., Djodari-Irani, A., Klein, J., & Sartorius, A. (2011). Pharmacological inhibition of the lateral habenula improves depressive-like behavior in an animal model of treatment resistant depression. *Behavioural Brain Research, 216*, 463–465.

Woolley, C. S., Gould, E., & McEwen, B. S. (1990). Exposure to excess glucocorticoids alters dendritic morphology of adult hippocampal pyramidal neurons. *Brain Research, 531*, 225–231.

Yilmaz, A., Schulz, D., Aksoy, A., & Canbeyli, R. (2002). Prolonged effect of an anesthetic dose of ketamine on behavioral despair. *Pharmacology Biochemistry and Behavior, 71*, 341–344.

Zalcman, S., Green-Johnson, J. M., Murray, L., Nance, D. M., Dyck, D., Anisman, H., & Greenberg, A. H. (1994). Cytokine-specific central monoamine alterations induced by interleukin-1, -2 and -6. *Brain Research, 643*, 40–49.

Zhou, D., Kusnecov, A. W., Shurin, M. R., DePaoli, M., & Rabin, B. S. (1993). Exposure to physical and psychological stressors elevates plasma interleukin 6: relationship to the activation of hypothalamic-pituitary-adrenal axis. *Endocrinology, 133*, 2523–2530.

Zubin, J., & Spring, B. (1977). Vulnerability—a new view of schizophrenia. *Journal of Abnormal Psychology, 86*, 103–126.

Chapter 5

Molecular Mechanisms of Depression

Artemis Varidaki, Hasan Mohammad, Eleanor T. Coffey
Turku Centre for Biotechnology, Åbo Akademi University and the University of Turku, Turku, Finland

INTRODUCTION

Molecular events in the synapse determine whether appropriate neuronal connections will form and be maintained. When deregulated, these mechanisms can lead to maladaptive responses to stress. Thus, chronic stress exposure promotes the rundown of synaptic activity and loss of neuroendocrine control. Several proteins that regulate synapse formation and maintenance are implicated in depression, including LIMK1, Rac1, RhoA, spinophilin, and kalirin-7. Similarly, protein kinases, in particular mechanistic target of rapamycin (mTOR), which controls translation of brain-derived neurotrophic factor (BDNF) mRNA, and glycogen synthase kinase 3 (GSK-3), which inhibits adult hippocampal neurogenesis, are important downstream targets of the antidepressant ketamine. Considerable diversity has been identified among the signaling molecules that are linked with depression, which may explain the heterogeneity of the disorder. A great deal of information has been obtained regarding the molecular mechanisms of depressive behavior from studies in animals and humans. These key findings are described in the following sections.

REGULATION OF SYNAPTIC PLASTICITY DURING DEPRESSION

Several human-based studies have indicated that structural and functional changes in brain regions such as the amygdala, prefrontal cortex (PFC), and hippocampus contribute to the pathophysiology of depression and anxiety (Campbell, Marriott, Nahmias, & MacQueen, 2004; Frodl et al., 2007; Savitz & Drevets, 2009; Stockmeier et al., 2004; Veer et al., 2010) (see also Chapter 10). More specifically, dendritic atrophy, leading to the loss of synaptic connections

Systems Neuroscience in Depression. http://dx.doi.org/10.1016/B978-0-12-802456-0.00005-4

in regions controlling emotions, contributes to the development of mood disorders (Cook & Wellman, 2004; Magarinos & McEwen, 1995).

Dendritic Anomalies in Depression

It is extensively reported that early life stress increases the risk for developing depressive-like symptoms in adult life (Frodl, Reinhold, Koutsouleris, Reiser, & Meisenzahl, 2010; Heim & Binder, 2012; Heim, Plotsky, & Nemeroff, 2004). Childhood and adolescence is a critical period for new synapse formation and maturation. Dendritic branching undergoes continuous structural changes, and stabilization is dependent on synaptic activity (Cline, 2001; Niell, Meyer, & Smith, 2004; Vaughn, 1989). In rodent models of depression, chronic immobilization stress or administration of the stress hormone corticosterone triggers the activation of molecular pathways, leading to reduced dendrite arborization in specific subregions. This was prominent in apical dendrites of the CA3 region of the hippocampus and in the medial PFC (Goldwater et al., 2009; Radley et al., 2004). Significantly, this remodeling is *N*-methyl-D-aspartate (NMDA)-dependent, as administration of an NMDA receptor antagonist prevented dendritic reorganization in both areas (Christian, Miracle, Wellman, & Nakazawa, 2011; Martin & Wellman, 2011). In contrast, however, in the basolateral amygdala (BLA), dendrite branching and length are increased following chronic immobilization stress (Vyas, Mitra, Shankaranarayana Rao, & Chattarji, 2002). Accordingly, in this region, chronic stress increases synaptic activity, associated with increased fear and anxiety responses (Zhang & Rosenkranz, 2012). Consistent with these findings in the chronic immobilization stress model of depression, chronic unpredictable mild stress also induced dendrite atrophy in the hippocampus and PFC (Sharma & Thakur, 2015). These studies served to highlight regional differences in dendritic arborization that occur upon stress exposure in animal models of depression. They consistently show dendrite atrophy in the CA3 and PFC with increased spine density in the BLA, associated with increased aversive stimuli and eliciting of a fear response. These structural changes reflect adaptive responses to stress in central loci that are deregulated in disease (Table 1).

Spine Anomalies in Depression

Many studies have shown that the density and morphology of dendritic spines are altered in animal models of anxiety and depression. Spines are classified as mushroom, thin, and stubby, depending on their morphology (Rochefort & Konnerth, 2012). Newly generated spines are dynamic; however, as synaptic circuits mature, spine dynamics decrease, and most immature thin spines undergo experience-dependent elimination (Trachtenberg et al., 2002) or are converted into mushroom like spines that form stronger synaptic connections (Holtmaat et al., 2005). In rodent models of depression, reduced spine density

TABLE 1 Regional Changes in Activity and Expression of Proteins/mRNAs (as Indicated) in Brains from Animal Models of Depression and Subjects That Suffered from MDD; the Corresponding References Are Indicated

Synapto-Dendritic Alterations			Brain Region	References
Morphological changes	Dendritic arborization	Decreased	Hippocampus	Magarinos and McEwen (1995)
		Decreased	PFC	Goldwater et al. (2009), Kim et al. (2014) and Radley et al. (2004)
		Increased	BLA	Kim et al. (2014), Mitra and Sapolsky (2008) and Vyas et al. (2002)
	Spine density	Decreased	Hippocampus	Martinez-Tellez et al. (2009) and Norrholm and Ouimet (2001)
		Decreased	PFC	Radley et al. (2006, 2008, 2013)
		Decreased	NAc	Martinez-Tellez et al. (2009)
		Increased	BLA	Maroun et al. (2013), Mitra and Sapolsky (2008), Mitra et al. (2005) and Vyas et al. (2002)
Synaptic activity	Glutamatergic synapses	Decreased	Hippocampus	Hajszan et al. (2009)
		Decreased	PFC	Kang et al. (2012)
	GABA levels	Decreased	PFC	Hasler et al. (2007)
Neurotrophic factors	BDNF	Decreased	Hippocampus	Elizalde et al. (2010), Pizarro et al. (2004), Rasmusson et al. (2002) and Smith et al. (1995)
		Decreased	PFC	Dwivedi et al. (2003)
		Decreased	BLA	Pizarro et al. (2004)

Continued

TABLE 1 Regional Changes in Activity and Expression of Proteins/mRNAs (as Indicated) in Brains from Animal Models of Depression and Subjects That Suffered from MDD; the Corresponding References Are Indicated—cont'd

Synapto-Dendritic Alterations			Brain Region	References
Receptors	FZD6	Decreased	Hippocampus	Voleti et al. (2012)
	NMDA receptor subunits	Increased	Hippocampus	Law and Deakin, (2001), Martisova et al. (2012) and Ryan et al. (2009)
		Decreased	PFC	Beneyto and Meador-Woodruff (2008), Feyissa et al. (2009) and Yuen et al. (2012)
	AMPA receptor subunits	Increased	Hippocampus	Martisova et al. (2012)
		Decreased	PFC	Li et al. (2011) and Yuen et al. (2012)
Synaptic proteins	Kalirin-7 protein	Decreased	Hippocampus	Li et al. (2010) and Qiao et al. (2014)
	Spinophilin mRNA	Decreased	Hippocampus	Law et al. (2004) and Willard et al. (2014)
	PSD-95/GRIP1/PICK1/SAP97	Decreased	PFC	Jourdi et al. (2003), Kang et al. (2012) and Li et al. (2011)
Cytoskeletal regulatory proteins	RhoA	Increased	Hippocampus	Chen et al. (2013)
	Rac1 mRNA	Decreased	NAc	Golden et al. (2013)
	LIMK1 mRNA	Decreased	PFC	Nakatani et al. (2004)
	MAP2	Decreased	Hippocampus	Bianchi et al. (2006) and Soetanto et al. (2010)
		Decreased	PFC	Kang et al. (2012) and Nakatani et al. (2004)
	Dystrophin-related protein 2	Increased	Hippocampus	Mallei et al. (2011)
	Stathmin	Increased	PFC	Mallei et al. (2011)
	ARC	Decreased	Hippocampus	Elizalde et al. (2010) and Ying et al. (2002)
		Decreased	PFC	Elizalde et al. (2010)

Vesicle/endocytic proteins	VGLUT1	Decreased	Hippocampus	Elizalde et al. (2010) and Zink et al. (2010)
	Clathrin light chain a and b	Decreased	Hippocampus	Hu et al. (2013)
	Dynamin I	Decreased	Hippocampus	Hu et al. (2013)
	RAB3a	Decreased	Hippocampus	Thakker-Varia et al. (2001)
		Decreased	PFC	Kang et al. (2012)
	Synapsin I and II	Decreased	PFC	Elizalde et al. (2010), Kang et al. (2012) and Li et al. (2011)
Kinases/phosphatases/protein	CamKII β protein	Increased	Lateral habenula	Li et al. (2013)
		Decreased	Hippocampus	Li et al. (2013)
	ERK	Decreased	Hippocampus	Duric et al. (2010), Dwivedi et al. (2001) and First et al. (2011)
		Decreased	PFC	Yuan et al. (2010)
	AKT	Decreased	Hippocampus	Hsiung et al. (2003)
		Decreased	PFC	Karege et al. (2007)
		Decreased	VTA	Krishnan et al. (2008)
	MKP-1	Increased	Hippocampus	Duric et al. (2010)
		Decreased	PFC	Malki et al. (2015)
	SGK1 mRNA	Increased	Hippocampus	Anacker et al. (2013)
	mTOR/p70S6K	Decreased	PFC	Jernigan et al. (2011) and Ota et al. (2014)
	RAF	Decreased	Hippocampus	Dwivedi, Rizavi, Conley, and Pandey (2005)
		Decreased	PFC	Dwivedi et al. (2005)
	Disheveled	Decreased	NAc	Wilkinson et al. (2011)
Nuclear proteins	CREB	Decreased	PFC	Dwivedi et al. (2003)
	Gata1 mRNA	Decreased	PFC	Kang et al. (2012)
	Redd1 mRNA	Increased	PFC	Ota et al. (2014)

is observed in subregions of the PFC (Radley, Anderson, Hamilton, Alcock, & Romig-Martin, 2013; Radley et al., 2006, 2008) and hippocampus (Hajszan et al., 2009; Martinez-Tellez, Hernandez-Torres, Gamboa, & Flores, 2009; Norrholm & Ouimet, 2001), while dendritic arborization and spine density in the BLA are typically increased, leading to a hyperactive firing state (Kim et al., 2014; Maroun et al., 2013; Mitra, Jadhav, McEwen, Vyas, & Chattarji, 2005; Mitra & Sapolsky, 2008; Sharma & Thakur, 2015) (Table 1). This reflects a loss of inhibitory input to the amygdala. Specifically, the chronic variable stress model of depression-reduced mushroom spine density in neurons that projected from the medial PFC (prelimbic) to the GABAergic neurons in the bed nuclei of the stria terminalis (Radley et al., 2013). These neurons negatively regulate corticotrophin-releasing hormone (CRH) production, thereby controlling hypothalamic–pituitary–adrenal (HPA) axis activity (Radley, Gosselink, & Sawchenko, 2009). Thus, the consequence of chronic variable stress was synaptic atrophy in the medial PFC and increased HPA axis/CRH activity. This observation is consistent with a general decrease in GABA levels in humans diagnosed with depression (Hasler et al., 2007) as well as a reduction in the number of GABAergic neurons in the PFC (Rajkowska, O'Dwyer, Teleki, Stockmeier, & Miguel-Hidalgo, 2007; Sanacora et al., 2004). Interestingly, GABAergic neurons express CRH receptor 1 in dendritic spines. Deletion of this receptor (Refojo et al., 2011) or activity blockade restored spine density and reduced anxiety-like behavior (Andres et al., 2013; Chen, Dube, Rice, & Baram, 2008). In agreement with this, optogenetic stimulation of neurons in the medial PFC has an antidepressant effect in mice undergoing the chronic social defeat paradigm (Covington et al., 2010). These findings demonstrate how synaptic atrophy can lead to hypersecretion of CRH, which will in turn promote symptoms of melancholia such as heightened anxiety, activation of the locus coeruleus norepinephrine system, and hypercortisolism (Gold, 2015).

A large body of evidence indicates that antidepressants act by regulating synaptic plasticity, i.e., by remodeling synapses and dendrites. In particular, desipramine and its precursor imipramine, which belong to the tricyclic antidepressants, reversed synapse loss that occurred following stress (Chen, Madsen, Wegener, & Nyengaard, 2010; Hajszan et al., 2009). Imipramine also increased the length of apical dendrites in the hippocampus and PFC (Bessa et al., 2009). Similarly, another tricyclic antidepressant, amitriptyline, enhanced spine density in hippocampal neurons (Norrholm & Ouimet, 2001). A chemically distinct drug, tianeptine, which works via modulation of glutamatergic signaling, reversed stress-induced plasticity changes in the amygdala (dendrite growth and anxiety behavior) while increasing NMDA and α-amino-3-hydroxy-5-methyl-4-isoxazolepropionic acid (AMPA) currents in the hippocampus (Pillai, Anilkumar, & Chattarji, 2012). Also, tianeptine blocks neuronal atrophy caused by stressful events in CA3 neurons (McEwen et al., 1997). In addition to the tricyclic antidepressants, a 4-week treatment with fluoxetine, a selective reuptake serotonin inhibitor (SSRI), increased dendritic spine density with an equivalent increase in mushroom spines (and decrease in thin spines) in the proximal dendrites of the CA1 hippocampus (McAvoy, Russo, Kim, Rankin, & Sahay, 2015; Rubio et al., 2013) and in a subregion of the cortex (Ampuero et al., 2010). Chronic fluoxetine is crucial for increasing dendritic arborization of immature granule neurons expressing doublecortin protein (Hajszan, MacLusky, & Leranth, 2005). In contrast, a 1-h treatment with fluoxetine did not alter spine density in the apical BLA, though it induced a paradoxical increase in anxiety behavior that is associated with acute fluoxetine treatment (Ravinder, Pillai, & Chattarji, 2011).

MOLECULAR SIGNALING AT THE SYNAPSE THAT IS RELEVANT TO DEPRESSION

Glutamatergic synapses are composed of a presynaptic exocytic compartment and a postsynaptic, electron-dense region hosting neurotransmitter receptors that initiate signaling events (Wilhelm et al., 2014). The postsynaptic terminal also contains recycling endosomes, which control the turnover of signaling molecules and receptors. Several molecules that are implicated in neuropsychiatric disorders act to regulate cytoskeletal machinery, which is essential for dendrite and spine homeostasis (Zhang & Benson, 2001)—particularly actin, but tubulin polymers are also crucial for spine plasticity (Hoogenraad & Akhmanova, 2010; Okamoto, Nagai, Miyawaki, & Hayashi, 2004; Star, Kwiatkowski, & Murthy, 2002). Experiments using photo switchable *green fluorescent protein* (GFP) fused to actin have revealed the importance of actin polymerization for spine head enlargement and maturation (Honkura, Matsuzaki, Noguchi, Ellis-Davies, & Kasai, 2008). Not surprisingly, therefore, proteins that regulate actin polymerization are enriched in spines and are critical for spine morphology changes (Lamprecht, Farb, Rodrigues, & LeDoux, 2006; Tada & Sheng, 2006) (Figure 1) as well as membrane trafficking and receptor endocytosis (Hotulainen et al., 2009). Several of these proteins have been shown to be deregulated in animal models of depression and in patients suffering from major depressive disorder (MDD) or bipolar disorder, as outlined below.

The **Rho family of GTPases** are key regulators of actin remodeling. In particular, Rac1 is an important modulator of spine volume, which it does by reorganizing filamentous-actin (F-actin). Active Rac1 induces activation of LIM kinase 1 (LIMK1). LIMK1 in turn inhibits cofilin, an actin depolymerizing factor (Niwa, Nagata-Ohashi, Takeichi, Mizuno, & Uemura, 2002) (Figure 1), leading to increased F-actin, spine head growth, and enhanced synaptic function (Meng et al., 2002). In the "learned helplessness" model of depression, Limk1 mRNA levels are dramatically decreased in the frontal cortex, and its expression is partially, though not fully, restored by imipramine or fluoxetine treatment (Nakatani, Aburatani, Nishimura, Semba, & Yoshikawa, 2004). Importantly, LIMK1 expression is repressed by mIR-134, which in turn is negatively regulated by BDNF (Schratt et al., 2006). Also, BDNF in dendrites exerts retrograde control of presynaptic glutamate release (Jakawich et al., 2010), providing an

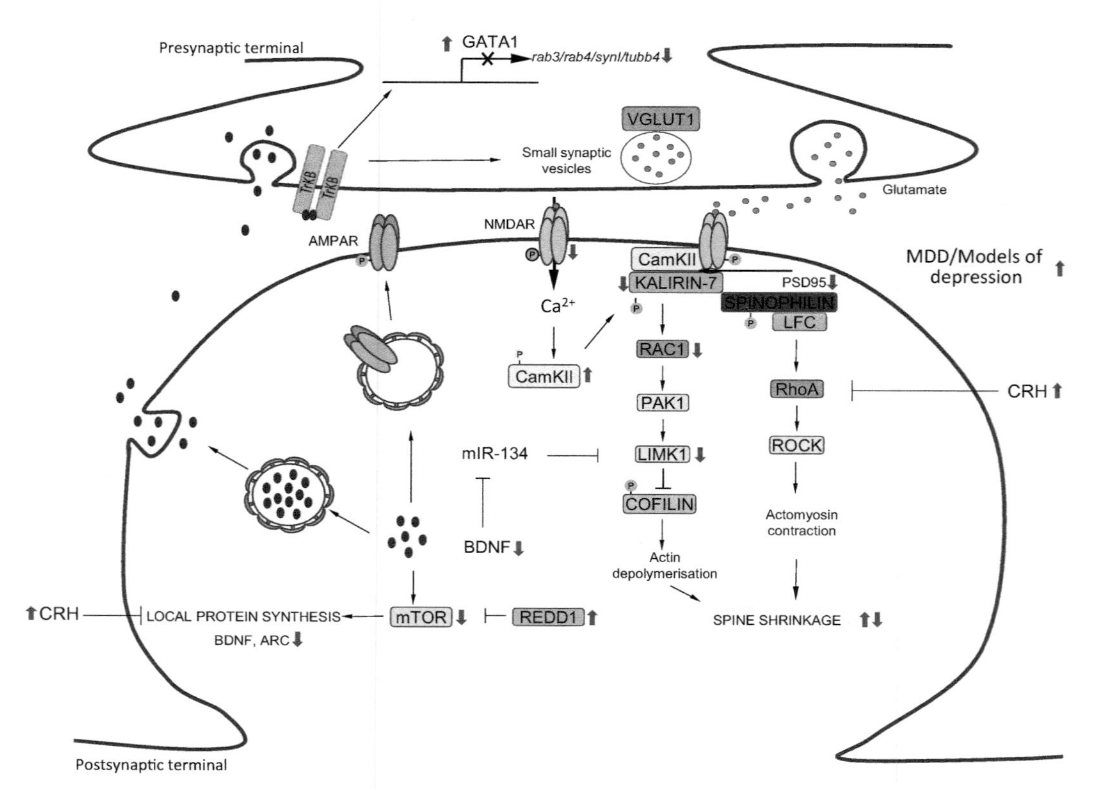

FIGURE 1 Molecular regulation of signaling events in the spine head leading to plasticity changes. Proteins that are activated (or upregulated) in rodent models of depression or in MDD patients, are marked with an upward arrow and those inhibited (or downregulated) with a downward arrow. The effect of increased CRH is also indicated.

additional loop whereby Rac1/LIMK1 activity can be regulated in the context of depression when BDNF levels are reduced (Figure 1).

The action of another small GTPase, RhoA, antagonizes that of Rac1 and causes spine head shrinkage by rho associated coiled-coil containing protein kinase (ROCK)-mediated, actomyosin-induced contractility (Murakoshi, Wang, & Yasuda, 2011) (Figure 1). RhoA also simplifies dendritic architecture (Nakayama, Harms, & Luo, 2000). Notably, application of the stress hormone CRH to hippocampal neurons increases RhoA activity and reduces spine density (Chen et al., 2013) (Figure 1). The activities of Rac1 and RhoA appear to be coordinated, because spine enlargement requires both increased activity of Rac1 and decreased activity of RhoA (Sin, Haas, Ruthazer, & Cline, 2002). Interestingly, Rac1 is epigenetically downregulated in the nucleus accumbens (NAc) of mice undergoing social defeat stress and results in increased stubby spine density (Golden et al., 2013). Moreover, overexpression of constitutively active Rac1 in the NAc reversed the depressive behavior following social defeat stress. Importantly, this study also reported a repressed chromatin state in the region surrounding the Rac1 promoter in the NAc of patients suffering from depression. These examples demonstrated how atypical regulation of gene expression and mRNA translation of synaptic proteins can evoke inappropriate plasticity, resulting in depressive behavior.

Spinophilin

Spinophilin is a dendritic spine protein (also known as neurabin) that acts as a scaffold and reorganizes actin (Allen, Ouimet, & Greengard, 1997). It does this via the activation of RhoA (Ryan et al., 2009) (Figure 1). Consistent with the repressive function of RhoA in spine formation, mice lacking spinophilin display increased spine density in the caudate-putamen, and in the hippocampus, these mice produce filopodial-like spines prematurely (Feng et al., 2000). In the elevated plus maze, spinophilin knock out mice spent increased time in the open arms, indicating a low anxiety phenotype. They also displayed an enhanced antidepressant response to desipramine (Cottingham, Li, & Wang, 2012). Specific knockdown of spinophilin in the anterior cingulate cortex (ACC) increased synaptic activity and lowered anxiety-like behavior. However, it increased the depressive phenotype in the forced swim test (Kim et al., 2011). Notably, inhibition of spinophilin reproduced the low anxiety phenotype but not the depressive phenotype, suggesting a tight association with anxiety behavior. These studies raised awareness of spinophilin as a regulator of synaptic activity and a potential target for anxiety disorders associated with the ACC.

The association of spinophilin with depressive behavior is less clear. Reduced spinophilin mRNA was observed in the hippocampus CA3, hilus, subiculum, and entorhinal cortex of human patients diagnosed with depression and schizophrenia (Law, Weickert, Hyde, Kleinman, & Harrison, 2004). However, in spontaneously depressed monkeys, spinophilin protein remained unchanged in the hippocampus CA3 but decreased in the CA1 (Willard, Hemby, Register, McIntosh, & Shively, 2014). Yet in rats, chronic restraint stress increased spinophilin protein levels in the hippocampus CA3 (Orlowski, Elfving, Muller, Wegener, & Bjarkam, 2012). These discrepancies may to some extent be explained by a lack of correlation between mRNA and protein. Additionally, whether spinophilin levels increase or decrease during depression or anxiety may simply reflect the extent of synaptic atrophy.

Glutamatergic Signaling in Depression

Mounting evidence suggests that the glutamatergic system is abnormally affected in MDD. Glutamate levels are increased in *cerebrospinal fluid*, blood, and brain tissue from patients with MDD, and this is countered by antidepressant treatment (Tokita, Yamaji, & Hashimoto, 2012). Also, atypical expression of glutamate receptors (NMDA, mGluR, and AMPA) are reported in postmortem brain (Tokita, Yamaji, & Hashimoto, 2012). Significantly, NMDA receptor antagonists, ketamine and riluzole, produce robust antidepressant effects in patients with MDD (Zarate et al., 2010; Zarate, Jr. et al., 2006). Studies in rodents have correlated chronic stressful events with region-specific changes in glutamate receptor expression (Table 1). For example, repeated subchronic stress in mice decreased the surface expression of GluR1 and NR1 and impaired neurotransmission in the PFC (Yuen et al., 2012). Interestingly, acute stress had the opposite effect and was associated with working memory (Yuen et al., 2009). In patients with MDD, NR1 mRNA (Law & Deakin, 2001) as well as NR2A and NR2B protein were also reduced in the PFC (Feyissa, Chandran, Stockmeier, & Karolewicz, 2009; Woo, Walsh, & Benes, 2004). Another study showed no change in NMDA receptor mRNA levels in bipolar patients but a significant decrease in expression of the synapse-associated protein (SAP) 102 scaffold that organizes downstream signaling (Beneyto & Meador-Woodruff, 2008). Other studies reported increased expression of NR1 and NR2A in the hippocampus of mice following maternal separation stress (Martisova et al., 2012; Ryan et al., 2009). Thus, it can be summarized that atypical expression of glutamate receptors and downstream signaling components is associated with depression in both animal models and human subjects, while the nature of the changes differ between stress models.

Inhibition of glutamatergic transmission occurs following repeated stressful and depressive events and can be attributed to the downregulation of presynaptic proteins, such as synapsin I and III, synaptophysin (Kang et al., 2012), and the vesicular glutamate transporter 1 (VGLUT1) (Garcia–Garcia et al., 2009; King et al., 2014; Zink, Vollmayr, Gebicke-Haerter, & Henn, 2010) (Figure 1). In rats, chronic mild stress downregulated several proteins involved in exocytosis and endocytosis (Hu et al., 2013). Among the endocytosis regulating proteins are clathrin light chain proteins and dynamin-I (Granseth, Odermatt, Royle, & Lagnado, 2006). In the learned helplessness paradigm, cytoskeletal regulatory proteins such as dystrophin-related protein 2 and stathmin were increased, suggesting the involvement of microtubule dynamics in the regulation of stress signaling (Curmi et al., 1997; Mallei et al., 2011). Moreover, microtubule-associated protein 2 (MAP2), which exerts a stabilizing effect on dendritic microtubules (Caceres, Payne, Binder, & Steward, 1983), was downregulated in the hippocampus of socially isolated mice (Bianchi et al., 2006). Mice lacking high-molecular-weight MAP2 display shorter apical dendrites in the hippocampus (Harada, Teng, Takei, Oguchi, & Hirokawa, 2002). This is most likely due to less stable microtubules as a direct consequence of reduced MAP2. Related to this, exogenous expression in the motor cortex of GFP-MAP2 that is pseudophosphorylated on the JNK1 sites increases dendrite length (Komulainen et al., 2014), further demonstrating that MAP2 phosphorylation defines dendritic architecture in vivo and is a possible mediator of dendritic irregularities, such as those that occur in neuropsychiatric disorders (Komulainen et al., 2014). MAP2 is also expressed in dendritic spines (Caceres et al., 1983), and loss of MAP2 has been associated with spine loss in schizophrenia (Shelton et al., 2015). Also, NMDA inhibits microtubule growth within the spine head (Kapitein & Hoogenraad, 2011), suggesting that the stability of microtubule polymers in dendrites and spines contributes to dendritic morphology changes and spine growth/atrophy. In patients with MDD, postsynaptic density protein 95 (PSD-95) was reduced in the PFC, suggesting synaptic loss (Kang et al., 2012), and a study of the dentate gyrus in MDD patients found that MAP2 levels were reduced (Duric et al., 2013; Soetanto et al., 2010).

Kalirin-7

Kalirin-7 is a brain-specific, guanine nucleotide exchange factor that is associated with a range of neuropsychiatric disorders. It is suggested that anchoring of Ca^{2+}/calmodulin-dependent protein kinase II (CamKII) to NR2B serves as a docking site for kalirin-7 at the postsynaptic density (Kiraly, Lemtiri-Chlieh, Levine, Mains, & Eipper, 2011). This allows phosphorylated kalirin-7 to activate Rac1 to promote actin polymerization and spine head growth (Xie et al., 2007) (Figure 1). Of note, mutation of an amino acid in the catalytic domain of kalirin-7 is associated with MDD (Russell et al., 2014), and chronic restraint stress in mice leads to decreased kalirin-7 expression in the hippocampus (Li, Li, & An, 2010; Qiao, An, Ren, & Ma, 2014). Kalirin-7 interacts with several PSD-95/discs large/ZO-1 homologous (PDZ) containing proteins including GluR1, PSD-95, and spinophilin, and decreased kalirin-7 mRNA in dendritic spines is associated with schizophrenia and attention deficit disorder as well as MDD (Mandela & Ma, 2012).

BDNF IN SYNAPTIC PLASTICITY AND DEPRESSION

Neurotrophic factors, especially BDNF, have been widely studied in the context of synaptic plasticity and in relation to depressive and anxiety disorders. Human patients suffering from depression display decreased serum levels of BDNF (Karege et al., 2002; Shimizu et al., 2003). In rodents, BDNF mRNA decreases following stress in the hippocampus (Pizarro et al., 2004; Rasmusson, Shi, & Duman, 2002; Smith, Makino, Kvetnansky, & Post, 1995) and PFC (Duman & Monteggia, 2006). BDNF is also downregulated upon treatment with corticosterone (Schaaf, de Jong, de Kloet, & Vreugdenhil, 1998). On the other hand, BDNF has also been seen to increase following stress (Fanous, Hammer, & Nikulina, 2010; Marmigere, Givalois, Rage, Arancibia, & Tapia-Arancibia, 2003) and is proposed to serve as a protective mechanism against the effects of future stressful events. Interestingly, specific deletion of the BDNF locus in the forebrain of female mice resulted in a depressive-like phenotype as compared to male mice with the same conditional gene deletion, suggesting that a gender difference in susceptibility may be explained by BDNF signaling in this region (Monteggia et al., 2007). Similarly, selective loss of BDNF in the dentate gyrus attenuated the efficacy of antidepressants (Adachi, Barrot, Autry, Theobald, & Monteggia, 2008), and knockdown of BDNF in the hippocampus induced anhedonia in rodents (Taliaz, Stall, Dar, & Zangen, 2010). Importantly, long-term resilience to chronic stress is reduced in the absence of BDNF (Taliaz et al., 2011).

A physiological consequence of elevated BDNF is an increase of the presynaptic pool of vesicles at active zones and an enhanced quantal neurotransmitter release (Figure 1) (Tyler & Pozzo-Miller, 2001). BDNF also upregulates expression of several synaptic proteins such as synapsin-I, PSD-95, and GluR1 (O'Leary, Wu, & Castren, 2009). A single nucleotide polymorphism in BDNF, Val66Met, confers susceptibility to depression in humans and rodents and is associated with anxiety disorders (Chen et al., 2006). This single amino acid alteration results in reduced BDNF in synapse-localized secretory granules as a result of impaired dendritic transport (Chiaruttini et al., 2009; Egan et al., 2003). Mice carrying the Val66Met mutation are irresponsive to fluoxetine (Chen et al., 2006). Human studies have correlated the Val66Met polymorphism with reduced hippocampal volume (Frodl et al., 2007), and carriers of this specific mutation who experienced early life stress exhibited a strong likelihood for developing depression later on (Gatt et al., 2009). BDNF blood levels were also found to be decreased in subjects who experienced early life stress, psychosocial stressors in particular, in those carrying the risk met-allele of the Val66Met BDNF gene (Elzinga et al., 2011). In mice carrying the Val66Met mutation, dendritic spine density is reduced in the medial PFC and BLA (Liu et al., 2012; Yu et al., 2012), with increased preponderance of thin spines in distal dendrites (Liu et al., 2012). These findings are consistent with studies showing reduced spine density after loss of BDNF (Hiester, Galati, Salinas, & Jones, 2013; Magarinos et al., 2011; Tolwani et al., 2002). Furthermore, overexpression of BDNF increased spine density in the BLA and decreased depressive-like behavior in mice, although it also increased anxiety-like behavior (Govindarajan et al., 2006).

BDNF is known to regulate intracellular pathways involved in anxiety and depression. For example, it induces expression of the immediate early gene Arc, which localizes in dendrites where it is regulated by synaptic activity (Lyford et al., 1995; Ying et al., 2002) (Figure 1). In spines, Arc regulates AMPA receptor recycling, and mice lacking Arc exhibit decreased AMPA receptor endocytosis (Chowdhury et al., 2006). Moreover, Arc is associated with the stabilization of dendritic spines (Bramham, 2008), and reduced Arc expression may contribute to depression-related behavior in mice (Elizalde et al., 2010). BDNF also positively regulates LIMK1 by abolishing the action of the brain-enriched miRNA 134, which inhibits translation of LIMK1 (Schratt et al., 2006) (Figure 1). LIMK in turn contributes to dendritic spine growth (Meng et al., 2002).

Studies in mice lacking BDNF have been informative in defining its molecular action. They show increased expression of glutamate receptor-interacting protein 1 (GRIP1), SAP97, and protein interacting with C kinase 1 (PICK1), PDZ-domain proteins that determine synaptic plasticity (Jourdi et al., 2003). Additionally, postsynaptic release of BDNF stimulates presynaptic Rab3a expression (Alder et al., 2005; Jakawich et al., 2010; Thakker-Varia, Alder, Crozier, Plummer, & Black, 2001) (Figure 1). Rab3a is a major monomeric G protein in the brain that is enriched on synaptic vesicles, and its expression is decreased in patients suffering from MDD (Kang et al., 2012). The overexpression of GATA binding protein 1 (GATA1), a transcription factor that represses Rab3a, evokes a depressive phenotype in the forced swim test. These mice exhibit reduced dendritic arborization and spine density. Consistent with this observation, GATA1 expression is increased in MDD patients (Kang et al., 2012). Impaired regulation of BDNF is inextricably linked with depressive behavior. As seen elsewhere in this chapter, BDNF triggers several major signaling pathways that have been individually shown to be critical for antidepressant responses in animal models and are deregulated in MDD. Among these are mitogen-activated protein kinases (MAPK) and CAMKII signaling and their downstream target cyclic-AMP response element-binding protein (CREB). The mTOR pathway is also associated with BDNF regulation (Sabatini, Erdjument-Bromage, Lui, Tempst, & Snyder, 1994) (Figures 1 and 2).

PROTEIN KINASES IMPLICATED IN DEPRESSION

mTOR Signaling and Its Relevance in Depression

mTOR regulates synaptic protein synthesis in response to activity leading to altered spine density and morphology (Leal, Comprido, & Duarte, 2014; Schratt, Nigh, Chen, Hu, & Greenberg, 2004) (Figure 1). In particular, chronic inhibition of mTOR was shown to reduce dendrite complexity and spine density in hippocampal neurons (Kumar, Zhang, Swank, Kunz, & Wu, 2005). Consistent with this, the activation of mTOR enhanced synaptic responses

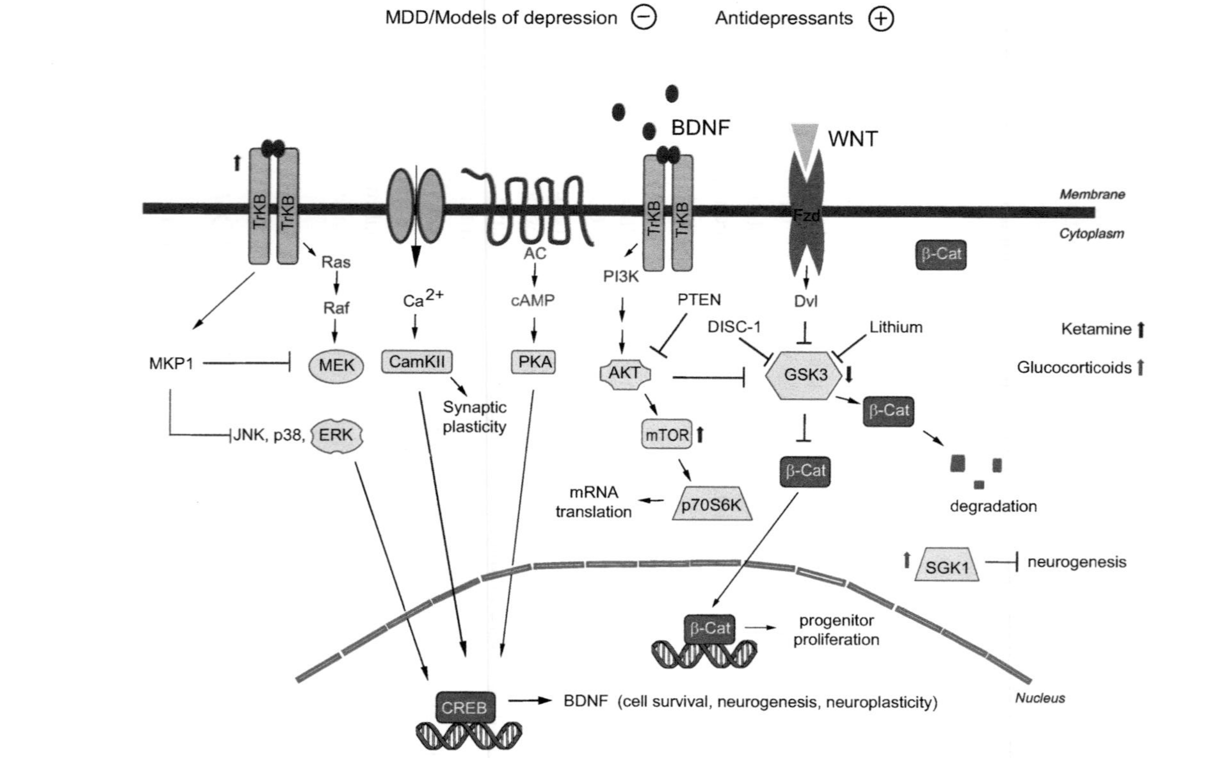

FIGURE 2 Signaling pathways that are negatively regulated in animal models of depression and in human subjects with MDD. Proteins that are activated (upward arrow) or inhibited (downward arrow) by ketamine are indicated in blue. Proteins that are increased by glucocorticoids are indicated with red arrows.

from both glutamatergic and GABAergic neurons by increasing the ready releasable pool of synaptic vesicles. mTOR activation also increased synapse number. Conversely, the inhibition of mTOR with rapamycin reduced glutamatergic transmission after 72 h, suggesting a requirement for de novo protein synthesis. Interestingly, rapamycin had no effect on inhibitory synaptic transmission. A possible cause may be that GABAergic transmission has a relatively lower demand for de novo protein synthesis (Weston, Chen, & Swann, 2012). Notably, localized synthesis of BDNF in the postsynaptic terminal is regulated by dendritic mTOR. This serves to induce a retrograde enhancement of neurotransmitter release from the apposed presynaptic terminal, providing a mechanism for homeostatic transsynaptic control of presynaptic function (Henry et al., 2012) (Figure 1).

p70S6K is a serine/threonine kinase substrate of mTOR. p70S6K responds to mTOR to promote protein synthesis. Interestingly, it was shown that viral mediated expression of an active form of p70S6K in the medial PFC had an antidepressant effect in the forced swim test and blocked anhedonia induced by chronic stress (Dwyer, Maldonado-Avilés, Lepack, DiLeone, & Duman, 2015). These results suggest that p70S6K is a critical player downstream of mTOR in regulating depressive behavior.

In patients with MDD, robust decreases in mTOR levels in the PFC have been reported from postmortem tissue (Abelaira, Réus, Neotti, & Quevedo, 2014; Jernigan et al., 2011). mTOR action in the context of psychiatric disorders seems to be modulated by regulated in development and DNA damage responses-1 (REDD1), a protein that is induced under conditions of stress, and whereupon it inhibits mTOR by stabilization of the mTOR inhibitory complex TSC1-TSC2 (Figure 1). Mice lacking REDD1 exhibit increased spine density, whereas REDD1 overexpression produced depressive-like behavior. Similarly, in MDD patients, REDD1 expression is elevated (Ota et al., 2014).

Ketamine and mTOR

A problem of unresponsiveness to treatment among a large proportion of patients has fueled a search for nonmonoaminergic mechanisms. The NMDA receptor antagonist ketamine was identified as one such drug with fast-acting antidepressant responses in treatment-resistant patients (Zarate et al., 2010; Zarate, Jr. et al., 2006). Intraperitoneal injection of ketamine in mice enhances glutamate levels and mTOR activity in the PFC within 30 min and in the striatum by 80 min postinjection (Li, Lee, et al., 2010; Moghaddam, Adams, Verma, & Daly, 1997). Thus the PFC is considered to be the primary site of GABAergic disinhibition in response to ketamine. At the molecular level, ketamine was shown to activate mTOR and stimulate production of synaptic proteins Arc, GluR1, PSD-95, and synapsin I in the PFC of rats (Li, Lee, et al., 2010) as well as BDNF (Autry et al., 2011). Consistent with mTOR activation, ketamine also increases the release of BDNF from the postsynaptic terminal

(Duman & Aghajanian, 2012; Henry et al., 2012). Moreover, it was shown that the blockade of mTOR prevented the synaptogenic and antidepressant action of ketamine (Li, Lee, et al., 2010), and in patients with MDD, the inhibition of mTOR with rapamycin blocked the antidepressant effect of ketamine (Yu et al., 2013). This finding is important as it identifies mTOR as an essential, proximal downstream effector of the antidepressant response to ketamine. Ketamine itself produces unwanted psychiatric side effects in approximately 20% of patients (Rasmussen, 2014), thus there is a drive to understand the downstream mechanisms so that alternative targets can be identified.

CamKII α and β

CamKII α and β are neuron-specific kinases, expressed at high molar concentrations in the brain (Brocke, Srinivasan, & Schulman, 1995). They are activated in response to glutamate-induced NMDA receptor activation, following which the receptor becomes permeable to calcium as well as Na^+ and K^+ (Malenka, Kauer, Zucker, & Nicoll, 1988). The elevation of intracellular calcium triggers the autophosphorylation of a threonine residue in the catalytic domain of CamKII, leading to its activation. This autophosphorylation is critical for synaptic plasticity (Lisman & Zhabotinsky, 2001) (Figures 1 and 2). In addition to phosphorylating proteins in the synapse, active CamKII phosphorylates CREB, leading to its transactivation (Wu, Deisseroth, & Tsien, 2001). CREB in turn upregulates the expression of BDNF, thereby controlling behavioral homeostasis, as discussed elsewhere in this chapter (Figure 2). It was revealed in a proteomics screen that CamKII β (but not α) was specifically upregulated in the habenula in rodent models of depression and downregulated following antidepressant treatment. Furthermore, blocking CamKII β activity reversed the depressive symptoms, indicating a causal role (Li et al., 2013). Given the central role of CamKII in regulating synaptic stability as well as gene transcription, it is not surprising that deregulation of this kinase will cause behavioral defects. Interestingly, a proteomics study using selected reaction monitoring reported that CamKII α, β, and γ expression are downregulated in postmortem anterior PFC from patients with MDD (Wesseling, Gottschalk, & Bahn, 2014).

MAPKs and Depression

The extracellular signal-regulated kinase (ERK) MAPK cascade is activated by extracellular signals such as BDNF leading to cellular growth and differentiation (Figure 2). Evidence suggests that ERK-dependent events are suppressed in animal models of depression (Dwivedi et al., 2001; Feng, Guan, Yang, & Fang, 2003; Gourley et al., 2007; Meller et al., 2003; Qi, Lin, Li, Pan, & Wang, 2006; Qi et al., 2008; Tiraboschi et al., 2004). Specifically, a reduction in ERK activity was shown in the hippocampus and frontal cortex of rats following chronic mild stress exposure, and this was reversed by treatment with the antidepressant fluoxetine (First et al., 2011). Moreover, infusion with U0126, an inhibitor of MAPK kinase (MEK) to ERK signaling, induced anhedonia and increased anxiety behavior in rats. Also, in postmortem studies of suicide victims, hippocampal levels of ERK and upstream pathway components MEK and Raf were decreased (Dwivedi et al., 2005; Dwivedi et al., 2001; Yuan et al., 2010).

The MAPK phosphatase-1 (MKP-1; also known as dual specificity phosphatase-1 [DUSP1] or CL100) reverses ERK, p38, and c-Jun N-terminal kinase (JNK) activation. MKP-1 expression is elevated in the hippocampus of depressed suicide subjects (Duric et al., 2010). It is classified as an immediate early gene that is induced by oxidative stress and hypoxia (Keyse & Emslie, 1992; Laderoute et al., 1999; Seta, Kim, Kim, Millhorn, & Beitner-Johnson, 2001). The expression of MKP-1 also increased after chronic mild stress in rodents, consistent with findings from postmortem studies of subjects with MDD, while it returned to normal levels upon antidepressant treatment (Duric et al., 2010). Notably, this study also showed that virus-mediated expression of MKP-1 in the hippocampus was sufficient to induce a depressive phenotype in mice, whereas the deletion of MKP-1 resulted in resilience to chronic stress. Consistent with this, BDNF was shown to increase MKP-1 expression, leading to inhibition of JNK, destabilization of microtubules, and axon branching (Jeanneteau, Deinhardt, Miyoshi, Bennett, & Chao, 2010). Whether this regulation of JNK is consequential in the context of depression is not known; however, a study reported that MKP-1 mRNA was decreased in the PFC of MDD patients (Malki et al., 2015). These findings correlate regional changes in MKP-1 expression with depressive outcome and support the idea that the blockade of ERK signaling contributes to depressive behavior.

V-Akt Murine Thymoma Viral Oncogene Homolog Signaling in Depression

V-akt murine thymoma viral oncogene homolog (AKT) is activated downstream of BDNF receptor binding and is also implicated in MDD (Carter, 2007; Detera-Wadleigh, 2001) (Figure 2). In depressed suicide subjects, levels of AKT and phospho-AKT (representing the catalytically active form) were decreased in the hippocampus and PFC (Hsiung et al., 2003; Karege et al., 2007). Decreased levels of AKT have also been reported in the ventral tegmental area (VTA) of mice subjected to chronic social defeat stress, and selective blockade of AKT activity in this region renders mice susceptible to depressive-like behavior (Krishnan et al., 2008). Reduced activity of AKT may result from increased expression of the phosphatase and tensin analog, which negatively regulates AKT by dephosphorylating its upstream regulator PI3K (Duric et al., 2010; Dwivedi et al., 2010). Interestingly, the blockade of PI3K/AKT signaling was sufficient to attenuate adult hippocampal neurogenesis following exercise (Bruel-Jungerman et al., 2009) and to prevent ketamine-induced antidepressant effects (Li, Lee, et al., 2010). The most potent downstream target of the PI3K/AKT pathway is GSK-3, which has gained considerable attention in the context of depression.

GSK-3 Signaling and Depression

Lithium, the gold standard treatment for bipolar disorder for over 50 decades, is an allosteric inhibitor of GSK-3α and β. The mood-stabilizing effect of lithium has been proposed to be mediated via GSK-3 inhibition, though GSK-3 is not the only target of this drug (Gould & Manji, 2005; Silva et al., 2008). GSK-3β has been shown to promote neuronal death responses (Hongisto, Vainio, Thompson, Courtney, & Coffey, 2008; Tanis & Duman, 2007). SSRI-based antidepressants inhibit GSK-3 by increasing phosphorylation of serine 9. This occurs via 5-HT1A receptors (Li & Jope, 2010). Also, pharmacological inhibition of GSK-3 activity produces antidepressant-like effects in animal models of depression (Gould, Einat, Bhat, & Manji, 2004). β-catenin is an important substrate of GSK-3, which upon phosphorylation undergoes proteosomal degradation (Henderson, 2000). Conversely, inhibition of GSK-3 increases levels of β-catenin in the cytosol, providing structural support at the plasma membrane and increased gene transcription in the nucleus (Figure 2).

GSK-3 inhibits adult hippocampal neurogenesis, which may contribute to anxiety and depressive behavior (Carlezon, Duman, & Nestler, 2005; Li, Lee, et al., 2010). Interestingly, disrupted in schizophrenia-1 (DISC-1) blocks GSK-3β through a direct interaction. This reduced neural progenitor proliferation leads to premature cell cycle exit, most likely via increased nuclear β-catenin activity (Figure 2). DISC-1 also reversed depressive behavior in the mice, as did pharmacological inhibition of GSK-3 (Mao et al., 2009). Importantly, in humans, GSK-3β SNPs have been associated with depression (Inkster et al., 2009; Saus et al., 2010). Not surprisingly, as GSK-3 is directly regulated by AKT, its activity is inhibited by ketamine. Accordingly, ketamine was shown to increase phosphorylation on serine-9 (GSK-3β) and serine 21 (GSK-3α; i.e., phosphorylation on the AKT sites). The role of serine-9 phosphorylation in ketamine responsiveness was tested in mice where serine-9 (GSK-3β) and serine 21 (GSK-3α) were mutated to alanine, leading to constitutively active GSK-3 that cannot be inhibited by AKT. These mice were irresponsive to ketamine in the learned helplessness model of depression, suggesting a requisite role for GSK-3 inhibition in mediating the behavioral response to ketamine (Beurel, Song, & Jope, 2011). Furthermore, GSK-3 phosphorylation in response to ketamine depends on AMPA receptor activation (Jourdi et al., 2009), as does the antidepressant action of ketamine (Maeng et al., 2008). It has even been proposed that GSK-3 inhibition may produce its antidepressant effects by deregulating AMPA receptor trafficking (Wei, Liu, & Yan, 2010). This may be achieved by GSK-3β phosphorylation of kinesin light chain, which has been shown to dissociate AMPA receptor (GluR1) cargo (Du et al., 2010). This study also showed that a peptide, which interfered with kinesin light chain phosphorylation, induced antidepressant-like activity in mice. Finally, there is evidence that GSK-3 may contribute to epigenetic regulation of depression (Aubry, Schwald, Ballmann, & Karege, 2009; Krishnan & Nestler, 2008).

Wnt Signaling and Depression

Wnts are extracellular ligands that act via the canonical Wnt pathway to negatively regulate GSK-3. While Wnts play essential roles during development, in adults Wnt regulates neuroplasticity (Ciani & Salinas, 2005; Nusse et al., 2008; Purro et al., 2008). Wnts are glycoproteins that bind to Frizzled receptors (Fzd) and inhibit GSK-3 through the activation of disheveled phosphoproteins (Kikuchi, Yamamoto, & Kishida, 2007; Logan & Nusse, 2004) (Figure 2). Wnt2 is increased upon several antidepressant treatments, including chronic electroconvulsive therapy in rats, and virus-mediated regional expression of Wnt2 in the hippocampus evokes antidepressant-like responses (Okamoto et al., 2010). Wnt3a expression is also increased following SSRI treatment in rats, whereupon it enhances adult hippocampal neurogenesis (Pinnock, Blake, Platt, & Herbert, 2010). Moreover, Fzd subtypes have been implicated in stress and antidepressant treatments. For example, the expression of Fzd6 was increased in the hippocampus after chronic electroconvulsive therapy in mice and decreased following the chronic unpredictable stress paradigm. Moreover, viral-mediated knockdown of Fzd6 in the hippocampus evokes anhedonic behavior and increased anxiety (Voleti, Tanis, Newton, & Duman, 2012). In a postmortem study of depressed suicide subjects and in animal models of social defeat stress, the mRNA and protein levels of Disheveled Segment Polarity Protein were significantly decreased in the NAc (Wilkinson et al., 2011). These findings consistently associate impaired Wnt signaling with depression-like behavior.

Serum and Glucocorticoid-Regulated Kinase 1 (SGK1): A Mediator of Glucocorticoid Signaling in Depression

In depressed patients and rodent stress models, the levels of glucocorticoid hormones are severely elevated due to impaired HPA axis feedback regulation. Increased glucocorticoid signaling activates glucocorticoid and mineralocorticoid receptors in the hippocampus and reduces neurogenesis, thereby impairing neuroplasticity (Solas, Gerenu, Gil-Bea, & Ramírez, 2013). SGK1, a serine threonine kinase target of the glucocorticoid receptor, was shown to be activated in response to glucocorticoid treatment in neurons (Anacker et al., 2011). Studies suggest that SGK1 acts downstream of glucocorticoids to inhibit adult hippocampal neurogenesis, neuronal excitability, and working memory in the PFC after acute stress (Anacker et al., 2013; Yuen et al., 2011).

The mRNA level of SGK1 was found to be correlated with hippocampal volumes in patients with depression. In particular, a lack of mRNA expression SGK1 in patients with MDD that seems to correspond to a blunted cortisol response is associated with smaller hippocampal cornu ammonis and dentate gyrus volumes (Frodl et al., 2014). Moreover, the significant increase in mRNA levels of SGK1 in the peripheral blood of drug-free depressed patients and in the hippocampus of rats subjected to unpredictable chronic mild stress (Anacker et al., 2013) or mice subjected to early life stress (Bockmühl et al., 2015) suggests its relevance in the pathophysiology of depression.

CREB and Depression

CREB is a transcription factor target of several signaling cascades relevant for the pathogenesis of depression. In a study of 26 suicide subjects, it was shown that CREB expression in the hippocampus and PFC was decreased (Dwivedi et al., 2003). Likewise, in animal models of depression, CREB activity was reduced, as discerned from phosphorylation of Ser-133 (Qi et al., 2008). Consistent with this, CREB levels are increased in the hippocampus following treatment with the antidepressant fluoxetine, leading in turn to augmentation of BDNF expression (Nibuya, Nestler, & Duman, 1996; Tiraboschi et al., 2004) (Figure 2). Indeed, the most common monoaminergic antidepressants converge on CREB (Carlezon et al., 2005). Also, protein kinase A, which activates CREB, was activated in response to chronic antidepressant treatment (Tardito et al., 2006), while ERK and calmodulin-dependent kinases activate CREB in a stimulus-dependent manner (Deisseroth & Tsien, 2002; Sheng, Thompson, & Greenberg, 1991; Tardito et al., 2006) (Figure 2). Thus, the convergence of antidepressant action on CREB is hardly surprising. In the nucleus, CREB binds gene targets to regulate neuroplasticity, cell survival, and cognition (Carlezon et al., 2005) (Figure 2). Importantly, BDNF is among these targets, and as discussed earlier, it protects both mice and humans from aversive stimuli and depressive-like symptoms. Together these findings correlate reduced CREB function with depressive outcomes.

Depression is a disease where the adaptive responses to stress, which normally provide resilience, have gone awry. One means for recovering homeostatic regulation is through the use of drugs that target critical mechanisms to regain normal function. Clearly, the perturbations in signaling pathways that occur in depressive disorders are diverse. Precisely for this reason, targeting of these signaling proteins offers an opportunity for developing stratified treatments. Continued research to understand the mechanistic basis of depression making use of systems-level approaches provides great potential for improved diagnostic methods and therapy.

ACKNOWLEDGMENT

We would like to thank Sylvia Ortega-Martinez for her helpful comments.

REFERENCES

Abelaira, H. M., Réus, G. Z., Neotti, M. V., & Quevedo, J. (2014). The role of mTOR in depression and antidepressant responses. *Life Sciences, 101*(1–2), 10–14. http://dx.doi.org/10.1016/j.lfs.2014.02.014.

Adachi, M., Barrot, M., Autry, A. E., Theobald, D., & Monteggia, L. M. (2008). Selective loss of brain-derived neurotrophic factor in the dentate gyrus attenuates antidepressant efficacy. *Biological Psychiatry, 63*(7), 642–649. http://dx.doi.org/10.1016/j.biopsych.2007.09.019.

Alder, J., Thakker-Varia, S., Crozier, R. A., Shaheen, A., Plummer, M. R., & Black, I. B. (2005). Early presynaptic and late postsynaptic components contribute independently to brain-derived neurotrophic factor-induced synaptic plasticity. *Journal of Neuroscience, 25*, 3080–3085. http://dx.doi.org/10.1523/jneurosci.2970-04.2005.

Allen, P. B., Ouimet, C. C., & Greengard, P. (1997). Spinophilin, a novel protein phosphatase 1 binding protein localized to dendritic spines. *Proceedings of the National Academy of Sciences of the United States of America, 94*(18), 9956–9961.

Ampuero, E., Rubio, F. J., Falcon, R., Sandoval, M., Diaz-Veliz, G., Gonzalez, R. E., … Wyneken, U. (2010). Chronic fluoxetine treatment induces structural plasticity and selective changes in glutamate receptor subunits in the rat cerebral cortex. *Neuroscience, 169*(1), 98–108. http://dx.doi.org/10.1016/j.neuroscience.2010.04.035.

Anacker, C., Cattaneo, A., Musaelyan, K., Zunszain, P. A., Horowitz, M., Molteni, R., … Pariante, C. M. (2013). Role for the kinase SGK1 in stress, depression, and glucocorticoid effects on hippocampal neurogenesis. *Proceedings of the National Academy of Sciences of the United States of America, 110*(21), 8708–8713. http://dx.doi.org/10.1073/pnas.1300886110.

Anacker, C., Zunszain, P. A., Cattaneo, A., Carvalho, L. A., Garabedian, M. J., Thuret, S., … Pariante, C. M. (2011). Antidepressants increase human hippocampal neurogenesis by activating the glucocorticoid receptor. *Molecular Psychiatry, 16*, 738–750. http://dx.doi.org/10.1038/mp.2011.26.

Andres, A. L., Regev, L., Phi, L., Seese, R. R., Chen, Y., Gall, C. M., & Baram, T. Z. (2013). NMDA receptor activation and calpain contribute to disruption of dendritic spines by the stress neuropeptide CRH. *Journal of Neuroscience, 33*(43), 16945–16960. http://dx.doi.org/10.1523/jneurosci.1445-13.2013.

Aubry, J. M., Schwald, M., Ballmann, E., & Karege, F. (2009). Early effects of mood stabilizers on the Akt/GSK-3beta signaling pathway and on cell survival and proliferation. *Psychopharmacology (Berl), 205*(3), 419–429. http://dx.doi.org/10.1007/s00213-009-1551-2.

Autry, A. E., Adachi, M., Nosyreva, E., Na, E. S., Los, M. F., Cheng, P. F., … Monteggia, L. M. (2011). NMDA receptor blockade at rest triggers rapid behavioural antidepressant responses. *Nature, 475*(7354), 91–95. http://dx.doi.org/10.1038/nature10130.

Beneyto, M., & Meador-Woodruff, J. H. (2008). Lamina-specific abnormalities of NMDA receptor-associated postsynaptic protein transcripts in the prefrontal cortex in schizophrenia and bipolar disorder. *Neuropsychopharmacology, 33*(9), 2175–2186. http://dx.doi.org/10.1038/sj.npp.1301604.

Bessa, J. M., Ferreira, D., Melo, I., Marques, F., Cerqueira, J. J., Palha, J. A., … Sousa, N. (2009). The mood-improving actions of antidepressants do not depend on neurogenesis but are associated with neuronal remodeling. *Molecular Psychiatry, 14*(8), 764–773. http://dx.doi.org/10.1038/mp.2008.119, 739.

Beurel, E., Song, L., & Jope, R. S. (2011). Inhibition of glycogen synthase kinase-3 is necessary for the rapid antidepressant effect of ketamine in mice. *Molecular Psychiatry, 16*, 1068–1070. England.

Bianchi, M., Fone, K. F., Azmi, N., Heidbreder, C. A., Hagan, J. J., & Marsden, C. A. (2006). Isolation rearing induces recognition memory deficits accompanied by cytoskeletal alterations in rat hippocampus. *European Journal of Neuroscience, 24*(10), 2894–2902. http://dx.doi.org/10.1111/j.1460-9568.2006.05170.x.

Bockmühl, Y., Patchev, A. V., Madejska, A., Hoffmann, A., Sousa, J. C., Sousa, N., ... Spengler, D. (2015). Methylation at the CpG island shore region upregulates Nr3c1 promoter activity after early-life stress. *Epigenetics, 10*(3), 247–257. http://dx.doi.org/10.1080/15592294.2015.1017199.

Bramham, C. R. (2008). Local protein synthesis, actin dynamics, and LTP consolidation. *Current Opinion in Neurobiology, 18*(5), 524–531. http://dx.doi.org/10.1016/j.conb.2008.09.013.

Brocke, L., Srinivasan, M., & Schulman, H. (1995). Developmental and regional expression of multifunctional Ca2+/calmodulin-dependent protein kinase isoforms in rat brain. *Journal of Neuroscience, 15*(10), 6797–6808.

Bruel-Jungerman, E., Veyrac, A., Dufour, F., Horwood, J., Laroche, S., & Davis, S. (2009). Inhibition of PI3K-Akt signaling blocks exercise-mediated enhancement of adult neurogenesis and synaptic plasticity in the dentate gyrus. *PLoS One, 4*(11), e7901. http://dx.doi.org/10.1371/journal.pone.0007901.

Caceres, A., Payne, M. R., Binder, L. I., & Steward, O. (1983). Immunocytochemical localization of actin and microtubule-associated protein MAP2 in dendritic spines. *Proceedings of the National Academy of Sciences of the United States of America, 80*(6), 1738–1742.

Campbell, S., Marriott, M., Nahmias, C., & MacQueen, G. M. (2004). Lower hippocampal volume in patients suffering from depression: a meta-analysis. *American Journal of Psychiatry, 161*(4), 598–607.

Carlezon, W. A., Jr., Duman, R. S., & Nestler, E. J. (2005). The many faces of CREB. *Trends in Neurosciences, 28*, 436–445. http://dx.doi.org/10.1016/j.tins.2005.06.005.

Carter, C. J. (2007). Multiple genes and factors associated with bipolar disorder converge on growth factor and stress activated kinase pathways controlling translation initiation: implications for oligodendrocyte viability. *Neurochemistry International, 50*(3), 461–490. http://dx.doi.org/10.1016/j.neuint.2006.11.009.

Chen, Y., Dube, C. M., Rice, C. J., & Baram, T. Z. (2008). Rapid loss of dendritic spines after stress involves derangement of spine dynamics by corticotropin-releasing hormone. *Journal of Neuroscience, 28*(11), 2903–2911. http://dx.doi.org/10.1523/jneurosci.0225-08.2008.

Chen, Z. Y., Jing, D., Bath, K. G., Ieraci, A., Khan, T., Siao, C. J., ... Lee, F. S. (2006). Genetic variant BDNF (Val66Met) polymorphism alters anxiety-related behavior. *Science, 314*, 140–143. http://dx.doi.org/10.1126/science.1129663.

Chen, Y., Kramar, E. A., Chen, L. Y., Babayan, A. H., Andres, A. L., Gall, C. M., ... Baram, T. Z. (2013). Impairment of synaptic plasticity by the stress mediator CRH involves selective destruction of thin dendritic spines via RhoA signaling. *Molecular Psychiatry, 18*(4), 485–496. http://dx.doi.org/10.1038/mp.2012.17.

Chen, F., Madsen, T. M., Wegener, G., & Nyengaard, J. R. (2010). Imipramine treatment increases the number of hippocampal synapses and neurons in a genetic animal model of depression. *Hippocampus, 20*(12), 1376–1384. http://dx.doi.org/10.1002/hipo.20718.

Chiaruttini, C., Vicario, A., Li, Z., Baj, G., Braiuca, P., Wu, Y., ... Tongiorgi, E. (2009). Dendritic trafficking of BDNF mRNA is mediated by translin and blocked by the G196A (Val66Met) mutation. *Proceedings of the National Academy of Sciences of the United States of America, 106*(38), 16481–16486. http://dx.doi.org/10.1073/pnas.0902833106.

Chowdhury, S., Shepherd, J. D., Okuno, H., Lyford, G., Petralia, R. S., Plath, N., ... Worley, P. F. (2006). Arc/Arg3.1 interacts with the endocytic machinery to regulate AMPA receptor trafficking. *Neuron, 52*(3), 445–459. http://dx.doi.org/10.1016/j.neuron.2006.08.033.

Christian, K. M., Miracle, A. D., Wellman, C. L., & Nakazawa, K. (2011). Chronic stress-induced hippocampal dendritic retraction requires CA3 NMDA receptors. *Neuroscience, 174*, 26–36. http://dx.doi.org/10.1016/j.neuroscience.2010.11.033.

Ciani, L., & Salinas, P. C. (2005). WNTs in the vertebrate nervous system: from patterning to neuronal connectivity. *Nature Reviews Neuroscience, 6*(5), 351–362. http://dx.doi.org/10.1038/nrn1665.

Cline, H. T. (2001). Dendritic arbor development and synaptogenesis. *Current Opinion In Neurobiology, 11*(1), 118–126.

Cook, S. C., & Wellman, C. L. (2004). Chronic stress alters dendritic morphology in rat medial prefrontal cortex. *Journal of Neurobiology, 60*(2), 236–248. http://dx.doi.org/10.1002/neu.20025.

Cottingham, C., Li, X., & Wang, Q. (2012). Noradrenergic antidepressant responses to desipramine in vivo are reciprocally regulated by arrestin3 and spinophilin. *Neuropharmacology, 62*(7), 2354–2362. http://dx.doi.org/10.1016/j.neuropharm.2012.02.011.

Covington, H. E., 3rd, Lobo, M. K., Maze, I., Vialou, V., Hyman, J. M., Zaman, S., ... Nestler, E. J. (2010). Antidepressant effect of optogenetic stimulation of the medial prefrontal cortex. *Journal of Neuroscience, 30*(48), 16082–16090. http://dx.doi.org/10.1523/jneurosci.1731-10.2010.

Curmi, P. A., Andersen, S. S., Lachkar, S., Gavet, O., Karsenti, E., Knossow, M., & Sobel, A. (1997). The stathmin/tubulin interaction in vitro. *Journal of Biological Chemistry, 272*(40), 25029–25036.

Deisseroth, K., & Tsien, R. W. (2002). Dynamic multiphosphorylation passwords for activity-dependent gene expression. *Neuron, 34*, 179–182. United States.

Detera-Wadleigh, S. D. (2001). Lithium-related genetics of bipolar disorder. *Annals of Medicine, 33*(4), 272–285.

Duman, R. S., & Aghajanian, G. K. (2012). Synaptic dysfunction in depression: potential therapeutic targets. *Science, 338*, 68–72. http://dx.doi.org/10.1126/science.1222939.

Duman, R. S., & Monteggia, L. M. (2006). A neurotrophic model for stress-related mood disorders. *Biological Psychiatry, 59*(12), 1116–1127. http://dx.doi.org/10.1016/j.biopsych.2006.02.013.

Duric, V., Banasr, M., Licznerski, P., Schmidt, H. D., Stockmeier, C. A., Simen, A. A., ... Duman, R. S. (2010). A negative regulator of MAP kinase causes depressive behavior. *Nature Medicine, 16*, 1328–1332. http://dx.doi.org/10.1038/nm.2219.

Duric, V., Banasr, M., Stockmeier, C. A., Simen, A. A., Newton, S. S., Overholser, J. C., ... Duman, R. S. (2013). Altered expression of synapse and glutamate related genes in post-mortem hippocampus of depressed subjects. *International Journal of Neuropsychopharmacology, 16*(1), 69–82. http://dx.doi.org/10.1017/s1461145712000016.

Du, J., Wei, Y., Liu, L., Wang, Y., Khairova, R., Blumenthal, R., ... Manji, H. K. (2010). A kinesin signaling complex mediates the ability of GSK-3beta to affect mood-associated behaviors. *Proceedings of the National Academy of Sciences of the United States of America, 107*(25), 11573–11578. http://dx.doi.org/10.1073/pnas.0913138107.

Dwivedi, Y., Rizavi, H. S., Conley, R. R., Roberts, R. C., Tamminga, C. A., & Pandey, G. N. (2003). Altered gene expression of brain-derived neurotrophic factor and receptor tyrosine kinase B in postmortem brain of suicide subjects. *Archives of General Psychiatry, 60*(8), 804–815.

Dwivedi, Y., Rizavi, H. S., Conley, R. R., & Pandey, G. N. (2005). ERK MAP kinase signaling in post-mortem brain of suicide subjects: differential regulation of upstream Raf kinases Raf-1 and B-Raf. *Molecular Psychiatry, 11*, 86–98. http://dx.doi.org/10.1038/sj.mp.4001744.

Dwivedi, Y., Rizavi, H. S., Roberts, R. C., Conley, R. C., Tamminga, C. A., & Pandey, G. N. (2001). Reduced activation and expression of ERK1/2 MAP kinase in the post-mortem brain of depressed suicide subjects. *Journal of Neurochemistry, 77*, 916–928.

Dwivedi, Y., Rizavi, H. S., Zhang, H., Roberts, R. C., Conley, R. R., & Pandey, G. N. (2010). Modulation in activation and expression of phosphatase and tensin homolog on chromosome ten, Akt1, and 3-phosphoinositide-dependent kinase 1: further evidence demonstrating altered phosphoinositide 3-kinase signaling in postmortem brain of suicide subjects. *Biological Psychiatry, 67*, 1017–1025. http://dx.doi.org/10.1016/j.biopsych.2009.12.031.

Dwyer, J. M., Maldonado-Avilés, J. G., Lepack, A. E., DiLeone, R. J., & Duman, R. S. (2015). Ribosomal protein S6 kinase 1 signaling in prefrontal cortex controls depressive behavior. *Proceedings of the National Academy of Sciences of the United States of America*. . http://dx.doi.org/10.1073/pnas.1505289112.

Egan, M. F., Kojima, M., Callicott, J. H., Goldberg, T. E., Kolachana, B. S., Bertolino, A., … Weinberger, D. R. (2003). The BDNF val66met polymorphism affects activity-dependent secretion of BDNF and human memory and hippocampal function. *Cell, 112*(2), 257–269.

Elizalde, N., Pastor, P. M., Garcia-Garcia, A. L., Serres, F., Venzala, E., Huarte, J., … Tordera, R. M. (2010). Regulation of markers of synaptic function in mouse models of depression: chronic mild stress and decreased expression of VGLUT1. *Journal of Neurochemistry, 114*(5), 1302–1314. http://dx.doi.org/10.1111/j.1471-4159.2010.06854.x.

Elzinga, B. M., Molendijk, M. L., Oude Voshaar, R. C., Bus, B. A., Prickaerts, J., Spinhoven, P., et al. (2011). The impact of childhood abuse and recent stress on serum brain-derived neurotrophic factor and the moderating role of BDNF Val66Met. *Psychopharmacology (Berl), 214*(1), 319–328.

Fanous, S., Hammer, R. P., Jr., & Nikulina, E. M. (2010). Short- and long-term effects of intermittent social defeat stress on brain-derived neurotrophic factor expression in mesocorticolimbic brain regions. *Neuroscience, 167*(3), 598–607. http://dx.doi.org/10.1016/j.neuroscience.2010.02.064.

Feng, P., Guan, Z., Yang, X., & Fang, J. (2003). Impairments of ERK signal transduction in the brain in a rat model of depression induced by neonatal exposure of clomipramine. *Brain Research, 991*, 195–205.

Feng, J., Yan, Z., Ferreira, A., Tomizawa, K., Liauw, J. A., Zhuo, M., … Greengard, P. (2000). Spinophilin regulates the formation and function of dendritic spines. *Proceedings of the National Academy of Sciences of the United States of America, 97*(16), 9287–9292.

Feyissa, A. M., Chandran, A., Stockmeier, C. A., & Karolewicz, B. (2009). Reduced levels of NR2A and NR2B subunits of NMDA receptor and PSD-95 in the prefrontal cortex in major depression. *Progress in Neuropsychopharmacology and Biological Psychiatry, 33*(1), 70–75. http://dx.doi.org/10.1016/j.pnpbp.2008.10.005.

First, M., Gil-Ad, I., Taler, M., Tarasenko, I., Novak, N., & Weizman, A. (2011). The effects of fluoxetine treatment in a chronic mild stress rat model on depression-related behavior, brain neurotrophins and ERK expression. *Journal of Molecular Neuroscience, 45*(2), 246–255. http://dx.doi.org/10.1007/s12031-011-9515-5.

Frodl, T., Reinhold, E., Koutsouleris, N., Reiser, M., & Meisenzahl, E. M. (2010). Interaction of childhood stress with hippocampus and prefrontal cortex volume reduction in major depression. *Journal of Psychiatric Research, 44*(13), 799–807. http://dx.doi.org/10.1016/j.jpsychires.2010.01.006.

Frodl, T., Schule, C., Schmitt, G., Born, C., Baghai, T., Zill, P., … Meisenzahl, E. M. (2007). Association of the brain-derived neurotrophic factor Val66Met polymorphism with reduced hippocampal volumes in major depression. *Archives of General Psychiatry, 64*(4), 410–416. http://dx.doi.org/10.1001/archpsyc.64.4.410.

Frodl, T., Skokauskas, N., Frey, E. M., Morris, D., Gill, M., & Carballedo, A. (2014). BDNF Val66Met genotype interacts with childhood adversity and influences the formation of hippocampal subfields. *Human Brain Mapping, 35*(12), 5776–5783. http://dx.doi.org/10.1002/hbm.22584

Garcia-Garcia, A. L., Elizalde, N., Matrov, D., Harro, J., Wojcik, S. M., Venzala, E., … Tordera, R. M. (2009). Increased vulnerability to depressive-like behavior of mice with decreased expression of VGLUT1. *Biological Psychiatry, 66*(3), 275–282. http://dx.doi.org/10.1016/j.biopsych.2009.02.027.

Gatt, J. M., Nemeroff, C. B., Dobson-Stone, C., Paul, R. H., Bryant, R. A., Schofield, P. R., … Williams, L. M. (2009). Interactions between BDNF Val66Met polymorphism and early life stress predict brain and arousal pathways to syndromal depression and anxiety. *Molecular Psychiatry, 14*(7), 681–695. http://dx.doi.org/10.1038/mp.2008.143.

Gold, P. W. (2015). The organization of the stress system and its dysregulation in depressive illness. *Molecular Psychiatry, 20*(1), 32–47. http://dx.doi.org/10.1038/mp.2014.163.

Golden, S. A., Christoffel, D. J., Heshmati, M., Hodes, G. E., Magida, J., Davis, K., … Russo, S. J. (2013). Epigenetic regulation of RAC1 induces synaptic remodeling in stress disorders and depression. *Nature Medicine, 19*(3), 337–344. http://dx.doi.org/10.1038/nm.3090.

Goldwater, D. S., Pavlides, C., Hunter, R. G., Bloss, E. B., Hof, P. R., McEwen, B. S., & Morrison, J. H. (2009). Structural and functional alterations to rat medial prefrontal cortex following chronic restraint stress and recovery. *Neuroscience, 164*(2), 798–808. http://dx.doi.org/10.1016/j.neuroscience.2009.08.053.

Gould, T. D., Einat, H., Bhat, R., & Manji, H. K. (2004). AR-A014418, a selective GSK-3 inhibitor, produces antidepressant-like effects in the forced swim test. *International Journal of Neuropsychopharmacology, 7*(4), 387–390. http://dx.doi.org/10.1017/s1461145704004535.

Gould, T. D., & Manji, H. K. (2005). Glycogen synthase kinase-3: a putative molecular target for lithium mimetic drugs. *Neuropsychopharmacology, 30*(7), 1223–1237. http://dx.doi.org/10.1038/sj.npp.1300731.

Gourley, S. L., Wu, F. J., Kiraly, D. D., Ploski, J. E., Kedves, A. T., Duman, R. S., & Taylor, J. R. (2007). Regionally specific regulation of ERK MAP kinase in a model of antidepressant-sensitive chronic depression. *Biological Psychiatry, 63*, 353–359. http://dx.doi.org/10.1016/j.biopsych.2007.07.016.

Govindarajan, A., Rao, B. S., Nair, D., Trinh, M., Mawjee, N., Tonegawa, S., & Chattarji, S. (2006). Transgenic brain-derived neurotrophic factor expression causes both anxiogenic and antidepressant effects. *Proceedings of the National Academy of Sciences of the United States of America, 103*(35), 13208–13213. http://dx.doi.org/10.1073/pnas.0605180103.

Granseth, B., Odermatt, B., Royle, S. J., & Lagnado, L. (2006). Clathrin-mediated endocytosis is the dominant mechanism of vesicle retrieval at hippocampal synapses. *Neuron, 51*(6), 773–786. http://dx.doi.org/10.1016/j.neuron.2006.08.029.

Hajszan, T., Dow, A., Warner-Schmidt, J. L., Szigeti-Buck, K., Sallam, N. L., Parducz, A., … Duman, R. S. (2009). Remodeling of hippocampal spine synapses in the rat learned helplessness model of depression. *Biological Psychiatry, 65*(5), 392–400. http://dx.doi.org/10.1016/j.biopsych.2008.09.031.

Hajszan, T., MacLusky, N. J., & Leranth, C. (2005). Short-term treatment with the antidepressant fluoxetine triggers pyramidal dendritic spine synapse formation in rat hippocampus. *European Journal of Neuroscience, 21*(5), 1299–1303. http://dx.doi.org/10.1111/j.1460-9568.2005.03968.x.

Harada, A., Teng, J., Takei, Y., Oguchi, K., & Hirokawa, N. (2002). MAP2 is required for dendrite elongation, PKA anchoring in dendrites, and proper PKA signal transduction. *Journal of Cell Biology, 158*(3), 541–549. http://dx.doi.org/10.1083/jcb.200110134.

Hasler, G., van der Veen, J. W., Tumonis, T., Meyers, N., Shen, J., & Drevets, W. C. (2007). Reduced prefrontal glutamate/glutamine and gamma-aminobutyric acid levels in major depression determined using proton magnetic resonance spectroscopy. *Archives of General Psychiatry, 64*(2), 193–200. http://dx.doi.org/10.1001/archpsyc.64.2.193.

Heim, C., & Binder, E. B. (2012). Current research trends in early life stress and depression: review of human studies on sensitive periods, gene-environment interactions, and epigenetics. *Experimental Neurology, 233*(1), 102–111. http://dx.doi.org/10.1016/j.expneurol.2011.10.032.

Heim, C., Plotsky, P. M., & Nemeroff, C. B. (2004). Importance of studying the contributions of early adverse experience to neurobiological findings in depression. *Neuropsychopharmacology, 29*(4), 641–648. http://dx.doi.org/10.1038/sj.npp.1300397.

Henderson, B. R. (2000). Nuclear-cytoplasmic shuttling of APC regulates beta-catenin subcellular localization and turnover. *Nature Cell Biology, 2*(9), 653–660. http://dx.doi.org/10.1038/35023605.

Henry, F. E., McCartney, A. J., Neely, R., Perez, A. S., Carruthers, C. J., Stuenkel, E. L., … Sutton, M. A. (2012). Retrograde changes in presynaptic function driven by dendritic mTORC1. *Journal of Neuroscience, 32*, 17128–17142. http://dx.doi.org/10.1523/jneurosci.2149-12.2012.

Hiester, B. G., Galati, D. F., Salinas, P. C., & Jones, K. R. (2013). Neurotrophin and Wnt signaling cooperatively regulate dendritic spine formation. *Molecular and Cellular Neuroscience, 56*, 115–127. http://dx.doi.org/10.1016/j.mcn.2013.04.006.

Holtmaat, A. J., Trachtenberg, J. T., Wilbrecht, L., Shepherd, G. M., Zhang, X., Knott, G. W., & Svoboda, K. (2005). Transient and persistent dendritic spines in the neocortex in vivo. *Neuron, 45*(2), 279–291. http://dx.doi.org/10.1016/j.neuron.2005.01.003.

Hongisto, V., Vainio, J. C., Thompson, R., Courtney, M. J., & Coffey, E. T. (2008). The Wnt pool of glycogen synthase kinase 3beta is critical for trophic-deprivation-induced neuronal death. *Molecular and Cellular Biology, 28*(5), 1515–1527. http://dx.doi.org/10.1128/MCB.02227-06 [pii].

Honkura, N., Matsuzaki, M., Noguchi, J., Ellis-Davies, G. C., & Kasai, H. (2008). The subspine organization of actin fibers regulates the structure and plasticity of dendritic spines. *Neuron, 57*(5), 719–729. http://dx.doi.org/10.1016/j.neuron.2008.01.013.

Hoogenraad, C. C., & Akhmanova, A. (2010). Dendritic spine plasticity: new regulatory roles of dynamic microtubules. *Neuroscientist, 16*(6), 650–661. http://dx.doi.org/10.1177/1073858410386357.

Hotulainen, P., Llano, O., Smirnov, S., Tanhuanpaa, K., Faix, J., Rivera, C., & Lappalainen, P. (2009). Defining mechanisms of actin polymerization and depolymerization during dendritic spine morphogenesis. *Journal of Cell Biology, 185*(2), 323–339. http://dx.doi.org/10.1083/jcb.200809046.

Hsiung, S. C., Adlersberg, M., Arango, V., Mann, J. J., Tamir, H., & Liu, K. P. (2003). Attenuated 5-HT1A receptor signaling in brains of suicide victims: involvement of adenylyl cyclase, phosphatidylinositol 3-kinase, Akt and mitogen-activated protein kinase. *Journal of Neurochemistry, 87*(1), 182–194.

Hu, Y., Zhou, J., Fang, L., Liu, H., Zhan, Q., Luo, D., … Xie, P. (2013). Hippocampal synaptic dysregulation of exo/endocytosis-associated proteins induced in a chronic mild-stressed rat model. *Neuroscience, 230*, 1–12. http://dx.doi.org/10.1016/j.neuroscience.2012.08.026.

Inkster, B., Nichols, T. E., Saemann, P. G., Auer, D. P., Holsboer, F., Muglia, P., & Matthews, P. M. (2009). Association of GSK3beta polymorphisms with brain structural changes in major depressive disorder. *Archives of General Psychiatry, 66*(7), 721–728. http://dx.doi.org/10.1001/archgenpsychiatry.2009.70.

Jakawich, S. K., Nasser, H. B., Strong, M. J., McCartney, A. J., Perez, A. S., Rakesh, N., … Sutton, M. A. (2010). Local presynaptic activity gates homeostatic changes in presynaptic function driven by dendritic BDNF synthesis. *Neuron, 68*, 1143–1158. http://dx.doi.org/10.1016/j.neuron.2010.11.034.

Jeanneteau, F., Deinhardt, K., Miyoshi, G., Bennett, A. M., & Chao, M. V. (2010). The MAP kinase phosphatase MKP-1 regulates BDNF-induced axon branching. *Nature Neuroscience, 13*(11), 1373–1379. http://dx.doi.org/10.1038/nn.2655.

Jernigan, C. S., Goswami, D. B., Austin, M. C., Iyo, A. H., Chandran, A., Stockmeier, C. A., & Karolewicz, B. (2011). The mTOR signaling pathway in the prefrontal cortex is compromised in major depressive disorder. *Progress in Neuropsychopharmacology and Biological Psychiatry, 35*, 1774–1779. http://dx.doi.org/10.1016/j.pnpbp.2011.05.010.

Jourdi, H., Hsu, Y. T., Zhou, M., Qin, Q., Bi, X., & Baudry, M. (2009). Positive AMPA receptor modulation rapidly stimulates BDNF release and increases dendritic mRNA translation. *Journal of Neuroscience, 29*(27), 8688–8697. http://dx.doi.org/10.1523/jneurosci.6078-08.2009.

Jourdi, H., Iwakura, Y., Narisawa-Saito, M., Ibaraki, K., Xiong, H., Watanabe, M., … Nawa, H. (2003). Brain-derived neurotrophic factor signal enhances and maintains the expression of AMPA receptor-associated PDZ proteins in developing cortical neurons. *Developmental Biology, 263*(2), 216–230.

Kang, H. J., Voleti, B., Hajszan, T., Rajkowska, G., Stockmeier, C. A., Licznerski, P., … Duman, R. S. (2012). Decreased expression of synapse-related genes and loss of synapses in major depressive disorder. *Nature Medicine, 18*, 1413–1417. http://dx.doi.org/10.1038/nm.2886.

Kapitein, L. C., & Hoogenraad, C. C. (2011). Which way to go? Cytoskeletal organization and polarized transport in neurons. *Molecular and Cellular Neuroscience, 46*(1), 9–20. http://dx.doi.org/10.1016/j.mcn.2010.08.015.

Karege, F., Perret, G., Bondolfi, G., Schwald, M., Bertschy, G., & Aubry, J. M. (2002). Decreased serum brain-derived neurotrophic factor levels in major depressed patients. *Psychiatry Research, 109*(2), 143–148.

Karege, F., Perroud, N., Burkhardt, S., Schwald, M., Ballmann, E., La Harpe, R., & Malafosse, A. (2007). Alteration in kinase activity but not in protein levels of protein kinase B and glycogen synthase kinase-3beta in ventral prefrontal cortex of depressed suicide victims. *Biological Psychiatry, 61*(2), 240–245. http://dx.doi.org/10.1016/j.biopsych.2006.04.036.

Keyse, S. M., & Emslie, E. A. (1992). Oxidative stress and heat shock induce a human gene encoding a protein-tyrosine phosphatase. *Nature, 359*(6396), 644–647. http://dx.doi.org/10.1038/359644a0.

Kikuchi, A., Yamamoto, H., & Kishida, S. (2007). Multiplicity of the interactions of Wnt proteins and their receptors. *Cellular Signalling, 19*(4), 659–671. http://dx.doi.org/10.1016/j.cellsig.2006.11.001.

Kim, S. S., Wang, H., Li, X. Y., Chen, T., Mercaldo, V., Descalzi, G., … Zhuo, M. (2011). Neurabin in the anterior cingulate cortex regulates anxiety-like behavior in adult mice. *Molecular Brain, 4*, 6. http://dx.doi.org/10.1186/1756-6606-4-6.

Kim, H., Yi, J. H., Choi, K., Hong, S., Shin, K. S., & Kang, S. J. (2014). Regional differences in acute corticosterone-induced dendritic remodeling in the rat brain and their behavioral consequences. *BMC Neuroscience, 15*, 65. http://dx.doi.org/10.1186/1471-2202-15-65.

King, M. V., Kurian, N., Qin, S., Papadopoulou, N., Westerink, B. H., Cremers, T. I., … Sharp, T. V. (2014). Lentiviral delivery of a vesicular glutamate transporter 1 (VGLUT1)-targeting short hairpin RNA vector into the mouse hippocampus impairs cognition. *Neuropsychopharmacology, 39*(2), 464–476. http://dx.doi.org/10.1038/npp.2013.220.

Kiraly, D. D., Lemtiri-Chlieh, F., Levine, E. S., Mains, R. E., & Eipper, B. A. (2011). Kalirin binds the NR2B subunit of the NMDA receptor, altering its synaptic localization and function. *Journal of Neuroscience, 31*(35), 12554–12565. http://dx.doi.org/10.1523/jneurosci.3143-11.2011.

Komulainen, E., Zdrojewska, J., Freemantle, E., Mohammad, H., Kulesskaya, N., Deshpande, P., … Coffey, E. T. (2014). JNK1 controls dendritic field size in L2/3 and L5 of the motor cortex, constrains soma size, and influences fine motor coordination. *Frontiers in Cellular Neuroscience, 8*, 272. http://dx.doi.org/10.3389/fncel.2014.00272.

Krishnan, V., Han, M. H., Mazei-Robison, M., Iniguez, S. D., Ables, J. L., Vialou, V., … Nestler, E. J. (2008). AKT signaling within the ventral tegmental area regulates cellular and behavioral responses to stressful stimuli. *Biological Psychiatry, 64*(8), 691–700. http://dx.doi.org/10.1016/j.biopsych.2008.06.003.

Krishnan, V., & Nestler, E. J. (2008). The molecular neurobiology of depression. *Nature, 455*(7215), 894–902. http://dx.doi.org/10.1038/nature07455.

Kumar, V., Zhang, M. X., Swank, M. W., Kunz, J., & Wu, G. Y. (2005). Regulation of dendritic morphogenesis by Ras-PI3K-Akt-mTOR and Ras-MAPK signaling pathways. *Journal of Neuroscience, 25*(49), 11288–11299. http://dx.doi.org/10.1523/jneurosci.2284-05.2005.

Laderoute, K. R., Mendonca, H. L., Calaoagan, J. M., Knapp, A. M., Giaccia, A. J., & Stork, P. J. (1999). Mitogen-activated protein kinase phosphatase-1 (MKP-1) expression is induced by low oxygen conditions found in solid tumor microenvironments. A candidate MKP for the inactivation of hypoxia-inducible stress-activated protein kinase/c-Jun N-terminal protein kinase activity. *Journal of Biological Chemistry, 274*(18), 12890–12897.

Lamprecht, R., Farb, C. R., Rodrigues, S. M., & LeDoux, J. E. (2006). Fear conditioning drives profilin into amygdala dendritic spines. *Nature Neuroscience, 9*(4), 481–483. http://dx.doi.org/10.1038/nn1672.

Law, A. J., & Deakin, J. F. (2001). Asymmetrical reductions of hippocampal NMDAR1 glutamate receptor mRNA in the psychoses. *Neuroreport, 12*(13), 2971–2974.

Law, A. J., Weickert, C. S., Hyde, T. M., Kleinman, J. E., & Harrison, P. J. (2004). Reduced spinophilin but not microtubule-associated protein 2 expression in the hippocampal formation in schizophrenia and mood disorders: molecular evidence for a pathology of dendritic spines. *American Journal of Psychiatry, 161*(10), 1848–1855. http://dx.doi.org/10.1176/appi.ajp.161.10.1848.

Leal, G., Comprido, D., & Duarte, C. B. (2014). BDNF-induced local protein synthesis and synaptic plasticity. *Neuropharmacology, 76*(Pt C), 639–656. http://dx.doi.org/10.1016/j.neuropharm.2013.04.005.

Li, X., & Jope, R. S. (2010). Is glycogen synthase kinase-3 a central modulator in mood regulation? *Neuropsychopharmacology, 35*(11), 2143–2154. http://dx.doi.org/10.1038/npp.2010.105.

Li, N., Lee, B., Liu, R. J., Banasr, M., Dwyer, J. M., Iwata, M., … Duman, R. S. (2010). mTOR-dependent synapse formation underlies the rapid antidepressant effects of NMDA antagonists. *Science, 329*, 959–964. http://dx.doi.org/10.1126/science.1190287.

Li, W., Li, Q. J., & An, S. C. (2010). Preventive effect of estrogen on depression-like behavior induced by chronic restraint stress. *Neuroscience Bulletin, 26*, 140–146. http://dx.doi.org/10.1007/s12264-010-0609-9.

Li, N., Liu, R. J., Dwyer, J. M., Banasr, M., Lee, B., Son, H., ... Duman, R. S. (2011). Glutamate N-methyl-D-aspartate receptor antagonists rapidly reverse behavioral and synaptic deficits caused by chronic stress exposure. *Biological Psychiatry, 69*(8), 754–761. http://dx.doi.org/10.1016/j.biopsych.2010.12.015.

Li, K., Zhou, T., Liao, L., Yang, Z., Wong, C., Henn, F., ... Hu, H. (2013). βCaMKII in lateral habenula mediates core symptoms of depression. *Science, 341*(6149), 1016–1020. http://dx.doi.org/10.1126/science.1240729.

Lisman, J. E., & Zhabotinsky, A. M. (2001). A model of synaptic memory: a CaMKII/PP1 switch that potentiates transmission by organizing an AMPA receptor anchoring assembly. *Neuron, 31*(2), 191–201.

Liu, R. J., Lee, F. S., Li, X. Y., Bambico, F., Duman, R. S., & Aghajanian, G. K. (2012). Brain-derived neurotrophic factor Val66Met allele impairs basal and ketamine-stimulated synaptogenesis in prefrontal cortex. *Biological Psychiatry, 71*(11), 996–1005. http://dx.doi.org/10.1016/j.biopsych.2011.09.030.

Logan, C. Y., & Nusse, R. (2004). The Wnt signaling pathway in development and disease. *Annual Review of Cell and Developmental Biology, 20*, 781–810. http://dx.doi.org/10.1146/annurev.cellbio.20.010403.113126.

Lyford, G. L., Yamagata, K., Kaufmann, W. E., Barnes, C. A., Sanders, L. K., Copeland, N. G., ... Worley, P. F. (1995). Arc, a growth factor and activity-regulated gene, encodes a novel cytoskeleton-associated protein that is enriched in neuronal dendrites. *Neuron, 14*, 433–445.

Maeng, S., Zarate, C. A., Jr., Du, J., Schloesser, R. J., McCammon, J., Chen, G., & Manji, H. K. (2008). Cellular mechanisms underlying the antidepressant effects of ketamine: role of alpha-amino-3-hydroxy-5-methylisoxazole-4-propionic acid receptors. *Biological Psychiatry, 63*(4), 349–352. http://dx.doi.org/10.1016/j.biopsych.2007.05.028.

Magarinos, A. M., Li, C. J., Gal Toth, J., Bath, K. G., Jing, D., Lee, F. S., & McEwen, B. S. (2011). Effect of brain-derived neurotrophic factor haploinsufficiency on stress-induced remodeling of hippocampal neurons. *Hippocampus, 21*(3), 253–264. http://dx.doi.org/10.1002/hipo.20744.

Magarinos, A. M., & McEwen, B. S. (1995). Stress-induced atrophy of apical dendrites of hippocampal CA3c neurons: comparison of stressors. *Neuroscience, 69*(1), 83–88.

Malenka, R. C., Kauer, J. A., Zucker, R. S., & Nicoll, R. A. (1988). Postsynaptic calcium is sufficient for potentiation of hippocampal synaptic transmission. *Science, 242*(4875), 81–84.

Malki, K., Pain, O., Tosto, M. G., Du Rietz, E., Carboni, L., & Schalkwyk, L. C. (2015). Identification of genes and gene pathways associated with major depressive disorder by integrative brain analysis of rat and human prefrontal cortex transcriptomes. *Translational Psychiatry, 5*, e519. http://dx.doi.org/10.1038/tp.2015.15.

Mallei, A., Giambelli, R., Gass, P., Racagni, G., Mathe, A. A., Vollmayr, B., & Popoli, M.00 (2011). Synaptoproteomics of learned helpless rats involve energy metabolism and cellular remodeling pathways in depressive-like behavior and antidepressant response. *Neuropharmacology, 60*(7–8), 1243–1253. http://dx.doi.org/10.1016/j.neuropharm.2010.12.012.

Mandela, P., & Ma, X. M. (2012). Kalirin, a key player in synapse formation, is implicated in human diseases. *Neural Plasticity, 2012*, 728161. http://dx.doi.org/10.1155/2012/728161.

Mao, Y., Ge, X., Frank, C. L., Madison, J. M., Koehler, A. N., Doud, M. K., ... Tsai, L. H. (2009). Disrupted in schizophrenia 1 regulates neuronal progenitor proliferation via modulation of GSK3beta/beta-catenin signaling. *Cell, 136*(6), 1017–1031. http://dx.doi.org/10.1016/j.cell.2008.12.044.

Marmigere, F., Givalois, L., Rage, F., Arancibia, S., & Tapia-Arancibia, L. (2003). Rapid induction of BDNF expression in the hippocampus during immobilization stress challenge in adult rats. *Hippocampus, 13*(5), 646–655. http://dx.doi.org/10.1002/hipo.10109.

Maroun, M., Ioannides, P. J., Bergman, K. L., Kavushansky, A., Holmes, A., & Wellman, C. L. (2013). Fear extinction deficits following acute stress associate with increased spine density and dendritic retraction in basolateral amygdala neurons. *European Journal of Neuroscience, 38*(4), 2611–2620. http://dx.doi.org/10.1111/ejn.12259.

Martin, K. P., & Wellman, C. L. (2011). NMDA receptor blockade alters stress-induced dendritic remodeling in medial prefrontal cortex. *Cerebral Cortex, 21*(10), 2366–2373. http://dx.doi.org/10.1093/cercor/bhr021.

Martinez-Tellez, R. I., Hernandez-Torres, E., Gamboa, C., & Flores, G. (2009). Prenatal stress alters spine density and dendritic length of nucleus accumbens and hippocampus neurons in rat offspring. *Synapse, 63*(9), 794–804. http://dx.doi.org/10.1002/syn.20664.

Martisova, E., Solas, M., Horrillo, I., Ortega, J. E., Meana, J. J., Tordera, R. M., & Ramirez, M. J. (2012). Long lasting effects of early-life stress on glutamatergic/GABAergic circuitry in the rat hippocampus. *Neuropharmacology, 62*(5–6), 1944–1953. http://dx.doi.org/10.1016/j.neuropharm.2011.12.019.

McAvoy, K., Russo, C., Kim, S., Rankin, G., & Sahay, A. (2015). Fluoxetine induces input-specific hippocampal dendritic spine remodeling along the septotemporal axis in adulthood and middle age. *Hippocampus*. . http://dx.doi.org/10.1002/hipo.22464.

McEwen, B. S., Conrad, C. D., Kuroda, Y., Frankfurt, M., Magarinos, A. M., & McKittrick, C. (1997). Prevention of stress-induced morphological and cognitive consequences. *European Neuropsychopharmacology, 7*(Suppl 3), S323–S328.

Meller, E., Shen, C., Nikolao, T. A., Jensen, C., Tsimberg, Y., Chen, J., & Gruen, R. J. (2003). Region-specific effects of acute and repeated restraint stress on the phosphorylation of mitogen-activated protein kinases. *Brain Research, 979*, 57–64.

Meng, Y., Zhang, Y., Tregoubov, V., Janus, C., Cruz, L., Jackson, M., … Jia, Z. (2002). Abnormal spine morphology and enhanced LTP in LIMK-1 knockout mice. *Neuron, 35*, 121–133.

Mitra, R., Jadhav, S., McEwen, B. S., Vyas, A., & Chattarji, S. (2005). Stress duration modulates the spatiotemporal patterns of spine formation in the basolateral amygdala. *Proceedings of the National Academy of Sciences of the United States of America, 102*(26), 9371–9376. http://dx.doi.org/10.1073/pnas.0504011102.

Mitra, R., & Sapolsky, R. M. (2008). Acute corticosterone treatment is sufficient to induce anxiety and amygdaloid dendritic hypertrophy. *Proceedings of the National Academy of Sciences of the United States of America, 105*(14), 5573–5578. http://dx.doi.org/10.1073/pnas.0705615105.

Moghaddam, B., Adams, B., Verma, A., & Daly, D. (1997). Activation of glutamatergic neurotransmission by ketamine: a novel step in the pathway from NMDA receptor blockade to dopaminergic and cognitive disruptions associated with the prefrontal cortex. *Journal of Neuroscience, 17*, 2921–2927.

Monteggia, L. M., Luikart, B., Barrot, M., Theobold, D., Malkovska, I., Nef, S., … Nestler, E. J. (2007). Brain-derived neurotrophic factor conditional knockouts show gender differences in depression-related behaviors. *Biological Psychiatry, 61*(2), 187–197. http://dx.doi.org/10.1016/j.biopsych.2006.03.021.

Murakoshi, H., Wang, H., & Yasuda, R. (2011). Local, persistent activation of Rho GTPases during plasticity of single dendritic spines. *Nature, 472*(7341), 100–104. http://dx.doi.org/10.1038/nature09823.

Nakatani, N., Aburatani, H., Nishimura, K., Semba, J., & Yoshikawa, T. (2004). Comprehensive expression analysis of a rat depression model. *Pharmacogenomics Journal, 4*(2), 114–126. http://dx.doi.org/10.1038/sj.tpj.6500234.

Nakayama, A. Y., Harms, M. B., & Luo, L. (2000). Small GTPases Rac and Rho in the maintenance of dendritic spines and branches in hippocampal pyramidal neurons. *Journal of Neuroscience, 20*(14), 5329–5338.

Nibuya, M., Nestler, E. J., & Duman, R. S. (1996). Chronic antidepressant administration increases the expression of cAMP response element binding protein (CREB) in rat hippocampus. *Journal of Neuroscience, 16*, 2365–2372.

Niell, C. M., Meyer, M. P., & Smith, S. J. (2004). In vivo imaging of synapse formation on a growing dendritic arbor. *Nature Neuroscience, 7*(3), 254–260. http://dx.doi.org/10.1038/nn1191.

Niwa, R., Nagata-Ohashi, K., Takeichi, M., Mizuno, K., & Uemura, T. (2002). Control of actin reorganization by Slingshot, a family of phosphatases that dephosphorylate ADF/cofilin. *Cell, 108*(2), 233–246.

Norrholm, S. D., & Ouimet, C. C. (2001). Altered dendritic spine density in animal models of depression and in response to antidepressant treatment. *Synapse, 42*(3), 151–163. http://dx.doi.org/10.1002/syn.10006.

Nusse, R., Fuerer, C., Ching, W., Harnish, K., Logan, C., Zeng, A., … Kalani, Y. (2008). Wnt signaling and stem cell control. *Cold Spring Harbor Symposia on Quantitative Biology, 73*, 59–66. http://dx.doi.org/10.1101/sqb.2008.73.035.

O'Leary, O. F., Wu, X., & Castren, E. (2009). Chronic fluoxetine treatment increases expression of synaptic proteins in the hippocampus of the ovariectomized rat: role of BDNF signalling. *Psychoneuroendocrinology, 34*(3), 367–381. http://dx.doi.org/10.1016/j.psyneuen.2008.09.015.

Okamoto, K., Nagai, T., Miyawaki, A., & Hayashi, Y. (2004). Rapid and persistent modulation of actin dynamics regulates postsynaptic reorganization underlying bidirectional plasticity. *Nature Neuroscience, 7*(10), 1104–1112. http://dx.doi.org/10.1038/nn1311.

Okamoto, H., Voleti, B., Banasr, M., Sarhan, M., Duric, V., Girgenti, M. J., … Duman, R. S. (2010). Wnt2 expression and signaling is increased by different classes of antidepressant treatments. *Biological Psychiatry, 68*(6), 521–527. http://dx.doi.org/10.1016/j.biopsych.2010.04.023.

Orlowski, D., Elfving, B., Muller, H. K., Wegener, G., & Bjarkam, C. R. (2012). Wistar rats subjected to chronic restraint stress display increased hippocampal spine density paralleled by increased expression levels of synaptic scaffolding proteins. *Stress, 15*(5), 514–523. http://dx.doi.org/10.3109/10253890.2011.643516.

Ota, K. T., Liu, R. J., Voleti, B., Maldonado-Aviles, J. G., Duric, V., Iwata, M., … Duman, R. S. (2014). REDD1 is essential for stress-induced synaptic loss and depressive behavior. *Nature Medicine, 20*(5), 531–535. http://dx.doi.org/10.1038/nm.3513.

Pillai, A. G., Anilkumar, S., & Chattarji, S. (2012). The same antidepressant elicits contrasting patterns of synaptic changes in the amygdala vs hippocampus. *Neuropsychopharmacology, 37*(12), 2702–2711. http://dx.doi.org/10.1038/npp.2012.135.

Pinnock, S. B., Blake, A. M., Platt, N. J., & Herbert, J. (2010). The roles of BDNF, pCREB and Wnt3a in the latent period preceding activation of progenitor cell mitosis in the adult dentate gyrus by fluoxetine. *PLoS One, 5*(10), e13652. http://dx.doi.org/10.1371/journal.pone.0013652.

Pizarro, J. M., Lumley, L. A., Medina, W., Robison, C. L., Chang, W. E., Alagappan, A., … Meyerhoff, J. L. (2004). Acute social defeat reduces neurotrophin expression in brain cortical and subcortical areas in mice. *Brain Research, 1025*(1–2), 10–20. http://dx.doi.org/10.1016/j.brainres.2004.06.085.

Purro, S. A., Ciani, L., Hoyos-Flight, M., Stamatakou, E., Siomou, E., & Salinas, P. C. (2008). Wnt regulates axon behavior through changes in microtubule growth directionality: a new role for adenomatous polyposis coli. *Journal of Neuroscience, 28*(34), 8644–8654. http://dx.doi.org/10.1523/jneurosci.2320-08.2008.

Qiao, H., An, S. C., Ren, W., & Ma, X. M. (2014). Progressive alterations of hippocampal CA3-CA1 synapses in an animal model of depression. *Behavioural Brain Research, 275*, 191–200. http://dx.doi.org/10.1016/j.bbr.2014.08.040.

Qi, X., Lin, W., Li, J., Li, H., Wang, W., Wang, D., & Sun, M. (2008). Fluoxetine increases the activity of the ERK-CREB signal system and alleviates the depressive-like behavior in rats exposed to chronic forced swim stress. *Neurobiology of Disease, 31*, 278–285. http://dx.doi.org/10.1016/j.nbd.2008.05.003.

Qi, X., Lin, W., Li, J., Pan, Y., & Wang, W. (2006). The depressive-like behaviors are correlated with decreased phosphorylation of mitogen-activated protein kinases in rat brain following chronic forced swim stress. *Behavioural Brain Research, 175*, 233–240. http://dx.doi.org/10.1016/j.bbr.2006.08.035.

Radley, J. J., Anderson, R. M., Hamilton, B. A., Alcock, J. A., & Romig-Martin, S. A. (2013). Chronic stress-induced alterations of dendritic spine subtypes predict functional decrements in an hypothalamo-pituitary-adrenal-inhibitory prefrontal circuit. *Journal of Neuroscience, 33*(36), 14379–14391. http://dx.doi.org/10.1523/jneurosci.0287-13.2013.

Radley, J. J., Gosselink, K. L., & Sawchenko, P. E. (2009). A discrete GABAergic relay mediates medial prefrontal cortical inhibition of the neuroendocrine stress response. *Journal of Neuroscience, 29*(22), 7330–7340. http://dx.doi.org/10.1523/jneurosci.5924-08.2009.

Radley, J. J., Rocher, A. B., Miller, M., Janssen, W. G., Liston, C., Hof, P. R., … Morrison, J. H. (2006). Repeated stress induces dendritic spine loss in the rat medial prefrontal cortex. *Cerebral Cortex, 16*(3), 313–320. http://dx.doi.org/10.1093/cercor/bhi104.

Radley, J. J., Rocher, A. B., Rodriguez, A., Ehlenberger, D. B., Dammann, M., McEwen, B. S., … Hof, P. R. (2008). Repeated stress alters dendritic spine morphology in the rat medial prefrontal cortex. *Journal of Comparative Neurology, 507*(1), 1141–1150. http://dx.doi.org/10.1002/cne.21588.

Radley, J. J., Sisti, H. M., Hao, J., Rocher, A. B., McCall, T., Hof, P. R., … Morrison, J. H. (2004). Chronic behavioral stress induces apical dendritic reorganization in pyramidal neurons of the medial prefrontal cortex. *Neuroscience, 125*(1), 1–6. http://dx.doi.org/10.1016/j.neuroscience.2004.01.006.

Rajkowska, G., O'Dwyer, G., Teleki, Z., Stockmeier, C. A., & Miguel-Hidalgo, J. J. (2007). GABAergic neurons immunoreactive for calcium binding proteins are reduced in the prefrontal cortex in major depression. *Neuropsychopharmacology, 32*(2), 471–482. http://dx.doi.org/10.1038/sj.npp.1301234.

Rasmussen, K. G. (2014). Psychiatric side effects of ketamine in hospitalized medical patients administered subanesthetic doses for pain control. *Acta Neuropsychiatrica, 26*(4), 230–233. http://dx.doi.org/10.1017/neu.2013.61.

Rasmusson, A. M., Shi, L., & Duman, R. (2002). Downregulation of BDNF mRNA in the hippocampal dentate gyrus after re-exposure to cues previously associated with footshock. *Neuropsychopharmacology, 27*(2), 133–142. http://dx.doi.org/10.1016/s0893-133x(02)00286-5.

Ravinder, S., Pillai, A. G., & Chattarji, S. (2011). Cellular correlates of enhanced anxiety caused by acute treatment with the selective serotonin reuptake inhibitor fluoxetine in rats. *Frontiers in Behavioral Neuroscience, 5*, 88. http://dx.doi.org/10.3389/fnbeh.2011.00088.

Refojo, D., Schweizer, M., Kuehne, C., Ehrenberg, S., Thoeringer, C., Vogl, A. M., … Deussing, J. M. (2011). Glutamatergic and dopaminergic neurons mediate anxiogenic and anxiolytic effects of CRHR1. *Science, 333*(6051), 1903–1907. http://dx.doi.org/10.1126/science.1202107.

Rochefort, N. L., & Konnerth, A. (2012). Dendritic spines: from structure to in vivo function. *EMBO Reports, 13*(8), 699–708. http://dx.doi.org/10.1038/embor.2012.102.

Rubio, F. J., Ampuero, E., Sandoval, R., Toledo, J., Pancetti, F., & Wyneken, U. (2013). Long-term fluoxetine treatment induces input-specific LTP and LTD impairment and structural plasticity in the CA1 hippocampal subfield. *Frontiers in Cellular Neuroscience, 7*, 66. http://dx.doi.org/10.3389/fncel.2013.00066.

Russell, T. A., Blizinsky, K. D., Cobia, D. J., Cahill, M. E., Xie, Z., Sweet, R. A., ... Penzes, P. (2014). A sequence variant in human KALRN impairs protein function and coincides with reduced cortical thickness. *Nature Communications, 5*, 4858. http://dx.doi.org/10.1038/ncomms5858.

Ryan, B., Musazzi, L., Mallei, A., Tardito, D., Gruber, S. H., El Khoury, A., ... Popoli, M. (2009). Remodelling by early-life stress of NMDA receptor-dependent synaptic plasticity in a gene-environment rat model of depression. *International Journal of Neuropsychopharmacology, 12*(4), 553–559. http://dx.doi.org/10.1017/s1461145708009607.

Sabatini, D. M., Erdjument-Bromage, H., Lui, M., Tempst, P., & Snyder, S. H. (1994). RAFT1: a mammalian protein that binds to FKBP12 in a rapamycin-dependent fashion and is homologous to yeast TORs. *Cell, 78*(1), 35–43.

Sanacora, G., Gueorguieva, R., Epperson, C. N., Wu, Y. T., Appel, M., Rothman, D. L., ... Mason, G. F. (2004). Subtype-specific alterations of gamma-aminobutyric acid and glutamate in patients with major depression. *Archives of General Psychiatry, 61*(7), 705–713. http://dx.doi.org/10.1001/archpsyc.61.7.705.

Saus, E., Soria, V., Escaramis, G., Crespo, J. M., Valero, J., Gutierrez-Zotes, A., ... Urretavizcaya, M. (2010). A haplotype of glycogen synthase kinase 3beta is associated with early onset of unipolar major depression. *Genes, Brain, and Behavior, 9*(7), 799–807. http://dx.doi.org/10.1111/j.1601-183X.2010.00617.x.

Savitz, J., & Drevets, W. C. (2009). Bipolar and major depressive disorder: neuroimaging the developmental-degenerative divide. *Neuroscience and Biobehavioral Reviews, 33*(5), 699–771. http://dx.doi.org/10.1016/j.neubiorev.2009.01.004.

Schaaf, M. J., de Jong, J., de Kloet, E. R., & Vreugdenhil, E. (1998). Downregulation of BDNF mRNA and protein in the rat hippocampus by corticosterone. *Brain Research, 813*(1), 112–120.

Schratt, G. M., Nigh, E. A., Chen, W. G., Hu, L., & Greenberg, M. E. (2004). BDNF regulates the translation of a select group of mRNAs by a mammalian target of rapamycin-phosphatidylinositol 3-kinase-dependent pathway during neuronal development. *Journal of Neuroscience, 24*(33), 7366–7377. http://dx.doi.org/10.1523/jneurosci.1739-04.2004.

Schratt, G. M., Tuebing, F., Nigh, E. A., Kane, C. G., Sabatini, M. E., Kiebler, M., & Greenberg, M. E. (2006). A brain-specific microRNA regulates dendritic spine development. *Nature, 439*(7074), 283–289. http://dx.doi.org/10.1038/nature04367.

Seta, K. A., Kim, R., Kim, H. W., Millhorn, D. E., & Beitner-Johnson, D. (2001). Hypoxia-induced regulation of MAPK phosphatase-1 as identified by subtractive suppression hybridization and cDNA microarray analysis. *Journal of Biological Chemistry, 276*(48), 44405–44412. http://dx.doi.org/10.1074/jbc.M103346200.

Sharma, H. R., & Thakur, M. K. (2015). Correlation of ERα/ERβ expression with dendritic and behavioural changes in CUMS mice. *Physiology and Behavior, 145*, 71–83. http://dx.doi.org/10.1016/j.physbeh.2015.03.041.

Shelton, M. A., Newman, J. T., Gu, H., Sampson, A. R., Fish, K. N., MacDonald, M. L., ... Sweet, R. A. (2015). Loss of microtubule-associated protein 2 immunoreactivity linked to dendritic spine loss in schizophrenia. *Biological Psychiatry*. . http://dx.doi.org/10.1016/j.biopsych.2014.12.029.

Sheng, M., Thompson, M. A., & Greenberg, M. E. (1991). CREB: a Ca(2+)-regulated transcription factor phosphorylated by calmodulin-dependent kinases. *Science, 252*(5011), 1427–1430.

Shimizu, E., Hashimoto, K., Okamura, N., Koike, K., Komatsu, N., Kumakiri, C., ... Iyo, M. (2003). Alterations of serum levels of brain-derived neurotrophic factor (BDNF) in depressed patients with or without antidepressants. *Biological Psychiatry, 54*(1), 70–75.

Silva, R., Mesquita, A. R., Bessa, J., Sousa, J. C., Sotiropoulos, I., Leao, P., ... Sousa, N. (2008). Lithium blocks stress-induced changes in depressive-like behavior and hippocampal cell fate: the role of glycogen-synthase-kinase-3beta. *Neuroscience, 152*(3), 656–669. http://dx.doi.org/10.1016/j.neuroscience.2007.12.026.

Sin, W. C., Haas, K., Ruthazer, E. S., & Cline, H. T. (2002). Dendrite growth increased by visual activity requires NMDA receptor and Rho GTPases. *Nature, 419*(6906), 475–480. http://dx.doi.org/10.1038/nature00987.

Smith, M. A., Makino, S., Kvetnansky, R., & Post, R. M. (1995). Stress and glucocorticoids affect the expression of brain-derived neurotrophic factor and neurotrophin-3 mRNAs in the hippocampus. *Journal of Neuroscience, 15*(3 Pt 1), 1768–1777.

Soetanto, A., Wilson, R. S., Talbot, K., Un, A., Schneider, J. A., Sobiesk, M., ... Arnold, S. E. (2010). Association of anxiety and depression with microtubule-associated protein 2- and synaptopodin-immunolabeled dendrite and spine densities in hippocampal CA3 of older humans. *Archives of General Psychiatry, 67*(5), 448–457. http://dx.doi.org/10.1001/archgenpsychiatry.2010.48.

Solas, M., Gerenu, G., Gil-Bea, F. J., & Ramírez, M. J. (2013). Mineralocorticoid receptor activation induces insulin resistance through c-Jun N-terminal kinases in response to chronic corticosterone: cognitive implications. *Journal of Neuroendocrinology, 25*(4), 350–356. http://dx.doi.org/10.1111/jne.12006.

Star, E. N., Kwiatkowski, D. J., & Murthy, V. N. (2002). Rapid turnover of actin in dendritic spines and its regulation by activity. *Nature Neuroscience, 5*(3), 239–246. http://dx.doi.org/10.1038/nn811.

Stockmeier, C. A., Mahajan, G. J., Konick, L. C., Overholser, J. C., Jurjus, G. J., Meltzer, H. Y., ... Rajkowska, G. (2004). Cellular changes in the postmortem hippocampus in major depression. *Biological Psychiatry, 56*(9), 640–650. http://dx.doi.org/10.1016/j.biopsych.2004.08.022.

Tada, T., & Sheng, M. (2006). Molecular mechanisms of dendritic spine morphogenesis. *Current Opinion in Neurobiology, 16*(1), 95–101. http://dx.doi.org/10.1016/j.conb.2005.12.001.

Taliaz, D., Loya, A., Gersner, R., Haramati, S., Chen, A., & Zangen, A. (2011). Resilience to chronic stress is mediated by hippocampal brain-derived neurotrophic factor. *Journal of Neuroscience, 31*(12), 4475–4483. http://dx.doi.org/10.1523/JNEUROSCI.5725-10.2011.

Taliaz, D., Stall, N., Dar, D. E., & Zangen, A. (2010). Knockdown of brain-derived neurotrophic factor in specific brain sites precipitates behaviors associated with depression and reduces neurogenesis. *Molecular Psychiatry, 15*(1), 80–92. http://dx.doi.org/10.1038/mp.2009.67.

Tanis, K. Q., & Duman, R. S. (2007). Intracellular signaling pathways pave roads to recovery for mood disorders. *Annals of Medicine, 39*, 531–544. http://dx.doi.org/10.1080/07853890701483270.

Tardito, D., Perez, J., Tiraboschi, E., Musazzi, L., Racagni, G., & Popoli, M. (2006). Signaling pathways regulating gene expression, neuroplasticity, and neurotrophic mechanisms in the action of antidepressants: a critical overview. *Pharmacological Reviews, 58*(1), 115–134. http://dx.doi.org/10.1124/pr.58.1.7.

Thakker-Varia, S., Alder, J., Crozier, R. A., Plummer, M. R., & Black, I. B. (2001). Rab3A is required for brain-derived neurotrophic factor-induced synaptic plasticity: transcriptional analysis at the population and single-cell levels. *Journal of Neuroscience, 21*, 6782–6790.

Tiraboschi, E., Tardito, D., Kasahara, J., Moraschi, S., Pruneri, P., Gennarelli, M., ... Popoli, M. (2004). Selective phosphorylation of nuclear CREB by fluoxetine is linked to activation of CaM kinase IV and MAP kinase cascades. *Neuropsychopharmacology, 29*, 1831–1840. http://dx.doi.org/10.1038/sj.npp.1300488.

Tokita, K., Yamaji, T., & Hashimoto, K. (2012). Roles of glutamate signaling in preclinical and/or mechanistic models of depression. *Pharmacology Biochemistry and Behavior, 100*(4), 688–704. http://dx.doi.org/10.1016/j.pbb.2011.04.016.

Tolwani, R. J., Buckmaster, P. S., Varma, S., Cosgaya, J. M., Wu, Y., Suri, C., & Shooter, E. M. (2002). BDNF overexpression increases dendrite complexity in hippocampal dentate gyrus. *Neuroscience, 114*(3), 795–805.

Trachtenberg, J. T., Chen, B. E., Knott, G. W., Feng, G., Sanes, J. R., Welker, E., & Svoboda, K. (2002). Long-term in vivo imaging of experience-dependent synaptic plasticity in adult cortex. *Nature, 420*(6917), 788–794. http://dx.doi.org/10.1038/nature01273.

Tyler, W. J., & Pozzo-Miller, L. D. (2001). BDNF enhances quantal neurotransmitter release and increases the number of docked vesicles at the active zones of hippocampal excitatory synapses. *Journal of Neuroscience, 21*(12), 4249–4258.

Vaughn, J. E. (1989). Fine structure of synaptogenesis in the vertebrate central nervous system. *Synapse, 3*(3), 255–285. http://dx.doi.org/10.1002/syn.890030312.

Veer, I. M., Beckmann, C. F., van Tol, M. J., Ferrarini, L., Milles, J., Veltman, D. J., … Rombouts, S. A. (2010). Whole brain resting-state analysis reveals decreased functional connectivity in major depression. *Frontiers in Systems Neuroscience, 4*. http://dx.doi.org/10.3389/fnsys.2010.00041.

Voleti, B., Tanis, K. Q., Newton, S. S., & Duman, R. S. (2012). Analysis of target genes regulated by chronic electroconvulsive therapy reveals role for Fzd6 in depression. *Biological Psychiatry, 71*(1), 51–58. http://dx.doi.org/10.1016/j.biopsych.2011.08.004.

Vyas, A., Mitra, R., Shankaranarayana Rao, B. S., & Chattarji, S. (2002). Chronic stress induces contrasting patterns of dendritic remodeling in hippocampal and amygdaloid neurons. *Journal of Neuroscience, 22*(15), 6810–6818. doi:20026655.

Wei, J., Liu, W., & Yan, Z. (2010). Regulation of AMPA receptor trafficking and function by glycogen synthase kinase 3. *Journal of Biological Chemistry, 285*(34), 26369–26376. http://dx.doi.org/10.1074/jbc.M110.121376.

Wesseling, H., Gottschalk, M. G., & Bahn, S. (2014). Targeted multiplexed selected reaction monitoring analysis evaluates protein expression changes of molecular risk factors for major psychiatric disorders. *International Journal of Neuropsychopharmacology, 18*(1). http://dx.doi.org/10.1093/ijnp/pyu015.

Weston, M. C., Chen, H., & Swann, J. W. (2012). Multiple roles for mammalian target of rapamycin signaling in both glutamatergic and GABAergic synaptic transmission. *Journal of Neuroscience, 32*(33), 11441–11452. http://dx.doi.org/10.1523/jneurosci.1283-12.2012.

Wilhelm, B. G., Mandad, S., Truckenbrodt, S., Krohnert, K., Schafer, C., Rammner, B., … Rizzoli, S. O. (2014). Composition of isolated synaptic boutons reveals the amounts of vesicle trafficking proteins. *Science, 344*(6187), 1023–1028. http://dx.doi.org/10.1126/science.1252884.

Wilkinson, M. B., Dias, C., Magida, J., Mazei-Robison, M., Lobo, M., Kennedy, P., … Nestler, E. J. (2011). A novel role of the WNT-dishevelled-GSK3beta signaling cascade in the mouse nucleus accumbens in a social defeat model of depression. *Journal of Neuroscience, 31*(25), 9084–9092. http://dx.doi.org/10.1523/jneurosci.0039-11.2011.

Willard, S. L., Hemby, S. E., Register, T. C., McIntosh, S., & Shively, C. A. (2014). Altered expression of glial and synaptic markers in the anterior hippocampus of behaviorally depressed female monkeys. *Neuroscience Letters, 563*, 1–5. http://dx.doi.org/10.1016/j.neulet.2014.01.012.

Woo, T. U., Walsh, J. P., & Benes, F. M. (2004). Density of glutamic acid decarboxylase 67 messenger RNA-containing neurons that express the N-methyl-D-aspartate receptor subunit NR2A in the anterior cingulate cortex in schizophrenia and bipolar disorder. *Archives of General Psychiatry, 61*(7), 649–657. http://dx.doi.org/10.1001/archpsyc.61.7.649.

Wu, G. Y., Deisseroth, K., & Tsien, R. W. (2001). Activity-dependent CREB phosphorylation: convergence of a fast, sensitive calmodulin kinase pathway and a slow, less sensitive mitogen-activated protein kinase pathway. *Proceedings of the National Academy of Sciences of the United States of America, 98*(5), 2808–2813. http://dx.doi.org/10.1073/pnas.051634198.

Xie, Z., Srivastava, D. P., Photowala, H., Kai, L., Cahill, M. E., Woolfrey, K. M., … Penzes, P. (2007). Kalirin-7 controls activity-dependent structural and functional plasticity of dendritic spines. *Neuron, 56*(4), 640–656. http://dx.doi.org/10.1016/j.neuron.2007.10.005.

Ying, S. W., Futter, M., Rosenblum, K., Webber, M. J., Hunt, S. P., Bliss, T. V., & Bramham, C. R. (2002). Brain-derived neurotrophic factor induces long-term potentiation in intact adult hippocampus: requirement for ERK activation coupled to CREB and upregulation of Arc synthesis. *Journal of Neuroscience, 22*, 1532–1540.

Yuan, P., Zhou, R., Wang, Y., Li, X., Li, J., Chen, G., … Manji, H. K. (2010). Altered levels of extracellular signal-regulated kinase signaling proteins in postmortem frontal cortex of individuals with mood disorders and schizophrenia. *Journao of Affective Disorders, 124*(1–2), 164–169. http://dx.doi.org/10.1016/j.jad.2009.10.017.

Yuen, E. Y., Liu, W., Karatsoreos, I. N., Feng, J., McEwen, B. S., & Yan, Z. (2009). Acute stress enhances glutamatergic transmission in prefrontal cortex and facilitates working memory. *Proceedings of the National Academy of Sciences of the United States of America, 106*(33), 14075–14079. http://dx.doi.org/10.1073/pnas.0906791106.

Yuen, E. Y., Liu, W., Karatsoreos, I. N., Ren, Y., Feng, J., McEwen, B. S., & Yan, Z. (2011). Mechanisms for acute stress-induced enhancement of glutamatergic transmission and working memory. *Molecular Psychiatry, 16*(2), 156–170. http://dx.doi.org/10.1038/mp.2010.50.

Yuen, E. Y., Wei, J., Liu, W., Zhong, P., Li, X., & Yan, Z. (2012). Repeated stress causes cognitive impairment by suppressing glutamate receptor expression and function in prefrontal cortex. *Neuron, 73*(5), 962–977. http://dx.doi.org/10.1016/j.neuron.2011.12.033.

Yu, H., Wang, D. D., Wang, Y., Liu, T., Lee, F. S., & Chen, Z. Y. (2012). Variant brain-derived neurotrophic factor Val66Met polymorphism alters vulnerability to stress and response to antidepressants. *Journal of Neuroscience, 32*(12), 4092–4101. http://dx.doi.org/10.1523/jneurosci.5048-11.2012.

Yu, J. J., Zhang, Y., Wang, Y., Wen, Z. Y., Liu, X. H., Qin, J., & Yang, J. L. (2013). Inhibition of calcineurin in the prefrontal cortex induced depressive-like behavior through mTOR signaling pathway. *Psychopharmacology (Berl), 225*(2), 361–372. http://dx.doi.org/10.1007/s00213-012-2823-9.

Zarate, C. A., Jr., Singh, J. B., Carlson, P. J., Brutsche, N. E., Ameli, R., Luckenbaugh, D. A., … Manji, H. K. (2006). A randomized trial of an N-methyl-D-aspartate antagonist in treatment-resistant major depression. *Archives of General Psychiatry, 63*(8), 856–864. http://dx.doi.org/10.1001/archpsyc.63.8.856.

Zarate, C., Machado-Vieira, R., Henter, I., Ibrahim, L., Diazgranados, N., & Salvadore, G. (2010). Glutamatergic modulators: the future of treating mood disorders? *Harvard Review of Psychiatry, 18*(5), 293–303. http://dx.doi.org/10.3109/10673229.2010.511059.

Zhang, W., & Benson, D. L. (2001). Stages of synapse development defined by dependence on F-actin. *Journal of Neuroscience, 21*(14), 5169–5181.

Zhang, W., & Rosenkranz, J. A. (2012). Repeated restraint stress increases basolateral amygdala neuronal activity in an age-dependent manner. *Neuroscience, 226*, 459–474. http://dx.doi.org/10.1016/j.neuroscience.2012.08.051.

Zink, M., Vollmayr, B., Gebicke-Haerter, P. J., & Henn, F. A. (2010). Reduced expression of glutamate transporters vGluT1, EAAT2 and EAAT4 in learned helpless rats, an animal model of depression. *Neuropharmacology, 58*(2), 465–473. http://dx.doi.org/10.1016/j.neuropharm.2009.09.005.

Part III

Systems Neuroscience

Chapter 6

Epigenetic Mechanisms in Depression

Melissa L. Levesque[1], Moshe Szyf[2], Linda Booij[3,4,5]
[1]*Campbell Family Mental Health Research Institute, Centre for Addiction and Mental Health, Toronto, ON, Canada;* [2]*Department of Pharmacology, McGill University, Montreal, QC, Canada;* [3]*Department of Psychology, Concordia University, Montreal, QC, Canada;* [4]*Sainte-Justine Hospital Research Centre, University of Montreal, Montreal, QC, Canada;* [5]*Department of Psychiatry, McGill University, Montreal, QC, Canada*

Major depressive disorder (MDD) is common, with a lifetime prevalence estimated at 16.2% (Kessler, Merikangas, & Wang, 2007). In fact, the World Health Organization estimated that MDD will be the leading cause of disease burden by 2030 (WHO, 2012). The median age of onset ranges between 20 and 25 in most countries (Andrade et al., 2003), and women are particularly vulnerable to mood disorders (Kessler et al., 2003). The rate of recurrence is high, with each episode raising the probability of a new one by 16% (Solomon et al., 2000). MDD thus represents a serious problem for the society as a whole. Our understanding of its etiology, however, is still limited.

Over the years, many alterations in neurobiological systems, alone or in combination, have been identified in association with MDD. However, the precise role(s) of specific neurobiological systems in terms of onset, relapse, and effective treatment are far from being understood. For instance, antidepressant medications acting on monoaminergic systems have been on the market for more than 30 years; however, nonresponse rates are still high, and factors associated with adequate or adverse responses are still not known (Booij, Tremblay, Szyf, & Benkelfat, 2015). Furthermore, it still seems impossible to identify those people at risk or those who are most vulnerable for relapse after recovery (Booij, Tremblay, et al., 2015).

Historically, alterations in the serotonin (5-HT) neurotransmitter system and in the hypothalamic–pituitary–adrenal (HPA) axis are perhaps the most commonly studied biological systems in relation to MDD (Booij, Wang, Levesque, Tremblay, & Szyf, 2013; Booij, Tremblay, et al., 2015). Not surprisingly, with the advancement of technologies, theories have been refined over the years. For instance, while it was initially postulated that an absolute

Systems Neuroscience in Depression. http://dx.doi.org/10.1016/B978-0-12-802456-0.00006-6

"deficit" in 5-HT was the cause of depression, later studies in at-risk populations (e.g., recovered patients, first-degree relatives) favored a less "deterministic" model, suggesting that, rather than an absolute deficit, low 5-HT neurotransmission may represent a risk factor for MDD. This risk is more likely to be expressed in the presence of environmental stressors ("diathesis-stress model"; Booij, Tremblay, et al., 2015). Likewise, dysregulation of the HPA axis activity has also been widely implicated in MDD (Hardeveld et al., 2014; Labonte et al., 2012), as demonstrated by an altered response to the dexamethasone/corticotropin-releasing hormone (CRH) challenge in some depressed patients (Hardeveld et al., 2014). Some studies also report alterations in HPA axis functioning in recovered depressed patients and at-risk individuals (Bhagwagar, Hafizi, & Cowen, 2003; Dienes, Hazel, & Hammen, 2013; Pintor et al., 2007), suggestive of a trait alteration contributing to a higher vulnerability to MDD (Booij et al., 2013). In addition, studies also show an association between brain-derived neurotrophic factor (BDNF) levels and depression (Dalton, Kolshus, & McLoughlin, 2014; Roth & Sweatt, 2011b). Notably, decreased BDNF mRNA levels are observed in animal models of depression as well as in depressed patients (Carlberg et al., 2014; D'Addario et al., 2013; Dalton et al., 2014; Dell'Osso et al., 2014; Dwivedi et al., 2003; Fuchikami et al., 2011; Keller et al., 2010; Roth & Sweatt, 2011b). Finally, studies also suggest that the immune system may be altered in patients with MDD (Christian, 2012; Felger & Lotrich, 2013; Irwin & Miller, 2007; Mills, Scott, Wray, Cohen-Woods, & Baune, 2013). Taken together, these findings suggest that the mechanisms underlying (risk for) MDD are complex and involve multiple systems. Although genetic and environmental interactions seem to be involved, the physiological mechanisms underlying (risk for) MDD have yet to be resolved.

Following the observation of epigenetic changes in the glucocorticoid receptor (GR) gene in rats exposed to less maternal care (Weaver et al., 2004) and in postmortem brains of victims of childhood abuse (McGowan et al., 2009), a number of studies have examined epigenetic processes in peripheral tissues and in association with early life adversity and MDD. In this chapter, we describe part of the growing body of evidence demonstrating the relevance of epigenetic processes for MDD. The overarching hypothesis is that environmental stressors reprogram biological systems through epigenetic processes, such as DNA methylation, by altering the expression of genes involved in the function of these systems ((Booij et al., 2013), and see Figure 1). These epigenetic changes are not stochastic but may reflect a global response to an environmental trigger (Booij et al., 2013). This, in turn, may provide an explanation for how genes and the environment physiologically interact in the development of MDD.

We will first describe gene by environment ($G \times E$) interaction models (diathesis-stress) in the context of risk for MDD. Next, we will review research on epigenetic mechanisms as potential underlying mechanisms of how genes

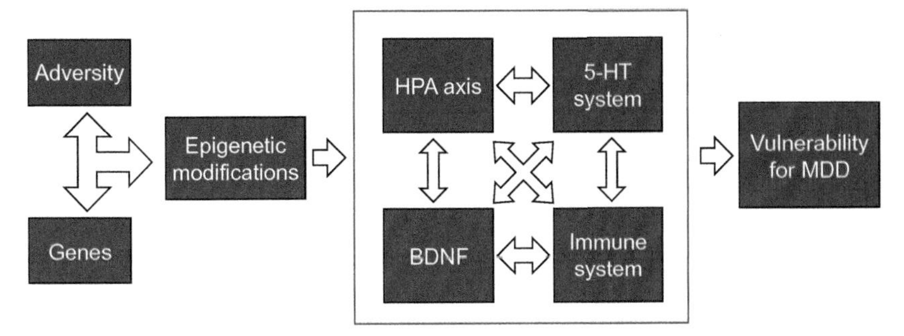

FIGURE 1 Proposed model of the serotonin (5-HT) system, HPA axis, BDNF, and the immune system and associated DNA methylation as well as its role in vulnerability to MDD.

and the environment interact. We will then describe evidence supporting the presence of epigenetic modifications in specific biological systems relevant for MDD. These include the 5-HT system, the HPA axis, BDNF, and the immune system. Furthermore, we will discuss the stability of epigenetic mechanisms and describe methodological issues and limitations that are encountered when studying epigenetic mechanisms in (living) humans. We will finish by discussing the possibility of eventually using epigenetic marks as a marker for depression onset, relapse, and therapeutic response.

DIATHESIS-STRESS MODEL

There is much support for both genetics and the environment being implicated in conveying (vulnerability to) MDD. Various studies have found associations between genotypes/alleles of polymorphisms of specific genes such as (but not limited to) the 5-HT transporter (SLC6A4; Booij et al., 2013; Kendler, Kuhn, Vittum, Prescott, & Riley, 2005), tryptophan hydroxylase 2 (TPH2; Berger et al., 2012; Chen & Miller, 2013), the GR (NR3C1) gene (Galecka et al., 2013; Szczepankiewicz et al., 2011), and the BDNF gene (Hosang, Shiles, Tansey, McGuffin, & Uher, 2014; Jiang & Salton, 2013). However, results are inconsistent across studies, and no single gene has been steadily associated with risk for MDD. Moreover, studies in monozygotic (MZ) twins, who share 100% of their genes, have shown that there is discordance of approximately 40% in depressive and related symptoms within MZ twins (Nivard et al., 2015). Following Bronfenbrenner's ecological model of development, which posits that a person's personality is influenced not only by his or her own characteristics (genes) but also by his/her immediate and more distant environment (Bronfenbrenner, 1979), numerous studies have demonstrated associations between adverse environmental events and mental health disorders (i.e., Pechtel & Pizzagalli, 2011), such as low socioeconomic status (SES; van Oort, van der Ende, Wadsworth, Verhulst, & Achenbach, 2011) and maternal stress (Walker et al., 2011). However, two people can be exposed to the same stressor and fare quite differently. It is thus becoming increasingly obvious that vulnerability is quite complex, with multiple factors and mechanisms interacting in conveying risk. The G × E model has been proposed (Karg & Sen, 2012; Prathikanti & Weinberger, 2005), which postulates that the impact of a specific environment on the phenotype varies, depending on genetics or, conversely, that the effect of a specific genotype depends on the environment (Karg & Sen, 2012).

One of the most well-studied genes in relation to MDD is the 5-HTTLPR polymorphism of the SLC6A4 gene. It has widely been shown that MDD is most prevalent in individuals with the short allele of the 5-HTTLPR when combined with life stress such as abuse during childhood (Caspi et al., 2003). Negative studies have, however, also been reported (e.g., see Risch et al., 2009). Although some of the negative studies appeared to be accounted by methodological differences (Karg, Burmeister, Shedden, & Sen, 2011; Uher & McGuffin, 2010), it has also been proposed that timing of adversity may be highly relevant for conveying vulnerability. Adverse exposures occurring during the prenatal and early postnatal period may have a particularly significant impact on development, in effect increasing vulnerability for later mental health disorders (Danese & McEwen, 2012; Shonkoff, Boyce, & McEwen, 2009; Shonkoff & Garner, 2012). A better understanding of these critical periods and the mechanisms by which these alterations occur may enable us to target interventions when and in whom they may be the most useful.

EPIGENETICS AS A POTENTIAL MECHANISM UNDERLYING DIATHESIS-STRESS

Epigenetic patterns are thought to be influenced by genetics, inherited transgenerationally from parents, and influenced by the environment (Szyf, 2015). Given that environmentally driven gene expression patterns are generally set up during the critical prenatal and early postnatal developmental periods (e.g., Booij, Tremblay, et al., 2015), adversity occurring during this time frame can lead to long-lasting modifications of genetic expression. These modifications are likely to occur through epigenetic alterations, which can be defined as any long-term change to gene expression, or the epigenome, which persists past the end of the trigger without a change in gene sequence or structure (McGowan & Szyf, 2010). The epigenome consists of DNA, chromatin, and other chemical compounds that attach to chromatin or DNA (Razin, 1998). Briefly, DNA wraps around chromatin, and a number of modifications to chromatin or DNA itself can affect how DNA is expressed. Histones in chromatin are responsible for packaging DNA and can be modified by phosphorylation, acetylation, and ubiquitination (McGowan & Szyf, 2010) as well as the less well-known SUMOylation and adenosine diphosphate ribosylation (Tsankova, Renthal, Kumar, & Nestler, 2007). Since DNA is wrapped around the nucleosomes, these covalent modifications promote an open or closed chromatin state, determine whether the transcription machinery has access to DNA, and thus turn on/off gene expression (McGowan & Szyf, 2010).

DNA METHYLATION

DNA methylation, since it affects the DNA molecule itself through the enzymatic addition of a methyl group to DNA, is of particular interest. The methyl group is provided by the donor S-adenosylmethionine and attached to the 5′ position of the cytosine in cytosine-phosphate-guanine (CpG) dinucleotide (McGowan & Szyf, 2010; Murgatroyd & Spengler, 2011b) through the action of the DNA methyltransferases (DNMTs) (McGowan & Szyf, 2010). Methylation is widely found in the genome with the exception of CpG islands. These are DNA patches of approximately 1000 base pairs rich in CpGs that are often associated with genes, particularly promoters and enhancers, which respectively initiate and enhance transcription and thus expression of a particular gene (Bird, 2002; Suzuki & Bird, 2008). These CpG islands represent approximately 1% of the genome and are largely unmethylated (Murgatroyd & Spengler, 2011b). Contrary to methylation occurring outside of promoters and enhancers, methylation within CpG sites in promoters and enhancers leads to permanent silencing of the gene, either directly by blocking transcriptional factors from binding to the DNA sequence or by attracting proteins to form corepressor complexes in order to indirectly suppress gene expression (Bird, 2002; Jones et al., 1998; Nan et al., 1998). Although no change has been made to the genome itself, DNA methylation can effectively silence a gene.

EARLY ENVIRONMENT AND DNA METHYLATION

DNA methylation patterns are shaped during gestation by a series of methylation and demethylation events (Benvenisty, Mencher, Meyuhas, Razin, & Reshef, 1985; Razin & Cedar, 1993). DNA methylation during this period is highly vulnerable to the environment. Many studies have shown that environmental events are associated with epigenetic modifications, including DNA methylation (Borghol et al., 2012; Essex et al., 2013; Filiberto et al., 2011; Labonte, Azoulay, Yerko, Turecki, & Brunet, 2014). For instance, one study showed that individuals who were exposed to the Dutch famine in the perinatal period had, six decades later, altered DNA methylation patterns as compared to their siblings (Heijmans, Kremer, Tobi, Boomsma, & Slagboom, 2007). In turn, later studies found associations between DNA methylation patterns and adversity during childhood, including parental stress (Essex et al., 2013), low SES (Borghol et al., 2012), peer victimization (Ouellet-Morin et al., 2012), and abuse (Suderman et al., 2014). From these (and other) studies, it can be concluded that early adverse exposures are associated with alterations in DNA methylation in a wide variety of human biological systems.

CONFOUNDS: GENETICS

A methodological issue in studying environmental factors that predispose an individual to MDD is the potential for G × E interactions. Indeed, studies have shown that some genetic variants may be more susceptible to alterations in gene expression following environmental manipulations. This implies that carriers of certain polymorphic variants may be more susceptible to methylation in varying environmental conditions. Some examples can be found in the literature, particularly in association with methylation at the 5-HTTLPR promoter. For instance, Kinnally et al. (2010) found higher mean methylation of 5-HTTLPR in carriers of the *s* allele. Williams et al. (2009) report that early life stress is associated with increased activation in the amygdala, ventromedial prefrontal cortex (PFC), and anterior cingulate but only in those with the *s* allele of the 5-HTTLPR. Pluess et al. (2011) report that infants carrying the *s* allele show more negative emotionality when mothers reported high levels of anxiety during pregnancy. In regard to childhood abuse, Beach, Brody, Todorov, Gunter, and Philibert (2010, 2011) found positive associations linking childhood sexual abuse with DNA methylation of the 5-HTTLPR promoter and methylation with symptoms of antisocial personality disorder in adulthood but only in carriers of the *s* allele. Finally, lower methylation is found in association with unresolved trauma or loss, specifically in the *s/s* genotype, while an inverse correlation was found in the *l/l* genotype (van IJzendoorn, Caspers, Bakermans-Kranenburg, Beach, & Philibert, 2010). On the other hand, other studies find no association between genotype and DNA methylation (i.e., Booij, Szyf, et al., 2015; Ouellet-Morin et al., 2012; Wang, Szyf, et al., 2012; Wankerl et al., 2014), and in a study in a large adolescent sample, the association between stressful life events and SLC6A4 methylation was stronger for carriers of the *l/l* genotype than for *s* carriers (van der Knaap et al., 2015). In addition to sample size, methodological differences such as differences in investigated CpG sites, sample characteristics, and the way adversity has been conceptualized and assessed may account for some of the discrepancies in results between studies.

The best method to overcome this methodological issue of the G × E interaction and assess the impact of the environment while controlling for genetics is to use MZ twins. Since MZ twins share 100% of their genes, within-pair epigenetic differences must be due to unique environmental experiences. Using this method, Ollikainen et al. (2010) demonstrated that even as newborns, variations in methylation can be detected across MZ twins, suggesting that epigenetic mechanisms have already started affecting development in utero. In a study by our group, we found that although within-pair correlations in DNA methylation were very high (96–99%), high within-pair variability is found in genes associated with development, cellular mechanisms, and psychological/neurological disorders (Levesque et al., 2014). Other studies have found similar results. Gordon et al. (2011) found that the most discordant DNA methylation sites across cotwins were associated with genes involved in the functioning of the immune system and responding to the environment. Similar findings of DNA methylation discordance on genes associated with immune function were found in MZ twins discordant for psoriasis (Gervin et al., 2012) and autoimmune inflammatory diseases (Javierre et al., 2010). Moreover, both Gordon et al. (2011) and Saffery et al. (2012) found that the most discordantly methylated genes from cord blood mononuclear cells and human umbilical vascular endothelial cells were those shown to be involved in responding to external signals, such as the immune response. Finally, studies have found within-pair DNA methylation discordance in twin pairs discordant for autism (Nguyen, Rauch, Pfeifer, & Hu, 2010; Wong et al., 2014), bipolar disorder (Kuratomi et al., 2008), risk-taking behavior (Kaminsky et al., 2008), Alzheimer disease (Mastroeni, McKee, Grover, Rogers, & Coleman, 2009), intestinal disease (Harris, Nagy-Szakal, & Kellermayer, 2013), diabetes (Rakyan et al., 2011; Ribel-Madsen et al., 2012; Stefan, Zhang, Concepcion, Yi, & Tomer, 2014), and birth weight (Gordon et al., 2012). Overall, studies show high similarity within twin pairs across tissues; however, differences are also found, particularly when phenotypes diverge across twins. More work is needed, particularly during childhood and adolescence, but it is already clear that epigenetic mechanisms are active in humans and affect development in early life.

SPECIFIC SYSTEMS INVOLVED IN DEPRESSION

An emerging number of studies have investigated the association between early adverse exposures and epigenetic processes in biological systems relevant for MDD, particularly genes related to the HPA axis, 5-HT system, and BDNF, and genes regulating the immune system. Some of these studies will be reviewed below.

HPA Axis

Research increasingly supports a prominent role of epigenetic mechanisms in altering the expression of genes involved in the regulation of the HPA axis, as reviewed in Booij et al. (2013). One of the most important, or at least the best known, is the gene encoding for the GR, NR3C1. The GR receptor is classified as a ligand-activated transcription factor, meaning that when a ligand binds to the receptor, this leads to a change in the configuration of the receptor, thereby allowing binding to specific sites on DNA and regulation of gene transcription. Various cofactors can associate with the GR, and thus the effect of binding can vary in nature, intensity, and direction (Meaney, 2010).

The early postnatal environment, for instance, normal variations in maternal care during the first week of life in rodents, programs DNA methylation patterns of NR3C1 and thus HPA activity (Roth & Sweatt, 2011a).

Early stress in mice results in decreased levels of NR3C1 mRNA in the hippocampus, paraventricular nucleus (PVN), and pituitary (Murgatroyd et al., 2009; Weaver et al., 2004), and this is accompanied by increased HPA axis reactivity and anxious behavior (Weaver et al., 2004). More specifically, early stressors, including low maternal care, are associated with altered methylation in specific cytosines. This altered methylation results in decreased expression of genes, notably the exon 1_7 promoter for NR3C1 specific to the hippocampus (Harris & Seckl, 2011; Murgatroyd & Spengler, 2011b; Weaver, 2007; Weaver et al., 2004) as well as decreased histone acetylation of the promoter (Weaver et al., 2004), with both changes persisting into adulthood. High levels of maternal care, on the other hand, are associated with histone acetylation and lower levels of methylation in the 5′ CpG dinucleotide of the transcriptional activator nerve growth factor-inducible protein A (NGFI-A) binding site of the NR3C1 exon 1_7 promoter and increased NGFI-A binding to 1_7 promoter, as compared to low maternal care. Importantly, the 3′ CpG site is only a few nucleotides away, and methylation is unaffected by maternal care in this region, which demonstrates the specificity of the mechanisms involved (Meaney, 2010). To validate that this is not a genetic effect, Weaver et al. (2004) showed that methylation of the NR3C1 promoter in low-care offspring could be reversed by cross-fostering (biological offspring of low-care mothers cared for by high-care mothers) with an associated decreased HPA response to stress. Furthermore, these changes emerged early in life and remained stable, supporting the importance of the methylation process. In summary, lower maternal care is linked with lower histone acetylation and higher methylation of a specific NR3C1 promoter, and this is associated with greater stress reactivity.

Early life parental care has also been shown to be associated with methylation of HPA axis-regulating genes in humans. For instance, maternal prenatally depressed mood has been associated with NR3C1 promoter methylation at the transcription factor NGFI-A-binding region in newborns and infants, accompanied by an altered stress response with enhanced salivary cortisol levels when measured at three months of age (Glover, O'Connor, & O'Donnell, 2010; Oberlander et al., 2008). A life history of childhood abuse has also been associated with the stress response of the HPA axis, and individuals who experienced childhood abuse are more vulnerable to depression and committing suicide as adolescents and adults (Brown, Cohen, Johnson, & Smailes, 1999; Heim & Nemeroff, 2001; Widom, DuMont, & Czaja, 2007). In postmortem brain samples from suicide victims with a history of early life abuse, enhanced DNA methylation in the NR3C1 exon 1_F promoter (the human homolog of the exon 1_7 promoter) and decreased NR3C1 mRNA expression were found in the hippocampus (McGowan et al., 2009). Suicide and depressed patients with no history of early life abuse, however, showed no epigenetic changes of the hippocampal GR (Murgatroyd & Spengler, 2011b). Another study elaborated on these findings, also demonstrating differential hippocampal NR3C1 expression of the 1_B, 1_C, and 1_H variants in this adult sample (Labonte et al., 2012). Moreover, Alt et al. (2010) found decreased NR3C1 expression in the amygdala, cingulate gyrus, and inferior PFC as well as decreased NGFI-A expression in the postmortem hippocampus of depressed patients as compared to controls. Results appear to be similar when NR3C1 methylation is measured from peripheral blood, with childhood maltreatment associated with increased NR3C1 methylation (Perroud et al., 2011). Studies with no association between epigenetic mechanisms and depression can also be found (e.g., Alt et al., 2010; Januar, Saffery, & Ryan, 2015). Overall, as in animals, low maternal care as well as childhood abuse has been linked with altered DNA methylation of the NR3C1 promoter in various brain regions later in life, and this is associated with a greater risk for depressive symptoms and suicide.

Maternal care during infancy is also associated with altered expression of the arginine vasopressin (Avp) gene. Studies show that 3 h of daily maternal separation during the first week of life in rodents is associated with lifelong hypomethylation of CpG residues in the CGI3 region of the Avp enhancer gene in the PVN of the hypothalamus, persistent upregulation of Avp expression in the PVN, and increased corticosterone both under basal condition and in response to acute stress (Murgatroyd et al., 2009; Roth & Sweatt, 2011b). Interestingly, early stress-induced hypomethylation of the Avp enhancer was further localized to a downstream Avp enhancer, which contains high-affinity context-dependent DNA-binding sites for methyl CpG binding protein 2 (MeCP2) (Harris & Seckl, 2011; Murgatroyd & Spengler, 2011a). Hypomethylation reduced MeCP2's ability to recruit and bind repressive histone complexes such as histone deacetylases and DNMTs (Murgatroyd & Spengler, 2011b). If repressor complexes cannot be bound to MeCP2, gene expression is not suppressed and the HPA axis is hyperactivated, which can increase vulnerability for MDD and other HPA axis-associated mental disorders.

Other genes also reported to be affected by early stress include the corticotropin-releasing factor (CRF) gene, the pro-opiomelanocortin (POMC) gene coding for adrenocorticotropic hormone (downstream target of AVP and CRH signaling), and the HSD11B2 gene. A chronic daily variable stress paradigm in mice was associated with altered methylation of specific cytosines within the regulatory region of the CRF gene in the hypothalamus and amygdala and was accompanied by depressive-type behaviors (Roth & Sweatt, 2011a). Furthermore, a number of studies have found altered methylation of the POMC promoter region, which leads to gene expression silencing following early stress (reviewed in Murgatroyd & Spengler, 2011b). Finally, Harris and Seckl (2011) demonstrated that methylation of CpG islands within the promoter and exon 1 region of the HSD11B2 gene is associated with differential expression levels of the HSD2 enzyme, hence modulating the amount of glucocorticoids reaching the fetus during pregnancy.

5-HT

Serotonin is the most widely studied neurotransmitter in relation to MDD. Among the many 5-HT proteins, the 5-HT transporter is perhaps the most widely studied (Booij et al., 2013). In nonhuman adult primates, Kinnally et al. (2010, 2011) found that the degree of methylation of the 5-HT transporter gene was positively associated with increased reactivity to early stress. Specifically, the *s* allele of the 5-HTTLPR was more methylated in those animals that had been nursery-reared than in those that were mother-reared. Furthermore, nursery-reared individuals with higher SLC6A4 CpG methylation displayed higher anxiety during maternal-social separation as compared to mother-reared infants (Kinnally et al., 2010). Finally, a study demonstrated that when embryonic hippocampal neurons are placed in culture and treated with 5-HT, which mimics the extracellular signal associated with maternal licking and grooming behaviors, the 5′ CpG site is demethylated, while no changes are found at the 3′ CpG site, as reviewed in Meaney (2010).

In humans, studies demonstrated that early stress, including maternal depression during pregnancy and a history of childhood abuse, was associated with altered levels of SLC6A4 promoter methylation, at birth and during adulthood (Beach et al., 2010, 2011; Devlin, Brain, Austin, & Oberlander, 2010; van IJzendoorn et al., 2010). In turn, we found that site-specific SLC6A4 promoter methylation in white blood cells and T cells of adults was associated with lower in vivo measures of brain 5-HT synthesis in the lateral orbitofrontal cortex bilaterally and higher childhood aggression (Wang, Szyf, et al., 2012). Moreover, the same CpG sites of the SLC6A4 gene were associated with childhood trauma, reduced hippocampal volume, and neural processing of negative stimuli in both MDD patients and controls (Booij, Szyf, et al., 2015; Frodl et al., 2015).

Although the association with early life adversity has been quite well demonstrated, the association with depressive symptomatology has been unclear. In a study in healthy twins, within-pair differences (thereby controlling for DNA sequence) in SLC6A4 methylation were associated with within-pair differences in depressive symptoms (Zhao, Goldberg, Bremner, & Vaccarino, 2013). In patient samples, SLC6A4 methylation did not correlate with MDD diagnosis or depressive symptomatology (Booij, Szyf, et al., 2015; Kang et al., 2013). However, greater SLC6A4 methylation has been shown to correlate with greater functional impairment and stress sensitivity (Kang et al., 2013) as well treatment responsiveness (Domschke et al., 2014; Okada et al., 2014). Though preliminary, taking these findings together suggests that levels of SLC6A4 methylation may reflect responses to adverse exposures and individual depression vulnerability but that other biological or social risk factors as well as protective factors need to be considered to account for the actual phenotype.

Very little is known regarding functioning of other neurotransmitter systems following early adversity. However, studies have shown that licking and grooming behavior in rodents is associated with the mesolimbic dopamine system in addition to the HPA axis (Champagne et al., 2006; Pena, Neugut, Calarco, & Champagne, 2014). In addition, prenatal restraint stress in rodents leads to altered γ-aminobutyric acid and glutamate receptor expression in the amygdala and hippocampus as well as anxious- and depressive-like phenotypes in these animals (Laloux et al., 2012; Zuena et al., 2008). Moreover, these phenotypes can be reversed by chronic treatment with the selective serotonin reuptake inhibitor (SSRI) fluoxetine (Marroco et al., 2014). More work will be needed to investigate epigenetic alterations to dopamine, glutamate, and other neurotransmitter-related genes and how these relate to depression.

BDNF

BDNF levels are also known to be decreased in depressed patients (Carlberg et al., 2014; Dell'Osso et al., 2014; Fuchikami et al., 2011; Song et al., 2014) and in animal models of depression (Roth & Sweatt, 2011b), although again, findings are not entirely consistent (see Januar et al., 2015). Specifically, some studies have shown decreased BDNF mRNA and protein levels in the hippocampus, amygdala, and PFC of adult mice exposed to bouts of infant isolation or abusive care (Dalton et al., 2014; Roth & Sweatt, 2011b). These mice also show increased anxiety-like behavior in adulthood (Roth & Sweatt, 2011b). Further examination of mice exposed to a stressed abusive caregiver in infancy revealed that 12 cytosine-guanine dinucleotide sites within the regulatory region of exon IV of the BDNF gene were hypermethylated as compared to control mice (Roth & Sweatt, 2011b). Notably, treatment with antidepressants as well as the demethylating agent zebularine for 7 days reversed the methylation and gene expression alterations seen following early adversity in these mice (Coppell, Pei, & Zetterstrom, 2003; Roth & Sweatt, 2011b). Similar results showed that chronic treatment with imipramine, a tricylic antidepressant, can also reverse the epigenetic alterations to BDNF following early defeat stress in mice (Tsankova et al., 2006).

In humans, postmortem studies have demonstrated increased methylation and a decrease in BDNF mRNA in regions such as the hippocampus and Wernicke's area in suicide victims postmortem (Dwivedi et al., 2003; Keller et al., 2010). Furthermore, alterations in BDNF methylation have been found in the blood of (living) MDD patients (Carlberg et al., 2014; D'Addario et al., 2013; Dell'Osso et al., 2014; Fuchikami et al., 2011). Specifically, studies have found increased levels of BDNF exon I promoter methylation in MDD patients as compared to healthy controls (Carlberg et al., 2014; Dell'Osso et al., 2014). In addition, Song et al. (2014) found altered methylation of the BDNF promoter region of exon 1 in saliva from depressed patients as compared to controls. Interestingly, BDNF methylation levels are also associated with treatment. Effective antidepressants have often been found to increase BDNF levels (Chen, Ernst, & Turecki, 2011; Lopez et al., 2013), while BDNF overexpression demonstrates antidepressant activity (Hellweg, Ziegenhorn, Heuser, & Deuschle, 2008). Studies therefore suggest that alterations in BDNF gene expression are associated with depression, as shown in patient samples as well as experimentally using animal models.

Immune System and Cytokines

Accumulating evidence also suggests that alterations to the immune system may be associated with mental health disorders, including mood disorders (Christian, 2012; Felger & Lotrich, 2013; Irwin & Miller, 2007; Mills et al., 2013). Cytokines, signaling molecules with immune-modulating activity that can be either pro- or anti-inflammatory, depending on their target, may be at the base of this association (Mills et al., 2013). Indeed, increased levels of circulating proinflammatory cytokines have been found in the context of high levels of stress and/or depression (Dowlati et al., 2010; Hassanain, Bhatt, Zalcman, & Siegel, 2005; Pesce et al., 2011). Antidepressants are also associated with cytokine levels, for instance, by suppressing proinflammatory cytokines such as interleukin (IL)-2 (Myint & Kim, 2003). Furthermore, studies in animal models have demonstrated a role for cytokines in brain development (Deverman & Patterson, 2009), and early environmental adversity is associated with altered functioning of the immune system, including cytokines (Barreau, Ferrier, Fioramonti, & Bueno, 2004; O'Connor, Moynihan, & Caserta, 2014; Powell et al., 2013).

This association between the early environment and immune system functioning is likely mediated by epigenetic mechanisms, since several studies have demonstrated an association between the early environment and DNA methylation in immune pathways (Borghol et al., 2012; Provencal et al., 2012). For instance, maternal deprivation in rhesus macaques has been associated with altered DNA methylation patterns in both the PFC and T cells (Provencal et al., 2012), and pathway analyses identified pathways associated with immune responses. Further support for an association between the environment and DNA methylation in immune pathways comes from a study conducted by our group. We found, in genetically identical MZ twins, variability in DNA methylation patterns in pathways associated with immune function, suggesting that within-pair discordance in DNA methylation is associated with discordance in external signals (Levesque et al., 2014). Finally, Powell et al. (2013) found that methylation at the CpG unit 5 of the IL-11 gene was associated with treatment response to antidepressants in depressed individuals. In addition, lower levels of methylation at the CpG unit 5 have been associated with a better response to antidepressants (Powell et al., 2013). Thus, evidence is accumulating that early stress is associated with methylation of genes involved in immune pathways, and that levels of cytokines are altered in the context of depression. Further studies will be necessary to assess whether the immune system may contribute to increased vulnerability for MDD through epigenetic modifications.

Associations between 5-HT Neurotransmission, the HPA Axis, Levels of BDNF, and Immune Function

It is clear that none of the systems described above function in isolation, and there is increasing evidence that monoamine transmission, the HPA axis, growth factors including BDNF, and immune function interact together to increase vulnerability for depressive disorders. For instance, it appears that early stress activates inflammatory signaling pathways in the brain, which results in altered neurotransmission of monoamines, including 5-HT, as well as altered HPA axis function and levels of growth factors (Felger & Lotrich, 2013; Mills et al., 2013). As proof-of-principle examples, cytokine administration has been shown to activate the HPA axis (Dunn, 2000) as well as 5-HT turnover (Capuron & Miller, 2011) through increased expression of the tryptophan-metabolizing enzyme indoleamine-2,3-dioxygenase (Capuron & Miller, 2011; Myint & Kim, 2003; Tsao, Lin, Chen, Bai, & Wu, 2006), thereby decreasing overall brain serotonin synthesis. In a study in rats, early life stress and the 5-HTTLPR genotype interacted to affect DNA methylation of the CRF gene promoter in the amygdala (van der Doelen et al., 2015). Furthermore, glucocorticoids can repress proinflammatory proteins and cytokine genes (Klengel et al., 2013). In addition, it was shown that BDNF levels are positively associated with IL-6 levels in MDD patients but not in healthy controls (Patas et al., 2014). Finally, polymorphisms of 5-HT genes and the BDNF Val66Met gene also interact to cause the depressive phenotype (Ignacio, Reus, Abelaira, & Quevedo, 2014). It was shown that the activation of the 5-HT2A receptor modulates expression of BDNF mRNA in the brain and that selective SSRI administration is associated with increased BDNF expression (reviewed in Ignacio et al., 2014). More work is needed to better understand how these systems interact with one another, but increasing evidence suggests that they may all be involved in increasing vulnerability to MDD following exposure to early adversity.

STABILITY OF DNA METHYLATION PATTERNS

So far we have reviewed literature showing that prenatal and early postnatal life events are associated with gene expression throughout life. A related question is whether DNA methylation is responsive to external factors throughout the lifespan. This is highly relevant for treatment and prognosis.

If the methylome is indeed dynamic throughout life, then differences in DNA methylation profiles in identical twins should increase through life, and they do. Although within-pair discordances are present from birth (Ollikainen et al., 2010), they increase with age (Alisch et al., 2012; Bjornsson et al., 2008; Boks et al., 2009; Booij, Szyf, et al., 2015; Fuke et al., 2004; Gomes et al., 2012; Gronniger et al., 2010; Hannum et al., 2013; Hernandez et al., 2011; Wang, Liu, et al., 2012), particularly when twins have divergent experiences

(Bell et al., 2012; Fraga et al., 2005; Langevin et al., 2011; Talens et al., 2012). Even in utero, differences in gene expression increase with gestational age (Novakovic et al., 2011). The direction of change is complex, with DNA methylation increasing with age at some sites and decreasing at others (Alisch et al., 2012; Heyn et al., 2012). Furthermore, DNA methylation does not vary with age in all genes equally, suggesting some specificity (Johansson, Enroth, & Gyllensten, 2013; Martino et al., 2013). Genes found to be associated with age are enriched for functions including DNA binding and regulation of transcription (Hernandez et al., 2011), molecular and cellular characteristics of skin tissue development (Gronniger et al., 2010), and aging-related conditions including Alzheimer disease, cancer, tissue degradation, DNA damage, and oxidative stress (Hannum et al., 2013). In our study in adolescent MZ twins (Levesque et al., 2014), we further addressed the question of stability by identifying gene pathways that are stable and dynamic over a period of 3–6 months. We found 258 sites distributed over 226 genes showing highly variable DNA methylation within twin pairs but stable methylation over several months. These sites were associated with 16 pathways involved in several diseases and disorders, development, and cellular mechanisms. We also found 47 sites across 46 genes that showed high within-pair variability but were dynamic over a period of 3–6 months. These sites were associated with three networks involved in development, cancer, and different disorders. We thus showed that while DNA methylation of some genes is stable across time, it is dynamic at some specific sites, and this information may eventually be used to identify targets for interventions in the treatment of mental health disorders, including depression.

METHODOLOGICAL ISSUES AND LIMITATIONS

An important methodological concern in assessing methylation in relation to mental health problems including MDD is that it is impossible to measure DNA methylation directly in the living human brain. What we can do is assess methylation from peripheral tissues such as blood and saliva. The question of whether DNA methylation patterns are tissue-specific or system-wide arises. Studies addressing this question have found that DNA methylation patterns of some genes are highly tissue-specific. Thus, methylation patterns of certain genes from peripheral cells may not reflect methylation patterns of those same genes in the brain. One study assessing DNA methylation profiles in the PFC and T cells of rhesus macaques that were deprived from their mothers in early life showed that a large part of DNA methylation alterations were tissue-specific (Provencal et al., 2012). However, studies have shown that although methylation patterns of some genes are highly tissue-specific, others may be similarly methylated in the brain and in peripheral cells (Gordon et al., 2012; Lokk et al., 2014; Muangsub, Samsuwan, Tongyoo, Kitkumthorn, & Mutirangura, 2014; Provencal et al., 2012; Rodriguez et al., 2014; Wu et al., 2014). Furthermore, studies have shown that methylation levels in peripheral tissues are associated with brain function. For instance, we have demonstrated that DNA methylation upstream of the SLC6A4 promoter in monocytes and in T cells is associated with in vivo measures of brain 5-HT synthesis in adult human males (Wang, Szyf, et al., 2012). Furthermore, studying the same CpG sites as in Wang, Szyf, et al. (2012) using whole blood DNA (Frodl et al., 2015), we found that the degree of methylation of the SLC6A4 promoter, as assessed in peripheral blood, is positively associated with activation in response to negative emotional stimuli in the anterior insula and negatively associated with activation in response to negative emotional stimuli in the posterior insula as well as inversely related to hippocampal volume (Booij, Szyf, et al., 2015). Finally, increased SLC6A4 methylation from both blood- and saliva-extracted DNA has also been associated with increased amygdala reactivity to threatening stimuli (Nikolova et al., 2014). Work in our laboratory is presently ongoing comparing the association between brain measures and DNA methylation as assessed in different tissues/cell types.

Another issue stems from sex differences. As seen in various animal model studies (Charil, Laplante, Vaillancourt, & King, 2010), males appear to be more sensitive to the effects of early stress than females. For instance, Pawluski et al. (2012) demonstrated that exposure to fluoxetine in newly born offspring of a dam stressed during pregnancy affects GR expression in the hippocampus in males but not in female offspring. Human studies have also shown discordant gene expression following early adversity across men and women. For instance, Charil et al. (2010) demonstrated that 11β-HSD2 expression and activity were significantly decreased in males but not in females following early stress, thereby exposing males to higher levels of cortisol than females. Furthermore, it was shown that SLC6A4 methylation levels are greater in males than females independent of early adversity (Booij, Szyf, et al., 2015). Moreover, Shalev et al. (2009) found that the association between a Val66Met polymorphism and cortisol increase following a social challenge is sex-specific. Given that many studies of early stress collapse both sexes together, further investigations are required to address sex differences.

Another important methodological consideration in this literature was brought up by Charil et al. (2010) and consists of the use of different early stress paradigms in animal models yielding divergent results. A similar issue has been raised in humans, given that there are multiple ways to measure stress. What constitutes stress and/or adversity varies across studies and may in part explain divergent results. Future work will be needed to investigate the long-term impact of different stressors, while considering stress type, duration, and subjective experience, in addition to timing. Furthermore, longitudinal research should be favored, as it allows for the investigation of causal relationships between methylation and depressive symptomatology.

Nevertheless, it has now been widely shown that early adversity is associated with DNA methylation patterns of several genes, and it appears likely that these modifications in gene expression have long-term, system-wide consequences for the functioning of an individual. It is possible that epigenetic marks may eventually be used as biomarkers of depression vulnerability and/or treatment response (reviewed in Frieling & Tadic, 2013). For instance, there is some preliminary evidence that response to treatment with antidepressants could be predicted from pretreatment increased SLC6A4 methylation rates (Domschke et al., 2014; Okada et al., 2014), from BDNF levels in plasma and serum during the first week of treatment (Dreimuller et al., 2012; Tadic et al., 2011), and from pretreatment DNA methylation levels of a site within the BDNF exon IV promoter region (Tadic et al., 2014). Also of interest are the findings of a study showing that the treatment response to cognitive behavioral therapy was associated with site-specific methylation in the 5-HT transporter in children with anxiety disorders (Roberts et al., 2014). Further studies in prospective longitudinal samples as well as high-risk and recovered MDD samples will be necessary to determine whether alterations in methylation patterns in specific biological systems relevant to MDD may be used to assess vulnerability to MDD onset and relapse and can be predictive of treatment response.

CONCLUSION

Even though our understanding of physiological mechanisms underlying the association between the early environment and MDD is still limited, progress is being made. We know that several biological systems, including the HPA axis, 5-HT neurotransmission, BDNF, and the immune system, are involved in the etiology of MDD. We know that epigenetic mechanisms, including DNA methylation, are involved in altering the functioning of these biological systems following early adverse environmental exposures and that these alterations are often long term. Future work will be necessary to further investigate the alterations to these systems and how they interact to underlie vulnerability for MDD. Eventually, our understanding of the biological mechanisms underlying vulnerability for MDD may be sufficient to develop more efficient treatments and improve prognosis. It may even be possible to utilize our understanding of vulnerability to develop interventions to prevent the development of MDD in targeted individuals.

This chapter is an adaptation of the following papers:

- Booij, L., Tremblay, R. E., Szyf, M., & Benkelfat, C. (2015). Genetic and early environmental influences on the serotonin system: consequences for brain development and risk for psychopathology. *Journal of Psychiatry and Neuroscience.*
- Booij, L., Wang, D., Lévesque, M. L., Tremblay, R. E, & Szyf, M. (2013). Looking beyond the DNA sequence: The relevance of DNA methylation processes for the stress-diathesis model of depression. *Philosophical Transactions of the Royal Society B-Biological Sciences, 368*(1615), 1–16.

REFERENCES

Alisch, R. S., Barwick, B. G., Chopra, P., Myrick, L. K., Satten, G. A., Conneely, K. N., & Warren, S. T. (2012). Age-associated DNA methylation in pediatric populations. *Genome Research, 22*(4), 623–632.

Alt, S. R., Turner, J. D., Klok, M. D., Meijer, O. C., Lakke, E. A., Derijk, R. H., & Muller, C. P. (2010). Differential expression of glucocorticoid receptor transcripts in major depressive disorder is not epigenetically programmed. *Psychoneuroendocrinology, 35*(4), 544–556.

Andrade, L., Caraveo-Anduaga, J. J., Berglund, P., Bijl, R. V., De Graaf, R., Vollebergh, W., … Wittchen, H. U. (2003). The epidemiology of major depressive episodes: results from the International Consortium of Psychiatric Epidemiology (ICPE) Surveys. *International Journal of Methods in Psychiatric Research, 12*(1), 3–21.

Barreau, F., Ferrier, L., Fioramonti, J., & Bueno, L. (2004). Neonatal maternal deprivation triggers long term alterations in colonic epithelial barrier and mucosal immunity in rats. *Gut, 53*(4), 501–506.

Beach, S. R., Brody, G. H., Todorov, A. A., Gunter, T. D., & Philibert, R. A. (2010). Methylation at SLC6A4 is linked to family history of child abuse: an examination of the Iowa Adoptee sample. *American Journal of Medical Genetics Part B, Neuropsychiatric Genetics, 153B*(2), 710–713.

Beach, S. R., Brody, G. H., Todorov, A. A., Gunter, T. D., & Philibert, R. A. (2011). Methylation at 5HTT mediates the impact of child sex abuse on women's antisocial behavior: an examination of the Iowa adoptee sample. *Psychosomatic Medicine, 73*(1), 83–87.

Bell, J. T., Tsai, P. C., Yang, T. P., Pidsley, R., Nisbet, J., Glass, D., … Deloukas, P. (2012). Epigenome-wide scans identify differentially methylated regions for age and age-related phenotypes in a healthy ageing population. *PLoS Genetics, 8*(4), e1002629.

Benvenisty, N., Mencher, D., Meyuhas, O., Razin, A., & Reshef, L. (1985). Sequential changes in DNA methylation patterns of the rat phosphoenolpyruvate carboxykinase gene during development. *Proceedings of the National Academy of Sciences of the United States of America, 82*(2), 267–271.

Berger, S. M., Weber, T., Perreau-Lenz, S., Vogt, M. A., Gartside, S. E., Maser-Gluth, C., … Bartsch, D. (2012). A functional Tph2 C1473G polymorphism causes an anxiety phenotype via compensatory changes in the serotonergic system. *Neuropsychopharmacology, 37*(9), 1986–1998.

Bhagwagar, Z., Hafizi, S., & Cowen, P. J. (2003). Increase in concentration of waking salivary cortisol in recovered patients with depression. *American Journal of Psychiatry, 160*(10), 1890–1891.

Bird, A. (2002). DNA methylation patterns and epigenetic memory. *Genes & Developement, 16*(1), 6–21.

Bjornsson, H. T., Sigurdsson, M. I., Fallin, M. D., Irizarry, R. A., Aspelund, T., Cui, H., … Feinberg, A. P. (2008). Intra-individual change over time in DNA methylation with familial clustering. *JAMA, 299*(24), 2877–2883.

Boks, M. P., Derks, E. M., Weisenberger, D. J., Strengman, E., Janson, E., Sommer, I. E., … Ophoff, R. A. (2009). The relationship of DNA methylation with age, gender and genotype in twins and healthy controls. *PLoS One, 4*(8), e6767.

Booij, L., Wang, D., Levesque, M. L., Tremblay, R. E., & Szyf, M. (2013). Looking beyond the DNA sequence: the relevance of DNA methylation processes for the stress-diathesis model of depression. *Philosophical Transactions of the Royal Society of London. Series B, Biological Sciences, 368*(1615), 20120251.

Booij, L., Tremblay, R. E., Szyf, M., & Benkelfat, C. (2015). Genetic and early environmental influences on the serotonin system: consequences for brain development and risk for psychopathology. *Journal of Psychiatry & Neuroscience, 40*(1), 5–18.

Booij, L., Szyf, M., Carballedo, A., Frey, E. M., Morris, D., Dymov, S., … Frodl, T. (2015). DNA methylation of the serotonin transporter gene in peripheral cells and stress-related changes in hippocampal volume: a study in depressed patients and healthy controls. *PLoS One, 10*(3), e0119061.

Borghol, N., Suderman, M., McArdle, W., Racine, A., Hallett, M., Pembrey, M., … Szyf, M. (2012). Associations with early-life socio-economic position in adult DNA methylation. *International Journal of Epidemiology, 41*(1), 62–74.

Bronfenbrenner, U. (1979). *The Ecology of human development: Experiments by nature and design.* Cambridge: Harvard University Press.

Brown, J., Cohen, P., Johnson, J. G., & Smailes, E. M. (1999). Childhood abuse and neglect: specificity of effects on adolescent and young adult depression and suicidality. *Journal of the American Academy of Child and Adolescence Psychiatry, 38*(12), 1490–1496.

Capuron, L., & Miller, A. H. (2011). Immune system to brain signaling: neuropsychopharmacological implications. *Pharmacology & Therapeutics, 130*(2), 226–238.

Carlberg, L., Scheibelreiter, J., Hassler, M. R., Schloegelhofer, M., Schmoeger, M., Ludwig, B., … Schosser, A. (2014). Brain-derived neurotrophic factor (BDNF)-epigenetic regulation in unipolar and bipolar affective disorder. *Journal of Affective Disorders, 168*, 399–406.

Caspi, A., Sugden, K., Moffitt, T. E., Taylor, A., Craig, I. W., Harrington, H., … Poulton, R. (2003). Influence of life stress on depression: moderation by a polymorphism in the 5-HTT gene. *Science, 301*(5631), 386–389.

Champagne, F. A., Chretien, P., Stevenson, C. W., Zhang, T. Y., Gratton, A., & Meaney, M. J. (2006). Variations in the nucleus accumbens dopamine associated with individual differences in maternal behavior in the rat. *Journal of Neuroscience, 24*, 4113–4123.

Charil, A., Laplante, D. P., Vaillancourt, C., & King, S. (2010). Prenatal stress and brain development. *Brain Research Reviews, 65*(1), 56–79.

Chen, E. S., Ernst, C., & Turecki, G. (2011). The epigenetic effects of antidepressant treatment on human prefrontal cortex BDNF expression. *International Journal of Neuropsychopharmacology, 14*(3), 427–429.

Chen, G. L., & Miller, G. M. (2013). Tryptophan hydroxylase-2: an emerging therapeutic target for stress disorders. *Biochemical Pharmacology, 85*(9), 1227–1233.

Christian, L. M. (2012). Psychoneuroimmunology in pregnancy: immune pathways linking stress with maternal health, adverse birth outcomes, and fetal development. *Neuroscience and Biobehavioral Reviews, 36*(1), 350–361.

Coppell, A. L., Pei, Q., & Zetterstrom, T. S. (2003). Bi-phasic change in BDNF gene expression following antidepressant drug treatment. *Neuropharmacology, 44*(7), 903–910.

D'Addario, C., Dell'Osso, B., Galimberti, D., Palazzo, M. C., Benatti, B., Di Francesco, A., … Maccarrone, M. (2013). Epigenetic modulation of BDNF gene in patients with major depressive disorder. *Biological Psychiatry, 73*(2), e6–e7.

Dalton, V. S., Kolshus, E., & McLoughlin, D. M. (2014). Epigenetics and depression: return of the repressed. *Journal of Affective Disorders, 155*, 1–12.

Danese, A., & McEwen, B. S. (2012). Adverse childhood experiences, allostasis, allostatic load, and age-related disease. *Physiology & Behavior, 106*(1), 29–39.

Dell'Osso, B., D'Addario, C., Carlotta Palazzo, M., Benatti, B., Camuri, G., Galimberti, D., … Altamura, A. C. (2014). Epigenetic modulation of BDNF gene: differences in DNA methylation between unipolar and bipolar patients. *Journal of Affective Disorders, 166*, 330–333.

Deverman, B. E., & Patterson, P. H. (2009). Cytokines and CNS development. *Neuron, 64*(1), 61–78.

Devlin, A. M., Brain, U., Austin, J., & Oberlander, T. F. (2010). Prenatal exposure to maternal depressed mood and the MTHFR C677T variant affect SLC6A4 methylation in infants at birth. *PLoS One, 5*(8), e12201.

Dienes, K. A., Hazel, N. A., & Hammen, C. L. (2013). Cortisol secretion in depressed, and at-risk adults. *Psychoneuroendocrinology, 38*(6), 927–940.

van der Doelen, R. H., Arnoldussen, I. A., Ghareh, H., van Och, L., Homberg, J. R., & Kozicz, T. (2015). Early life adversity and serotonin transporter gene variation interact to affect DNA methylation of the corticotropin-releasing factor gene promoter region in the adult rat brain. *Development and Psychopathology, 27*(1), 123–135.

Domschke, K., Tidow, N., Schwarte, K., Deckert, J., Lesch, K. P., Arolt, V., ... Baune, B. T. (2014). Serotonin transporter gene hypomethylation predicts impaired antidepressant treatment response. *International Journal of Neuropsychopharmacology, 17*(8), 1167–1176.

Dowlati, Y., Herrmann, N., Swardfager, W., Liu, H., Sham, L., Reim, E. K., & Lanctot, K. L. (2010). A meta-analysis of cytokines in major depression. *Biological Psychiatry, 67*(5), 446–457.

Dreimuller, N., Schlicht, K. F., Wagner, S., Peetz, D., Borysenko, L., Hiemke, C., ... Tadic, A. (2012). Early reactions of brain-derived neurotrophic factor in plasma (pBDNF) and outcome to acute antidepressant treatment in patients with major depression. *Neuropharmacology, 62*(1), 264–269.

Dunn, A. J. (2000). Cytokine activation of the HPA axis. *Annals of the New York Academy of Sciences, 917*, 608–617.

Dwivedi, Y., Rizavi, H. S., Conley, R. R., Roberts, R. C., Tamminga, C. A., & Pandey, G. N. (2003). Altered gene expression of brain-derived neurotrophic factor and receptor tyrosine kinase B in postmortem brain of suicide subjects. *Archives of General Psychiatry, 60*(8), 804–815.

Essex, M. J., Thomas Boyce, W., Hertzman, C., Lam, L. L., Armstrong, J. M., Neumann, S. M., & Kobor, M. S. (2013). Epigenetic vestiges of early developmental adversity: childhood stress exposure and DNA methylation in adolescence. *Child Development, 84*(1), 58–75.

Felger, J. C., & Lotrich, F. E. (2013). Inflammatory cytokines in depression: neurobiological mechanisms and therapeutic implications. *Neuroscience, 246*, 199–229.

Filiberto, A. C., Maccani, M. A., Koestler, D., Wilhelm-Benartzi, C., Avissar-Whiting, M., Banister, C. E., ... Marsit, C. J. (2011). Birthweight is associated with DNA promoter methylation of the glucocorticoid receptor in human placenta. *Epigenetics, 6*(5), 566–572.

Fraga, M. F., Ballestar, E., Paz, M. F., Ropero, S., Setien, F., Ballestar, M. L., ... Esteller, M. (2005). Epigenetic differences arise during the lifetime of monozygotic twins. *Proceedings of the National Academy of Sciences of the United States of America, 102*(30), 10604–10609.

Frieling, H., & Tadic, A. (2013). Value of genetic and epigenetic testing as biomarkers of response to antidepressant treatment. *International Review of Psychiatry(Abingdon, England), 25*(5), 572–578.

Frodl, T., Szyf, M., Carballedo, A., Ly, V., Dymov, S., Vaisheva, F., ... Booij, L. (2015). DNA methylation of serotonin transporter gene (SLC6A4) associates with brain function involved in processing of emotional stimuli. *Journal of Psychiatry & Neuroscience, 40*(2), 140180.

Fuchikami, M., Morinobu, S., Segawa, M., Okamoto, Y., Yamawaki, S., Ozaki, N., ... Terao, T. (2011). DNA methylation profiles of the brain-derived neurotrophic factor (BDNF) gene as a potent diagnostic biomarker in major depression. *PLoS One, 6*(8), e23881.

Fuke, C., Shimabukuro, M., Petronis, A., Sugimoto, J., Oda, T., Miura, K., ... Jinno, Y. (2004). Age related changes in 5-methylcytosine content in human peripheral leukocytes and placentas: an HPLC-based study. *Annals of Human Genetics, 68*(Pt 3), 196–204.

Galecka, E., Szemraj, J., Bienkiewicz, M., Majsterek, I., Przybylowska-Sygut, K., Galecki, P., & Lewinski, A. (2013). Single nucleotide polymorphisms of NR3C1 gene and recurrent depressive disorder in population of Poland. *Molecular Biology Reports, 40*(2), 1693–1699.

Gervin, K., Vigeland, M. D., Mattingsdal, M., Hammero, M., Nygard, H., Olsen, A. O., ... Lyle, R. (2012). DNA methylation and gene expression changes in monozygotic twins discordant for psoriasis: identification of epigenetically dysregulated genes. *PLoS Genetics, 8*(1), e1002454.

Glover, V., O'Connor, T. G., & O'Donnell, K. (2010). Prenatal stress and the programming of the HPA axis. *Neuroscience and Biobehavioral Reviews, 35*(1), 17–22.

Gomes, M. V., Toffoli, L. V., Arruda, D. W., Soldera, L. M., Pelosi, G. G., Neves-Souza, R. D., ... Marquez, A. S. (2012). Age-related changes in the global DNA methylation profile of leukocytes are linked to nutrition but are not associated with the MTHFR C677T genotype or to functional capacities. *PLoS One, 7*(12), e52570.

Gordon, L., Joo, J. H., Andronikos, R., Ollikainen, M., Wallace, E. M., Umstad, M. P., ... Craig, J. M. (2011). Expression discordance of monozygotic twins at birth: effect of intrauterine environment and a possible mechanism for fetal programming. *Epigenetics, 6*(5), 579–592.

Gordon, L., Joo, J. E., Powell, J. E., Ollikainen, M., Novakovic, B., Li, X., ... Saffery, R. (2012). Neonatal DNA methylation profile in human twins is specified by a complex interplay between intrauterine environmental and genetic factors, subject to tissue-specific influence. *Genome Research, 22*(8), 1395–1406.

Gronniger, E., Weber, B., Heil, O., Peters, N., Stab, F., Wenck, H., ... Lyko, F. (2010). Aging and chronic sun exposure cause distinct epigenetic changes in human skin. *PLoS Genetics, 6*(5), e1000971.

Hannum, G., Guinney, J., Zhao, L., Zhang, L., Hughes, G., Sadda, S., ... Zhang, K. (2013). Genome-wide methylation profiles reveal quantitative views of human aging rates. *Molecular Cell, 49*(2), 359–367.

Hardeveld, F., Spijker, J., Vreeburg, S. A., Graaf, R. D., Hendriks, S. M., Licht, C. M., ... Beekman, A. T. (2014). Increased cortisol awakening response was associated with time to recurrence of major depressive disorder. *Psychoneuroendocrinology, 50*, 62–71.

Harris, R. A., Nagy-Szakal, D., & Kellermayer, R. (2013). Human metastable epiallele candidates link to common disorders. *Epigenetics, 8*(2), 157–163.

Harris, A., & Seckl, J. (2011). Glucocorticoids, prenatal stress and the programming of disease. *Hormones and Behavior, 59*(3), 279–289.

Hassanain, M., Bhatt, S., Zalcman, S., & Siegel, A. (2005). Potentiating role of interleukin-1beta (IL-1beta) and IL-1beta type 1 receptors in the medial hypothalamus in defensive rage behavior in the cat. *Brain Research, 1048*(1–2), 1–11.

Heijmans, B. T., Kremer, D., Tobi, E. W., Boomsma, D. I., & Slagboom, P. E. (2007). Heritable rather than age-related environmental and stochastic factors dominate variation in DNA methylation of the human IGF2/H19 locus. *Human Molecular Genetics, 16*(5), 547–554.

Heim, C., & Nemeroff, C. B. (2001). The role of childhood trauma in the neurobiology of mood and anxiety disorders: preclinical and clinical studies. *Biological Psychiatry, 49*(12), 1023–1039.

Hellweg, R., Ziegenhorn, A., Heuser, I., & Deuschle, M. (2008). Serum concentrations of nerve growth factor and brain-derived neurotrophic factor in depressed patients before and after antidepressant treatment. *Pharmacopsychiatry, 41*(2), 66–71.

Hernandez, D. G., Nalls, M. A., Gibbs, J. R., Arepalli, S., van der Brug, M., Chong, S., ... Singleton, A. B. (2011). Distinct DNA methylation changes highly correlated with chronological age in the human brain. *Human Molecular Genetics, 20*(6), 1164–1172.

Heyn, H., Li, N., Ferreira, H. J., Moran, S., Pisano, D. G., Gomez, A., ... Esteller, M. (2012). Distinct DNA methylomes of newborns and centenarians. *Proceedings of the National Academy of Sciences of the United States of America, 109*(26), 10522–10527.

Hosang, G. M., Shiles, C., Tansey, K. E., McGuffin, P., & Uher, R. (2014). Interaction between stress and the BDNF Val66Met polymorphism in depression: a systematic review and meta-analysis. *BMC Medicine, 12*, 7.

Ignacio, Z. M., Reus, G. Z., Abelaira, H. M., & Quevedo, J. (2014). Epigenetic and epistatic interactions between serotonin transporter and brain-derived neurotrophic factor genetic polymorphism: insights in depression. *Neuroscience, 275*, 455–468.

van IJzendoorn, M. H., Caspers, K., Bakermans-Kranenburg, M. J., Beach, S. R., & Philibert, R. (2010). Methylation matters: interaction between methylation density and serotonin transporter genotype predicts unresolved loss or trauma. *Biological Psychiatry, 68*(5), 405–407.

Irwin, M. R., & Miller, A. H. (2007). Depressive disorders and immunity: 20 years of progress and discovery. *Brain, Behavior, and Immunity, 21*(4), 374–383.

Januar, V., Saffery, R., & Ryan, J. (2015). Epigenetics and depressive disorders: a review of current progress and future directions. *International Journal of Epidemiology, 44*(4), 1364–1387.

Javierre, B. M., Fernandez, A. F., Richter, J., Al-Shahrour, F., Martin-Subero, J. I., Rodriguez-Ubreva, J., ... Ballestar, E. (2010). Changes in the pattern of DNA methylation associate with twin discordance in systemic lupus erythematosus. *Genome Research, 20*(2), 170–179.

Jiang, C., & Salton, S. R. (2013). The role of Neurotrophins in major depressive disorder. *Translational Neuroscience, 4*(1), 46–58.

Johansson, A., Enroth, S., & Gyllensten, U. (2013). Continuous aging of the human DNA methylome throughout the human lifespan. *PLoS One, 8*(6), e67378.

Jones, P. L., Veenstra, G. J., Wade, P. A., Vermaak, D., Kass, S. U., Landsberger, N., ... Wolffe, A. P. (1998). Methylated DNA and MeCP2 recruit histone deacetylase to repress transcription. *Nature Genetics, 19*(2), 187–191.

Kaminsky, Z., Petronis, A., Wang, S. C., Levine, B., Ghaffar, O., Floden, D., & Feinstein, A. (2008). Epigenetics of personality traits: an illustrative study of identical twins discordant for risk-taking behavior. *Twin Research and Human Genetics, 11*(1), 1–11.

Kang, H. J., Kim, J. M., Stewart, R., Kim, S. Y., Bae, K. Y., Kim, S. W., ... Yoon, J. S. (2013). Association of SLC6A4 methylation with early adversity, characteristics and outcomes in depression. *Progress in Neuro-psychopharmacology & Biological Psychiatry, 44*, 23–28.

Karg, K., Burmeister, M., Shedden, K., & Sen, S. (2011). The serotonin transporter promoter variant (5-HTTLPR), stress, and depression meta-analysis revisited: evidence of genetic moderation. *Archives of General Psychiatry, 68*(5), 444–454.

Karg, K., & Sen, S. (2012). Gene × Environment interaction models in psychiatric genetics. *Current Topics in Behavioral Neurosciences, 12*, 441–462.

Keller, S., Sarchiapone, M., Zarrilli, F., Videtic, A., Ferraro, A., Carli, V., ... Chiariotti, L. (2010). Increased BDNF promoter methylation in the Wernicke area of suicide subjects. *Archives of General Psychiatry, 67*(3), 258–267.

Kendler, K. S., Kuhn, J. W., Vittum, J., Prescott, C. A., & Riley, B. (2005). The interaction of stressful life events and a serotonin transporter polymorphism in the prediction of episodes of major depression: a replication. *Archives General Psychiatry, 62*(5), 529–535.

Kessler, R. C., Berglund, P., Demler, O., Jin, R., Koretz, D., Merikangas, K. R., ... Wang, P. S., &, National Comorbidity Survey, Replication. (2003). The epidemiology of major depressive disorder: results from the National Comorbidity Survey Replication (NCS-R). *JAMA, 289*(23), 3095–3105.

Kessler, R. C., Merikangas, K. R., & Wang, P. S. (2007). Prevalence, comorbidity, and service utilization for mood disorders in the United States at the beginning of the twenty-first century. *Annual Review Clinical Psychology, 3*, 137–158.

Kinnally, E. L., Capitanio, J. P., Leibel, R., Deng, L., LeDuc, C., Haghighi, F., & Mann, J. J. (2010). Epigenetic regulation of serotonin transporter expression and behavior in infant rhesus macaques. *Genes, Brain, and Behavior, 9*(6), 575–582.

Kinnally, E. L., Feinberg, C., Kim, D., Ferguson, K., Leibel, R., Coplan, J. D., & John Mann, J. (2011). DNA methylation as a risk factor in the effects of early life stress. *Brain, Behavior, Immunity, 25*(8), 1548–1553.

Klengel, T., Mehta, D., Anacker, C., Rex-Haffner, M., Pruessner, J. C., Pariante, C. M., ... Binder, E. B. (2013). Allele-specific FKBP5 DNA demethylation mediates gene-childhood trauma interactions. *Nature Neuroscience, 16*(1), 33–41.

van der Knapp, L. J., Riese, H., Hudziak, J. J., Verbiest, M. M. P. J., Verhulst, F. C., Oldehinkel, A. J., & van Oort, F. V. A. (2015). Adverse life events and allele-specific methylation of the serotonin transporter (SLC6A4) in adolescents: the TRAILS study. *Psychosomatic Medicine, 77*(3), 246–255. Advance online publication.

Kuratomi, G., Iwamoto, K., Bundo, M., Kusumi, I., Kato, N., Iwata, N., ... Kato, T. (2008). Aberrant DNA methylation associated with bipolar disorder identified from discordant monozygotic twins. *Molecular Psychiatry, 13*(4), 429–441.

Labonte, B., Yerko, V., Gross, J., Mechawar, N., Meaney, M. J., Szyf, M., & Turecki, G. (2012). Differential glucocorticoid receptor exon 1(B), 1(C), and 1(H) expression and methylation in suicide completers with a history of childhood abuse. *Biological Psychiatry, 72*(1), 41–48.

Labonte, B., Azoulay, N., Yerko, V., Turecki, G., & Brunet, A. (2014). Epigenetic modulation of glucocorticoid receptors in posttraumatic stress disorder. *Translational Psychiatry, 4*, e368.

Laloux, C., Mairesse, J., van Camp, G., Giovine, G., Branchi, I., Bouret, S., ... Maccari, S. (2012). Anxiety-like behaviour and associated neurochemical and endocronological alterations in male pups exposed to prenatal stress. *Psychoneuroendocrinology, 37*(10), 1646–1658.

Langevin, S. M., Houseman, E. A., Christensen, B. C., Wiencke, J. K., Nelson, H. H., Karagas, M. R., ... Kelsey, K. T. (2011). The influence of aging, environmental exposures and local sequence features on the variation of DNA methylation in blood. *Epigenetics, 6*(7), 908–919.

Levesque, M. L., Casey, K. F., Szyf, M., Ismaylova, E., Ly, V., Verner, M.-P., ... Booij, L. (2014). Genome-wide DNA methylation variability in adolescent monozygotic twins followed since birth. *Epigenetics, 9*(10), 1410–1421.

Lokk, K., Modhukur, V., Rajashekar, B., Martens, K., Magi, R., Kolde, R., ... Tonisson, N. (2014). DNA methylome profiling of human tissues identifies global and tissue-specific methylation patterns. *Genome Biology, 15*(4), r54.

Lopez, J. P., Mamdani, F., Labonte, B., Beaulieu, M. M., Yang, J. P., Berlim, M. T., ... Turecki, G. (2013). Epigenetic regulation of BDNF expression according to antidepressant response. *Molecular Psychiatry, 18*(4), 398–399.

Marrocco, J., Reynaert, M. L., Gatta, E., Gabriel, C., Mocaer, E., di Prisco, S., ... Mairesse, J. (2014). The effects of antidepressant treatment in prenatally stressed rats support the glutamatergic hypothesis of stress-related disorders. *Journal of Neuroscience, 34*(6), 2015–2024.

Martino, D., Loke, Y. J., Gordon, L., Ollikainen, M., Cruickshank, M. N., Saffery, R., & Craig, J. M. (2013). Longitudinal, genome-scale analysis of DNA methylation in twins from birth to 18 months of age reveals rapid epigenetic change in early life and pair-specific effects of discordance. *Genome Biology, 14*(5), R42.

Mastroeni, D., McKee, A., Grover, A., Rogers, J., & Coleman, P. D. (2009). Epigenetic differences in cortical neurons from a pair of monozygotic twins discordant for Alzheimer's disease. *PLoS One, 4*(8), e6617.

McGowan, P. O., Sasaki, A., D'Alessio, A. C., Dymov, S., Labonte, B., Szyf, M., ... Meaney, M. J. (2009). Epigenetic regulation of the glucocorticoid receptor in human brain associates with childhood abuse. *Nature Neuroscience, 12*(3), 342–348.

McGowan, P. O., & Szyf, M. (2010). The epigenetics of social adversity in early life: implications for mental health outcomes. *Neurobiology of Disease, 39*(1), 66–72.

Meaney, M. J. (2010). Epigenetics and the biological definition of gene × environment interactions. *Child Development, 81*(1), 41–79.

Mills, N. T., Scott, J. G., Wray, N. R., Cohen-Woods, S., & Baune, B. T. (2013). Research review: the role of cytokines in depression in adolescents: a systematic review. *Journal of Child Psychology and Psychiatry, and allied disciplines, 54*(8), 816–835.

Muangsub, T., Samsuwan, J., Tongyoo, P., Kitkumthorn, N., & Mutirangura, A. (2014). Analysis of methylation microarray for tissue specific detection. *Gene, 553*(1), 31–41.

Murgatroyd, C., Patchev, A. V., Wu, Y., Micale, V., Bockmuhl, Y., Fischer, D., ... Spengler, D. (2009). Dynamic DNA methylation programs persistent adverse effects of early-life stress. *Nature Neuroscience, 12*(12), 1559–1566.

Murgatroyd, C., & Spengler, D. (2011a). Epigenetic programming of the HPA axis: early life decides. *Stress, 14*(6), 581–589.

Murgatroyd, C., & Spengler, D. (2011b). Epigenetics of early child development. *Frontiers in Psychiatry, 2*, 16.

Myint, A. M., & Kim, Y. K. (2003). Cytokine-serotonin interaction through IDO: a neurodegeneration hypothesis of depression. *Medical Hypotheses, 61*(5–6), 519–525.

Nan, X., Ng, H. H., Johnson, C. A., Laherty, C. D., Turner, B. M., Eisenman, R. N., & Bird, A. (1998). Transcriptional repression by the methyl-CpG-binding protein MeCP2 involves a histone deacetylase complex. *Nature, 393*(6683), 386–389.

Nguyen, A., Rauch, T. A., Pfeifer, G. P., & Hu, V. W. (2010). Global methylation profiling of lymphoblastoid cell lines reveals epigenetic contributions to autism spectrum disorders and a novel autism candidate gene, RORA, whose protein product is reduced in autistic brain. *FASEB Journal, 24*(8), 3036–3051.

Nikolova, Y. S., Koenen, K. C., Galea, S., Wang, C. M., Seney, M. L., Sibille, E., ... Hariri, A. R. (2014). Beyond genotype: serotonin transporter epigenetic modification predicts human brain function. *Nature Neuroscience, 17*(9), 1153–1155.

Nivard, M. G., Dolan, C. V., Kendler, K. S., Kan, K. J., Willemsen, G., van Beijsterveldt, C. E., ... Boomsma, D. I. (2015). Stability in symptoms of anxiety and depression as a function of genotype and environment: a longitudinal twin study from ages 3 to 63 years. *Psychological Medicine, 45*(5), 1039–1049.

Novakovic, B., Yuen, R. K., Gordon, L., Penaherrera, M. S., Sharkey, A., Moffett, A., ... Saffery, R. (2011). Evidence for widespread changes in promoter methylation profile in human placenta in response to increasing gestational age and environmental/stochastic factors. *BMC Genomics, 12*, 529.

O'Connor, T. G., Moynihan, J. A., & Caserta, M. T. (2014). Annual research review: the neuroinflammation hypothesis for stress and psychopathology in children—developmental psychoneuroimmunology. *Journal of Child Psychology and Psychiatry, and Allied Disciplines, 55*(6), 615–631.

Oberlander, T. F., Weinberg, J., Papsdorf, M., Grunau, R., Misri, S., & Devlin, A. M. (2008). Prenatal exposure to maternal depression, neonatal methylation of human glucocorticoid receptor gene (NR3C1) and infant cortisol stress responses. *Epigenetics, 3*(2), 97–106.

Okada, S., Morinobu, S., Fuchikami, M., Segawa, M., Yokomaku, K., Kataoka, T., ... Mimura, M. (2014). The potential of SLC6A4 gene methylation analysis for the diagnosis and treatment of major depression. *Journal of Psychiatric Research, 53*, 47–53.

van Oort, F. V., van der Ende, J., Wadsworth, M. E., Verhulst, F. C., & Achenbach, T. M. (2011). Cross-national comparison of the link between socioeconomic status and emotional and behavioral problems in youths. *Social Psychiatry and Psychiatric Epidemiology, 46*(2), 167–172.

Ollikainen, M., Smith, K. R., Joo, E. J., Ng, H. K., Andronikos, R., Novakovic, B., ... Craig, J. M. (2010). DNA methylation analysis of multiple tissues from newborn twins reveals both genetic and intrauterine components to variation in the human neonatal epigenome. *Human Molecular Genetics, 19*(21), 4176–4188.

World Health Organization. (2012). *Depression: A global Crisis*. Retrieved 2015 from http://www.who.int/mediacentre/factsheets/fs369/en/.

Ouellet-Morin, I., Wong, C. C., Danese, A., Pariante, C. M., Papadopoulos, A. S., Mill, J., & Arseneault, L. (2012). Increased serotonin transporter gene (SERT) DNA methylation is associated with bullying victimization and blunted cortisol response to stress in childhood: a longitudinal study of discordant monozygotic twins. *Psychological Medicine*, 1–11.

Patas, K., Penninx, B. W., Bus, B. A., Vogelzangs, N., Molendijk, M. L., Elzinga, B. M., ... Oude Voshaar, R. C. (2014). Association between serum brain-derived neurotrophic factor and plasma interleukin-6 in major depressive disorder with melancholic features. *Brain, Behavior, and Immunity, 36*, 71–79.

Pawluski, J. L., Rayen, I., Niessen, N. A., Kristensen, S., van Donkelaar, E. L., Balthazart, J., ... Charlier, T. D. (2012). Developmental fluoxetine exposure differentially alters central and peripheral measures of the HPA system in adolescent male and female offspring. *Neuroscience, 220*, 131–141.

Pechtel, P., & Pizzagalli, D. A. (2011). Effects of early life stress on cognitive and affective function: an integrated review of human literature. *Psychopharmacology, 214*(1), 55–70.

Pena, C. J., Neugut, Y. D., Calarco, C. A., & Champagne, F. A. (2014). Effects of maternal care on the development of midbrain dopamine pathways and reward-directed behavior in female offspring. *The European Journal of Neuroscience, 39*, 946–956.

Perroud, N., Paoloni-Giacobino, A., Prada, P., Olie, E., Salzmann, A., Nicastro, R., ... Malafosse, A. (2011). Increased methylation of glucocorticoid receptor gene (NR3C1) in adults with a history of childhood maltreatment: a link with the severity and type of trauma. *Translational Psychiatry, 1*, e59.

Pesce, M., Speranza, L., Franceschelli, S., Ialenti, V., Patruno, A., Febo, M. A., ... Grilli, A. (2011). Biological role of interleukin-1beta in defensive-aggressive behaviour. *Journal of Biological Regulators and Homeostatic Agents, 25*(3), 323–329.

Pintor, L., Torres, X., Navarro, V., Martinez de Osaba, M. A., Matrai, S., & Gasto, C. (2007). Corticotropin-releasing factor test in melancholic patients in depressed state versus recovery: a comparative study. *Progress in Neuro-psychopharmacology & Biological Psychiatry, 31*(5), 1027–1033.

Pluess, M., Velders, F. P., Belsky, J., van, IJzendoorn, M. H., Bakermans-Kranenburg, M. J., Jaddoe, V. W., ... Tiemeier, H. (2011). Serotonin transporter polymorphism moderates effects of prenatal maternal anxiety on infant negative emotionality. *Biological Psychiatry, 69*(6), 520–525.

Powell, T. R., Smith, R. G., Hackinger, S., Schalkwyk, L. C., Uher, R., McGuffin, P., ... Tansey, K. E. (2013). DNA methylation in interleukin-11 predicts clinical response to antidepressants in GENDEP. *Translational Psychiatry, 3*, e300.

Prathikanti, S., & Weinberger, D. R. (2005). Psychiatric genetics—the new era: genetic research and some clinical implications. *British Medical Bulletin, 73-74*, 107–122.

Provencal, N., Suderman, M. J., Guillemin, C., Massart, R., Ruggiero, A., Wang, D., ... Szyf, M. (2012). The signature of maternal rearing in the methylome in rhesus macaque prefrontal cortex and T cells. *Journal of Neuroscience, 32*(44), 15626–15642.

Rakyan, V. K., Beyan, H., Down, T. A., Hawa, M. I., Maslau, S., Aden, D., ... Leslie, R. D. (2011). Identification of type 1 diabetes-associated DNA methylation variable positions that precede disease diagnosis. *PLoS Genetics, 7*(9), e1002300.

Razin, A. (1998). CpG methylation, chromatin structure and gene silencing-a three-way connection. *EMBO Journal, 17*(17), 4905–4908.

Razin, A., & Cedar, H. (1993). DNA methylation and embryogenesis. *Experimentia Supplementum, 64*, 343–357.

Ribel-Madsen, R., Fraga, M. F., Jacobsen, S., Bork-Jensen, J., Lara, E., Calvanese, V., & Poulsen, P. (2012). Genome-wide analysis of DNA methylation differences in muscle and fat from monozygotic twins discordant for type 2 diabetes. *PLoS One, 7*(12), e51302.

Risch, N., Herrell, R., Lehner, T., Liang, K. Y., Eaves, L., Hoh, J., ... Merikangas, K. R. (2009). Interaction between the serotonin transporter gene (5-HTTLPR), stressful life events, and risk of depression: a meta-analysis. *JAMA, 301*(23), 2462–2471.

Roberts, S., Lester, K. J., Hudson, J. L., Rapee, R. M., Creswell, C., Cooper, P. J., ... Eley, T. C. (2014). Serotonin tranporter methylation and response to cognitive behaviour therapy in children with anxiety disorders. *Translational Psychiatry, 4*, e444.

Rodriguez, E., Baurecht, H., Wahn, A. F., Kretschmer, A., Hotze, M., Zeilinger, S., ... Weidinger, S. (2014). An integrated epigenetic and transcriptomic analysis reveals distinct tissue-specific patterns of DNA methylation associated with atopic dermatitis. *Journal of Investigative Dermatology, 134*(7), 1873–1883.

Roth, T. L., & Sweatt, J. D. (2011a). Annual Research Review: epigenetic mechanisms and environmental shaping of the brain during sensitive periods of development. *Journal of Child Psychology and Psychiatry, and Allied Disciplines, 52*(4), 398–408.

Roth, T. L., & Sweatt, J. D. (2011b). Epigenetic marking of the BDNF gene by early-life adverse experiences. *Hormones and Behavior, 59*(3), 315–320.

Saffery, R., Morley, R., Carlin, J. B., Joo, J. H., Ollikainen, M., Novakovic, B., ... Craig, J. M. (2012). Cohort profile: the peri/post-natal epigenetic twins study. *International Journal of Epidemiology, 41*(1), 55–61.

Shalev, I., Lerer, E., Israel, S., Uzefovsky, F., Gritsenko, I., Mankuta, D., ... Kaitz, M. (2009). BDNF Val66Met polymorphism is associated with HPA axis reactivity to psychological stress characterized by genotype and gender interactions. *Psychoneuroendocrinology, 34*(3), 382–388.

Shonkoff, J. P., Boyce, W. T., & McEwen, B. S. (2009). Neuroscience, molecular biology, and the childhood roots of health disparities: building a new framework for health promotion and disease prevention. *JAMA, 301*(21), 2252–2259.

Shonkoff, J. P., & Garner, A. S. (2012). The lifelong effects of early childhood adversity and toxic stress. *Pediatrics, 129*(1), e232–246.

Solomon, D. A., Keller, M. B., Leon, A. C., Mueller, T. I., Lavori, P. W., Shea, M. T., ... Endicott, J. (2000). Multiple recurrences of major depressive disorder. *The American Journal of Psychiatry, 157*(2), 229–233.

Song, Y., Miyaki, K., Suzuki, T., Sasaki, Y., Tsutsumi, A., Kawakami, N., ... Shimbo, T. (2014). Altered DNA methylation status of human brain derived neurotrophis factor gene could be useful as biomarker of depression. *American Journal of Medical Genetics Part B, Neuropsychiatric Genetics, 165B*(4), 357–364.

Stefan, M., Zhang, W., Concepcion, E., Yi, Z., & Tomer, Y. (2014). DNA methylation profiles in type 1 diabetes twins point to strong epigenetic effects on etiology. *Journal of Autoimmunity, 50*, 33–37.

Suderman, M., Borghol, N., Pappas, J. J., Pinto Pereira, S. M., Pembrey, M., Hertzman, C., ... Szyf, M. (2014). Childhood abuse is associated with methylation of multiple loci in adult DNA. *BMC Medical Genomics, 7*, 13.

Suzuki, M. M., & Bird, A. (2008). DNA methylation landscapes: provocative insights from epigenomics. *Nature Reviews. Genetics, 9*(6), 465–476.

Szczepankiewicz, A., Leszczynska-Rodziewicz, A., Pawlak, J., Rajewska-Rager, A., Dmitrzak-Weglarz, M., Wilkosc, M., ... Hauser, J. (2011). Glucocorticoid receptor polymorphism is associated with major depression and predominance of depression in the course of bipolar disorder. *Journal of Affective Disorders, 134*(1–3), 138–144.

Szyf, M. (2015). Nongenetic inheritance and transgenerational epigenetics. *Trends in Molecular Medicine, 21*(2), 134–144.

Tadic, A., Wagner, S., Schlicht, K. F., Peetz, D., Borysenko, L., Dreimuller, N., ... Lieb, K. (2011). The early non-increase of serum BDNF predicts failure of antidepressant treatment in patients with major depression: a pilot study. *Progress in Neuro-psychopharmacology & Biological Psychiatry, 35*(2), 415–420.

Tadic, A., Muller-Engling, L., Schlicht, K. F., Kotsiari, A., Dreimuller, N., Kleimann, A., ... Frieling, H. (2014). Methylation of the promoter of brain-derived neurotrophic factor exon IV and antidepressant response in major depression. *Molecular Psychiatry, 19*(3), 281–283.

Talens, R. P., Christensen, K., Putter, H., Willemsen, G., Christiansen, L., Kremer, D., ... Heijmans, B. T. (2012). Epigenetic variation during the adult lifespan: cross-sectional and longitudinal data on monozygotic twin pairs. *Aging Cell, 11*(4), 694–703.

Tsankova, N. M., Berton, O., Renthal, W., Kumar, A., Neve, R. L., & Nestler, E. J. (2006). Sustained hippocampal chromatin regulation in a mouse model of depression and antidepressant action. *Nature Neuroscience, 9*(4), 519–525.

Tsankova, N., Renthal, W., Kumar, A., & Nestler, E. J. (2007). Epigenetic regulation in psychiatric disorders. *Nature Reviews Neurosciences, 8*(5), 355–367.

Tsao, C. W., Lin, Y. S., Chen, C. C., Bai, C. H., & Wu, S. R. (2006). Cytokines and serotonin transporter in patients with major depression. *Progress in Neuro-psychopharmacology & Biological Psychiatry, 30*(5), 899–905.

Uher, R., & McGuffin, P. (2010). The moderation by the serotonin transporter gene of environmental adversity in the etiology of depression: 2009 update. *Molecular Psychiatry, 15*(1), 18–22.

Walker, S. P., Wachs, T. D., Grantham-McGregor, S., Black, M. M., Nelson, C. A., Huffman, S. L., ... Richter, L. (2011). Inequality in early childhood: risk and protective factors for early child development. *Lancet, 378*(9799), 1325–1338.

Wang, D., Liu, X., Zhou, Y., Xie, H., Hong, X., Tsai, H. J., ... Wang, X. (2012). Individual variation and longitudinal pattern of genome-wide DNA methylation from birth to the first two years of life. *Epigenetics, 7*(6), 594–605.

Wang, D., Szyf, M., Benkelfat, C., Provencal, N., Turecki, G., Caramaschi, D., ... Booij, L. (2012b). Peripheral SLC6A4 DNA methylation is associated with in vivo measures of human brain serotonin synthesis and childhood physical aggression. *PLoS One, 7*(6), e39501.

Wankerl, M., Miller, R., Kirschbaum, C., Hennig, J., Stalder, T., & Alexander, N. (2014). Effects of genetic and early environmental risk factors for depression on serotonin transporter expression and methylation profiles. *Translational Psychiatry, 4*, e402.

Weaver, I. C. (2007). Epigenetic programming by maternal behavior and pharmacological intervention. Nature versus nurture: let's call the whole thing off. *Epigenetics, 2*(1), 22–28.

Weaver, I. C., Cervoni, N., Champagne, F. A., D'Alessio, A. C., Sharma, S., Seckl, J. R., ... Meaney, M. J. (2004). Epigenetic programming by maternal behavior. *Nature Neuroscience, 7*(8), 847–854.

Widom, C. S., DuMont, K., & Czaja, S. J. (2007). A prospective investigation of major depressive disorder and comorbidity in abused and neglected children grown up. *Archives of General Psychiatry, 64*(1), 49–56.

Williams, L. M., Gatt, J. M., Schofield, P. R., Olivieri, G., Peduto, A., & Gordon, E. (2009). 'Negativity bias' in risk for depression and anxiety: brain-body fear circuitry correlates, 5-HTT-LPR and early life stress. *NeuroImage, 47*(3), 804–814.

Wong, C. C., Meaburn, E. L., Ronald, A., Price, T. S., Jeffries, A. R., Schalkwyk, L. C., ... Mill, J. (2014). Methylomic analysis of monozygotic twins discordant for autism spectrum disorder and related behavioural traits. *Molecular Psychiatry, 19*(4), 495–503.

Wu, H. C., Wang, Q., Chung, W. K., Andrulis, I. L., Daly, M. B., John, E. M., ... Terry, M. B. (2014). Correlation of DNA methylation levels in blood and saliva DNA in young girls of the LEGACY Girls study. *Epigenetics, 9*(7), 929–933.

Zhao, J., Goldberg, J., Bremner, J. D., & Vaccarino, V. (2013). Association between promoter methylation of serotonin transporter gene and depressive symptoms: a monozygotic twin study. *Psychosomatic Medicine, 75*(6), 523–529.

Zuena, A. R., Mairesse, J., Casolini, P., Cinque, C., Alema, G. S., Morley-Fletcher, S., ... Maccari, S. (2008). Prenatal restraint stress generates two distinct behavioral and neurochemical profiles in male and female rats. *PloS One, 3*(5), e2170.

Chapter 7

Identifying Large-Scale Neural Networks Using fMRI

Peter C. Mulders[1,2], **Philip F. van Eijndhoven**[1,2], **Christian F. Beckmann**[2,3,4]

[1]*Department of Psychiatry, Radboud University Medical Centre, Nijmegen, The Netherlands;* [2]*Department of Cognitive Neuroscience, Radboud University Medical Centre, Nijmegen, The Netherlands;* [3]*Donders Institute for Brain, Cognition and Behaviour, Centre for Neuroscience, Nijmegen, The Netherlands;* [4]*Oxford University Centre for Functional MRI of the Brain (FMRIB), Nuffield Department of Clinical Neurosciences, University of Oxford, Oxford, UK*

INTRODUCTION

The field of neuroimaging has advanced considerably, both in terms of data acquisition and analysis methodology, and many new developments have been applied to studying the underlying biology of major depressive disorder (MDD). While earlier neuroimaging techniques primarily investigated focal structural and functional changes (Alcaro, Panksepp, Witczak, Hayes, & Northoff, 2010; Hamilton, Chen, & Gotlib, 2013; Koenigs & Grafman, 2009; Northoff, Wiebking, Feinberg, & Panksepp, 2011; Price & Drevets, 2012), depression is increasingly understood as a disorder of distributed effects among interacting brain regions (Drevets, Savitz, & Trimble, 2008; Hamilton et al., 2013; Mayberg, 1997). Within such a conceptual framework, different brain regions are dynamically organized into functional networks of interconnected areas (or "nodes") that interact and thereby enable specific functions (Bressler, 1995). With the advance of network-based research in system-level neurosciences in general, new techniques now allow us to identify these large-scale brain networks. In this chapter, we will discuss popular methods of investigating these networks, their advantages, and disadvantages and summarize the current understanding of network dysfunction in depression.

METHODS

The current leading belief in systems-level neuroscience is that no function is localized exclusively within a single brain area. Instead, each action, decision, or emotion is thought to be generated by the interaction of multiple regions

Systems Neuroscience in Depression. http://dx.doi.org/10.1016/B978-0-12-802456-0.00007-8

across the brain. Areas that routinely cooperate to create a specific outcome are organized within "functional networks," and there have been considerable advances in techniques that attempt to characterize these networks in terms of their organization and interactivity. Understanding how these different techniques inform us on the different aspects of brain function requires some basic understanding of both imaging acquisition and analysis, which we will briefly discuss.

Basics of (Functional) Image Acquisition

The most commonly applied methods of network research make use of functional magnetic resonance imaging (fMRI). This type of imaging obtains a T2*-weighted blood oxygen level-dependent (BOLD) signal, which captures the shift in magnetic susceptibility that occurs when the ratio of oxygenated and deoxygenated hemoglobin changes (Ogawa & Lee, 1990). Given that neuronal activity requires oxygen, the change in BOLD-signal is strongly related to the activity of the surrounding cells, making it an indirect, yet reliable, noninvasive way to assess functional activity (Siero et al., 2014). Even during periods of rest, the brain is a highly active organ, accounting for approximately 20% of energy consumed, while during strong activation the increase in BOLD-signal relative to baseline is around 5% (Raichle et al., 2001; Shulman et al., 1997). Because this increase is small as compared to the high and subject-specific baseline activity, it is not yet feasible to use the raw BOLD-signal to evaluate disease-related patterns in a single individual. However, this type of data contains many other interesting features that bypass these limitations.

One aspect fundamental to network analysis is the fact that the BOLD-signal is far from constant. There are significant spontaneous fluctuations in signal activity during both activation and rest, and these signals can be interpreted as a combination of outputs from various signal sources, such as cardiovascular and respiratory effects and movement as well as the spontaneous fluctuations in neuronal activity. By measuring the BOLD-signal within one session at multiple points in time, a four-dimensional (4D) image is obtained, where each three-dimensional (3D) volumetric element (or "voxel," typically of size $\sim 2 \times 2 \times 2$ mm) in the brain has a (BOLD-signal) value for its activity at that particular time. With sufficient measurements, the signal intensity for a particular voxel over time (it's "time course") can be used to detect any similarities in temporal dynamics across the brain between voxels. Because of the different frequencies of the signals that are mixed within the BOLD-signal, it is important to pay attention to the different acquisition parameters when obtaining 4D fMRI. Besides the spatial resolution (which determines the size of each voxel), aspects like temporal resolution (time between each 3D image), total number of volumes, and total scanning time need to be sufficient to describe the temporal dynamics. As scanning systems and sequences are improved, higher spatial resolution and signal-to-noise ratio and a shorter temporal resolution will further help correctly identify even subtle variations within brain networks.

Extracting Networks

Initially, networks were defined based on regions that showed a significant activation pattern during a specific condition. This approach is quite similar to earlier studies using positron emission topography (PET), which used more invasive means to assess brain activity. As mentioned earlier, BOLD is a relative, i.e., non-quantitative measure of activation. This means that translating the data into discernible networks requires either a contrast when compared to baseline activity or investigation of the synchronization of the spontaneous signal fluctuations during rest. An important methodological development regarding the latter was the finding that even during rest, areas that share functionality exhibit similar temporal dynamics, leading to the discovery of the so-called "resting-state networks." The nature of the resting state, being independent of task-based paradigms, offers an important advantage of being a method that is reproducible across different populations and study settings (Buckner, 2012; Snyder & Raichle, 2012). While "resting-state network" is commonly used to describe the large functional networks in the brain, these networks are not exclusively tied to the resting state but instead are ever dynamic in adapting to changing environments or stimuli.

In contrast to the resting-state, task-based activation studies commonly use a block- or event-related design to assess the changes in BOLD-signal when compared to a baseline signal (rest). This contrast then represents the task-related changes within the context. Given that functional connectivity is dynamic by nature (as opposed to structural connectivity), networks will interact to perform specific tasks, and disease-related changes in their interaction can become more apparent during a specific setting as compared to the resting state. Unfortunately, designing a task in such a way as to highlight the specific function relevant to the research question while minimizing the effect of confounds, such as variations in intelligence or cognitive deficit, can be problematic. The results obtained might not be exclusively related to the function under investigation, especially in a clinical population, where the execution of a task can be influenced by a large variety of patient-bound factors. Instead, the assumption that a task only elicits task-related responses (the principle of "pure insertion") may not be optimal for functional neuroimaging, as it does not take the interactions of different cognitive components into account (Friston et al., 1996).

Although all of the methods of identifying large-scale neural networks use a similar type of imaging data, there are large differences in their base assumptions and complexity and therefore ensuing interpretations. Broadly speaking, there are two distinct approaches to analyzing fMRI. The first uses correlations of the time series within the 4D image to create a map of functional connectivity, which can be defined as the temporal correlation of two distant neurophysiological effects. The second approach uses the fMRI image to calculate a certain connectivity-related variable for each voxel in the brain separately to inform network hypotheses. Within the first approach, similar patterns of activation and/or deactivation (similar "time courses") in areas will lead to high correlations or high "functional connectivity." This represents a strong functional relation and increases the likelihood of them belonging to the same network. Both seed-based correlational analysis (SCA) and independent component analysis (ICA) use this approach, each with their advantages and limitations. Other means of using the temporal dynamics in the BOLD-signal such as "regional homogeneity" (ReHo) and "amplitude of low-frequency fluctuations" (ALFF or fALFF) instead opt for the voxel-wise calculation of a derived measure to inform local coherence or the power of certain frequencies. Finally, Graph theory analysis can be seen as a more downstream technique that can be used to convert the information obtained from any method into models that describe the different aspects of a network and its nodes (Salvador, Suckling, Schwarzbauer, & Bullmore, 2005; Wang, Zuo, & He, 2010). We will now discuss these different approaches in more detail.

Seed-Based Correlation Analysis

SCA was the first measure of functional connectivity to be widely employed in a clinical setting. SCA uses a predefined "seed" region that is hypothesized to be important within the study context. The reference time course of this seed region (e.g., average BOLD-signal within a spherical region of interest around a certain location) is then correlated to another region of interest or, more commonly, to the time courses of all other voxels in the brain. This can be used to generate an image where each voxel has a value corresponding to that voxel's connectivity strength with the seed region. By means of thresholding, a map can then be obtained containing all voxels that are strongly (positively or negatively) correlated to the seed, representing the seed region's functional connectivity network. If the seed region used is one of the core regions of a wider system, then this approach is a reasonably reliable method for identifying networks, e.g., many authors investigating networks in depression have used a seed region in the posterior cingulate cortex to extract the so-called "Default Mode Network,", which is a system encompassing medial prefrontal and posterior areas believed to reflect self-generated thought. To test for changes in connectivity in a disorder, the difference in acquired connectivity maps between two groups is used to interpret the changed functional relation of the seed region.

Interpreting the results obtained from SCA is straightforward: an increase in connectivity reflects an increase in functional synchronization between two areas. It is a highly sensitive method and suitable to test a clear a priori hypothesis. The drawback, however, is that this method relies heavily on accurately localizing the seed, and a small variance in seed selection can lead to significantly different results (Cole, Smith, & Beckmann, 2010). Also, as some authors have found that the boundaries of networks are altered in specific states and disorders, such as depression (Greicius, Flores, & Menon, 2007; Zhou et al., 2010), drawing inference about the state of a network as a whole on the basis of time courses obtained from a single area may not always be reliable. Furthermore, correctly modeling the background noise within a small area is challenging, and results of whole-brain SCA can be confounded by similarities of the noise within the seed region to noise in the rest of the brain (Cole et al., 2010).

In summary, SCA is highly sensitive to connectivity changes within the seed region, but the reliability of the results depends largely on having the right model and hypothesis. While results obtained from SCA are straightforward as far as the seed region is concerned, they should be treated with some caution when interpreting neural networks as a whole.

Independent Component Analysis

In contrast to SCA, ICA does not require a specific a priori hypothesis but instead uses all of the information available within the data to drive its network estimations. It uses a multivariate approach to decompose the entire data set into temporally coherent, spatially independent "components" (corresponding to a large-scale network as obtained through other methods) and associated time courses, which describe the temporal evolution of the entire component map (Beckmann, DeLuca, Devlin, & Smith, 2005) (see Figure 1). Essentially,

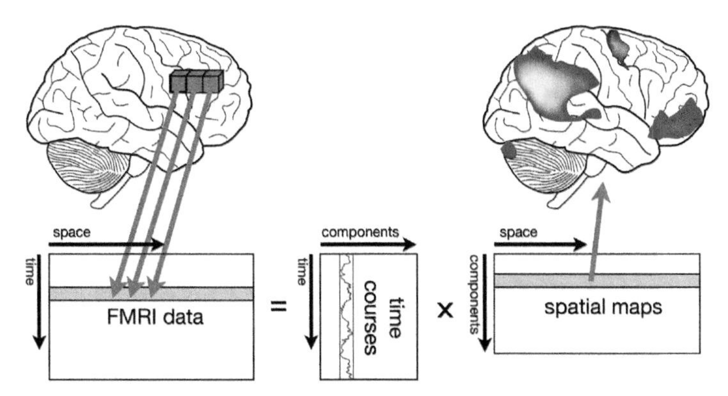

FIGURE 1 Overview of independent component analysis. The 4D-fMRI data is represented as a "time × space" 2D matrix, which is decomposed into a "time × component" and "space × component" matrix. The "space × component" matrix can then be used to derive the spatial maps representing the neural network.

a 4D fMRI image can be observed as a two-dimensional (2D) "time × space" matrix where the value at each voxel is the result of a mixture of different signal sources. By optimizing for maximal statistical independence to estimate the (non-Gaussian) underlying signal sources, ICA decomposes this matrix into a number of components with similar time courses that are spatially independent.

The number of components N is referred to as the "model order" and represents the (calculated or selected) number of components that the data is decomposed into. While a low model order (20–30) is often sufficient to identify large-scale networks, a higher model order can be used to investigate the more fine-grained within-network or subnetwork dynamics (Smith et al., 2009). Importantly, since background noise in fMRI data has different temporal characteristics when compared to "real" functional networks, ICA can also be used during preprocessing to identify physiological noise and movement artefacts. These "noise components" can be manually inspected and removed, although ICA-based denoising algorithms have been developed to distinguish "true" components from noise using their frequency or spatial characteristics (Kundu, Inati, Evans, Luh, & Bandettini, 2012; Pruim, Mennes, Buitelaar, & Beckmann, 2015; Pruim, Mennes, van Rooij, et al., 2015; Salimi-Khorshidi et al., 2014).

The final output from ICA is one map per component, where each voxel represents the association of that voxel with the component. Using these maps, ICA can inform us about a voxel's connection within its network and about (changes in) the relationship between different networks as well as how the configuration of different networks change under a specific condition, for instance, during a depressive episode or as a result of successful treatment. For group analysis, component maps are usually estimated at the group level and then reconstructed within the individual space, e.g., using "dual regression" (Filippini et al., 2009). This generates subject-specific maps for each component that can then be compared across groups using nonparametric hypothesis testing. ICA has repeatedly been found to be a highly reliable method of identifying networks, and as it does not require a predefined area of interest, makes maximum use of the abundance of information within the collected data.

Regional Homogeneity

Regional homogeneity (ReHo) provides the means to assess activation patterns in a model-free, localized manner (Zang, Jiang, Lu, He, & Tian, 2004). It differs from SCA and ICA in that it does not look at whole-brain temporal synchronization but instead calculates "Kendall's coefficient of concordance" (KCC) for each voxel independently. The KCC represents the degree to which a voxel's time course is similar to that of its neighboring voxels. By calculating this value for each voxel individually, a whole-brain map is generated, reflecting local coherence for each voxel in the brain. One assumed advantage of ReHo is that it might be less sensitive to regional differences in noise (as the noise within a voxel is expected to be similar to the noise in surrounding voxels). The initial paper showcasing this method found that during a functional task (finger movement), local synchronization tends to increase in motor areas, both on the contralateral and ipsilateral sides (Zang et al., 2004). Although they related it to patterns of activation seen in similar studies, a synchronized deactivation within an area could provide similarly increased ReHo. As such, the exact positioning of ReHo in relation to other measures of activity or connectivity is not yet well defined. While it has its merits in being a new way to interpret local changes in activation patterns, its focus on local coherence makes it unsuited for identifying large-scale networks of activity throughout the brain.

Amplitude of Low-Frequency Fluctuations

Instead of basing the analyses on the time course of a signal, some researchers opt to extract information from functional imaging data based on the amplitude of the signal power per frequency. This "amplitude of low-frequency fluctuations" (ALFF) is calculated by averaging the square root of the power spectrum of time courses standardized to the global mean. Here, the assumption is that "interesting" fluctuations are due to hemodynamically convolved neuronal activity that (due to the property of the hemodynamic convolution) occupies a specific frequency range within the power spectral density of the BOLD-signal. Artifactual signal components, on the other hand, have rather unspecific underlying frequency content. Because of its sensitivity to noise, a later modification to this analysis instead uses the ratio of the power at each frequency to the total power within the entire frequency range (Cole et al., 2010; Zang et al., 2007; Zuo et al., 2008). Instead of representing connectivity, ALFF instead is a means to interpret the BOLD-signal itself, i.e., it characterizes how dominated a voxel's time course is by low-frequency signals relative to higher frequency contributions. In essence, it is indicative of how BOLD-dominated a measured time course is. Given that the distribution of frequencies is directly influenced by the different frequencies that are (sufficiently) represented within the MRI data, which depends heavily on the characteristics of the MRI sequence that were used to obtain the data (e.g., spin- versus gradient-echo imaging), findings in specific frequencies are not easily compared across different study designs. The (fractional) ALFF measures have been used as a basis for calculating connectivity and identifying networks (Taylor, Gohel, Di, Walter, & Biswal, 2012; Zhang et al., 2011), but data are limited on how this compares to connectivity maps obtained using direct correlation of time courses.

Graph Theory Analysis

Graph theory analysis (GTA) is a method that originated in mathematics and sociology and has since been applied in numerous different fields. In neuroscience, as opposed to the previous methods, it uses information generated using another method to inform a predefined model. This model contains both regions of interest (called "nodes" or "vertices") and their connections (called "edges"). In the brain, these nodes are key anatomical regions, while the edges represent connectivity. GTA then uses the data it has given to inform us on various characteristics of the nodes and edges within the network, for instance the number of relevant connections among all connections in the model or how strongly one node is related to several others. The flexibility within this method allows for interpreting both structural (e.g., mean diffusivity as obtained by diffusion tensor imaging) and functional connectivity measures, making it an interesting downstream approach to network research.

Correlations and Causality

After delineating networks within a data set and testing for group effects, it is common practice to subsequently try and characterize the difference in networks and/or connectivity by correlating these changes in estimated network characteristics to a variety of patient- or disease-related variables. When describing group differences in depression, it is important to consider that any significant group difference could be either state-related, trait-related, or even just an effect of current or past treatment. Importantly, correlations in general do not imply causality from a clinical perspective, but increased connectivity also just describes a relation between two areas without further informing directionality. Some researchers have attempted to infer this directionality using methods such as "dynamic causal modelling" or "Granger causal modelling" (Friston, Harrison, & Penny, 2003; Goebel, Roebroeck, Kim, & Formisano, 2003). Unfortunately, causal inference from BOLD-fMRI data is still hindered by a number of key issues such as the indirectness of measuring BOLD as representing neural activation, variance in regions of interest between individuals, and variance in BOLD-delay across the brain (Ramsey et al., 2010).

General Considerations

When considering methods investigating large-scale brain networks, it is important to distinguish the key characteristics and to relate them to the actual hypothesis that is under investigation. Unfortunately, results obtained using model-free whole-brain exploratory analyses are often presented as if answering empirical questions with a strong prior assumption. Therefore, as a reader it is important to understand if the method used fits the question.

Model-free methods are suitable for those interested in either whole-brain changes or changes within the network configurations on a larger scale, while certain highly sensitive methods (SCA) may be better suited to detecting small differences in areas that have been shown to be highly relevant. Another important consideration is the reliability of the measure in question. A review investigating this showed that some analysis methods used were significantly more consistent than others (Zuo & Xing, 2014), with SCA versus ICA comparisons being in favor of the latter in cases where large-scale network identifications are desired. One way to approach network investigations is to use model-free methods to explore large-scale changes in the first instance, while using specific sensitive measures to confirm and expand upon any relevant findings in other, similar populations.

CHANGES IN LARGE-SCALE NETWORKS IN DEPRESSION

In MDD most findings in both task-based and resting-state fMRI implicate one of three major neural networks: the default mode network (DMN), the cognitive-executive network (CEN), and the salience network (SN) (Figure 2) (Hamilton et al., 2013; Menon, 2011; Raichle et al., 2001; Seeley et al., 2007). The focus on these networks is not surprising, since they represent the brain's functionality during rest, cognition, and emotional processes, all of which are

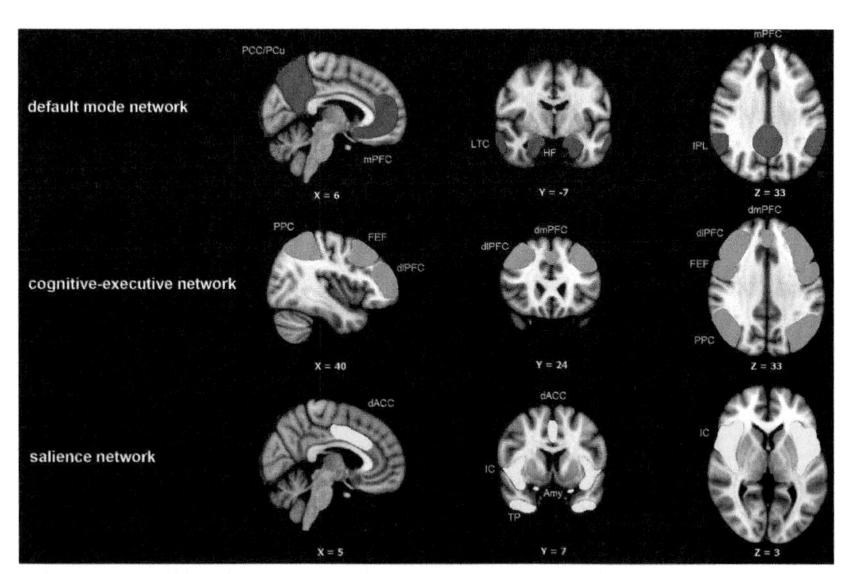

FIGURE 2 Major resting-state networks relevant in MDD. Representation of the major resting-state networks relevant in MDD. The DMN consists of two core regions: the mPFC and the PCC/PCu. Associated areas include the LTC, HF, and IPL. The CEN is centered on the dlPFC and PPC and also includes the dmPFC and FEF. The SN consists of the insular cortex, dACC, temporal pole, and Amygdala.

abnormal in the depressed state. In addition to changes within these networks, researchers have also called attention to the interactions between them and what drives the shift in network activity in response to external events. Here, we will discuss the most relevant networks for depression and how they are changed during a depressive episode. This summarizes all available recent literature of SCA and ICA papers, as these are the most consistent methods used and are best suited to inform functional networks.

Default Mode Network

The idea that during rest the mind is far from restful has been present since ancient times. Although the existence of brain areas with increased blood flow during passive states was already common knowledge early in the twentieth century, it was not until recently that the functional relevance of the brain at rest has really taken flight. When studying brain activity using PET, it was common practice to collect "resting" images as a contrast to discern task-related activation. These analyses often reported on task-related deactivations, which occurred in similar areas between studies. Research by Shulman and colleagues (among others) then began to focus on deactivation patterns, leading up to their discovery of the "default mode network" in 2001 as a highly consistent large-scale network that is most active during rest and deactivates during demanding cognitive tasks. The inception of such a network, which was hypothesized to account for the complex functions that arise during rest, has marked the start of an immense amount of research into the specific functioning of its different regions. Parallel to this, the emergence of the resting-state fMRI has helped push this field of research into the spotlight. This combination has also led to investigations into this network in disorders that clinically show differences in what patients experience during rest, of which depression is one of the most significant.

The DMN is often divided into an anterior subnetwork that centers on the medial prefrontal cortex (mPFC) and a posterior subnetwork that centers on the posterior cingulate cortex (PCC) and the precuneus cortex (PCu) (Andrews-Hanna, Reidler, & Sepulcre, 2010; Buckner, Andrews-Hanna, & Schacter, 2008). In addition to these core regions, associated DMN areas include the inferior parietal cortex (IPL), the hippocampal formation (HF), and the lateral temporal cortex (LTC) (Buckner et al., 2008; Greicius, Krasnow, Reiss, & Menon, 2003). As the DMN is located between the "primal" limbic system and the more lateral "higher cognitive" areas, it is ideally placed to integrate both emotional and cognitive processes. The split of the DMN into two smaller subnetworks has received support from functional studies showing that while these areas share similar temporal dynamics, they differ in regard to their specific function (Andrews-Hanna et al., 2010; Andrews-Hanna, Smallwood, Spreng, 2014). Another subdivision that has been suggested is of a "core" system, including the mPFC and PCC/PCu, and two related subsystems: a dorsal medial subsystem and a medial temporal subsystem (Andrews-Hanna et al., 2014). Although the exact within-DMN configuration on a smaller scale is still unclear, how this could prove highly relevant in studying depression will be discussed later.

Functionally, both the anterior and posterior DMN are related to spontaneous or self-generated cognition. The anterior DMN, with the mPFC as its main "hub", has strong structural connections to limbic areas such as the amygdala. Activity within this area has been related to both emotional control and self-referential activity, i.e., relating external emotionally salient stimuli to the self. In this regard, it has also been linked to internal mentation and investigated in disorders of theory of mind such as autism. By contrast, the posterior DMN, with its main hub in the PCC and the PCu, has been mostly implicated in memory-related processes through its relation to the hippocampal formation. Interestingly, activity within the PCC is also related to different levels of awareness, which relates to the DMN's other function as a "sentinel" network that observes and processes incoming stimuli without a specific focus, i.e., a "broad" attention that filters out relevant stimuli (Andrews-Hanna et al., 2014; Cavanna & Trimble, 2006; Leech & Sharp, 2014).

Independent Component Analysis

All of the ICA studies investigating the DMN in MDD reported an increase in connectivity within several nodes in the anterior part of the DMN (Greicius et al., 2007; Li et al., 2013; Manoliu et al., 2013; Zhu, Lu, et al., 2012), while the majority of papers also indicate a decrease in connectivity between the anterior and posterior subnetworks of the DMN. Changes within the posterior DMN differ between studies, which could be related to either the involvement of the PCu in networks other than the DMN (Manoliu et al., 2013) or age-linked differences in PCu connectivity (Yang et al., 2014). Additionally, studies looking at both the DMN and the CEN consistently report decreased connectivity between the posterior DMN and the CEN. Scales of clinical severity correlated with connectivity in different nodes of the DMN, while antidepressant treatment was shown to differentially affect connectivity within the anterior and posterior DMN and also to affect the connectivity between these subnetworks (Abbott et al., 2013; Li et al., 2013; Zhu, Wang, et al., 2012)

Greicius et al. (2007) were the first to investigate the role of the DMN in depression using ICA and reported increased connectivity within the anterior DMN in a group of MDD patients, most of whom were on antidepressant medication. Increased connectivity was found in the subgenual anterior cingulate cortex (ACC) (sgACC) and the orbitofrontal cortex (OFC) and accompanied by increased connectivity in the posterior hub of the DMN (precuneus) and the thalamus. A prominent node in the anterior DMN for the depressed patients, the sgACC was not part of the DMN in healthy control subjects, indicating that MDD affects the configuration of the DMN. The authors note that while the other changes in connectivity have been reported in other neuropsychiatric disorders, increased connectivity in the sgACC was unique for MDD, which was further emphasized by the finding that increased connectivity of the sgACC correlated with the duration of the depressive episode (Greicius et al., 2007). Zhu and colleagues investigated a group of first-episode treatment-naïve patients who also demonstrated increased connectivity within the anterior DMN (dorsomedial PFC, ventromedial PFC, ventral ACC, and medial OFC), which was related to levels of rumination (Zhu, Wang, et al., 2012). In contrast to Greicius, however, this study found decreased connectivity in the posterior DMN (PCC/PCu and angular gyrus), which was related to measures of overgeneralizing memory. The results indicate that anterior and posterior parts of the DMN may be differentially affected in MDD, which is further substantiated by the other two ICA papers (Li et al., 2013; Manoliu et al., 2013). Using a limited number of components (20), Li and colleagues identified the anterior and posterior section of the DMN, which showed increased connectivity within their respective core nodes (mPFC and PCu), but the subnetworks were spatially independent and asynchronous from one another (Li et al., 2013). Moreover, antidepressant treatment normalized connectivity in the posterior but not in the anterior DMN. These findings of changed within-network integrity in MDD are further supported by those of Guo and colleagues, who used ICA to identify the DMN but instead of connectivity reported on network homogeneity within the DMN (Guo et al., 2011). They found that network homogeneity within the DMN was increased in the anterior (dorsomedial prefrontal cortex; dmPFC) but decreased in the posterior DMN (inferior temporal gyrus). Together, these papers indicate that connectivity of the DMN in depression is primarily driven by activity in the anterior DMN, while connectivity between anterior and posterior segments of the DMN is decreased.

Consistent with these findings, Abbott and colleagues investigated between-network connectivity in a group of treatment-resistant depressed patients prior to electroconvulsive therapy and found decreased connectivity between the anterior and posterior DMN and between the posterior DMN and CEN (dorsolateral prefrontal cortex; dlPFC) at baseline (Abbott et al., 2013). The between-network connectivity changed from negative to positive following electroconvulsive therapy, demonstrating the importance of between-network interactions in addition to the within-network findings. A decrease in connectivity between the posterior DMN and the CEN was also reported by Manoliu et al. (2013). They used ICA to examine both within- and between-network changes in the DMN, CEN, and SN. Using a higher model order ICA (75), they found that the networks split into several smaller subnetworks (anterior/inferior-posterior/superior-posterior DMN, dorsal/left ventral/right ventral CEN and SN). Within-network connectivity was increased for the bilateral ACC within the anterior DMN and the SN. For the PCu, connectivity was increased within the inferior-posterior DMN, decreased within the left ventral CEN, and both increased and decreased within the superior-posterior DMN. In contrast with the decrease in connectivity between posterior DMN and the dorsal CEN, there was an increase in connectivity between the inferior-posterior DMN and SN. Importantly, connectivity of the right anterior insula within the SN not only correlated positively with depression severity but also correlated with changes in connectivity between the DMN and the CEN. This suggests that abnormal DMN−CEN interactions are driven by impaired right anterior insula-mediated control of network interactions, which is in line with other findings' involvement of the insula in switching between the DMN and the CEN (Goulden et al., 2014; Hamilton, Furman, & Chang, 2011; Sridharan, Levitin, & Menon, 2008).

The final two papers on the DMN using ICA have investigated a mixed group of MDD patients in a mild and remitted state, which limits a direct comparison with the abovementioned findings (Sexton et al., 2014; Veer et al., 2010). Sexton and colleagues identified the DMN (anterior and posterior), CEN, and SN and found no changes in mostly remitted elderly patients (Sexton et al., 2014). Veer and colleagues used 20-component ICA and did a full-brain comparison but report no changes in the DMN in a group of remitted and mildly depressed patients (Veer et al., 2010).

Seed-Based Correlation Analysis

Based on earlier work defining the DMN and findings from resting-state activation studies, the most commonly used seed regions in SCA studies investigating DMN connectivity are either the sgACC or the PCC (Greicius et al., 2003, 2007). Importantly, the sgACC is not always considered to be part of the DMN (Sheline, Price, Yan, & Mintun, 2010), although several authors point toward its inclusion in the DMN in depressed samples (Greicius et al., 2007; Zhou et al., 2010). A study by Zhou et al. (2010) on this issue found that the recruitment of the sgACC within the DMN is specific for MDD and not present in healthy controls, which is an important finding but for obvious reasons limits the use of this area as a seed region to study the DMN as a whole when comparing different subject groups. Regardless, a large number of studies have found connectivity of this region to be changed in relation to the DMN in depression, and it has been implicated both by structural and functional activation studies as highly relevant for MDD (Drevets et al., 2008; Hamani et al., 2011). In a similar fashion, the dmPFC has also been repeatedly identified as an important node in the context of depression (Northoff et al., 2011; Phillips, Drevets, Rauch, & Lane, 2003). Most seed-based papers focusing on the anterior DMN report changes in DMN connectivity similar to the ICA-based papers. Most notably, they show an increase in connectivity between different nodes in the anterior DMN (mainly with the sgACC) and between different parts of the ACC and the SN (amygdala, insula) as well as a decrease in connectivity between the anterior and posterior DMN. In contrast, researchers using seeds in the posterior DMN (PCC or PCu) show inconsistent changes in connectivity between anterior and posterior nodes of the DMN, although these studies do indicate that treatment influences connectivity within the DMN. The inconsistencies in findings from SCA in the posterior DMN could be related to the relatively large area from which to select the seed voxel, as variances in seed selection have been shown to significantly influence the representation of the DMN and thus any differences between groups (Cole et al., 2010).

Cognitive-Executive Network

Contrary to the DMN, the CEN, also known as the "cognitive-control" or "task-positive" network, is most active during cognitive tasks and is implicated in cognitive functioning, including attention and working memory. The DMN and the CEN are often seen as opposing networks, and task-related interactions between these networks are changed in MDD (Hamilton et al., 2011; Sheline et al., 2009). The CEN includes the lateral prefrontal cortex, the posterior parietal cortex (PPC), the frontal eye fields (FEF), and part of the dmPFC (Corbetta & Shulman, 2002; Rogers et al., 2004). While originally defined on the basis of several regions that all activated during (visual) attention tasks, the CEN is also consistently found using network-based analyses, for example, using ICA, where it is often referred to as the "fronto-parietal network." Depression is often hypothesized to be the result of a loss of top-down "cognitive control" over emotional processes and often is associated with cognitive deficits. These facts emphasize the importance of the CEN in relation to depression. Illustrating this, the dlPFC, one of the core regions of the CEN, was consistently shown to be less active during cognitive tasks in depressed patients and is the primary region of repetitive transcranial magnetic stimulation (rTMS), an effective treatment option for depressive episodes.

Independent Component Analysis

Only two papers thus far have used the ICA to investigate functional connectivity in the CEN (Abbott et al., 2013; Manoliu et al., 2013). As described above, they both reported decreased connectivity between the posterior DMN and the dorsal part of the CEN. Veer and colleagues also reported decreased connectivity of the frontal pole within a network consisting of areas related to attention and working memory (Veer et al., 2010).

Seed-Based Correlation Analysis

SCA studies on connectivity of the CEN have exclusively used the dlPFC as a seed region (Alexopoulos et al., 2012; Liston et al., 2014; Lui et al., 2011; Sheline et al., 2010; Van Tol et al., 2013; Ye et al., 2012). The majority of these papers report decreased connectivity between different nodes of the CEN in depression, while one paper found no changes in connectivity for a part of the dlPFC with reduced cortical thickness in MDD (Van Tol et al., 2013). As a popular target area for rTMS, Liston and colleagues specifically investigated how depression and rTMS influences the interaction of the DMN and the CEN. Besides the earlier mentioned increased connectivity of the sgACC with the DMN, they found that dlPFC connectivity was reduced with various other CEN regions (premotor cortex, posterior parietal cortex, bilateral cerebellum, and other lateral prefrontal cortical areas). They also found that rTMS of the dlPFC induced anticorrelations between dlPFC and the medial prefrontal regions of the DMN. As these networks are also anticorrelated in healthy individuals (Fox et al., 2005) and increased connectivity within the anterior DMN is found consistently in depression, the restoration of network interactions could indicate a possible mechanism of action for rTMS.

Salience Network

The SN, also known as the "affective network," is the network that is most involved in emotion regulation and salience processing. Through work in disentangling the SN from the CEN (which overlap in some settings), Seeley and colleagues used both SCA and ICA to inform their definition of the SN (Seeley et al., 2007). The SN has since been replicated by numerous other authors, although not all researchers agree on the exact boundaries. Most typically it consists of the fronto-insular cortex, the dorsal ACC, the amygdala, and the temporal poles, and activates in response to various salient stimuli, including acute stress (Hermans, Henckens, Joëls, & Fernández, 2014; Seeley et al., 2007). It is believed to reflect paralimbic emotional processing and to play a central role in emotional control through its extensive subcortical connectivity. Moreover, it has been implicated in switching between network dominance (Goulden et al., 2014; Sridharan et al., 2008), where the SN, and more specifically the anterior insula, is responsible for orchestrating the balance between the DMN and CEN.

The most investigated seed regions involved in the SN are the insular cortex (Avery et al., 2013; Horn et al., 2010; Lui et al., 2011) and the amygdala (Anand et al., 2005a; Lui et al., 2011; Pannekoek et al., 2014; Ramasubbu et al., 2014; Tahmasian et al., 2013; Tang et al., 2013; Yue et al., 2013), while three other papers used seeds in the hippocampus (Cao et al., 2012; Lui et al., 2011; Tahmasian et al., 2013). Compared to the DMN and the CEN, the SN is less well defined during the resting state, although many authors report changes in networks related to emotional processing. Possibly related to this, changes in the SN appear to be less consistent and more node-dependent in comparison to the DMN and CEN. The insular cortex is found repeatedly to be related to changes in between-network interaction, showing increased connectivity to nodes of the anterior DMN but a decoupling from other regions of emotional control in the SN and the CEN. The amygdala also shows mainly decreased connectivity with various brain regions both within and outside of the CEN, which likely reflects a loss of top-down control.

Independent Component Analysis

Two papers have investigated emotion-regulating networks using ICA (Manoliu et al., 2013; Veer et al., 2010). As described above, Manoliu and colleagues identified the SN and found increased connectivity within the bilateral ACC and decreased connectivity within the bilateral anterior insula. Notably, the right anterior insula connectivity correlated with depression severity and connectivity between the DMN and the CEN (Manoliu et al., 2013). This is indicative of a role for the anterior insula and SN in regulating DMN and CEN interactions, as proposed by several other papers (Goulden et al., 2014; Menon, 2011; Sridharan et al., 2008). Veer and colleagues used whole-brain ICA in a group of mildly depressed patients and also report a decoupling of the amygdala and insula within a network involved in emotional processing (Veer et al., 2010).

Seed-Based Correlation Analysis

As mentioned earlier, SCA studies investigating the SN show a large variance in their findings, which could be related to the relatively small areas to select your seed from. Overall, there appears to be a trend of increased connectivity of the SN with the anterior DMN, similar to the (limited) evidence from ICA studies. SCA of the insula also reports a relation to disease severity (Avery et al., 2013) and treatment response (Lui et al., 2011). Another important node in the SN and highly relevant for MDD (Phillips et al., 2003; Price & Drevets, 2010), papers using the amygdala as a seed region typically find decreased connectivity with various brain regions, which is in line with the uncoupling of the amygdala and insula from the SN, as found using ICA (Manoliu et al., 2013; Veer et al., 2010).

DISCUSSION

Due to the relative ease of collecting resting-state fMRI scans and the availability of new techniques such as ICA to accommodate the large amount of information involved, a large body of investigations into resting-state functional connectivity in depression is available. Our review of the available data yielded several findings that were consistent across different methods:

1. increased connectivity within the anterior DMN,
2. increased connectivity between the anterior DMN and the SN,
3. changed connectivity between the anterior and the posterior DMN, and
4. decreased connectivity between the posterior DMN and the CEN.

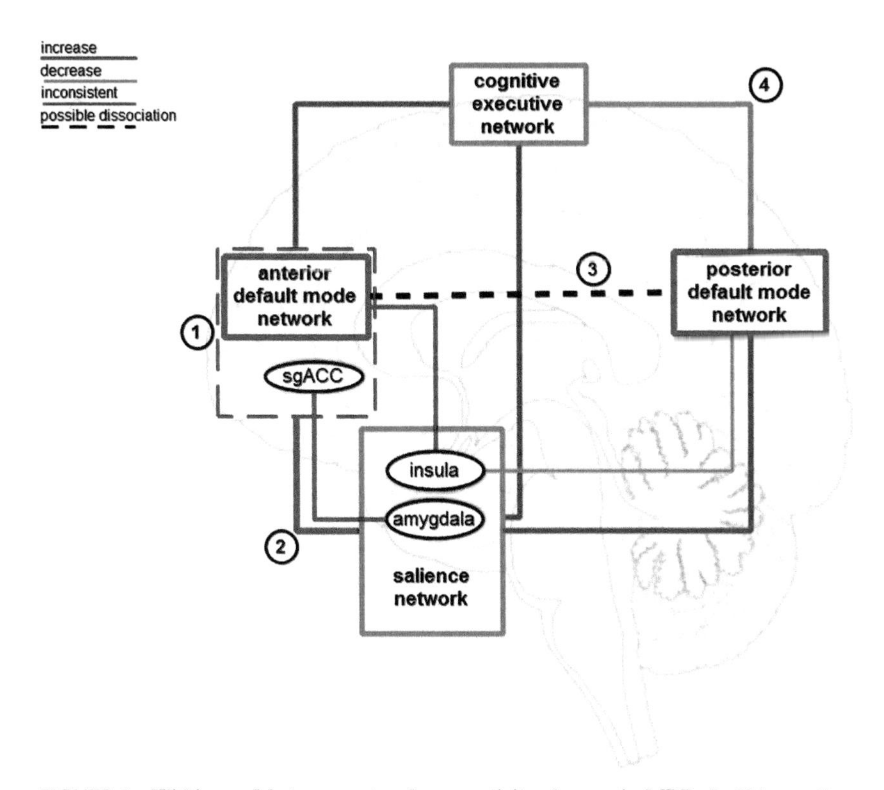

FIGURE 3 Within- and between-network connectivity changes in MDD. Red/blue outlines represent a within-network increase/decrease in connectivity; red/blue lines between networks represent a between-network increase/decrease in connectivity. Black ellipses represent key nodes related to connectivity. Numbers represent main findings: (1): increase in anterior DMN connectivity and inclusion of sgACC within the anterior DMN; (2): increased connectivity between anterior DMN and SN; (3): changed connectivity between anterior and posterior DMN; and (4): decreased connectivity between posterior DMN and CEN.

A summary of these findings is represented in Figure 3. Below, we will discuss the support for these findings and their implications for our current understanding of MDD and its treatment.

Increased Connectivity within the Anterior Default Mode Network and Its Relation to the Salience Network

The most consistent finding across all studies is increased connectivity in the anterior DMN, both within the DMN as a whole and between the different anterior nodes. In depression, gray matter volume in this area is typically decreased, while activity is increased both during rest (Drevets et al., 2008; Hamani et al., 2011; Mayberg et al., 2005; Rodríguez-Cano et al., 2014) and in response to emotionally salient stimuli (Gotlib et al., 2005; Sheline et al., 2009). Adding to this, the attenuation of anterior DMN activity during cognitive tasks (Gusnard, Akbudak, Shulman, & Raichle, 2001) is impaired in the depressed state (Rodríguez-Cano et al., 2014; Sheline et al., 2009). Hyperactivity in the sgACC has been an especially consistent finding (Hamani et al., 2011), and activity within this node serves as a marker of treatment response (Pizzagalli, 2011). Because of this, the change in DMN configuration to incorporate the sgACC in depression is intriguing (Greicius et al., 2007; Zhou et al., 2010), especially considering its extensive structural connectivity to both the DMN and the SN. Although not as often reported as the increase in anterior DMN connectivity, several papers indicate an increase in functional connectivity between the anterior DMN and the SN, which is consistent with reports of increased structural connectivity (Fang et al., 2012). Hyperactivity of the amygdala, especially in response to negative stimuli, is common in depressed patients and has been hypothesized to interact with the mPFC to underlie the negativity bias in MDD (Murray, Wise, & Drevets, 2011; Price & Drevets, 2012). The repeat finding of increased connectivity between the amygdala and sgACC, in addition to a report on its relation to disease onset (Davey et al., 2014), further highlights the relevance of this connection in MDD. However, whether these changes in connectivity constitute an increase in top-down modulation of limbic hyperactivity, bottom-up interference of self-processing regions, or both is unclear. Surprisingly, connectivity of the amygdala with other brain regions implicated in emotional control is decreased in MDD, which could indicate an inability to control amygdala hyperactivity by regions other than the sgACC. This is in line with a review by Rive and colleagues, who showed both differential functioning of ACC regions and an inability to recruit additional prefrontal resources in emotional control in depressed patients (Rive et al., 2013). The decreased connectivity of the amygdala with more lateral brain regions (insula and lateral PFC) is also consistent with limbic hyperactivity and lateral hypoactivity, as found in resting-state activation studies in depression (Northoff et al., 2011).

Changed Connectivity within the Default Mode Network and Its Relation to the CEN

Within the DMN, changes in connectivity between its anterior and posterior nodes seem to differ in MDD. Studies using ICA and SCA with anterior DMN seeds mainly report evidence of a dissociation of the DMN, while in the SCA investigations using PCC seeds, the majority report increased connectivity between anterior and posterior nodes. As described earlier, an anterior and posterior subnetwork within the DMN has been identified in healthy subjects (Andrews-Hanna et al., 2010; Buckner et al., 2008), and they were shown to contribute to different aspects of self-generated thought. However, the implications of changes in functional connectivity between the anterior and posterior subnetworks are not well understood. A paper by Leech and Sharp hypothesized that an increase in PCC connectivity with anterior DMN regions would relate to an increase in internally directed attention (Leech & Sharp, 2014), which is in line with its correlation to rumination scores in depression (Berman et al., 2011). In depressed patients, findings of decreased connectivity between the anterior and posterior DMN are supported by a decrease in structural connectivity in an sgACC-posterior DMN-based network (Korgaonkar, Fornito, Williams, & Grieve, 2014). Considering that using a higher model order in ICA reliably splits the DMN into its subnetworks (Abbott et al., 2013; Manoliu et al., 2013), a possible explanation for the inconsistent findings is that MDD merely accentuates a normal functional distinction already present within the DMN in healthy subjects. Regardless, it is worth noting that while findings are inconsistent, several authors have reported treatment for depression to selectively affect parts of the DMN, which underlines the relevance of the distinction between anterior and posterior subnetworks.

Another interesting finding concerning the posterior DMN specifically is its decreased connectivity with the CEN. In line with the role the posterior DMN has in awareness and directed attention (Leech & Sharp, 2014) and the role of the CEN in higher cognitive functioning (Corbetta & Shulman, 2002), the change in their interaction could underlie a difficulty in switching from a "default state," in which the DMN is dominant and which is directed internally, to an "executive state," in which the CEN is dominant and attention is directed toward outward stimuli (Hamilton et al., 2013). Several authors have indicated that the insular cortex might be crucial for this shift in network dominance (Goulden et al., 2014; Hamilton et al., 2011; Manoliu et al., 2013; Sridharan et al., 2008), which is supported by the increased connectivity of the insula with the anterior DMN and decreased connectivity with other networks.

Clinical Correlates and Treatment Effects

Many authors have attempted to correlate clinical measures to changes in connectivity in MDD. In agreement with the most consistent regions of changed connectivity, nearly all findings of significant correlation between connectivity and clinical scores are either within the DMN or the interaction between the DMN and the SN. In short, the severity of disease was related to connectivity of the sgACC (Connolly et al., 2013; Davey, Harrison, Yücel, & Allen, 2012; De Kwaasteniet et al., 2013; Salomons et al., 2014); the dmPFC (Sheline et al., 2010); the dorsal anterior cingulate cortex (dACC) (Pannekoek et al., 2014); the dorsal caudate (Furman, Hamilton, & Gotlib, 2011); and the insula (Avery et al., 2013; Manoliu et al., 2013). Connectivity of the sgACC was further related to disease duration (Greicius et al., 2007) and rumination (Berman et al., 2011; Connolly et al., 2013; Zhu, Wang, et al., 2012), while connectivity within the posterior DMN was related to overgeneralized memory (Zhu, Wang, et al., 2012).

Adding to this, numerous authors have reported changes in connectivity through various treatment modalities. Longitudinal studies found antidepressant medication to affect the anterior and posterior DMN differently, but whether the effect is specific for the sgACC (Kozel et al., 2011), the anterior DMN (Wu et al., 2011), posterior DMN (Li et al., 2013), their interaction (Andreescu et al., 2013), or the interaction between parts of the DMN and the SN (Anand et al., 2005b; McCabe et al., 2011) is unclear. However, as most of these changes were in subjects with a baseline difference in connectivity within the DMN, this signifies that effective antidepressant treatment restores aberrant network configuration within the DMN. Consistent with this, other treatment modalities find similar effects with TMS normalizing an increase in sgACC–DMN connectivity (Baeken et al., 2014; Liston et al., 2014; Salomons et al., 2014) and electroconvulsive therapy (ECT) also targeting disrupted connectivity between the anterior and posterior DMN (Abbott et al., 2013; Beall et al., 2012) and between the DMN and the CEN (Abbott et al., 2013; Beall et al., 2012; Perrin et al., 2012). Studies looking at differences between treatment-sensitive and treatment-responsive patients in a longitudinal setting also found that responders showed lower baseline connectivity of the PCC with the striatum (Andreescu et al., 2013) and higher connectivity of the insula with DMN nodes (Lui et al., 2011). Extending the clinical relevance of connectivity measures, a number of authors have been able to predict treatment response using functional connectivity. A positive response to rTMS was predicted by high baseline connectivity of the sgACC with the DMN (Liston et al., 2014; Salomons et al., 2014) or strong anticorrelation of the sgACC with the stimulation site in the dlPFC (Fox, Buckner, White, Greicius, & Pascual-Leone, 2012). A paper by Van Waarde et al. (2014) showed that connectivity patterns in two networks centered on the dmPFC and ACC could identify responders to ECT with high reliability, which corresponds with our findings that the majority of consistent changes in connectivity center on networks related to the anterior DMN.

Importantly, due to the differences in methods and results, there is no compelling evidence for differentiating different subtypes of depression based on connectivity measures. As subtypes of depression respond differently to treatment strategies, this is a line of research that could prove to be promising in the future (Bühler, Seemüller, & Läge, 2014; Gili et al., 2012; Rush, 2007).

Methodological Considerations and Limitations

Although the primary findings as presented above expand our current understanding of MDD, it is important to also address the shortcomings of current connectivity investigations. Especially for seed-based approaches, even the most consistent findings are often not reproduced or connectivity changes are only found using one of the regions as a seed. For instance, a number of studies find increased connectivity for the sgACC with the amygdala when using the sgACC as a seed region, while none of the papers using the amygdala as a seed region report increased connectivity with the sgACC. This heterogeneity in findings could reflect heterogeneity within the disorder itself, which is inherent to the symptom-based classification used to diagnose depression. However, it could also reflect the methodological difficulty in correct seed selection and the problems introduced when correlating one or a few seeds to all other brain voxels. This is illustrated when papers use several similar seed voxels close to one another and find large differences in the emerging connectivity pattern (Cole et al., 2010; Connolly et al., 2013). So, while seed-based approaches are very sensitive to changes in connectivity of the seed region under investigation, this high sensitivity could also induce spurious findings with no biological meaning, for instance, by merely reflecting similarity in noise within the two regions. In contrast to the SCA-based papers, studies using ICA to identify networks appear to be much more consistent in their findings, as in fact all studies using a depressed group found increased connectivity within the anterior DMN, and nearly all report changes in the connectivity of anterior and posterior DMN regions. A limitation of ICA, however, is that there is no clear consensus on a "correct" number of components to identify, and this directly influences the possible outcomes. However, different model orders could also be used to further our understanding of subnetworks within networks. For example, a limited number of components could be used to look at large-scale networks (e.g., the DMN as a whole), while a high number of components in the same data would allow us to look at decompositions into subnetworks and investigate the underlying changes in network configuration (e.g., interaction between anterior and posterior DMN). A limitation of investigating neural networks in general is that there is no consensus on the boundaries of these networks. Additionally, while key nodes are identified consistently, associated regions may vary in their connectivity to any given network.

Future Directions

Based on the current findings and methodological properties of the different techniques, there are several recommendations for future researchers. The decreased bias and increased consistency of findings using ICA over SCA leads us to propose that ICA should be used to give a model-free estimation of regions of change in depression. Consequently, SCA should be used to further expand upon these significant findings by being more sensitive to specific changes less related to within-network connectivity. As mentioned above, the number of components used in ICA analysis could help to delineate within-network configurations and subnetworks. Furthermore, large-scale networks are not static but instead change over time, even during rest (Gonzalez-Castillo et al., 2014; Zalesky, Fornito, Cocchi, Gollo, & Breakspear, 2014). More insight into these network dynamics, for instance, by looking at causal interactions and directionality of influence between the different nodes or networks, could inform us as to how the activation of networks is coordinated and possibly how network control is changed under various conditions such as the depressed state. Finally, relating changes in connectivity to distinct symptom profiles and neurocognitive domains would help us tackle the issue of large heterogeneity within MDD and improve treatment strategies for specific depression subtypes.

CONCLUSION

Research into large-scale neural networks and their dysfunction in depression is an exciting new field of research, especially considering the large advances in different methods of acquisition and analyses of the past decades. Connectivity studies in MDD expand upon activation studies by reporting increased connectivity within nodes of the anterior DMN and between the anterior DMN and the SN, changed connectivity between the anterior and posterior DMN, and decreased connectivity between the posterior DMN and the CEN. This could reflect a state of increased interaction between self-referential and emotional networks and the dominance of negative self-referential over cognitive processing, which corresponds to the clinical symptoms of depression. The consistent differences found in the interaction between nodes and networks as well as the clinical potential in predicting treatment response highlight the importance of functional connectivity in furthering our understanding and treatment of depression. Investigations into how large-scale networks consist of subnetworks and the interactions between different networks are important challenges for current researchers in the field.

REFERENCES

Abbott, C. C., Lemke, N. T., Gopal, S., Thoma, R. J., Bustillo, J., Calhoun, V. D., & Turner, J. A. (2013). Electroconvulsive therapy response in major depressive disorder: a pilot functional network connectivity resting state FMRI investigation. *Frontiers in Psychiatry, 4*, 10.

Alcaro, A., Panksepp, J., Witczak, J., Hayes, D. J., & Northoff, G. (2010). Is subcortical-cortical midline activity in depression mediated by glutamate and GABA? A cross-species translational approach. *Neuroscience and Biobehavioral Reviews, 34*, 592–605.

Alexopoulos, G., Hoptman, M., Kanellopoulos, D., Murphy, C., Lim, K., & Gunning, F. (2012). Functional connectivity in the cognitive control network and the default mode network in late-life depression. *Journal of Affective Disorders, 139*, 56–65.

Anand, A., Li, Y., Wang, Y., Wu, J., Gao, S., Bukhari, L., ... Lowe, M. J. (2005a). Activity and connectivity of brain mood regulating circuit in depression: a functional magnetic resonance study. *Biological Psychiatry, 57*, 1079–1088.

Anand, A., Li, Y., Wang, Y., Wu, J., Gao, S., Bukhari, L., ... Lowe, M. J. (2005b). Antidepressant effect on connectivity of the mood-regulating circuit: an FMRI study. *Neuropsychopharmacology, 30*, 1334–1344.

Andreescu, C., Tudorascu, D. L., Butters, M. A., Tamburo, E., Patel, M., Price, J., ... Aizenstein, H. (2013). Resting state functional connectivity and treatment response in late-life depression. *Psychiatry Research, 214*, 313–321.

Andrews-Hanna, J., Reidler, J., & Sepulcre, J. (2010). Functional-anatomic fractionation of the brain's default network. *Neuron, 65*, 550–562.

Andrews-Hanna, J. R., Smallwood, J., & Spreng, R. N. (2014). The default network and self-generated thought: component processes, dynamic control, and clinical relevance. *Annals of the New York Academy of Sciences, 1316*, 29–52.

Avery, J. A., Drevets, W. C., Moseman, S. E., Bodurka, J., Barcalow, J. C., & Simmons, W. K. (2013). Major depressive disorder is associated with abnormal interoceptive activity and functional connectivity in the insula. *Biological Psychiatry*, 1–9.

Baeken, C., Marinazzo, D., Wu, G.-R., Van Schuerbeek, P., De Mey, J., Marchetti, I., ... De Raedt, R. (2014). Accelerated HF-rTMS in treatment-resistant unipolar depression: insights from subgenual anterior cingulate functional connectivity. *World Journal of Biological Psychiatry*, 1–12.

Beall, E. B., Malone, D. A., Dale, R. M., Muzina, D. J., Koenig, K. A., Bhattacharrya, P. K., ... Lowe, M. J. (2012). Effects of electroconvulsive therapy on brain functional activation and connectivity in depression. *Journal of ECT, 28*, 234–241.

Beckmann, C. F., DeLuca, M., Devlin, J. T., & Smith, S. M. (2005). Investigations into resting-state connectivity using independent component analysis. *Philosophical Transactions of the Royal Society of London B Biological Sciences, 360*, 1001–1013.

Berman, M. G., Peltier, S., Nee, D. E., Kross, E., Deldin, P. J., & Jonides, J. (2011). Depression, rumination and the default network. *Social Cognitive and Affective Neuroscience, 6*, 548–555.

Bressler, S. L. (1995). Large-scale cortical networks and cognition. *Brain Research Reviews, 20*, 288–304.

Buckner, R. L. (2012). The serendipitous discovery of the brain's default network. *NeuroImage, 64*(2), 1137–1145.

Buckner, R. L., Andrews-Hanna, J. R., & Schacter, D. L. (2008). The brain's default network: anatomy, function, and relevance to disease. *Annals of the New York Academy of Sciences, 1124*, 1–38.

Bühler, J., Seemüller, F., & Läge, D. (2014). The predictive power of subgroups: an empirical approach to identify depressive symptom patterns that predict response to treatment. *Journal of Affective Disorders, 163*, 81–87.

Cao, X., Liu, Z., Xu, C., Li, J., Gao, Q., Sun, N., ... Zhang, K. (2012). Disrupted resting-state functional connectivity of the hippocampus in medication-naïve patients with major depressive disorder. *Journal of Affective Disorders, 141*, 194–203.

Cavanna, A. E., & Trimble, M. R. (2006). The precuneus: a review of its functional anatomy and behavioural correlates. *Brain, 129*, 564–583.

Cole, D. M., Smith, S. M., & Beckmann, C. F. (2010). Advances and pitfalls in the analysis and interpretation of resting-state FMRI data. *Frontiers in Systems Neuroscience, 4*, 8.

Connolly, C. G., Wu, J., Ho, T. C., Hoeft, F., Wolkowitz, O., Eisendrath, S., ... Yang, T. T. (2013). Resting-state functional connectivity of subgenual anterior cingulate cortex in depressed adolescents. *Biological Psychiatry, 74*, 898–907.

Corbetta, M., & Shulman, G. L. (2002). Control of goal-directed and stimulus-driven attention in the brain. *Nature Reviews Neuroscience, 3*, 201–215.

Davey, C. G., Harrison, B. J., Yücel, M., & Allen, N. B. (2012). Regionally specific alterations in functional connectivity of the anterior cingulate cortex in major depressive disorder. *Psychologie Medicale, 42*, 2071–2081.

Davey, C. G., Whittle, S., Harrison, B. J., Simmons, J. G., Byrne, M. L., Schwartz, O. S., & Allen, N. B. (2014). Functional brain-imaging correlates of negative affectivity and the onset of first-episode depression. *Psychologie Medicale*, 1–9.

De Kwaasteniet, B., Ruhe, E., Caan, M., Rive, M., Olabarriaga, S., Groefsema, M., ... Denys, D. (2013). Relation between structural and functional connectivity in major depressive disorder. *Biological Psychiatry, 74*, 40–47.

Drevets, W., Savitz, J., & Trimble, M. (2008). The subgenual anterior cingulate cortex in mood disorders. *CNS Spectrums, 13*, 663–681.

Fang, P., Zeng, L.-L., Shen, H., Wang, L., Li, B., Liu, L., & Hu, D. (2012). Increased cortical-limbic anatomical network connectivity in major depression revealed by diffusion tensor imaging. *PLoS One, 7*, e45972.

Filippini, N., MacIntosh, B. J., Hough, M. G., Goodwin, G. M., Frisoni, G. B., Smith, S. M., ... Mackay, C. E. (2009). Distinct patterns of brain activity in young carriers of the APOE-epsilon4 allele. *Proceedings of the National Academy of Sciences of the United States of America, 106*, 7209–7214.

Fox, M. D., Buckner, R. L., White, M. P., Greicius, M. D., & Pascual-Leone, A. (2012). Efficacy of transcranial magnetic stimulation targets for depression is related to intrinsic functional connectivity with the subgenual cingulate. *Biological Psychiatry, 72*, 595–603.

Fox, M. D., Snyder, A. Z., Vincent, J. L., Corbetta, M., Van Essen, D. C., & Raichle, M. E. (2005). The human brain is intrinsically organized into dynamic, anticorrelated functional networks. *Proceedings of the National Academy of Sciences of the United States of America, 102*, 9673–9678.

Friston, K. J., Harrison, L., & Penny, W. (2003). Dynamic causal modeling. *NeuroImage, 19*, 1273–1302.

Friston, K. J., Price, C. J., Fletcher, P., Moore, C., Frackowiak, R. S., & Dolan, R. J. (1996). The trouble with cognitive subtraction. *NeuroImage, 4*, 97–104.

Furman, D. J., Hamilton, J. P., & Gotlib, I. H. (2011). Frontostriatal functional connectivity in major depressive disorder. *Biology of Mood and Anxiety Disorders, 1*, 11.

Gili, M., Roca, M., Armengol, S., Asensio, D., Garcia-Campayo, J., & Parker, G. (2012). Clinical patterns and treatment outcome in patients with melancholic, atypical and non-melancholic depressions. *PLoS One, 7*, e48200.

Goebel, R., Roebroeck, A., Kim, D. S., & Formisano, E. (2003). Investigating directed cortical interactions in time-resolved fMRI data using vector autoregressive modeling and Granger causality mapping. *Magnetic Resonance in Imaging, 21*, 1251–1261.

Gonzalez-Castillo, J., Handwerker, D. A., Robinson, M. E., Hoy, C. W., Buchanan, L. C., Saad, Z. S., & Bandettini, P. A. (2014). The spatial structure of resting state connectivity stability on the scale of minutes. *Frontiers in Neuroscience, 8*, 138.

Gotlib, I. H., Sivers, H., Gabrieli, J. D. E., Whit, S., Goldin, P., & Minor, K. L. (2005). Subgenual anterior cingulate activation to valenced emotional stimuli in major depression. *Neuroreport, 16*, 1731–1734.

Goulden, N., Khusnulina, A., Davis, N. J., Bracewell, R. M., Bokde, A. L., McNulty, J. P., & Mullins, P. G. (2014). The salience network is responsible for switching between the default mode network and the central executive network: replication from DCM. *NeuroImage, 99C*, 180–190.

Greicius, M., Flores, B., & Menon, V. (2007). Resting-state functional connectivity in major depression: abnormally increased contributions from subgenual cingulate cortex and thalamus. *Biological Psychiatry, 62*, 429–437.

Greicius, M. D., Krasnow, B., Reiss, A. L., & Menon, V. (2003). Functional connectivity in the resting brain: a network analysis of the default mode hypothesis. *Proceedings of the National Academy of Sciences of the United States of America, 100*, 253–258.

Guo, W., Sun, X., Liu, L., Xu, Q., Wu, R., Liu, Z., ... Zhao, J. (2011). Disrupted regional homogeneity in treatment-resistant depression: a resting-state fMRI study. *Progress in Neuropsychopharmacology and Biological Psychiatry, 35*, 1297–1302.

Gusnard, D. A., Akbudak, E., Shulman, G. L., & Raichle, M. E. (2001). Medial prefrontal cortex and self-referential mental activity: relation to a default mode of brain function. *Proceedings of the National Academy of Sciences of the United States of America, 98*, 4259–4264.

Hamani, C., Mayberg, H., Stone, S., Laxton, A., Haber, S., & Lozano, A. M. (2011). The subcallosal cingulate gyrus in the context of major depression. *Biological Psychiatry, 69*, 301–308.

Hamilton, J. P., Chen, M. C., & Gotlib, I. H. (2013). Neural systems approaches to understanding major depressive disorder: an intrinsic functional organization perspective. *Neurobiology of Disease, 52*, 4–11.

Hamilton, J., Furman, D., & Chang, C. (2011). Default-mode and task-positive network activity in major depressive disorder: implications for adaptive and maladaptive rumination. *Biological Psychiatry, 70*, 327–333.

Hermans, E. J., Henckens, M. J., Joëls, M., & Fernández, G. (2014). Dynamic adaptation of large-scale brain networks in response to acute stressors. *Trends in Neurosciences, 37*, 304–314.

Horn, D. I., Yu, C., Steiner, J., Buchmann, J., Kaufmann, J., Osoba, A., ... Walter, M. (2010). Glutamatergic and resting-state functional connectivity correlates of severity in major depression - the role of pregenual anterior cingulate cortex and anterior insula. *Frontiers in Systems Neuroscience, 4*, 1–10.

Koenigs, M., & Grafman, J. (2009). The functional neuroanatomy of depression: distinct roles for ventromedial and dorsolateral prefrontal cortex. *Behavioural Brain Research, 201*, 239–243.

Korgaonkar, M. S., Fornito, A., Williams, L. M., & Grieve, S. M. (2014). Abnormal structural networks characterize major depressive disorder: a connectome analysis. *Biological Psychiatry, 76*, 567–574.

Kozel, F. A., Rao, U., Lu, H., Nakonezny, P. A., Grannemann, B., McGregor, T., ... Trivedi, M. H. (2011). Functional connectivity of brain structures correlates with treatment outcome in major depressive disorder. *Frontiers in Psychiatry, 2*, 7.

Kundu, P., Inati, S. J., Evans, J. W., Luh, W. M., & Bandettini, P. A. (2012). Differentiating BOLD and non-BOLD signals in fMRI time series using multi-echo EPI. *NeuroImage, 60*, 1759–1770.

Leech, R., & Sharp, D. J. (2014). The role of the posterior cingulate cortex in cognition and disease. *Brain, 137*, 12–32.

Li, B., Liu, L., Friston, K. J., Shen, H., Wang, L., Zeng, L.-L., & Hu, D. (2013). A treatment-resistant default mode subnetwork in major depression. *Biological Psychiatry, 74*, 48–54.

Liston, C., Chen, A. C., Zebley, B. D., Drysdale, A. T., Gordon, R., Leuchter, B., ... Dubin, M. J. (2014). Default mode network mechanisms of transcranial magnetic stimulation in depression. *Biological Psychiatry, 76*, 517–526.

Lui, S., Wu, Q., Qiu, L., Yang, X., Kuang, W., Chan, R. C. K., ... Gong, Q. (2011). Resting-state functional connectivity in treatment-resistant depression. *American Journal of Psychiatry, 168*, 642–648.

Manoliu, A., Meng, C., Brandl, F., Doll, A., Tahmasian, M., Scherr, M., ... Sorg, C. (2013). Insular dysfunction within the salience network is associated with severity of symptoms and aberrant inter-network connectivity in major depressive disorder. *Frontiers in Human Neuroscience, 7*, 930.

Mayberg, H. (1997). Limbic-cortical dysregulation: a proposed model of depression. *Journal of Neuropsychiatry and Clinical Neurosciences*, 471–481.

Mayberg, H. S., Lozano, A. M., Voon, V., McNeely, H. E., Seminowicz, D., Hamani, C., … Kennedy, S. H. (2005). Deep brain stimulation for treatment-resistant depression. *Neuron, 45*, 651–660.

McCabe, C., Mishor, Z., Filippini, N., Cowen, P. J., Taylor, M. J., & Harmer, C. J. (2011). SSRI administration reduces resting state functional connectivity in dorso-medial prefrontal cortex. *Molecular Psychiatry, 16*, 592–594.

Menon, V. (2011). Large-scale brain networks and psychopathology: a unifying triple network model. *Trends in Cognitive Science, 15*, 483–506.

Murray, E. A., Wise, S. P., & Drevets, W. C. (2011). Localization of dysfunction in major depressive disorder: prefrontal cortex and amygdala. *Biological Psychiatry, 69*, e43–54.

Northoff, G., Wiebking, C., Feinberg, T., & Panksepp, J. (2011). The "resting-state hypothesis" of major depressive disorder-a translational subcortical-cortical framework for a system disorder. *Neuroscience and Biobehavioral Reviews, 35*, 1929–1945.

Ogawa, S., & Lee, T. (1990). Brain magnetic resonance imaging with contrast dependent on blood oxygenation. *Proceedings of National Academy of Sciences, 87*, 9868–9872.

Pannekoek, J. N., van der Werff, S. J. A., Meens, P. H. F., van den Bulk, B. G., Jolles, D. D., Veer, I. M., … Vermeiren, R. R. (2014). Aberrant resting-state functional connectivity in limbic and salience networks in treatment-naïve clinically depressed adolescents. *Journal of Child Psychology and Psychiatry, 55*(12), 1317–1327.

Perrin, J. S., Merz, S., Bennett, D. M., Currie, J., Steele, D. J., Reid, I. C., & Schwarzbauer, C. (2012). Electroconvulsive therapy reduces frontal cortical connectivity in severe depressive disorder. *Proceedings of the National Academy of Sciences of the United States of America, 109*, 5464–5468.

Phillips, M. L., Drevets, W. C., Rauch, S. L., & Lane, R. (2003). Neurobiology of emotion perception II: implications for major psychiatric disorders. *Biological Psychiatry, 54*, 515–528.

Pizzagalli, D. (2011). Frontocingulate dysfunction in depression: toward biomarkers of treatment response. *Neuropsychopharmacology, 36*, 183–206.

Price, J. L., & Drevets, W. C. (2010). Neurocircuitry of mood disorders. *Neuropsychopharmacology, 35*, 192–216.

Price, J. L., & Drevets, W. C. (2012). Neural circuits underlying the pathophysiology of mood disorders. *Trends in Cognitive Sciences, 16*, 61–71.

Pruim, R. H. R., Mennes, M., Buitelaar, J. K., & Beckmann, C. F. (2015). Evaluation of ICA-AROMA and alternative strategies for motion artifact removal in resting-state fMRI. *Neuroimage, 112*, 278–287.

Pruim, R. H. R., Mennes, M., van, Rooij D., Llera, Arenas A., Buitelaar, J. K., & Beckmann, C. F. (2015). ICA-AROMA: a robust ICA-based strategy for removing motion artifact from fMRI data. *Neuroimage, 112*, 267–277.

Raichle, M. E., MacLeod, A. M., Snyder, A. Z., Powers, W. J., Gusnard, D. A., & Shulman, G. L. (2001). A default mode of brain function. *Proceedings of the National Academy of Sciences of the United States of America, 98*, 676–682.

Ramasubbu, R., Konduru, N., Cortese, F., Bray, S., Gaxiola-Valdez, I., & Goodyear, B. (2014). Reduced intrinsic connectivity of amygdala in adults with major depressive disorder. *Frontiers in Psychiatry, 5*, 17.

Ramsey, J. D., Hanson, S. J., Hanson, C., Halchenko, Y. O., Poldrack, R. A., & Glymour, C. (2010). Six problems for causal inference from fMRI. *NeuroImage, 49*, 1545–1558.

Rive, M. M., van Rooijen, G., Veltman, D. J., Phillips, M. L., Schene, A. H., & Ruhé, H. G. (2013). Neural correlates of dysfunctional emotion regulation in major depressive disorder. A systematic review of neuroimaging studies. *Neuroscience & Biobehavioral Reviews, 37*, 2529–2553.

Rodríguez-Cano, E., Sarró, S., Monté, G. C., Maristany, T., Salvador, R., McKenna, P. J., & Pomarol-Clotet, E. (2014). Evidence for structural and functional abnormality in the subgenual anterior cingulate cortex in major depressive disorder. *Psychologie Medicale, 44*, 3263–3273.

Rogers, M. A., Kasai, K., Koji, M., Fukuda, R., Iwanami, A., Nakagome, K., ... Kato, N. (2004). Executive and prefrontal dysfunction in unipolar depression: a review of neuropsychological and imaging evidence. *Neuroscience Research, 50*, 1–11.

Rush, A. (2007). The varied clinical presentations of major depressive disorder. *Journal of Clinical Psychiatry, 68*(Suppl.), 4–10.

Salimi-Khorshidi, G., Douaud, G., Beckmann, C. F., Glasser, M. F., Griffanti, L., & Smith, S. M. (2014). Automatic denoising of functional MRI data: combining independent component analysis and hierarchical fusion of classifiers. *NeuroImage, 90*, 449–468.

Salomons, T. V., Dunlop, K., Kennedy, S. H., Flint, A., Geraci, J., Giacobbe, P., & Downar, J. (2014). Resting-state cortico-thalamic-striatal connectivity predicts response to dorsomedial prefrontal rTMS in major depressive disorder. *Neuropsychopharmacology, 39*, 488–498.

Salvador, R., Suckling, J., Schwarzbauer, C., & Bullmore, E. (2005). Undirected graphs of frequency-dependent functional connectivity in whole brain networks. *Philosophical Transactions of the Royal Society of London B Biological Sciences, 360*, 937–946.

Seeley, W. W., Menon, V., Schatzberg, A. F., Keller, J., Glover, G. H., Kenna, H., ... Greicius, M. D. (2007). Dissociable intrinsic connectivity networks for salience processing and executive control. *Journal of Neuroscience, 27*, 2349–2356.

Sexton, C. E., Allan, C. L., Le Masurier, M., Bradley, K. M., Mackay, C. E., & Ebmeier, K. P. (2014). Magnetic resonance imaging in late-life depression: multimodal examination of network disruption. *Archives of General Psychiatry, 69*, 680–689.

Sheline, Y. I., Barch, D. M., Price, J. L., Rundle, M. M., Vaishnavi, S. N., Snyder, A. Z., ... Raichle, M. E. (2009). The default mode network and self-referential processes in depression. *Proceedings of the National Academy of Sciences of the United States of America, 106*, 1942–1947.

Sheline, Y. I., Price, J. L., Yan, Z., & Mintun, M. A. (2010). Resting-state functional MRI in depression unmasks increased connectivity between networks via the dorsal nexus. *Proceedings of the National Academy of Sciences of the United States of America, 107*, 11020–11025.

Shulman, G. L., Fiez, J. A., Corbetta, M., Buckner, R. L., Miezin, F. M., Raichle, M. E., & Petersen, S. E. (1997). Common blood flow changes across visual tasks: II. Decreases in cerebral cortex. *Journal of Cognitive Neuroscience, 9*, 648–663.

Siero, J. C. W., Hermes, D., Hoogduin, H., Luijten, P. R., Ramsey, N. F., & Petridou, N. (2014). BOLD matches neuronal activity at the mm scale: a combined 7T fMRI and ECoG study in human sensorimotor cortex. *NeuroImage, 101*, 177–184.

Smith, S. M., Fox, P. T., Miller, K. L., Glahn, D. C., Fox, P. M., Mackay, C. E., ... Beckmann, C. F. (2009). Correspondence of the brain's functional architecture during activation and rest. *Proceedings of the National Academy of Sciences of the United States of America, 106*, 13040–13045.

Snyder, A. Z., & Raichle, M. E. (2012). A brief history of the resting state: the Washington University perspective. *NeuroImage, 62*(2), 902–910.

Sridharan, D., Levitin, D. J., & Menon, V. (2008). A critical role for the right fronto-insular cortex in switching between central-executive and default-mode networks. *Proceedings of the National Academy of Sciences of the United States of America, 105*, 12569–12574.

Tahmasian, M., Knight, D. C., Manoliu, A., Schwerthöffer, D., Scherr, M., Meng, C., ... Sorg, C. (2013). Aberrant intrinsic connectivity of hippocampus and amygdala overlap in the fronto-insular and dorsomedial-prefrontal cortex in major depressive disorder. *Frontiers in Human Neuroscience, 7*, 639.

Tang, Y., Kong, L., Wu, F., Womer, F., Jiang, W., Cao, Y., ... Wang, F. (2013). Decreased functional connectivity between the amygdala and the left ventral prefrontal cortex in treatment-naive patients with major depressive disorder: a resting-state functional magnetic resonance imaging study. *Psychologie Medicale, 43*, 1921–1927.

Taylor, P. A., Gohel, S., Di, X., Walter, M., & Biswal, B. B. (2012). Functional covariance networks: obtaining resting-state networks from intersubject variability. *Brain Connect, 2*, 203–217.

Van Tol, M.-J., Li, M., Metzger, C. D., Hailla, N., Horn, D. I., Li, W., ... Walter, M. (2013). Local cortical thinning links to resting-state disconnectivity in major depressive disorder. *Psychologie Medicale*, 1–13.

Van Waarde, J. A., Scholte, H. S., van Oudheusden, L. J. B., Verwey, B., Denys, D., & van Wingen, G. A. (2014). A functional MRI marker may predict the outcome of electroconvulsive therapy in severe and treatment-resistant depression. *Molecular Psychiatry*, 1–6.

Veer, I. M., Beckmann, C. F., van Tol, M.-J., Ferrarini, L., Milles, J., Veltman, D. J., ... Rombouts, S. A. R. B. (2010). Whole brain resting-state analysis reveals decreased functional connectivity in major depression. *Frontiers in Systems Neuroscience, 4*.

Wang, J., Zuo, X., & He, Y. (2010). Graph-based network analysis of resting-state functional MRI. *Frontiers in Systems Neuroscience, 4*, 16.

Wu, Q.-Z., Li, D.-M., Kuang, W.-H., Zhang, T.-J., Lui, S., Huang, X.-Q., ... Gong, Q.-Y. (2011). Abnormal regional spontaneous neural activity in treatment-refractory depression revealed by resting-state fMRI. *Human Brain Mapping, 32*, 1290–1299.

Yang, Z., Chang, C., Xu, T., Jiang, L., Handwerker, D. A., Castellanos, F. X., ... Zuo, X.-N. (2014). Connectivity trajectory across lifespan differentiates the precuneus from the default network. *NeuroImage, 89*, 45–56.

Ye, T., Peng, J., Nie, B., Gao, J., Liu, J., Li, Y., ... Shan, B. (2012). Altered functional connectivity of the dorsolateral prefrontal cortex in first-episode patients with major depressive disorder. *European Journal of Radiology, 81*, 4035–4040.

Yue, Y., Yuan, Y., Hou, Z., Jiang, W., Bai, F., & Zhang, Z. (2013). Abnormal functional connectivity of amygdala in late-onset depression was associated with cognitive deficits. *PLoS One, 8*, e75058.

Zalesky, A., Fornito, A., Cocchi, L., Gollo, L. L., & Breakspear, M. (2014). Time-resolved resting-state brain networks. *Proceedings of the National Academy of Sciences of the United States of America, 111*(28), 10341–10346.

Zang, Y.-F., He, Y., Zhu, C.-Z., Cao, Q.-J., Sui, M.-Q., Liang, M., ... Wang, Y.-F. (2007). Altered baseline brain activity in children with ADHD revealed by resting-state functional MRI. *Brain and Development, 29*, 83–91.

Zang, Y., Jiang, T., Lu, Y., He, Y., & Tian, L. (2004). Regional homogeneity approach to fMRI data analysis. *NeuroImage, 22*, 394–400.

Zhang, Z., Liao, W., Zuo, X. N., Wang, Z., Yuan, C., Jiao, Q., ... Liu, Y. (2011). Resting-state brain organization revealed by functional covariance networks. *PLoS One, 6*.

Zhou, Y., Yu, C., Zheng, H., Liu, Y., Song, M., Qin, W., ... Jiang, T. (2010). Increased neural resources recruitment in the intrinsic organization in major depression. *Journal of Affective Disorders, 121*, 220–230.

Zhu, Z., Lu, Q., Meng, X., Jiang, Q., Peng, L., & Wang, Q. (2012). Spatial patterns of intrinsic neural activity in depressed patients with vascular risk factors as revealed by the amplitude of low-frequency fluctuation. *Brain Research, 1483*, 82–88.

Zhu, X., Wang, X., Xiao, J., Liao, J., Zhong, M., Wang, W., & Yao, S. (2012). Evidence of a dissociation pattern in resting-state default mode network connectivity in first-episode, treatment-naive major depression patients. *Biological Psychiatry, 71*, 611–617.

Zuo, X.-N., & Xing, X.-X. (2014). Test-retest reliabilities of resting-state FMRI measurements in human brain functional connectomics: a systems neuroscience perspective. *Neuroscience and Biobehavioral Reviews, 45*, 100–118.

Zuo, Q.-H., Zhu, C.-Z., Yang, Y., Zuo, X.-N., Long, X.-Y., Cao, Q.-J., ... Zang, Y.-F. (2008). An improved approach to detection of amplitude of low- frequency fluctuation (ALFF) for resting-state fMRI: fractional ALFF. *Journal of Neuroscience Methods, 172*, 137–141.

Chapter 8

Imaging Functional Systems in Depression

Christina B. Young[1,2,3], Bernhard Mueller[4], Indira Tendolkar[2,3,4]

[1]*Department of Psychology, Northwestern University, Evanston, IL, USA;* [2]*Donders Institute for Brain, Cognition and Behaviour, Nijmegen, The Netherlands;* [3]*Department of Psychiatry, Radboud University Medical Centre, Nijmegen, The Netherlands;* [4]*Department of Psychiatry and Psychotherapy, University Hospital Essen, Essen, Germany*

As highlighted in previous chapters, novel methods in systems neuroscience have identified numerous functional systems in the brain. These methods are being applied in increasing frequency to uncover neural aberrations underlying various psychiatric disorders. Furthermore, there has been a recent impetus to consider specific symptoms of mental illness instead of examining categorical disorders at large (Insel et al., 2010). In the context of major depressive disorder (MDD), research has shown that abnormalities in neural activity and functional connectivity are related to common symptoms in MDD. Accordingly, we highlight the relationship between neural abnormalities in regions within functional systems and common depression symptoms in this chapter. More specifically, we focus on mood-congruent processing, attention biases, memory biases, long-term memory deficits, rumination, and anhedonia.

MOOD-CONGRUENT PROCESSING IN DEPRESSION

The onset and maintenance of depression is affected by abnormal emotion processing. Numerous psychological studies have described mood-congruent processing in which individuals are biased toward stimuli that match their moods. In other words, individuals in a positive mood are biased toward positive stimuli, whereas those in a negative mood are biased toward negative stimuli. This is particularly important in the context of MDD, as depression is characterized by extended periods of negative mood. Indeed, depressed mood in MDD often causes one of the most pronounced cognitive distortions in MDD—a greater sensitivity to negative information. Increased sensitivity to negative, mood-congruent information then leads to deeper and more elaborate

Systems Neuroscience in Depression. http://dx.doi.org/10.1016/B978-0-12-802456-0.00008-X

processing of this content, which in turn influences subsequent attitudes, interpretations of events, and mood states (Ingram, 1984).

Neurally, the amygdala is central to mood-congruent biases. The amygdala is involved in the early detection of salient information at both conscious and subconscious levels (Bordi & LeDoux, 1992; Liddell et al., 2005; Morris, Ohman, & Dolan, 1999), and therefore helps identify the emotional significance of a stimulus and produces affective states (Phillips, Drevets, Rauch, & Lane, 2003). Additionally, the amygdala has numerous connections with regions important for emotion processing, including the orbitofrontal cortex (OFC) and other subcortical temporal structures, such as the hippocampus. Together, this temporo-amygdala orbitofrontal network modulates neural and behavioral responses to emotional stimuli (Drevets, Price, & Furey, 2008; Ongur, Ferry, & Price, 2003; Ongur & Price, 2000), as well as integrates visceral and emotional states with cognition and behavior (Catani, Dell'acqua, & Thiebaut de Schotten, 2013; Mesulam, 2000). Therefore, many of the depression neuroimaging studies examining mood-congruent biases have focused on or revealed abnormalities in amygdala activity and connectivity.

Socially relevant emotional stimuli such as faces have been used in many of the initial neuroimaging papers investigating mood-congruent processing biases. One of the most common findings in studies examining facial emotion processing in depression is greater amygdala responsiveness to negative stimuli (Stuhrmann, Suslow, & Dannlowski, 2011). Overtly presented sad (Fu et al., 2004; Surguladze et al., 2005) and fearful faces (Sheline et al., 2001) have been shown to elicit amygdala hyperactivation in MDD. Similarly, fearful and angry expressions combined in a face-matching task also elicited greater amygdala response in depressed patients relative to healthy controls (Peluso et al., 2009; Zhong et al., 2011). Amygdala hyperactivity within MDD has also been positively correlated with depression severity (Peluso et al., 2009). Furthermore, two studies subliminally presented facial expressions to examine whether aberrant amygdala activity related to mood-congruent processing occurs even at an early automatic processing stage. These studies showed that in comparison to healthy controls, depressed individuals have an increased and decreased amygdala response to sad and happy faces, respectively (Suslow et al., 2010; Victor, Furey, Fromm, Ohman, & Drevets, 2010). Thus, negative mood-congruent stimuli reliably elicits amygdala hyperactivity in MDD.

While these studies support the idea that pathological amygdala hyperactivity could affect the onset and maintenance of emotional disorders by eliciting dysfunctional negative biases, behavioral measures of biases are missing in the studies mentioned above. Dannlowski et al. (2006) addressed this gap by directly linking brain activity to behavioral biases in 35 medicated inpatients (24 women and 11 men) suffering from acute major depression. This study used an affective priming task that displayed either neutral target faces or no-face comparisons preceded by a fast presentation of a sad, angry, happy, or neutral face. A comparison of amygdala activity in response to masked emotional faces versus the no-face baseline revealed significant bilateral amygdala activation in response to masked sad faces as well as activation of the right amygdala in response to masked happy faces. Thus, similar to the studies mentioned above, depressed patients showed amygdala hyperactivity in response to negative stimuli. Participants in this study also repeated the affective priming task after scanning to rate each emotional expression in order to obtain bias scores. For each emotion, a bias score was calculated by subtracting mean ratings for neutral target faces primed by neutral faces from mean ratings for neutral target faces primed by emotional faces. Thus, a subject who responded more to a negative prime had a greater negative bias score. Using this bias score, the authors showed that negative biases were associated with stronger amygdala responses to masked angry and sad faces but not happy faces. Therefore, those who had stronger amygdala responses when presented with covert negative expressions also had stronger negative biases even during an automatic stage of affective information processing.

In addition to the studies that examined neural activity in response to facial expressions, a few studies also used facial expressions to probe aberrations in functional connectivity related to mood-congruent processing. These studies have primarily revealed that in response to negative faces, depressed individuals show reduced functional connectivity of the amygdala with other regions important for emotion processing. For example, in response to sad faces, depressed patients showed reduced functional connectivity between bilateral amygdala and several other limbic, temporal, and frontal regions, including the hippocampus; parahippocampal gyrus; inferior, middle, and superior temporal cortices; and inferior and middle frontal cortices, in comparison to healthy controls (Chen et al., 2008). Similarly, reduced functional connectivity of the amygdala with the dorsal anterior cingulate cortex (ACC) and dorsolateral prefrontal cortex (DLPFC) was seen in response to negative emotional faces in MDD patients versus healthy controls (Dannlowski et al., 2009). Taken together, Chen et al. (2008) and Dannlowski et al. (2009) showed reduced amygdala functional connectivity with numerous regions across the whole brain; however, a study by Matthews, Strigo, Simmons, Yang, and Paulus (2008) suggests that aberrations in amygdala connectivity may be more nuanced. Matthews et al. (2008) examined amygdala and cingulate connectivity specifically and demonstrated that depressed patients had increased functional connectivity between amygdala and subgenual ACC but decreased functional connectivity between amygdala and supragenual ACC. Furthermore, those with greater severe depression showed more reduced amygdala and supragenual ACC connectivity. Thus, amygdala functional connectivity appears to be generally reduced in MDD, but closer examinations of large regions may reveal locations with increased amygdala functional connectivity.

Other studies have also focused on OFC connectivity as the OFC works closely with the amygdala to modulate neural and behavioral responses to emotional stimuli (Drevets et al., 2008; Frodl et al., 2010; Ongur & Price, 2000;

Ongur et al., 2003). In response to sad and angry faces, medication-free depressed patients showed increased OFC functional connectivity with the DLPFC, which may reflect greater neural response to negative stimuli (Frodl et al., 2010). The same study also found reduced OFC connectivity with the precuneus and cingulate gyrus, and the authors suggested that this may be associated with problems in regulating self-schemas. Two studies have also used effective connectivity methods to assess the direction of connectivity. In the first study, depressed patients showed decreased top-down left OFC–amygdala connectivity in response to happy and sad facial expressions (Almeida et al., 2009). However, the second study also showed decreased bottom-up amygdala–OFC effective connectivity in depressed patients (Carballedo et al., 2011), suggesting deficits in both top-down and bottom-up processing. In addition to aberrant amygdala–OFC connectivity, depressed patients have also exhibited reduced amygdala–ACC and amygdala–prefrontal cortex connectivity in the right hemisphere as well as greater prefrontal–OFC effective connectivity in the left hemisphere (Carballedo et al., 2011). Taken together, these studies highlight the role of connectivity between the amygdala, temporal, and frontal regions in mood-congruent processing.

Importantly, treatment studies have shown that aberrant amygdala connectivity and activity can be normalized after treatment. In the only study examining the effects of antidepressant treatment on amygdala connectivity during face processing, Chen et al. (2008) showed that antidepressant treatment significantly increased amygdala functional connectivity with regions that previously showed reduced amygdala functional connectivity before treatment. With respect to aberrant amygdala activity, Sheline et al. (2001) conducted one of the first functional magnetic resonance imaging (fMRI) studies in acutely depressed patients and showed that antidepressant treatment decreased previously hyperactive amygdala response to fearful (and happy) faces. The same group of investigators later replicated the finding of normalization of exaggerated response to fearful faces in depressed patients (Fales et al., 2009). The reversal of abnormal amygdala responses after antidepressant treatment has also been shown in another study that combined cross-sectional and longitudinal design. Prior to treatment, Victor et al. (2010) demonstrated that both currently depressed and remitted depressed individuals had greater amygdala activity when processing subliminally presented sad faces in comparison to happy faces; in contrast, healthy controls had increased amygdala activity when processing happy faces in comparison to sad or neutral faces. After antidepressant treatment, this pattern reversed such that a bias towards happy faces developed in treated MDD individuals. However, the relevance of this finding to the antidepressant mechanism of action is unclear, since the authors also reported exaggerated responses to sad faces in unmedicated patients in remission. Indeed, the inconsistency of antidepressant effects is highlighted by Suslow et al. (2010), who demonstrated that currently depressed patients receiving antidepressant treatment do not have normalized amygdala responses but instead show hyperactive amygdala responses to sad faces and hypoactive responses to happy faces. Nevertheless, the studies by Sheline et al. (2001), Victor et al. (2010), and Chen et al. (2008) provide evidence that an elevated amygdala response to sad faces and its aberrant connectivity can be normalized with antidepressant treatment. Similarly, Fu et al. (2008) reported increased and normalized amygdala responses to sad faces in unmedicated, depressed patients after cognitive behavioral therapy. Taken together, these studies suggest that the amygdala may be part of a final common pathway of successful treatment effects.

In summary and as depicted in Figure 1(a), abnormal activity and connectivity of the amygdala are central to the mood-congruent biasing effects seen in depression. Studies using negative facial expressions have robustly elicited amygdala hyperactivity in those with depression in comparison to healthy controls. Negative biases in depression have also been directly linked

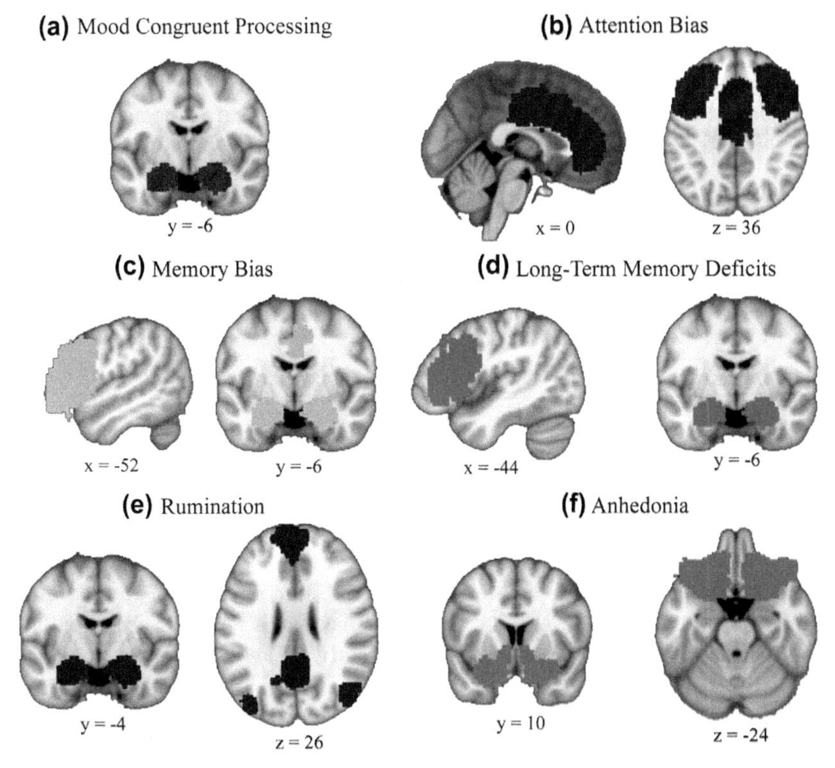

FIGURE 1 Neural regions related to common symptoms of depression. (a) Mood congruent processing is associated with the amygdala, (b) attention biases towards negative stimuli are related to rostral anterior cingulate cortex and lateral prefrontal cortex abnormalities, (c) memory biases are related to amygdala and frontal regions, (d) long-term memory deficits are associated with the hippocampus, (e) rumination is related to the amgydala and default-mode network, and (f) anhedonia is associated with the reward network.

to amygdala hyperactivity even during an automatic stage of affective information processing. Studies examining functional and effective connectivity have also shown that amygdala connectivity appears to be generally reduced in depression, although regions showing increased amygdala connectivity have also been found. Additionally, abnormal OFC and prefrontal connectivity in response to negative facial expressions have also been identified in depression. Finally, treatment studies suggest that aberrant amygdala activity and connectivity can be normalized with treatment, and the amygdala may be centrally involved in successful treatment effects.

BIASED ATTENTION TOWARD NEGATIVE STIMULI

In addition to overall increased sensitivity to negative stimuli, negative cognitive biases in depression can also be seen in specific cognitive domains (Beck, 1967). Two of the most well-studied cognitive domains are attention and memory. Behavioral evidence has shown that depressed individuals pay more attention to negative stimuli and less attention to positive stimuli in comparison to healthy controls (Kellough, Beevers, Ellis, & Wells, 2008). For example, MDD has been associated with a bias to attend to negative words (Bradley, Mogg, & Lee, 1997; Donaldson, Lam, & Mathews, 2007; Mogg, Bradley, & Williams, 1995) and faces (Gotlib, Krasnoperova, Yue, & Joormann, 2004; Joormann & Gotlib, 2007; Leyman, De Raedt, Schacht, & Koster, 2007). A more detailed examination of this negative attention bias has shown that patients with MDD engage more strongly and maintain attention to angry faces, whereas healthy controls not only direct attention away from but also more rapidly disengage from angry faces (Leyman et al., 2007). Thus, particular difficulty disengaging from negative stimuli may be characteristic of depression.

The negative attention bias is thought to involve both a bottom-up amygdala-centered neural system as well as a top-down control system involving the ACC and lateral prefrontal cortices (Browning, Holmes, Murphy, Goodwin, & Harmer, 2010). In the first neural system, amygdala hyperactivity discussed in the context of mood-congruent processing may put more brain systems on alert for negative stimuli (Whalen, Shin, Somerville, McLean, & Kim, 2002), thereby biasing attention. In the second neural system, the top-down control system may lead to difficulty disengaging from negative stimuli. The ACC and lateral prefrontal cortices that are part of the top-down control system are involved in detecting and resolving processing conflicts as well as maintaining attention on relevant stimuli. Thus, aberrations in these regions may lead to abnormal maintenance of attention on negative stimuli. Indeed, perturbations in both of these systems have been found in depression.

Studies on attentional biases in MDD have implicated the rostral ACC, which is typically recruited during inhibition (Bush, Luu, & Posner, 2000; Shafritz, Collins, & Blumberg, 2006). Using a negative affective priming task,

Eugène, Joormann, Cooney, Atlas, and Gotlib (2010) showed that inhibition of negative, but not positive words, was associated with rostral ACC activity in MDD patients. In contrast, rostral ACC activity was associated with inhibition of positive, but not negative words in healthy controls. Mitterschiffthaler et al. (2008) showed similar effects in one of the first studies using fMRI to study the emotional Stroop effect in MDD. In this study, the emotional Stroop effect was stronger in currently depressed patients such that reaction times were longer for sad words in the currently depressed group in comparison to healthy controls, but equivalent between groups for neutral words. Relating this behavioral effect to rostral ACC activity, the reaction times for negative words were positively correlated with activity in the rostral ACC in MDD patients. Thus, difficulty inhibiting responses to negative stimuli is related to rostral ACC activity in depression. Additional evidence of rostral ACC dysfunction can also be seen with the emotional go/no-go task. Using this task, Elliott, Rubinsztein, Sahakian, and Dolan (2002) showed that MDD patients had increased and decreased rostral ACC responses to sad and happy targets, respectively. In contrast, healthy controls showed the opposite activation profile. Taken together, these studies suggest that individuals with MDD require more cognitive effort to disengage from negative stimuli and this difficulty is reflected in increased rostral ACC activity.

Studies of attention biases in depression have also implicated the lateral prefrontal cortex, which is critical for cognitive control especially during inhibition of competing responses (Aron & Poldrack, 2005; Nee, Wager, & Jonides, 2007). In a study that involved shifting attention away from invalid emotional cues, depressed patients showed reduced activation in lateral prefrontal cortex and parietal regions (Beevers, Clasen, Stice, & Schnyer, 2010). Similarly, impaired top-down mechanisms were identified in the DLPFC, such that healthy controls, but not depressed patients, showed increased activity in the right DLPFC when fearful distractors were presented in an attentional interference task (Fales et al., 2009). These studies suggest that depression is characterized by difficulty engaging neural regions important for cognitive control of emotional information. However, studies have also shown increased activity in lateral prefrontal cortex in depression. Using a paradigm that embedded oddball target stimuli in blocks of sad or neutral images, Dichter, Felder, and Smoski (2009) showed greater activation in lateral prefrontal cortex and ACC during sad blocks in MDD patients, suggesting that more prefrontal activation was required to disengage from sad images and respond to target events. In a similar paradigm, Wang et al. (2008) showed that MDD patients were slower to respond to target-after-sad than target-versus-neutral stimuli. Furthermore, this difference in response time was positively correlated with inferior frontal gyrus activity during target-after-sad versus target-after-neutral conditions in MDD; this relationship was not seen in healthy controls. Therefore, activity in the inferior frontal gyrus, a region within the lateral prefrontal cortex, reflected effortful inhibition processing in depressed individuals who had difficulty disengaging from negative information. Taken together, these studies suggest that depression may be characterized by reduced activity in the lateral prefrontal cortex during the initial phase of attentional interference, but increased activity in this region when actively inhibiting the attention bias. However, this hypothesis needs to be directly tested.

In summary, depressed individuals not only orient their attention toward negative stimuli but also have particular difficulty disengaging from negative stimuli. The attention bias effects involve both amygdala hyperactivity, which relates to increased reactivity to negative stimuli, and aberrant functioning of the rostral ACC and lateral prefrontal cortex (Figure 1(b)), which relates to an inability to disengage from negative stimuli. To date, only one study has examined antidepressant effects on neural activity directly related to attention biases in MDD. Fales et al. (2009) used a fear attention task and showed increased DLPFC activity after treatment. Despite the scarcity of literature on treatment effects on attention biases in depression, studies on anxiety disorders and healthy controls have led to the hypothesis that pharmacological treatments influence the deployment of attention through the amygdala-based system, whereas psychological treatments influence control of attention through the lateral prefrontal cortex (Browning et al., 2010). However, whether this is true in depression remains to be seen.

MEMORY BIASES FOR NEGATIVE INFORMATION

Selective memory for negative information in depression contributes to the duration and severity of depressive episodes (Ingram, 1984; Teasdale, 1983) and has been proposed to be a cognitive endophenotype of depression (Hasler, Drevets, Manji, & Charney, 2004). Whereas healthy controls typically show an enhancement of memory for positive information (Bradley, Mogg, & Williams, 1995; Denny & Hunt, 1992), depressed patients have a tendency to recall negatively toned information (Bradley, Mogg, & Millar, 1996; Clark & Teasdale, 1982; Lloyd & Lishman, 1975). Furthermore, these memory biases in depression may be at least partly unconscious (Watkins, Vache, Verney, & Mathews, 1996). One of the contributing factors for the negative memory bias in depression is thought to be a negative mood at the time of encoding (Teasdale & Dent, 1987). Accordingly, the amygdala is thought to be central to the memory bias effect as the amygdala bolsters memory for emotional material (Cahill, Babinsky, Markowitsch, & McGaugh, 1995; Cahill & McGaugh, 1998).

Ramel et al. (2007) were the first to use fMRI to show that under a sad mood induction, bilateral amygdala response during encoding of valenced words predicted an increased recall of negative self-referent words in 14 remitted depressed patients in comparison to 14 healthy controls. Hamilton and Gotlib (2008) examined 14 acutely depressed patients as compared to

12 healthy controls and found similar results. In their study, depressed individuals remembered negative stimuli better than healthy controls but had equivalent memory for positive stimuli. Imaging results showed that in depressed patients, the right amygdala was more active and showed greater functional connectivity with the hippocampus and caudate-putamen during encoding of successfully remembered negative but not neutral or positive stimuli. Right amygdala activity in response to subsequently remembered negative stimuli was also significantly correlated with depression severity in MDD patients. Additional evidence of the influence of the amygdala in memory comes from an fMRI study investigating neutral memory in depression (van Eijndhoven et al., 2011). In this study, first-episode depressed patients, remitted patients, and healthy controls viewed and remembered neutral scenery images. Currently and remitted depressed patients showed increased activity in the amygdala during memory encoding when compared to controls. Thus, amygdala hyperactivity was involved in the memory of negative information and interfered with demanding memory tasks that were not affectively laden.

Whether memory biases are state- or trait-based characteristics is an important question. van Wingen et al. (2010) addressed this question by investigating medication-free currently depressed, recovered depressed, and healthy individuals as they memorized and retrieved happy and neutral faces. During memory encoding, state-based effects were found in the amygdala such that this region was involved in memory formation of both happy and neutral faces in healthy controls but only in memory formation of neutral faces in those with current depression. During memory retrieval, state-based effects were seen in the amygdala and posterior cingulate cortex (PCC) such that currently depressed patients, but not recovered patients or healthy controls, recruited these regions during retrieval of happy faces. In contrast, trait-based effects during both memory encoding and retrieval were found in the fusiform gyrus. Additionally, the OFC and inferior frontal gyrus showed trait-based effects during memory retrieval of happy faces. These findings led the authors to conclude that memory processing biases of positive information in the amygdala and PCC are state-based, while memory processing biases in the fusiform gyrus and prefrontal cortex are trait-based. This study, however, did not include negative stimuli, so it is unclear whether the neural biases are specific to happy faces or can be extended to emotional faces in general.

A study by Arnold et al. (2011) also examined memory formation biases in 14 medication-free women remitted from unipolar MDD and 14 matched controls. In this study, participants learned negative, positive, and neutral words and were later tested with free recall. The remitted MDD and healthy control groups had equivalent memory performance and activity patterns during memory formation of neutral and negative words. However, remitted MDD patients remembered fewer positive words than healthy controls. For positive words, remitted MDD patients showed greater activity in the amygdala/anterior hippocampus for later forgotten words in comparison to later remembered words, whereas controls showed the opposite pattern. Similarly, remitted patients exhibited greater activity in the cingulate gyrus, right inferior-, and left medial-frontal gyrus during successful encoding of positive words in comparison to healthy controls. When considered with the study by van Wingen et al. (2010), memory biases in the frontal regions appear to be trait-based and overreactivity of brain regions involved in emotion processing may lead to biased processing even during remission.

In summary, depressed individuals show strong memory biases for negative information. The amygdala is critically involved in negative memory biases (Figure 1(c)), and hyperactivity in this region interferes with memory of both affective and nonaffective information. Studies including currently depressed, remitted depressed, and healthy control groups have identified both state- and trait-based neural effects related to memory biases. Initial evidence suggests that memory processing biases of positive information in the amygdala may be state-based, while biases in frontal regions may be trait-based. Longitudinal treatment studies are still needed to better understand the persistence of neural abnormalities related to negative memory biases in depression.

LONG-TERM MEMORY DEFICITS

Depressed patients display deficits in specific facets of long-term memory (Burt, Zembar, & Niederehe, 1995). Studies of human memory have revealed a fundamental distinction between two ways of gaining access to past experiences: recollection and familiarity (Mandler, 1980; Yonelinas, 2001). Recollection involves the recognition of a stimulus, including the retrieval of contextual information that accompanied the original encounter. In contrast, familiarity refers to a feeling that a stimulus has been encountered before without contextual retrieval (Mandler, 1980). MDD is characterized by a selective deficit in recollection with preserved familiarity (Degl'Innocenti & Bäckman, 1999; Drakeford et al., 2010; MacQueen, Gakway, Hay, Young, & Joffe, 2002; Ramponi, Barnard, & Nimmo-Smith, 2004), and patients with several past episodes are more likely to display deficits in recollection (Basso & Bornstein, 1999; Fossati et al., 2004; Gorwood, Corruble, Falissard, & Goodwin, 2008; MacQueen et al., 2002). Depressed subjects also show a reduction in autobiographical memory specificity (Williams & Scott, 1988), which is thought to be part of a broader memory deficit in recollecting specific details about the context in which the memory was acquired (Raes et al., 2006; Ramponi et al., 2004). Long-term memory depends on the integrity of the medial temporal lobe (Eichenbaum, Yonelinas, & Ranganath, 2007), and this has motivated a line of research investigating the role of the hippocampus in the pathophysiology of depression (Bremner, Vythilingam, Vermetten, Vaccarino, & Charney, 2004; Campbell, Marriott, Nahmias, & MacQueen, 2004; MacQueen et al., 2003).

Using a paradigm that specifically engages hippocampal but not amygdala activation in healthy controls (Tendolkar et al., 2007), van Eijndhoven et al. (2011) investigated memory formation with and without episodic context in 20 medication-naïve patients with a first depressive episode, 20 medication-free patients recovered from a first episode, and 20 healthy controls. Patients in a first episode of MDD showed increased activity related to episodic memory formation in the frontolimbic network. These state-related activations may be a compensatory mechanism as memory performance was equivalent between groups. Additionally, although there were no differences in hippocampal activity between groups, both patient groups showed stronger subsequent memory effects in the amygdala when compared to controls. Thus, in MDD patients, amygdala activation specifically interacts with hippocampal-dependent memory formation of neutral stimuli. In a subsequent study, van Eijndhoven et al. (2013) dissociated the neural correlates of recollection- and familiarity-based memory retrieval. Notably, there were no changes in hippocampal activation between first-episode MDD patients and controls, supporting the idea that changes in hippocampal structure and function may rather be related to the later course of MDD.

In the same study examining recollection and familiarity-based memory retrieval, comparisons between first-episode MDD patients and controls revealed that during the acute state of depression, there was an increase in left prefrontal activity related to attempted recollection and no differences in activity during successful recollection. van Eijndhoven et al. (2013) concluded that in the early course of depression, depressive state is associated with increased left prefrontal activation during the attempt to recollect source information, suggesting an increased need for executive control during recollection in MDD. Indeed, a variety of studies investigating MDD patients during working memory tasks have shown that depressed patients require a greater recruitment of left prefrontal regions to maintain similar levels of task performance during conditions of increased cognitive load (Harvey et al., 2005; Matsuo et al., 2007; Walsh et al., 2007; Walter, Wolf, Spitzer, & Vasic, 2007).

The need for greater left prefrontal activity may be caused by maladaptive functional interactions in limbic-cortical regions (Harvey et al., 2005; Mayberg, 1997) since MDD patients may use their attentional resources primarily to focus on task-irrelevant or depression-relevant thoughts (Hartlage, Alloy, Vazquez, & Dykman, 1993). Increased left frontal activation serves to resolve this interference during cognitively demanding tasks and may compensate for a lack of deactivation of limbic structures such as the medial prefrontal cortex and the amygdala (Harvey et al., 2005). Although van Eijndhoven et al. (2013) did not find evidence for deactivation within limbic or other brain structures, it remains possible that increased activation of the left prefrontal cortex may effectively prevent limbic overactivity during recollection attempts. This claim is supported by work from Dichter et al.

(2009) who found amygdala overreactivity in MDD patients during a passive viewing condition but not during a cognitive control task, which was instead characterized by increased left prefrontal activation. Hence, problems during memory retrieval may be partially related to the high level of experienced effort of MDD patients during cognitively demanding tasks such as recollection.

In summary, depressed individuals show specific deficits in recollection of long-term memory. Although long-term memory is subserved by the hippocampus (Figure 1(d)), studies examining first-episode depression show that changes in hippocampal structure and function may be related to the later course of MDD. Instead of uncovering abnormal hippocampal activity in first-episode depression, these studies have identified aberrant amygdala activity and shown that amygdala activation interacts with hippocampal-dependent memory formation of neutral stimuli. Furthermore, increased activity in the left prefrontal cortex during memory recollection has been identified in depression, which is thought to be required in order to maintain similar levels of task performance. Thus, a need for high levels of effort during cognitively demanding tasks such as memory recollection may be involved in memory retrieval problems in MDD.

FUNCTIONAL CORRELATES OF RUMINATION

The attention and memory biases discussed thus far can predispose or lead to rumination, which is a form of self-referential processing or relating information to oneself. Rumination refers to repetitive thinking and focus on negative mood states (Nolen-Hoeksema, Morrow, & Fredrickson, 1993), and the relationship between negative biases and increased rumination in MDD is intuitive. Importantly, rumination has been associated with the severity and duration of a depressive episode (Nolen-Hoeksema, Wisco, & Lyubomirsky, 2008), and its presence increases the risk of relapse in remitted patients (Roberts, Gilboa, & Gotlib, 1998). Neuroimaging studies of rumination in MDD typically involve tasks that require participants to focus on oneself or involve resting state, during which participants are not engaged in any particular cognitive task.

Because rumination involves a persistent focus on negative mood states, one study examined activity patterns in the amygdala in response to negative stimuli over time (Siegle, Steinhauer, Thase, Stenger, & Carter, 2002). In this study, positive or negative emotional stimuli were first presented and then followed by nonemotional trials. When negative emotional stimuli were presented before nonemotional trials, depressed individuals showed sustained amygdala responses that lasted 25 s later throughout the following nonemotional processing trial. In contrast, amygdala responses in healthy controls decayed within 10 s. Additionally, the difference in sustained amygdala activity in response to negative and positive words moderately related to self-reported rumination. That is, those who showed a greater difference between amygdala response to happy and sad stimuli had lower rumination. Thus, rumination appears to involve an inability to modulate amygdala activity in response to negative stimuli.

Rumination also often involves emotional self-attributions, and this facet was examined in a study that required participants to judge the self-relatedness of positive and negative emotional stimuli (Grimm et al., 2009). In this study, depressed patients showed higher degrees of self-relatedness for negative emotional stimuli than healthy controls. When judging the self-relatedness of stimuli in contrast to passive viewing, depressed patients showed lower activity in the subcortical and cortical midline structures, including the dorsomedial prefrontal cortex, supragenual ACC, precuneus, ventral striatum (VS), and dorsomedial thalamus, than healthy controls. Furthermore, dorsomedial prefrontal cortex activity negatively correlated with depression severity, negative self-concept, and anhedonia/inhibition. From these results, the authors suggested that depression is characterized by increased negative self-attribution that is mediated by subcortical and cortical midline structures.

However, another study found opposite results and showed that ruminative self-focus was associated with overrecruitment of subcortical and cortical midline structures as well as prefrontal regions in MDD (Cooney, Joormann, Eugène, Dennis, & Gotlib, 2010). This study contrasted rumination (e.g., "Think about what people notice about your personality"), abstract distraction (e.g., "Think about what contributes to team spirit"), and concrete distraction (e.g., "Think about a row of shampoo bottles on display") in depressed and healthy individuals. Similar to rumination, abstract distraction is open-ended and less reliant on visual imagery, which is required by the concrete distraction condition. In comparison to healthy controls, depressed individuals showed greater activity in the amygdala, rostral ACC/medial prefrontal cortex, DLPFC, posterior cingulate, and parahippocampal gyrus during rumination versus abstract distraction. MDD individuals also showed increased activation in the OFC, subgenual ACC, and DLPFC in comparison to controls during rumination in contrast to concrete distraction. Taken together, whether midline structures show hypo- or hyperactivity may depend on the specific task used to probe rumination.

Although these task-based studies provide insight into the neural correlates of rumination, rumination is not typically prompted by specifically negative events or by instructions to reflect on oneself. Instead, negative introspection often occurs spontaneously and is unintentional. This closely relates to research showing that a network of brain regions called the default-mode network (DMN) underlies passive, self-referential mental activity (Menon, 2011). The DMN consists of an anterior portion that covers the medial prefrontal cortex, including parts of the ACC, and a posterior portion including the PCC, precuneus, inferior parietal cortex, hippocampal area, and lateral temporal cortex (Buckner, Andrews-Hanna, & Schacter, 2008). The DMN has been related to self-reference (Spreng, Mar, & Kim, 2009), emotion regulation (Etkin, Egner, & Kalisch, 2011), episodic memory (Sestieri, Corbetta, Romani, & Shulman, 2011), and autobiographical memory (Dastjerdi et al., 2011; Gusnard, Akbudak, Shulman, & Raichle, 2001; Spreng et al., 2009). Furthermore, the DMN is most active during rest, and activity in the DMN is decreased during demanding cognitive tasks. As a result, the DMN is typically examined during resting state or wakeful rest. Indeed, a wealth of research shows that aberrant DMN activity is related to rumination in depression.

In a study that examined DMN connectivity in depressed individuals and healthy controls, depressed individuals were shown to have more DMN connectivity with the subgenual anterior cingulate cortex (sgACC) than controls (Berman et al., 2010). This hyperconnectivity was only seen during rest and not during working memory task engagement. This is consistent with previous research that not only showed greater sgACC connectivity in depression but also that sgACC connectivity positively correlated with the length of a current depressive episode in MDD (Greicius et al., 2007). Critically, a closer examination of sgACC connectivity with the PCC, another key node of the DMN, revealed that sgACC-PCC connectivity positively correlated with rumination in both depressed and healthy individuals (Berman et al., 2010). This study provided the first direct evidence that DMN connectivity, particularly sgACC connectivity, is related to rumination.

Studies have continued to investigate the role of the DMN in rumination. In particular, one study examined the relationship between the DMN and the task-positive network (TPN), a set of regions that are engaged during active cognitive processing (Hamilton et al., 2011b). Here, they found that DMN dominance over TPN during rest was associated with greater depressive rumination and lower self-reflection in depressed individuals only. No association between DMN dominance over TPN and rumination was found in healthy controls. Another study by the same group used Granger causality analysis, an effective connectivity method, to examine patterns of neural cause and effect in the various components recruited during resting state in depression (Hamilton, Chen, Thomason, Schwartz, & Gotlib, 2011). Here, Hamilton et al. (2011b) found that the hippocampus initiates maladaptive overresponding in the sgACC and that there is aberrant mutually reinforcing excitation between sgACC and medial prefrontal cortex in depression. Furthermore, this medial prefrontal cortex and sgACC connectivity was positively correlated with depressive rumination, suggesting that the stronger the excitation of the sgACC to the medial prefrontal cortex, the greater the reported depressive rumination. Together, these results highlight the importance of the DMN, and the sgACC in particular, in rumination.

Although several studies have investigated treatment effects on the DMN, no direct relation with rumination has been made. For example, Abbott et al. (2013) investigated between-network connectivity in an elderly group of treatment-resistant depressed patients and found that electroconvulsive therapy reversed the relationship between the posterior DMN with the dorsomedial PFC and left dorsolateral PFC from negative to positive. Electroconvulsive therapy did not change any anterior DMN relationships, suggesting that treatment affects the anterior and posterior DMN differently. Similarly, another study in a small group of elderly depressed patients reported DMN-related PCC connectivity changes after treatment, which also suggests local differences in connectivity between anterior and posterior DMN. More specifically, the PCC showed increased connectivity with the dorsomedial PFC and OFC but decreased connectivity with the sgACC (Wu et al., 2011). Andreescu et al. (2013) also investigated a large group of elderly patients and showed treatment effects in the posterior part of the DMN after 12 weeks of antidepressant treatment, although these effects did not survive correction for white matter hyperintensities. Taken together, depression treatment does appear to affect activity in the DMN, although whether or not these changes relate to rumination specifically is still unknown.

In summary, rumination is an important symptom of depression that has been examined using both task-based and resting-state fMRI. Task-based studies have shown that rumination involves an inability to modulate amygdala activity in response to negative stimuli. Rumination tasks have also implicated midline brain structures, although whether these regions show hypo- or hyperactivity seems to depend on the specific task used. As a result, some studies have used resting state to probe rumination. These studies have shown that DMN connectivity is positively correlated with rumination and that the sgACC may be particularly important for this symptom (Figure 1(e)). Finally, although there are many studies examining the effects of treatment on DMN activity and connectivity, no studies to date have directly examined DMN-related treatment effects in relation to rumination.

FUNCTIONAL CORRELATES OF ANHEDONIA

One particularly important symptom in depression is anhedonia, defined as the reduced ability to experience pleasure from previously pleasurable stimuli (American Psychiatric Association, 2000). Anhedonia is commonly experienced in depression (Lewinsohn, Pettit, Joiner, & Seeley, 2003; Pelizza & Ferrari, 2009) and is a predictor of poor treatment response (Spijker, Bijl, de Graaf, & Nolen, 2001). Additionally, anhedonia is thought to be related to abnormalities in reward processing (Treadway, Bossaller, Shelton, & Zald, 2012) and, in general, fMRI studies have found that depression is characterized by reduced responsivity in the brain's cortical-basal ganglia reward network.

The cortical-basal ganglia reward network is anchored in the VS, ventral tegmental area, as well as substantia nigra (VTA/SN), OFC, and the ACC, including the ventromedial prefrontal cortex (vmPFC; Haber & Knutson, 2010). Both human and rodent research have demonstrated a critical role of the cortical-basal ganglia reward network in reward responsivity, incentive-based learning, estimation of reward probability, prediction error, and goal-directed behavior. fMRI studies have consistently shown increased VTA/SN and VS activity in response to a variety of rewards, including pleasant tastes (Small, Zatorre, Dagher, Evans, & Jones-Gotman, 2001), music (Koelsch, Fritz, Cramon, Müller, & Friederici, 2006; Menon & Levitin, 2005), humor (Mobbs, Greicius, Abdel-Azim, Menon, & Reiss, 2003), and money (Knutson, Westdorp, Kaiser, & Hommer, 2000). Additionally, the OFC and other forebrain regions have been shown to be important for more complex processes, such as determining value, magnitude, and probability of reward (Haber & Knutson, 2010), and the medial OFC in particular is important for processing reward valence (Berridge & Kringelbach, 2008). Importantly, a large meta-analysis of healthy reward processing demonstrated that the VS, vmPFC, anterior insula, and thalamus are engaged by both primary rewards, which have innate value, and secondary rewards, which gain value through learned associations with basic rewards (Sescousse, Caldú, Segura, & Dreher, 2013). Thus, these regions are involved in reward processing irrespective of the type of reward used.

In the depression and reward literature, reduced responsivity in core subcortical reward regions has been proposed to be a possible endophenotype of MDD as it is consistently found in both current and remitted depression. Using positive words and images as well as monetary rewards, a number of studies have demonstrated reduced activity in MDD in subcortical regions such as the VS, caudate, and putamen that are within the cortical-basal ganglia reward network (although see Knutson, Bhanji, Cooney, Atlas, & Gotlib, 2008; Kumari et al., 2003 for null as well as opposite findings, respectively). For example, in response to positive words (Epstein et al., 2006), pleasant pictures (Heller et al., 2009), and favorite music pieces (Osuch et al., 2009), depressed individuals showed reduced VS activity in comparison to healthy controls. Similarly, with monetary rewards, depressed individuals showed reduced reactivity in the caudate and NAc, a region within the VS, during reward outcome (Pizzagalli et al., 2009) as well as reduced caudate activity during reward selection and anticipation (Smoski et al., 2009). Reduced putamen activity in response to unexpected rewards was also found during reward reversal learning in depression (Robinson, Cools, Carlisi, Sahakian, & Drevets, 2012). This pattern of blunted activity is even present in remitted depression (McCabe, Cowen, & Harmer, 2009). Indeed, a meta-analysis of reward processing in MDD further demonstrated that depression is associated with blunted activity in basal ganglia regions (Zhang, Chang, Guo, Zhang, & Wang, 2013). One possible mechanism explaining the reduced subcortical reward activity seen in depression is that depressed individuals may be unable to sustain activity in the NAc over time when encountering positive stimuli (Heller et al., 2009).

In contrast to the fairly consistent finding of reduced subcortical activity in depression, findings in frontal cortical regions have been more varied.

For example, in response to favorite music pieces, depressed individuals showed reduced activity in the medial OFC in comparison to controls (Osuch et al., 2009). Reduced activity in the OFC was also seen during the cross-modal integration of the flavor and sight of chocolate (McCabe et al., 2009) as well as during monetary rewards (Dichter, Kozink, McClernon, & Smoski, 2012) in those with remitted depression. However, other studies have shown a greater OFC response in depressed individuals versus controls during the selection of stimuli leading to possible monetary reward (Smoski et al., 2009) as well as greater OFC activity in response to positive picture–caption pairs in individuals with treatment-resistant depression (Kumari et al., 2003).

One possible explanation for the discrepant OFC results is that the OFC has a variety of functions. The OFC has both a medial–lateral distinction, such that the medial regions are related to tracking reward value, whereas lateral regions are related to the evaluation of punishers, and a posterior–anterior distinction, such that complex rewards are further represented in more anterior regions than more basic rewards (Kringelbach, 2005). A better understanding of the functions of specific OFC regions as well as increased precision of the reporting of OFC results may help reconcile some of the discrepancies seen in the depression and reward literature.

Similarly, within the ACC, studies have also shown both increased and decreased activity in depressed versus healthy adults. In a study that included both pleasant images and monetary rewards, depressed individuals showed lower activation than healthy controls in the ACC in anticipation of pleasant images as well as money (Smoski, Rittenberg, & Dichter, 2011). Treatment-resistant depression has also been shown to be associated with reduced activity in the ACC (Kumari et al., 2003), and a meta-analysis of pleasant and monetary rewards in depression showed that depression is associated with reduced activity in the ACC (Zhang et al., 2013). However, greater dorsal ACC activation during monetary reward anticipation (Knutson et al., 2008) and increased activity in subgenual ACC has been shown in response to both happy and sad facial expressions in depression (Gotlib et al., 2005). Even in remitted depression, one study found increased ACC activity in response to monetary rewards (Dichter et al., 2012) while another found reduced ACC activity in response to the sight and taste of chocolate (McCabe et al., 2009).

Finally, with respect to the vmPFC, one study found a double dissociation between depressed and healthy controls as well as happy and sad stimuli. In response to sad facial expressions and autobiographical memory prompts, vmPFC activity was decreased in depressed but increased in healthy individuals (Keedwell, Andrew, Williams, Brammer, & Phillips, 2005). This pattern was reversed in response to happy stimuli; specifically, healthy individuals showed decreased vmPFC activity while depressed individuals showed increased vmPFC activity (Keedwell et al., 2005). However, in another study that used the sight and flavor of pleasant chocolate instead of happy stimuli, a reduced vmPFC response was found in those with remitted depression (McCabe et al., 2009). In general, conflicting results within prefrontal regions may be caused by varying definitions of the OFC, ACC, and vmPFC. Prefrontal regions have vastly different and complex anatomical connections to regions, including other prefrontal areas, hippocampus, amygdala, thalamus, and striatum (Haber, 2011). Thus, depression may be characterized by increased activity in certain prefrontal regions as well as decreased activity in others. Indeed, different vmPFC regions have been proposed to underlie positive and negative affect (Myers-Schulz & Koenigs, 2012).

Although the direction of abnormality is inconsistent in some regions within the cortical-basal ganglia network, there is compelling evidence that reduced activity in this network in general is associated with anhedonia. In a study that examined healthy adults, trait anhedonia in a healthy population was shown to be negatively correlated with activity in a number of cortical-basal ganglia reward regions, including NAc, OFC, ACC, and vmPFC in response to pleasant music (Keller et al., 2013). In clinical samples, Stoy et al. (2012) used the monetary reward to examine reward anticipation before and after selective serotonin reuptake inhibitor (SSRI) treatment. Depressed individuals showed reduced VS activity during reward and loss anticipation in comparison to healthy controls before treatment but equivalent activity after treatment. Furthermore, VS hypoactivity was associated with depression severity and anhedonic symptoms in particular. Similarly, Epstein et al. (2006) demonstrated that VS activity was negatively correlated with anhedonia in response to positive versus neutral words. Wacker, Dillon, and Pizzagalli (2009) also demonstrated that anhedonia, but not other depression or anxiety symptoms, was associated with reduced NAc activity during reward outcome, decreased resting activity in rostral ACC, and reduced NAc volume. Anhedonia has also been shown to be associated with reduced caudate volume in major depression (Pizzagalli et al., 2009) and in healthy adults (Harvey, Pruessner, Czechowska, & Lepage, 2007).

In addition to affecting activity in particular brain areas, anhedonia has also been associated with reduced connectivity between core cortical-basal ganglia network regions. Rodent models of depression have shown that appetitive conditioning and reward-related hedonic behavior involve afferent tracks from the mPFC and ventral hippocampus to the NAc (Deisseroth, 2014). In humans, Heller et al. (2009) demonstrated that depression was associated with reduced connectivity between the NAc and mPFC, and this may have contributed to the inability of sustaining NAc activity in depression. In healthy adults, trait anhedonia was examined in relation to connectivity of the NAc alone as well as the conjoint of the NAc and VTA/SN (Keller et al., 2013). This study did not find any significant trait anhedonia-related connectivity when examining the NAc alone but did show that trait anhedonia was negatively related to conjoint NAc and VTA/SN connectivity to the OFC and other paralimbic regions that regulate emotional reactivity to hedonic stimuli. Thus it may be important to examine core cortical-basal ganglia regions together in order to uncover the correlates of anhedonia.

Critically, targeted treatment to the NAc, a core region of the cortical basal-ganglia circuit, has led to the alleviation of anhedonia symptoms. In the first study of deep-brain stimulation to the NAc, spontaneous reports of increased interest were reported after treatment administration in three individuals with treatment-resistant depression (Schlaepfer et al., 2007). In a larger study of deep-brain stimulation to the NAc, the number of hedonic activities increased even after 12 months after the start of treatment (Bewernick et al., 2010). A longitudinal assessment of the sustained effects of deep-brain stimulation to the NAc also showed that after 8−12 months of treatment, the anti-anhedonic effect measured as number of positive activities remained stable (Bewernick, Kayser, Sturm, & Schlaepfer, 2012). Taken together, these studies provide direct evidence that the NAc is related to anhedonia and that treatments targeting this circuit can alleviate anhedonia symptoms.

Although the extant literature implicates the cortical-basal ganglia network in depression and anhedonia specifically, there is much that remains unknown. A variety of rewards have been used to assess network function, and although there are core brain regions that are consistently impacted in depression, there are also discrepant results. Understanding the generalizability of results will be important for conceptualizing depression. Research has also begun to focus on identifying network dysfunctions related to specific symptoms of depression, such as anhedonia, as opposed to the disorder in general. However, many of these studies assess for anhedonia in isolation, and it will be important to disentangle the effects of anhedonia from other symptoms of depression as well as depression severity more broadly. Furthermore, anhedonia itself has been proposed to consist of various types, such as anticipatory and consummatory anhedonia (Treadway & Zald, 2011), and a more refined understanding of how the cortical-basal ganglia network is related to specific subtypes may be particularly fruitful for advancing treatment.

In summary, anhedonia has been linked to deficits in the cortical-basal ganglia reward network, and reduced responsivity in core regions of this network has been proposed to be a possible endophenotype of MDD (Figure 1(f)). Studies using a variety of rewarding stimuli have consistently shown that depression is characterized by reduced activity in core subcortical regions such as the NAc and caudate when responding to positive stimuli. However, findings in frontal regions have been more discrepant, possibly due to varying broad definitions of OFC, ACC, and vmPFC regions. Despite these discrepancies, studies have begun to relate hypoactivity in the reward network specifically to anhedonia. Furthermore, reduced connectivity within the reward network has also been related to anhedonia. Thus, examining core cortical-basal ganglia regions together may be needed to elucidate the neural correlates of anhedonia. Finally, deep-brain stimulation treatment studies directly stimulating the NAc have shown that targeting this neural circuit can alleviate anhedonia symptoms.

CONCLUSIONS

In this chapter we highlighted the relationship between common symptoms of MDD and the activity and connectivity of various neural systems. More specifically, we linked (1) mood-congruent processing to hyperactivity in the amygdala, (2) attention biases toward negative stimuli to the rostral ACC and lateral prefrontal cortex, (3) memory biases for negative material to the amygdala and frontal regions, (4) long-term memory deficits to the hippocampus, (5) rumination to the amygdala and DMN, and (6) anhedonia to the reward network. Focused assessments of depression symptoms combined with network-level analyses have increased our understanding of the neural correlates underlying MDD behavioral abnormalities. Future studies continuing to use this approach will refine our knowledge of depression heterogeneity and aid in the development of more targeted treatments for this common and debilitating illness.

REFERENCES

Abbott, C. C., Lemke, N. T., Gopal, S., Thoma, R. J., Bustillo, J., Calhoun, V. D., & Turner, J. A. (2013). Electroconvulsive therapy response in major depressive disorder: a pilot functional network connectivity resting state fMRI investigation. *Frontiers in Psychiatry, 4*, 10. http://dx.doi.org/10.3389/fpsyt.2013.00010.

Almeida, J. R., Versace, A., Mechelli, A., Hassel, S., Quevedo, K., Kupfer, D. J., & Phillips, M. L. (2009). Abnormal amygdala-prefrontal effective connectivity to happy faces differentiates bipolar from major depression. *Biological Psychiatry, 66*(5), 451–459. http://dx.doi.org/10.1016/j.biopsych.2009.03.024.

American Psychiatric Association. (2000). *Diagnostic and statistical manual of mental disorders* (4th ed.) (Washington DC).

Andreescu, C., Tudorascu, D. L., Butters, M. A., Tamburo, E., Patel, M., Price, J., … Aizenstein, H. (2013). Resting state functional connectivity and treatment response in late-life depression. *Psychiatry Research, 214*(3), 313–321. http://dx.doi.org/10.1016/j.pscychresns.2013.08.007.

Aron, A. R., & Poldrack, R. A. (2005). The cognitive neuroscience of response inhibition: relevance for genetic research in attention-deficit/hyperactivity disorder. *Biological Psychiatry, 57*(11), 1285–1292. http://dx.doi.org/10.1016/j.biopsych.2004.10.026.

Arnold, J. F., Fitzgerald, D. A., Fernandez, G., Rijpkema, M., Rinck, M., Eling, P. A., … Tendolkar, I. (2011). Rose or black-coloured glasses? Altered neural processing of positive events during memory formation is a trait marker of depression. *Journal of Affective Disorders, 131*(1–3), 214–223. http://dx.doi.org/10.1016/j.jad.2010.12.011.

Basso, M. R., & Bornstein, R. A. (1999). Relative memory deficits in recurrent versus first-episode major depression on a word-list learning task. *Neuropsychology, 13*(4), 557–563.

Beck, A. T. (1967). *Depression: Clinical, experimental, and theoretical aspects*. New York: Hoeber Medical Division, Harper & Row.

Beevers, C. G., Clasen, P., Stice, E., & Schnyer, D. (2010). Depression symptoms and cognitive control of emotion cues: a functional magnetic resonance imaging study. *Neuroscience, 167*(1), 97–103. http://dx.doi.org/10.1016/j.neuroscience.2010.01.047.

Berman, M. G., Peltier, S., Nee, D. E., Kross, E., Deldin, P. J., & Jonides, J. (2011). Depression, rumination and the default network. *Social Cognitive and Affective Neuroscience, 6*(5), 548–555. http://dx.doi.org/10.1093/scan/nsq080.

Berridge, K. C., & Kringelbach, M. L. (2008). Affective neuroscience of pleasure: reward in humans and animals. *Psychopharmacology, 199*(3), 457–480. http://dx.doi.org/10.1007/s00213-008-1099-6.

Bewernick, B. H., Hurlemann, R., Matusch, A., Kayser, S., Grubert, C., Hadrysiewicz, B., ... Schlaepfer, T. E. (2010). Nucleus accumbens deep brain stimulation decreases ratings of depression and anxiety in treatment-resistant depression. *Biological Psychiatry, 67*(2), 110–116. http://dx.doi.org/10.1016/j.biopsych.2009.09.013.

Bewernick, B. H., Kayser, S., Sturm, V., & Schlaepfer, T. E. (2012). Long-term effects of nucleus accumbens deep brain stimulation in treatment-resistant depression: evidence for sustained efficacy. *Neuropsychopharmacology, 37*(9), 1975–1985.

Bordi, F., & LeDoux, J. (1992). Sensory tuning beyond the sensory system: an initial analysis of auditory response properties of neurons in the lateral amygdaloid nucleus and overlying areas of the striatum. *Journal of Neuroscience, 12*(7), 2493–2503.

Bradley, B. P., Mogg, K., & Williams, R. (1995). Implicit and explicit memory for emotion-congruent information in clinical depression and anxiety. *Behavior Research and Therapy, 33*(7), 755–770.

Bradley, B. P., Mogg, K., & Millar, N. (1996). Implicit memory bias in clinical and non-clinical depression. *Behaviour Research and Therapy, 34*(11–12), 865–879. http://dx.doi.org/10.1016/S0005-7967(96)00074-5.

Bradley, B. P., Mogg, K., & Lee, S. C. (1997). Attentional biases for negative information in induced and naturally occurring dysphoria. *Behavior Research and Therapy, 35*(10), 911–927.

Bremner, J. D., Vythilingam, M., Vermetten, E., Vaccarino, V., & Charney, D. S. (2004). Deficits in hippocampal and anterior cingulate functioning during verbal declarative memory encoding in midlife major depression. *American Journal of Psychiatry, 161*(4), 637–645.

Browning, M., Holmes, E. A., Murphy, S. E., Goodwin, G. M., & Harmer, C. J. (2010). Lateral prefrontal cortex mediates the cognitive modification of attentional bias. *Biological Psychiatry, 67*(10), 919–925. http://dx.doi.org/10.1016/j.biopsych.2009.10.031.

Buckner, R. L., Andrews-Hanna, J. R., & Schacter, D. L. (2008). The brain's default network: anatomy, function, and relevance to disease. *Annals of the New York Academy of Sciences, 1124*, 1–38. http://dx.doi.org/10.1196/annals.1440.011.

Burt, D. B., Zembar, M. J., & Niederehe, G. (1995). Depression and memory impairment: a meta-analysis of the association, its pattern, and specificity. *Psychological Bulletin, 117*(2), 285–305.

Bush, G., Luu, P., & Posner, M. I. (2000). Cognitive and emotional influences in anterior cingulate cortex. *Trends in Cognitive Sciences, 4*(6), 215–222.

Cahill, L., Babinsky, R., Markowitsch, H. J., & McGaugh, J. L. (1995). The amygdala and emotional memory. *Nature, 377*(6547), 295–296. http://dx.doi.org/10.1038/377295a0.

Cahill, L., & McGaugh, J. L. (1998). Mechanisms of emotional arousal and lasting declarative memory. *Trends in Neurosciences, 21*(7), 294–299. http://dx.doi.org/10.1016/S0166-2236(97)01214-9.

Campbell, S., Marriott, M., Nahmias, C., & MacQueen, G. M. (2004). Lower hippocampal volume in patients suffering from depression: a meta-analysis. *American Journal of Psychiatry, 161*(4), 598–607.

Carballedo, A., Scheuerecker, J., Meisenzahl, E., Schoepf, V., Bokde, A., Moller, H. J., … Frodl, T. (2011). Functional connectivity of emotional processing in depression. *Journal of Affective Disorders, 134*(1–3), 272–279. http://dx.doi.org/10.1016/j.jad.2011.06.021.

Catani, M., Dell'acqua, F., & Thiebaut de Schotten, M. (2013). A revised limbic system model for memory, emotion and behaviour. *Neuroscience & Biobehavioral Reviews, 37*(8), 1724–1737. http://dx.doi.org/10.1016/j.neubiorev.2013.07.001.

Chen, C. H., Suckling, J., Ooi, C., Fu, C. H., Williams, S. C., Walsh, N. D., … Bullmore, E. (2008). Functional coupling of the amygdala in depressed patients treated with antidepressant medication. *Neuropsychopharmacology, 33*(8), 1909–1918. http://dx.doi.org/10.1038/sj.npp.1301593.

Clark, D. M., & Teasdale, J. D. (1982). Diurnal variation in clinical depression and accessibility of memories of positive and negative experiences. *Journal of Abnormal Psychology, 91*(2), 87–95. http://dx.doi.org/10.1037/0021-843X.91.2.87.

Cooney, R. E., Joormann, J., Eugène, F., Dennis, E. L., & Gotlib, I. H. (2010). Neural correlates of rumination in depression. *Cognitive, Affective, & Behavioral Neuroscience, 10*(4), 470–478. http://dx.doi.org/10.3758/CABN.10.4.470.

Dannlowski, U., Kersting, A., Donges, U. S., Lalee-Mentzel, J., Arolt, V., & Suslow, T. (2006). Masked facial affect priming is associated with therapy response in clinical depression. *European Archives of Psychiatry & Clinical Neuroscience, 256*(4), 215–221. http://dx.doi.org/10.1007/s00406-005-0628-0.

Dannlowski, U., Ohrmann, P., Konrad, C., Domschke, K., Bauer, J., Kugel, H., … Suslow, T. (2009). Reduced amygdala-prefrontal coupling in major depression: association with MAOA genotype and illness severity. *International Journal of Neuropsychopharmacology, 12*(1), 11–22. http://dx.doi.org/10.1017/S1461145708008973.

Dastjerdi, M., Foster, B. L., Nasrullah, S., Rauschecker, A. M., Dougherty, R. F., Townsend, J. D., … Parvizi, J. (2011). Differential electrophysiological response during rest, self-referential, and non-self-referential tasks in human posteromedial cortex. *Proceedings of the National Academy of Sciences of the United States of America, 108*(7), 3023–3028. http://dx.doi.org/10.1073/pnas.1017098108.

Degl'Innocenti, A., & Bäckman, L. (1999). Source memory in major depression. *Journal of Affective Disorders, 54*(1–2), 205–209. http://dx.doi.org/10.1016/S0165-0327(98)00167-0.

Deisseroth, K. (2014). Circuit dynamics of adaptive and maladaptive behaviour. *Nature, 505*(7483), 309–317. http://dx.doi.org/10.1038/nature12982.

Denny, E. B., & Hunt, R. R. (1992). Affective valence and memory in depression: dissociation of recall and fragment completion. *Journal of Abnormal Psychology, 101*(3), 575–580.

Dichter, G. S., Felder, J. N., & Smoski, M. J. (2009). Affective context interferes with cognitive control in unipolar depression: an fMRI investigation. *Journal of Affective Disorders, 114*(1–3), 131–142. http://dx.doi.org/10.1016/j.jad.2008.06.027.

Dichter, G. S., Kozink, R. V., McClernon, F. J., & Smoski, M. J. (2012). Remitted major depression is characterized by reward network hyperactivation during reward anticipation and hypoactivation during reward outcomes. *Journal of Affective Disorders, 136*(3), 1126–1134. http://dx.doi.org/10.1016/j.jad.2011.09.048.

Donaldson, C., Lam, D., & Mathews, A. (2007). Rumination and attention in major depression. *Behaviour Research and Therapy, 45*(11), 2664–2678. http://dx.doi.org/10.1016/j.brat.2007.07.002.

Drakeford, J. L., Edelstyn, N. M., Oyebode, F., Srivastava, S., Calthorpe, W. R., & Mukherjee, T. (2010). Recollection deficiencies in patients with major depressive disorder. *Psychiatry Research, 175*(3), 205–210. http://dx.doi.org/10.1016/j.psychres.2008.08.010.

Drevets, W. C., Price, J. L., & Furey, M. L. (2008). Brain structural and functional abnormalities in mood disorders: implications for neurocircuitry models of depression. *Brain Structure & Function, 213*(1–2), 93–118. http://dx.doi.org/10.1007/s00429-008-0189-x.

Eichenbaum, H., Yonelinas, A. P., & Ranganath, C. (2007). The medial temporal lobe and recognition memory. *Annual Review of Neuroscience, 30*, 123–152. http://dx.doi.org/10.1146/annurev.neuro.30.051606.094328.

Elliott, R., Rubinsztein, J. S., Sahakian, B. J., & Dolan, R. J. (2002). The neural basis of mood-congruent processing biases in depression. *Archives of General Psychiatry, 59*(7), 597–604.

van Eijndhoven, P., van Wingen, G., Fernández, G., Rijpkema, M., Verkes, R. J., Buitelaar, J., & Tendolkar, I. (2011). Amygdala responsivity related to memory of emotionally neutral stimuli constitutes a trait factor for depression. *Neuroimage, 54*(2), 1677–1684. http://dx.doi.org/10.1016/j.neuroimage.2010.08.040.

van Eijndhoven, P., van Wingen, G., Fernandez, G., Rijpkema, M., Pop-Purceleanu, M., Verkes, R. J., … Tendolkar, I. (2013). Neural basis of recollection in first-episode major depression. *Human Brain Mapping, 34*(2), 283–294. http://dx.doi.org/10.1002/hbm.21439.

Epstein, J., Pan, H., Kocsis, J. H., Yang, Y., Butler, T., Chusid, J., … Silbersweig, D. A. (2006). Lack of ventral striatal response to positive stimuli in depressed versus normal subjects. *American Journal of Psychiatry, 163*(10), 1784–1790. http://dx.doi.org/10.1176/appi.ajp.163.10.1784.

Etkin, A., Egner, T., & Kalisch, R. (2011). Emotional processing in anterior cingulate and medial prefrontal cortex. *Trends in Cognitive Sciences, 15*(2), 85–93. http://dx.doi.org/10.1016/j.tics.2010.11.004.

Eugène, F., Joormann, J., Cooney, R. E., Atlas, L. Y., & Gotlib, I. H. (2010). Neural correlates of inhibitory deficits in depression. *Psychiatry Research, 181*(1), 30–35. http://dx.doi.org/10.1016/j.pscychresns.2009.07.010.

Fales, C. L., Barch, D. M., Rundle, M. M., Mintun, M. A., Mathews, J., Snyder, A. Z., & Sheline, Y. I. (2009). Antidepressant treatment normalizes hypoactivity in dorsolateral prefrontal cortex during emotional interference processing in major depression. *Journal of Affective Disorders, 112*(1–3), 206–211. http://dx.doi.org/10.1016/j.jad.2008.04.027.

Fossati, P., Harvey, P. O., Le Bastard, G., Ergis, A. M., Jouvent, R., & Allilaire, J. F. (2004). Verbal memory performance of patients with a first depressive episode and patients with unipolar and bipolar recurrent depression. *Journal of Psychiatric Research, 38*(2), 137–144. http://dx.doi.org/10.1016/j.jpsychires.2003.08.002.

Frodl, T., Bokde, A. L., Scheuerecker, J., Lisiecka, D., Schoepf, V., Hampel, H., … Meisenzahl, E. (2010). Functional connectivity bias of the orbitofrontal cortex in drug-free patients with major depression. *Biological Psychiatry, 67*(2), 161–167. http://dx.doi.org/10.1016/j.biopsych.2009.08.022.

Fu, C. H., Williams, S. C., Cleare, A. J., Brammer, M. J., Walsh, N. D., Kim, J., … Bullmore, E. T. (2004). Attenuation of the neural response to sad faces in major depression by antidepressant treatment: a prospective, event-related functional magnetic resonance imaging study. *Archives of General Psychiatry, 61*(9), 877–889. http://dx.doi.org/10.1001/archpsyc.61.9.877.

Fu, C. H. Y., Williams, S. C. R., Cleare, A. J., Scott, J., Mitterschiffthaler, M. T., Walsh, N. D., … Murray, R. M. (2008). Neural responses to sad facial expressions in major depression following cognitive behavioral therapy. *Biological Psychiatry, 64*(6), 505–512. http://dx.doi.org/10.1016/j.biopsych.2008.04.033.

Gorwood, P., Corruble, E., Falissard, B., & Goodwin, G. M. (2008). Toxic effects of depression on brain function: impairment of delayed recall and the cumulative length of depressive disorder in a large sample of depressed outpatients. *American Journal of Psychiatry, 165*(6), 731–739. http://dx.doi.org/10.1176/appi.ajp.2008.07040574.

Gotlib, I. H., Krasnoperova, E., Yue, D. N., & Joormann, J. (2004). Attentional biases for negative interpersonal stimuli in clinical depression. *Journal of Abnormal Psychology, 113*(1), 121–135. http://dx.doi.org/10.1037/0021-843X.113.1.121.

Gotlib, I. H., Sivers, H., Gabrieli, J. D., Whitfield-Gabrieli, S., Goldin, P., Minor, K. L., & Canli, T. (2005). Subgenual anterior cingulate activation to valenced emotional stimuli in major depression. *Neuroreport, 16*(16), 1731–1734.

Greicius, M. D., Flores, B. H., Menon, V., Glover, G. H., Solvason, H. B., Kenna, H., ... Schatzberg, A. F. (2007). Resting-state functional connectivity in major depression: abnormally increased contributions from subgenual cingulate cortex and thalamus. *Biological Psychiatry, 62*(5), 429–437. http://dx.doi.org/10.1016/j.biopsych.2006.09.020.

Grimm, S., Ernst, J., Boesiger, P., Schuepbach, D., Hell, D., Boeker, H., & Northoff, G. (2009). Increased self-focus in major depressive disorder is related to neural abnormalities in subcortical-cortical midline structures. *Human Brain Mapping, 30*(8), 2617–2627. http://dx.doi.org/10.1002/hbm.20693.

Gusnard, D. A., Akbudak, E., Shulman, G. L., & Raichle, M. E. (2001). Medial prefrontal cortex and self-referential mental activity: relation to a default mode of brain function. *Proceedings of the National Academy of Sciences of the United States of America, 98*(7), 4259–4264. http://dx.doi.org/10.1073/pnas.071043098.

Haber, S. N. (2011). Neuroanatomy of reward: a View from the ventral striatum. In J. A. Gottfried (Ed.), *Neurobiology of sensation and reward*. Boca Raton (FL).

Haber, S. N., & Knutson, B. (2010). The reward circuit: linking primate anatomy and human imaging. *Neuropsychopharmacology, 35*(1), 4–26.

Hamilton, J. P., Chen, G., Thomason, M. E., Schwartz, M. E., & Gotlib, I. H. (2011a). Investigating neural primacy in Major Depressive Disorder: multivariate Granger causality analysis of resting-state fMRI time-series data. *Molecular Psychiatry, 16*(7), 763–772. http://dx.doi.org/10.1038/mp.2010.46.

Hamilton, J. P., Furman, D. J., Chang, C., Thomason, M. E., Dennis, E., & Gotlib, I. H. (2011b). Default-mode and task-positive network activity in major depressive disorder: implications for adaptive and maladaptive rumination. *Biological Psychiatry, 70*(4), 327–333. http://dx.doi.org/10.1016/j.biopsych.2011.02.003.

Hamilton, J. P., & Gotlib, I. H. (2008). Neural substrates of increased memory sensitivity for negative stimuli in major depression. *Biological Psychiatry, 63*(12), 1155–1162. http://dx.doi.org/10.1016/j.biopsych.2007.12.015.

Hartlage, S., Alloy, L. B., Vazquez, C., & Dykman, B. (1993). Automatic and effortful processing in depression. *Psychological Bulletin, 113*(2), 247–278.

Harvey, P. O., Fossati, P., Pochon, J. B., Levy, R., LeBastard, G., Lehéricy, S., ... Dubois, B. (2005). Cognitive control and brain resources in major depression: an fMRI study using the n-back task. *Neuroimage, 26*(3), 860–869. http://dx.doi.org/10.1016/j.neuroimage.2005.02.048.

Harvey, P. O., Pruessner, J., Czechowska, Y., & Lepage, M. (2007). Individual differences in trait anhedonia: a structural and functional magnetic resonance imaging study in non-clinical subjects. *Molecular Psychiatry, 12*(8), 703, 767–775. http://dx.doi.org/10.1038/sj.mp.4002021.

Hasler, G., Drevets, W. C., Manji, H. K., & Charney, D. S. (2004). Discovering endophenotypes for major depression. *Neuropsychopharmacology, 29*(10), 1765–1781.

Heller, A. S., Johnstone, T., Shackman, A. J., Light, S. N., Peterson, M. J., Kolden, G. G., ... Davidson, R. J. (2009). Reduced capacity to sustain positive emotion in major depression reflects diminished maintenance of fronto-striatal brain activation. *Proceedings of the National Academy of Sciences, 106*(52), 22445–22450. http://dx.doi.org/10.1073/pnas.0910651106.

Ingram, R. E. (1984). Toward an information-processing analysis of depression. *Cognitive Therapy and Research, 8*(5), 443–477. http://dx.doi.org/10.1007/BF01173284.

Insel, T., Cuthbert, B., Garvey, M., Heinssen, R., Pine, D. S., Quinn, K., ... Wang, P. (2010). Research domain criteria (RDoC): toward a new classification framework for research on mental disorders. *American Journal of Psychiatry, 167*(7), 748–751. http://dx.doi.org/10.1176/appi.ajp.2010.09091379.

Joormann, J., & Gotlib, I. H. (2007). Selective attention to emotional faces following recovery from depression. *Journal Abnormal Psychology, 116*(1), 80–85. http://dx.doi.org/10.1037/0021-843X.116.1.80.

Keedwell, P. A., Andrew, C., Williams, S. C., Brammer, M. J., & Phillips, M. L. (2005). A double dissociation of ventromedial prefrontal cortical responses to sad and happy stimuli in depressed and healthy individuals. *Biological Psychiatry, 58*(6), 495–503. http://dx.doi.org/10.1016/j.biopsych.2005.04.035.

Keller, J., Young, C. B., Kelley, E., Prater, K., Levitin, D. J., & Menon, V. (2013). Trait anhedonia is associated with reduced reactivity and connectivity of mesolimbic and paralimbic reward pathways. *Journal of Psychiatric Research, 47*(10), 1319–1328. http://dx.doi.org/10.1016/j.jpsychires.2013.05.015.

Kellough, J. L., Beevers, C. G., Ellis, A. J., & Wells, T. T. (2008). Time course of selective attention in clinically depressed young adults: an eye tracking study. *Behavior Research and Therapy, 46*(11), 1238–1243. http://dx.doi.org/10.1016/j.brat.2008.07.004.

Knutson, B., Bhanji, J. P., Cooney, R. E., Atlas, L. Y., & Gotlib, I. H. (2008). Neural responses to monetary incentives in major depression. *Biological Psychiatry, 63*(7), 686–692. http://dx.doi.org/10.1016/j.biopsych.2007.07.023.

Knutson, B., Westdorp, A., Kaiser, E., & Hommer, D. (2000). FMRI visualization of brain activity during a monetary incentive delay task. *Neuroimage, 12*(1), 20–27. http://dx.doi.org/10.1006/nimg.2000.0593.

Koelsch, S., Fritz, T., V Cramon, D. Y., Müller, K., & Friederici, A. D. (2006). Investigating emotion with music: an fMRI study. *Human Brain Mapping, 27*(3), 239–250. http://dx.doi.org/10.1002/hbm.20180.

Kringelbach, M. L. (2005). The human orbitofrontal cortex: linking reward to hedonic experience. *Nature Reviews Neuroscience, 6*(9), 691–702. http://dx.doi.org/10.1038/nrn1747.

Kumari, V., Mitterschiffthaler, M. T., Teasdale, J. D., Malhi, G. S., Brown, R. G., Giampietro, V., ... Sharma, T. (2003). Neural abnormalities during cognitive generation of affect in treatment-resistant depression. *Biological Psychiatry, 54*(8), 777–791.

Lewinsohn, P. M., Pettit, J. W., Joiner, T. E., Jr., & Seeley, J. R. (2003). The symptomatic expression of major depressive disorder in adolescents and young adults. *Journal of Abnormal Psychology, 112*(2), 244–252.

Leyman, L., De Raedt, R., Schacht, R., & Koster, E. H. (2007). Attentional biases for angry faces in unipolar depression. *Psychology Medicine, 37*(3), 393–402. http://dx.doi.org/10.1017/S003329170600910X.

Liddell, B. J., Brown, K. J., Kemp, A. H., Barton, M. J., Das, P., Peduto, A., ... Williams, L. M. (2005). A direct brainstem-amygdala-cortical 'alarm' system for subliminal signals of fear. *Neuroimage, 24*(1), 235–243. http://dx.doi.org/10.1016/j.neuroimage.2004.08.016.

Lloyd, G. G., & Lishman, W. A. (1975). Effect of depression on the speed of recall of pleasant and unpleasant experiences. *Psychological Medicine, 5*(02), 173–180. http://dx.doi.org/10.1017/S0033291700056440.

MacQueen, G. M., Gakway, T. M., Hay, J., Young, L. T., & Joffe, R. T. (2002). Recollection memory deficits in patients with major depressive disorder predicted by past depressions but not current mood state or treatment status. *Psychological Medicine, 32*(2), 251–258. http://dx.doi.org/10.1017/S0033291701004834.

MacQueen, G. M., Campbell, S., McEwen, B. S., Macdonald, K., Amano, S., Joffe, R. T., ... Young, L. T. (2003). Course of illness, hippocampal function, and hippocampal volume in major depression. *Proceedings of the National Academy of Sciences of the United States of America, 100*(3), 1387–1392. http://dx.doi.org/10.1073/pnas.0337481100.

Mandler, G. (1980). *Recognizing: The judgment of previous occurrence*. American Psychological Association.

Matsuo, K., Glahn, D. C., Peluso, M. A., Hatch, J. P., Monkul, E. S., Najt, P., ... Soares, J. C. (2007). Prefrontal hyperactivation during working memory task in untreated individuals with major depressive disorder. *Molecular Psychiatry, 12*(2), 158–166. http://dx.doi.org/10.1038/sj.mp.4001894.

Matthews, S. C., Strigo, I. A., Simmons, A. N., Yang, T. T., & Paulus, M. P. (2008). Decreased functional coupling of the amygdala and supragenual cingulate is related to increased depression in unmedicated individuals with current major depressive disorder. *Journal of Affective Disorders, 111*(1), 13–20. http://dx.doi.org/10.1016/j.jad.2008.05.022.

Mayberg, H. S. (1997). Limbic-cortical dysregulation: a proposed model of depression. *Journal of Neuropsychiatry and Clinical Neurosciences, 9*(3), 471–481.

McCabe, C., Cowen, P. J., & Harmer, C. J. (2009). Neural representation of reward in recovered depressed patients. *Psychopharmacology (Berlin), 205*(4), 667–677. http://dx.doi.org/10.1007/s00213-009-1573-9.

Menon, V. (2011). Large-scale brain networks and psychopathology: a unifying triple network model. *Trends in Cognitive Sciences, 15*(10), 483–506. http://dx.doi.org/10.1016/j.tics.2011.08.003.

Menon, V., & Levitin, D. J. (2005). The rewards of music listening: response and physiological connectivity of the mesolimbic system. *Neuroimage, 28*(1), 175–184. http://dx.doi.org/10.1016/j.neuroimage.2005.05.053.

Mesulam, M. (2000). Behavioural neuroanatomy: large-scale networks, association cortex, frontal syndromes, the limbic system, and the hemispheric specializations. In M. Mesulam (Ed.), *Principles of behavioural and cognitive neurology* (pp. 1–120).

Mitterschiffthaler, M. T., Williams, S. C., Walsh, N. D., Cleare, A. J., Donaldson, C., Scott, J., & Fu, C. H. (2008). Neural basis of the emotional Stroop interference effect in major depression. *Psychological Medicine, 38*(2), 247–256. http://dx.doi.org/10.1017/S0033291707001523.

Mobbs, D., Greicius, M. D., Abdel-Azim, E., Menon, V., & Reiss, A. L. (2003). Humor modulates the mesolimbic reward centers. *Neuron, 40*(5), 1041–1048. http://dx.doi.org/10.1016/S0896-6273(03)00751-7.

Mogg, K., Bradley, B. P., & Williams, R. (1995). Attentional bias in anxiety and depression: the role of awareness. *British Journal of Clinical Psychology, 34*(1), 17–36. http://dx.doi.org/10.1111/j.2044-8260.1995.tb01434.x.

Morris, J. S., Ohman, A., & Dolan, R. J. (1999). A subcortical pathway to the right amygdala mediating "unseen" fear. *Proceedings of the National Academy of Sciences of the United States of America, 96*(4), 1680–1685.

Myers-Schulz, B., & Koenigs, M. (2012). Functional anatomy of ventromedial prefrontal cortex: Implications for mood and anxiety disorders. *Molecular Psychiatry, 17*(2), 132–141. http://dx.doi.org/10.1038/mp.2011.88.

Nee, D. E., Wager, T. D., & Jonides, J. (2007). Interference resolution: insights from a meta-analysis of neuroimaging tasks. *Cognitive, Affective & Behavioral Neuroscience, 7*(1), 1–17.

Nolen-Hoeksema, S., Morrow, J., & Fredrickson, B. L. (1993). Response styles and the duration of episodes of depressed mood. *Journal of Abnormal Psychology, 102*(1), 20–28.

Nolen-Hoeksema, S., Wisco, B. E., & Lyubomirsky, S. (2008). Rethinking rumination. *Perspectives on Psychological Science, 3*(5), 400–424. http://dx.doi.org/10.1111/j.1745-6924.2008.00088.x.

Öngur, D., Ferry, A. T., & Price, J. L. (2003). Architectonic subdivision of the human orbital and medial prefrontal cortex. *Journal of Comparative Neurology, 460*(3), 425–449. http://dx.doi.org/10.1002/cne.10609.

Ongur, D., & Price, J. L. (2000). The organization of networks within the orbital and medial prefrontal cortex of rats, monkeys and humans. *Cerebral Cortex, 10*(3), 206–219.

Osuch, E. A., Bluhm, R. L., Williamson, P. C., Theberge, J., Densmore, M., & Neufeld, R. W. (2009). Brain activation to favorite music in healthy controls and depressed patients. *Neuroreport, 20*(13), 1204–1208. http://dx.doi.org/10.1097/WNR.0b013e32832f4da3.

Pelizza, L., & Ferrari, A. (2009). Anhedonia in schizophrenia and major depression: state or trait? *Annals of General Psychiatry, 8*, 22. http://dx.doi.org/10.1186/1744-859X-8-22.

Peluso, M. A., Glahn, D. C., Matsuo, K., Monkul, E. S., Najt, P., Zamarripa, F., ... Soares, J. C. (2009). Amygdala hyperactivation in untreated depressed individuals. *Psychiatry Research, 173*(2), 158–161. http://dx.doi.org/10.1016/j.pscychresns.2009.03.006.

Phillips, M. L., Drevets, W. C., Rauch, S. L., & Lane, R. (2003). Neurobiology of emotion perception II: Implications for major psychiatric disorders. *Biological Psychiatry, 54*(5), 515–528.

Pizzagalli, D. A., Holmes, A. J., Dillon, D. G., Goetz, E. L., Birk, J. L., Bogdan, R., ... Fava, M. (2009). Reduced caudate and nucleus accumbens response to rewards in unmedicated individuals with major depressive disorder. *American Journal of Psychiatry, 166*(6), 702–710. http://dx.doi.org/10.1176/appi.ajp.2008.08081201.

Raes, F., Hermans, D., Williams, J. M., Beyers, W., Brunfaut, E., & Eelen, P. (2006). Reduced autobiographical memory specificity and rumination in predicting the course of depression. *Journal of Abnormal Psychology, 115*(4), 699–704. http://dx.doi.org/10.1037/0021-843X.115.4.699.

Ramel, W., Goldin, P. R., Eyler, L. T., Brown, G. G., Gotlib, I. H., & McQuaid, J. R. (2007). Amygdala reactivity and mood-congruent memory in individuals at risk for depressive relapse. *Biological Psychiatry, 61*(2), 231–239. http://dx.doi.org/10.1016/j.biopsych.2006.05.004.

Ramponi, C., Barnard, P., & Nimmo-Smith, I. (2004). Recollection deficits in dysphoric mood: an effect of schematic models and executive mode? *Memory, 12*(5), 655–670. http://dx.doi.org/10.1080/09658210344000189.

Roberts, J. E., Gilboa, E., & Gotlib, I. H. (1998). Ruminative response style and vulnerability to episodes of dysphoria: gender, neuroticism, and episode duration. *Cognitive Therapy and Research, 22*(4), 401–423. http://dx.doi.org/10.1023/A:1018713313894.

Robinson, O. J., Cools, R., Carlisi, C. O., Sahakian, B. J., & Drevets, W. C. (2012). Ventral striatum response during reward and punishment reversal learning in unmedicated major depressive disorder. *American Journal of Psychiatry, 169*(2), 152–159.

Schlaepfer, T. E., Cohen, M. X., Frick, C., Kosel, M., Brodesser, D., Axmacher, N., ... Sturm, V. (2007). Deep brain stimulation to reward circuitry alleviates anhedonia in refractory major depression. *Neuropsychopharmacology, 33*(2), 368–377.

Sescousse, G., Caldú, X., Segura, B., & Dreher, J. C. (2013). Processing of primary and secondary rewards: a quantitative meta-analysis and review of human functional neuroimaging studies. *Neuroscience & Biobehavioral Reviews, 37*(4), 681–696. http://dx.doi.org/10.1016/j.neubiorev.2013.02.002.

Sestieri, C., Corbetta, M., Romani, G. L., & Shulman, G. L. (2011). Episodic memory retrieval, parietal cortex, and the default mode network: functional and topographic analyses. *Journal of Neuroscience, 31*(12), 4407–4420. http://dx.doi.org/10.1523/JNEUROSCI.3335-10.2011.

Shafritz, K. M., Collins, S. H., & Blumberg, H. P. (2006). The interaction of emotional and cognitive neural systems in emotionally guided response inhibition. *Neuroimage, 31*(1), 468–475. http://dx.doi.org/10.1016/j.neuroimage.2005.11.053.

Sheline, Y. I., Barch, D. M., Donnelly, J. M., Ollinger, J. M., Snyder, A. Z., & Mintun, M. A. (2001). Increased amygdala response to masked emotional faces in depressed subjects resolves with antidepressant treatment: an fMRI study. *Biological Psychiatry, 50*(9), 651–658.

Siegle, G. J., Steinhauer, S. R., Thase, M. E., Stenger, V. A., & Carter, C. S. (2002). Can't shake that feeling: event-related fMRI assessment of sustained amygdala activity in response to emotional information in depressed individuals. *Biological Psychiatry, 51*(9), 693–707. http://dx.doi.org/10.1016/S0006-3223(02)01314-8.

Small, D. M., Zatorre, R. J., Dagher, A., Evans, A. C., & Jones-Gotman, M. (2001). Changes in brain activity related to eating chocolate: from pleasure to aversion. *Brain, 124*(Pt 9), 1720–1733.

Smoski, M. J., Felder, J., Bizzell, J., Green, S. R., Ernst, M., Lynch, T. R., & Dichter, G. S. (2009). fMRI of alterations in reward selection, anticipation, and feedback in major depressive disorder. *Journal of Affective Disorders, 118*(1–3), 69–78. http://dx.doi.org/10.1016/j.jad.2009.01.034.

Smoski, M. J., Rittenberg, A., & Dichter, G. S. (2011). Major depressive disorder is characterized by greater reward network activation to monetary than pleasant image rewards. *Psychiatry Research, 194*(3), 263–270. http://dx.doi.org/10.1016/j.pscychresns.2011.06.012.

Spijker, J., Bijl, R. V., de Graaf, R., & Nolen, W. A. (2001). Determinants of poor 1-year outcome of DSM-III-R major depression in the general population: results of the Netherlands Mental Health Survey and Incidence Study (NEMESIS). *Acta Psychiatriac Scandinavica, 103*(2), 122–130.

Spreng, R. N., Mar, R. A., & Kim, A. S. (2009). The common neural basis of autobiographical memory, prospection, navigation, theory of mind, and the default mode: a quantitative meta-analysis. *Journal of Cognitive Neurosciences, 21*(3), 489–510. http://dx.doi.org/10.1162/jocn.2008.21029.

Stoy, M., Schlagenhauf, F., Sterzer, P., Bermpohl, F., Hagele, C., Suchotzki, K., ... Strohle, A. (2012). Hyporeactivity of ventral striatum towards incentive stimuli in unmedicated depressed patients normalizes after treatment with escitalopram. *Journal of Psychopharmacology, 26*(5), 677–688. http://dx.doi.org/10.1177/0269881111416686.

Stuhrmann, A., Suslow, T., & Dannlowski, U. (2011). Facial emotion processing in major depression: a systematic review of neuroimaging findings. *Biology of Mood & Anxiety Disorders, 1*(1), 10. http://dx.doi.org/10.1186/2045-5380-1-10.

Surguladze, S., Brammer, M. J., Keedwell, P., Giampietro, V., Young, A. W., Travis, M. J., ... Phillips, M. L. (2005). A differential pattern of neural response toward sad versus happy facial expressions in major depressive disorder. *Biological Psychiatry, 57*(3), 201–209. http://dx.doi.org/10.1016/j.biopsych.2004.10.028.

Suslow, T., Konrad, C., Kugel, H., Rumstadt, D., Zwitserlood, P., Schöning, S., ... Dannlowski, U. (2010). Automatic mood-congruent amygdala responses to masked facial expressions in major depression. *Biological Psychiatry, 67*(2), 155–160. http://dx.doi.org/10.1016/j.biopsych.2009.07.023.

Teasdale, J. D. (1983). Negative thinking in depression: cause, effect, or reciprocal relationship. *Advances in Behaviour Research and Therapy, 5*(1), 3–25. http://dx.doi.org/10.1016/0146-6402(83)90013-9.

Teasdale, J. D., & Dent, J. (1987). Cognitive vulnerability to depression: an investigation of two hypotheses. *British Journal of Clinical Psychology, 26*(Pt 2), 113–126.

Tendolkar, I., Arnold, J., Petersson, K. M., Weis, S., Anke, B. D., van Eijndhoven, P., ... Fernández, G. (2007). Probing the neural correlates of associative memory formation: a parametrically analyzed event-related functional MRI study. *Brain Research, 1142*(0), 159–168. http://dx.doi.org/10.1016/j.brainres.2007.01.040.

Treadway, M. T., Bossaller, N. A., Shelton, R. C., & Zald, D. H. (2012). Effort-based decision-making in major depressive disorder: a translational model of motivational anhedonia. *Journal of Abnormal Psychology, 121*(3), 553–558. http://dx.doi.org/10.1037/a0028813.

Treadway, M. T., & Zald, D. H. (2011). Reconsidering anhedonia in depression: lessons from translational neuroscience. *Neuroscience and Biobehavioral Reviews, 35*(3), 537–555. http://dx.doi.org/10.1016/j.neubiorev.2010.06.006.

van Wingen, G. A., van Eijndhoven, P., Cremers, H. R., Tendolkar, I., Verkes, R. J., Buitelaar, J. K., & Fernandez, G. (2010). Neural state and trait bases of mood-incongruent memory formation and retrieval in first-episode major depression. *Journal of Psychiatric Research, 44*(8), 527–534. http://dx.doi.org/10.1016/j.jpsychires.2009.11.009.

Victor, T. A., Furey, M. L., Fromm, S. J., Ohman, A., & Drevets, W. C. (2010). Relationship between amygdala responses to masked faces and mood state and treatment in major depressive disorder. *Archives of General Psychiatry, 67*(11), 1128–1138. http://dx.doi.org/10.1001/archgenpsychiatry.2010.144.

Wacker, J., Dillon, D. G., & Pizzagalli, D. A. (2009). The role of the nucleus accumbens and rostral anterior cingulate cortex in anhedonia: integration of resting EEG, fMRI, and volumetric techniques. *Neuroimage, 46*(1), 327–337. http://dx.doi.org/10.1016/j.neuroimage.2009.01.058.

Walsh, N. D., Williams, S. C. R., Brammer, M. J., Bullmore, E. T., Kim, J., Suckling, J., ... Fu, C. H. Y. (2007). A longitudinal functional magnetic resonance imaging study of verbal working memory in depression after antidepressant therapy. *Biological Psychiatry, 62*(11), 1236–1243. http://dx.doi.org/10.1016/j.biopsych.2006.12.022.

Walter, H., Wolf, R. C., Spitzer, M., & Vasic, N. (2007). Increased left prefrontal activation in patients with unipolar depression: an event-related, parametric, performance-controlled fMRI study. *Journal of Affective Disorders, 101*(1–3), 175–185. http://dx.doi.org/10.1016/j.jad.2006.11.017.

Wang, L., LaBar, K. S., Smoski, M., Rosenthal, M. Z., Dolcos, F., Lynch, T. R., ... McCarthy, G. (2008). Prefrontal mechanisms for executive control over emotional distraction are altered in major depression. *Psychiatry Research, 163*(2), 143–155. http://dx.doi.org/10.1016/j.pscychresns.2007.10.004.

Watkins, P. C., Vache, K., Verney, S. P., & Mathews, A. (1996). Unconscious mood-congruent memory bias in depression. *Journal of Abnormal Psychology, 105*(1), 34–41. http://dx.doi.org/10.1037/0021-843X.105.1.34.

Whalen, P. J., Shin, L. M., Somerville, L. H., McLean, A. A., & Kim, H. (2002). Functional neuroimaging studies of the amygdala in depression. *Seminars in Clinical Neuropsychiatry, 7*(4), 234–242.

Williams, J. M. G., & Scott, J. (1988). Autobiographical memory in depression. *Psychological Medicine, 18*(03), 689–695. http://dx.doi.org/10.1017/S0033291700008370.

Wu, M., Andreescu, C., Butters, M. A., Tamburo, R., Reynolds, C. F., 3rd, & Aizenstein, H. (2011). Default-mode network connectivity and white matter burden in late-life depression. *Psychiatry Research, 194*(1), 39–46. http://dx.doi.org/10.1016/j.pscychresns.2011.04.003.

Yonelinas, A. P. (2001). Components of episodic memory: the contribution of recollection and familiarity. *Philosophical Transactions of the Royal Society of London B Biological Sciences, 356*(1413), 1363–1374. http://dx.doi.org/10.1098/rstb.2001.0939.

Zhang, W. N., Chang, S. H., Guo, L. Y., Zhang, K. L., & Wang, J. (2013). The neural correlates of reward-related processing in major depressive disorder: a meta-analysis of functional magnetic resonance imaging studies. *Journal of Affective Disorders, 151*(2), 531–539. http://dx.doi.org/10.1016/j.jad.2013.06.039.

Zhong, M., Wang, X., Xiao, J., Yi, J., Zhu, X., Liao, J., ... Yao, S. (2011). Amygdala hyperactivation and prefrontal hypoactivation in subjects with cognitive vulnerability to depression. *Biological Psychology, 88*(2–3), 233–242. http://dx.doi.org/10.1016/j.biopsycho.2011.08.007.

Chapter 9

Integrating the Stress Systems and Neuroimaging in Depression

Thomas Frodl[1,2], Veronica O'Keane[2]
[1]*Department of Psychiatry and Psychotherapy, Otto von Guericke University, Magdeburg, Germany;* [2]*Department of Psychiatry, Trinity College Dublin, Ireland*

INTRODUCTION

Integrated Neuroimaging in the Context of System Neuroscience.

Major depressive disorder (MDD) has until recently been conceptualized as an episodic disorder associated with "chemical imbalances" in the central nervous system but no permanent brain changes. Evidence has emerged suggesting that MDD is associated with smaller hippocampal volumes that may occur in the context of changes in other discrete regions, such as the cingulate cortex, in the absence of global morphological abnormalities. Evidence from neuroimaging, neuropathological, and lesion analysis studies further implicates limbic-cortical-striatal-pallidal-thalamic circuits, including prefrontal cortex, amygdala, ventromedial striatum, mediodorsal and midline thalamic nuclei, and ventral pallidum, in the pathophysiology of mood disorders (Miller, Saint Marie, Breier, & Swerdlow, 2010).

Changes in Brain Structure in MDD

Many structural imaging studies have reported that the hippocampus is small in patients with MDD. A meta-analysis including 9533 patients with MDD and 8846 controls found increased lateral ventricle and *cerebrospinal fluid* volume and a reduced volume of the basal ganglia, thalamus, hippocampus, frontal lobe, orbitofrontal cortex, and gyrus rectus in patients with MDD as compared to healthy controls (Kempton et al., 2011). Other meta-analyses were specific to the hippocampus and confirmed that patients had hippocampus volumes that were approximately 4–6% smaller than those of matched control subjects on the left and the right (Campbell, Marriott,

Systems Neuroscience in Depression. http://dx.doi.org/10.1016/B978-0-12-802456-0.00009-1

Nahmias, & MacQueen, 2004; McKinnon, Yucel, Nazarov, & MacQueen, 2009; Videbech & Ravnkilde, 2004). There are a few postmortem studies that have examined the hippocampus specifically in patients with MDD. In one study of 19 subjects with MDD and 21 normal control subjects, Stockmeier et al. (2004) examined the density of pyramidal neurons, dentate granule cell neurons, glia, and the size of the neuronal soma area. They reported that the density of granula cells and glia in the dentate gyrus and pyramidal neurons and glia in all cornu ammonis/hippocampal subfields was significantly increased by 30–35% in MDD, while the average soma size of pyramidal neurons was significantly decreased in patients with MDD. They suggested that a reduction in neuropil in MDD may account for decreased hippocampus volume detected by neuroimaging.

Changes in Functional Circuits in MDD

As a recurrent psychiatric disorder, functional magnetic resonance imaging (fMRI) is particularly important in order to investigate neural circuits involved in depressive states and those associated with vulnerability for MDD. Changes in brain function on a system neuroscience level are presented in a special chapter of the book by Tendolkar et al.

Resting-state fMRI that assesses neural activity during wakeful rest has been introduced as a candidate diagnostic biomarker in MDD. Network-based neuroimaging approaches thus far have mainly been limited to identifying four main networks in order to model possible changes in anxiety disorders and depression. The default mode network (DMN) consists of an anterior part, which covers the medial prefrontal cortex, including parts of the anterior cingulate cortex (ACC), and a posterior part, including the posterior cingulate cortex, precuneus, inferior parietal cortex, hippocampal area, and lateral temporal cortex (Buckner, Andrews-Hanna, & Schacter, 2008). Though most active during rest, activity in the DMN has been shown to be decreased during demanding cognitive tasks, and the DMN has been related to spontaneous cognition. The DMN was found to be related to emotion regulation, self-reference, and obsessive ruminations (Gusnard, Akbudak, Shulman, & Raichle, 2001). Its anterior regions are mostly related to self-referential activity and have strong connections to limbic regions such as the amygdala. The posterior cingulate cortex/precuneus is linked to arousal and awareness, and the conscious retrieval of past events is related to connections from here to the hippocampus and lateral temporal cortex.

The salience network (SN) consists of the frontoinsular cortex, dorsal ACC, and temporal poles and activates to various salient stimuli. It's hypothesized to play a central role in emotional control due to extensive subcortical connectivity (Seeley et al., 2007).

The affective network contains integrated regions of the affective subdivision of the ACC, amygdala, nucleus accumbens, hypothalamus, anterior insula, hippocampus, and orbitofrontal cortex (OFC) with reciprocal connections to autonomic, visceromotor, and endocrine systems (Bush, Luu, & Posner, 2000; Öngür, Ferry, & Price, 2003; Sheline, Price, Yan, & Mintun, 2010). This network is involved in emotional regulation and monitoring the salience of motivational stimuli (Fox, Corbetta, Snyder, Vincent, & Raichle, 2006; Sheline et al., 2010; Bush et al., 2000) and thus is linked to the SN.

Finally, often considered to be the opposite of the DMN, the cognitive-executive network (CEN) includes the lateral prefrontal cortex, lateral parietal cortex, and a part of the dorsomedial prefrontal cortex. This network is most active during cognitive tasks and is implicated in cognitive functioning, including attention and working memory. It is anticorrelated to the DMN, which means that when one is active the other is inactivated.

The majority of task-related fMRI studies in depression have targeted emotion processing brain networks through the use of affect-laden stimuli like the emotional face matching task. A "top-down" and "bottom-up" dysregulated prefrontal-subcortical circuitry has been suggested to underlie the failure to regulate mood in patients with MDD (Ochsner et al., 2004). Research on functional connectivity is in its infancy as a research tool. We have demonstrated frontolimbic functional dysconnections during emotional face matching in medication-free patients with MDD as compared to controls (Frodl, Bokde, et al., 2010), and these changes were found to be associated with treatment response (Lisiecka et al., 2011). Other studies have demonstrated that in depression increased levels of physiological activity within the DMN might be present (Price & Drevets, 2012). The CEN shows reduced functional connectivity in MDD (Brzezicka, 2013; Sylvester et al., 2013), and the SN was also found to be altered in MDD (Peterson, Thome, Frewen, & Lanius, 2014). The main findings in a meta-analysis, including 32 resting state fMRI studies, were hyperactivity/hyperconnectivity in midline structures related to the DMN and in lateral frontal and parietal areas related to the CEN (Sundermann, Olde Lutke Beverborg, & Pfleiderer, 2014).

Structural Connectivity

Some of the functional changes may be a consequence of structural changes, in particular changes in white matter fiber systems. Magnetic resonance diffusion tensor imaging (DTI) is a novel neuroimaging technique that can evaluate both the orientation and the diffusion characteristics of white matter tracts in vivo (Sexton, MacKay, & Ebmeier, 2009), and tractography methods allow the extraction of white matter fiber bundles relevant to MDD. Fractional anisotropy (FA) is a measure of the longitudinal diffusion, in relation to the perpendicular diffusion, in white matter tracts, often related to neural integrity. Interestingly, a significant positive correlation has been found between the average FA value of the cingulum tract and the level of functional connectivity within the DMN (Van Den Heuvel, Mandl, Luigjes, & Hulshoff POL, 2008). Thus changes in structural connectivity may provide extra parameters in the attempt to form connectivity signatures.

A meta-analysis of DTI studies of patients with MDD identified decreased fractional anisotropy in the white matter fascicles connecting the prefrontal cortex within frontal, temporal, and occipital lobes and amygdala and hippocampus (Liao et al., 2013). We reviewed the literature in DTI imaging in MDD and found, in our meta-analysis, a significant reduction in FA in the left superior longitudinal fasciculus in MDD (Murphy & Frodl, 2011). Moreover, childhood adversity measures were statistically related to the cingulum, uncinated fasciculus, and fronto-occipital fasciculus white matter tracts (Ugwu, Amico, Carballedo, Fagan, & Frodl, 2014).

Overall, however, findings from structural connectivity analyses show heterogeneous results without a clear definition of white matter pathways. Therefore, more research is needed, in particular to identify those altered white matter connections associated with common and more distinct psychopathology.

Neurometabolics

Finally, some of the biological underpinnings of altered functional connectivity might also be explored by identifying their associations with neuronal metabolic (neurochemical) and glial-related changes measured using MR spectroscopy. Quantitative in vivo MR spectroscopy allows for the assessment of neuronal and glial markers. Glutamate, glutamine, *N*-acetylaspartate (NAA), and gabaergic levels seem to impact blood oxygen level-dependent (BOLD) responses (Walter et al., 2009). Most interestingly, the glutamine deficit that has been reported in melancholic/high-anhedonia MDD seems to be related to smaller BOLD responses in pregenual ACC (pgACC; Walter et al., 2009), a key player within the DMN. Furthermore, Horn et al. (2010) reported that glutamatergic reductions in pgACC were related to depression severity and to deviant functional connectivity between the DMN and the SN, namely, between the pgACC and anterior insula. Importantly, these changes in functional connectivity have not yet been related to metabolites in the insula itself. While there is converging evidence for reduced NAA levels and total concentration of glutamate and glutamine in the pgACC (Ende, Demirakca, & Tost, 2006), y-aminobutyric acid levels remain unchanged in MDD (Hasler et al., 2007). A meta-analysis found that ACC reduction in glutamate in patients with MDD was of sufficient effect size to discriminate them from bipolar patients (Taylor, 2014).

How Can Biological System Changes and Neuroimaging Be Integrated?

Early life stress leads to epigenetic modifications in chromatin regulatory sites. The epigenetic changes alter gene expression and functions, affecting synaptic plasticity and glial cell functioning. These changes form the basis for large-scale functional disruptions of neuronal networks and metabolic brain changes in

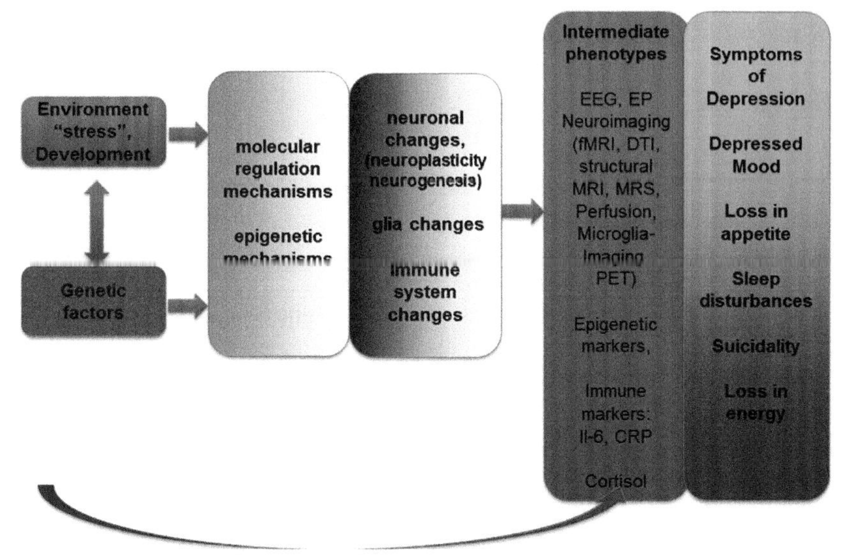

FIGURE 1 Early life stress leads to epigenetic modifications in chromatin regulatory sites. The epigenetic changes alter gene expression and functions, affecting synaptic plasticity and glial cell functioning. These changes form the basis for large-scale functional disruptions of neuronal networks and metabolic brain changes in humans with neuropsychiatric disorders.

humans with neuropsychiatric disorders (Figure 1). However, it is unknown which molecular and epigenetic changes are associated with specific or unspecific disruptions of brain function and metabolics in psychiatric pathologies, rendering diagnostic and prognostic exploitation of potential biomarkers from imaging and molecular specimens difficult. Therefore, we will review the stress system, inflammation system, neurotrophic system, and neurotransmitter system and their association with functional and structural brain changes.

DOES THE HYPOTHALAMUS-PITUITARY-ADRENAL (HPA) AXIS STRESS SYSTEM MEDIATE THE MORPHOLOGICAL AND FUNCTIONAL BRAIN CHANGES?

Stress is believed to play an important role in the pathogenesis of MDD. MDD is clinically recognized as a highly stress-sensitive illness (Kessler, 1997), and the hippocampus, which is altered in MDD, is a highly stress-sensitive brain region (Thomas, Hotsenpiller, & Peterson, 2007). There is much debate about the effect of early life adversity and its possible associations with the specific course of stress-related and depressive illnesses, long-lasting emotional problems (Carboni et al., 2010; Heim & Nemeroff, 2001; Mann & Currier, 2010), and hippocampal volumetric changes (Chen, Hamilton, & Gotlib, 2010).

Chronic stress can increase rates of depression in susceptible individuals, but the detailed pathophysiology underlying this process remains unknown (Tsankova et al., 2006).

The HPA Axis and Homeostasis

To a large extent, how our bodies respond to stressors determines our overall health. Physiological function is altered in response to acute stressors but with the ultimate outcome of returning to homeostatic, or baseline, control of these systems. This adaptive process is known as allostasis. If the stressor is excessive, or our stress systems are faulty, the strain exerted on our physiological systems by chronic stress prevents a return to a healthy state of homeostasis. The pathophysiological alteration in our stress-sensitive neuroendocrine, cardiovascular, immune, and neural systems brought about by this excessive strain is known as the allostatic load (Juster, McEwen, & Lupien, 2010). It is important to note that in order to keep our physiological environment relatively constant stress responses are necessary to continually adapt to an ever-changing environment.

The brain is the main regulatory organ for stress responses. The HPA axis, the major stress system in the body, is a neuroendocrine system involved in the production of the stress hormone cortisol by the adrenal glands. Cortisol is a glucocorticoid, so-called because it alters the function of numerous tissues in order to mobilize energy and change systems functionally to meet the demands of the stress challenge (De Kloet, Joëls, & Holsboer, 2005). Among the many processes affected by cortisol are glucose and fat metabolism, bone metabolism, cardiovascular responsiveness, and immune function. Glucorticoids also modify brain functions via two nuclear receptors that also function as transcriptional factors: the high-affinity glucocorticoid receptor (GR) in the hippocampus and the low-affinity mineralocorticoid receptor (MR) distributed throughout the brain (De Kloet, Joëls, & Holsboer, 2005; Heim & Nemeroff, 2001; Oberlander et al., 2008, Weaver et al., 2004).

Most of our knowledge about the chronic effects of high levels of cortisol on the brain is inferred from human studies of HPA axis function in MDD. Cortisol is released in a pulsatile ultradian pattern that varies in amplitude under different conditions of stress. Abnormal patterns of cortisol secretion occur in depression, with increased secretion of cortisol and blunting of the normal dip that should occur in the evening, leading to increased 24-h production of cortisol (Pariante & Lightman, 2008).

Measuring HPA Axis Function

A test that is frequently employed to examine abnormal cortisol secretory patterns in depression is the cortisol response to awakening (CAR). A study involving over 1500 participants in the Netherlands, 701 of whom were depressed, reported that basal cortisol levels and the CAR tend to be increased in those with depression (Vreeburg et al., 2009). Evening cortisol levels were also measured in this group and were higher in those who were depressed, but levels returned to normal with remission of the depression. The CAR did not alter with remission of depression. That the increased CAR may be genetically determined is suggested by the finding from this group that unaffected individuals with a parental history of depression have responses similar to affected individuals (Vreeburg et al., 2010). Thus, diurnal secretion of cortisol may be a state marker and likely to be impaired in state depression, whereas the CAR may be a trait marker of a depressive diathesis. This suggests poorer negative feedback of cortisol at brain MR and/or GR.

The cause of the impaired shutdown of the HPA axis in depression is not known, but there is much animal work indicating that negative feedback control on Corticotropin-releasing hormone (CRH) secretion may be impaired because of altered GR function in the hypothalamus (Claes, 2009). Recent human genetic work suggests that specific GR polymorphisms may be associated with a vulnerability to depression (Lahti et al., 2011). The relative resistance of brain glucocorticoid feedback mechanisms is thought to explain the reduced cortisol suppression to dexamethasone found in many patients suffering from depression (Paslakis et al., 2011; Pariante & Lightman, 2008). As a synthetic GR agonist, dexamethasone shuts down the HPA axis via negative feedback, and this is reflected in the lowered cortisol concentrations 12–16 h after dexamethasone administration. Dexamethasone nonsuppression, although not a specific biological marker for depression, is a widely used and simple test for demonstrating abnormal HPA axis dynamics in depression. Dexamethasone nonsuppression is more likely to occur in more familial types of depression and in more severely ill patients (Rush et al., 1996). MR are also involved in hypothalamic/hippocampal feedback inhibition, and altered MR function may occur in more severe forms of depression resistant to first-line treatments (Juruena et al., 2009). MR receptor function is tested by examining cortisol suppression following the administration of prednisolone, an MR receptor agonist.

The dexamethasone/CRH (dex/CRH) test combines the oral administration of dexamethasone with a CRH bolus (usually 100 μgm) typically given 16 h later. The exaggerated cortisol response found in those with depression is thought to reflect the underlying overactivity of the HPA axis when cortisol inhibition is removed (Ising et al., 2007). In a dex/CRH study comparing (1) healthy individuals with a positive family history for depression versus (2) healthy individuals with no family history and (3) individuals suffering from acute depression, the levels of cortisol were greatest in the depressed group, lowest in the healthy controls, and at an intermediate level in individuals with a positive family history, indicating an increased risk for developing this disease (Modell et al., 1998). Consistent with this, in the Munich Vulnerability Study, Holsboer found that individuals who do not suffer from a psychiatric condition but have a positive family history for depression show abnormal results in the dex/CRH test (Holsboer, 2000).

Glucocorticoids and Hippocampal Damage

The associations between glucocorticoids and stress and neuronal damage in the hippocampus indicate that the neurotoxic effects of glucocorticoids on the hippocampus can be visualized in terms of overall volume changes. The hippocampus has a role in memory processes, learning, and motivation (O'Mara, Sanchez-Vives, Brotons-Mas, & O'Hare, 2009), all of which can be altered in MDD (Zakzanis, Leach, & Kaplan, 1998).

The hippocampus is vulnerable to stress, particularly during the early developmental period (Teicher et al., 2003). This damage may be species-specific and time-dependent. While the separation of monkeys from their siblings or mothers seems not to affect hippocampal volumes (Lyons, Yang, Sawyer-Glover, Moseley, & Schatzberg, 2001; Spinelli et al., 2009), chronic stress exposure in rats or tree shrews does result in reduced hippocampal volumes (Lee, Jarome, Li, Kim, & Helmstetter, 2009; Ohl, Michaelis, Vollmann-Honsdorf, Kirschbaum, & Fuchs, 2000). Transient mild stress may enhance hippocampal function (Luine, Martinez, Villegas, Magarinos, & McEwen, 1996), but chronic or severe stress disrupts hippocampus-dependent memory in experimental animals, as reviewed in Sapolsky (2003).

Effects of hypercortisolemia on brain structure and function are difficult to separate from the effects of CRH and/or arginine vasopressin, as both are probably raised in MDD. Cushing syndrome is characterized by hypercortisolemia and low levels of hypothalamic adrenocorticotropic hormone (ACTH) secretagogs and provides a model for understanding the selective effects of high cortisol levels in the brain. Depression occurs in about 60% of patients with Cushing syndrome, tends to be atypical in type, and the majority remit when cortisol levels are normalized (Kelly, Kelly, & Faragher, 1996). Premature cortical atrophy and cognitive impairments also occur in Cushing syndrome (Simmons, Do, Lipper, & Laws, 2000).

Human studies demonstrate that impairments in learning and memory occur where cortisol levels are artificially acutely elevated or reduced, below or above average ranges (Lupien et al., 2002). A similar impairment has been observed with artificial, but chronic, elevation of cortisol levels (De Quervain, Roozendaal, Nitsch, McGaugh, & Hock, 2000; Newcomer et al., 1999; Young, Sahakian, Robbins, & Cowen, 1999). Rat work demonstrates that the impairments in cognitive function brought about by excessive exposure of the hippocampus to glucocorticoids result from glucocorticoid-induced changes in synaptic plasticity, reduction in neurogenesis, retraction of the apical dendrites of hippocampal pyramidal cells, and, in some situations, neuronal atrophy and cell death (Goosens & Sapolsky, 2007). A reduction in the amount of neuropil without frank cell loss has also been observed, a finding that appears to be consistent with observations from postmortem studies of the hippocampus in patients with MDD (Stockmeier et al., 2004).

The hippocampal damage brought about by excessive exposure to glucocorticoids should cause a reduction in the feedback inhibition mediated by cortisol, via the hippocampus, on CRH secretion; this results in further excessive cortisol secretion, creating a cascade of hippocampal damage (Sapolsky, Krey, & McEwen, 1986). The "Glucocorticoid Cascade Hypothesis" was formulated in 1986 and is now widely accepted as a pathophysiological pathway leading to brain changes associated with severe and enduring stress. This mechanism explains how chronic stress can lead to depression-like effects, brain changes that result in a dysfunctional central control of the HPA axis, and subsequent depression. It should be noted that in normal senescence a similar process occurs, leading to a reduced ability of the hippocampus to control the HPA axis and a gradual reduction in cognitive tasks mediated by the hippocampus.

Longitudinal Studies in Depression and Neuroimaging Findings

Longitudinal imaging studies should inform us in relation to the chronic effects of depression on the hippocampus. In a longitudinal study we demonstrated that a negative clinical outcome with more relapses and a chronic course was found to be associated with hippocampal, amygdala, ACC, and dorsomedial prefrontal cortex volume decline during a 3-year follow-up period (Frodl, Koutsouleris, et al., 2008). Moreover, during successful treatment, brain structures like the left inferior frontal cortex, right fusiform gyrus, and right cerebellum might increase in size (Lai & Hsu, 2011). The corollary that smaller hippocampal volumes have been found to be predictive for a bad clinical outcome in 1- and 3-year follow-up studies has also been demonstrated (Frodl, Meisenzahl, Zetzsche, et al., 2004).

Childhood Maltreatment and Neuroimaging Findings

Childhood adversity such as childhood maltreatment plays a significant role in a number of multifactorial mental disorders. Studies reveal that certain aspects of stress-related mental disorders result from maladaptive, stress-induced neuroplastic changes in specific neural circuits (Krishnan & Nestler, 2008). The contribution of childhood maltreatment to abnormal stress systems and responses is now supported by much evidence from multiple viewpoints.

Hippocampus

Studies have analyzed the effects that childhood maltreatment and trauma have on the hippocampus (Table 1). Bremner et al. (1997) showed that adult survivors of severe childhood abuse had a 12% smaller left hippocampus volume when compared to comparison subjects. In keeping with this, women who reported having been sexually abused in childhood were found to have significantly reduced left-sided hippocampal volume as compared to nonvictimized women (Stein, Koverola, Hanna, Torchia, & McClarty, 1997). The finding of smaller left

TABLE 1 Structural MRI Studies on the Effect of Childhood Maltreatment in Humans

Study	Population	Method	Results	Comment
Studies in Adults				
Bremner et al. (1997)	**17 adults**, childhood physical or sexual abuse, **17 healthy controls** matched by age, sex, race, handedness, education, body size, and alcohol	1.5 T MRI Five sections of a mid-hippocampal segment were analyzed Early trauma inventory	12% smaller left hippocampal volume in abused subjects No significant difference in the right hippocampus	Left hippocampus smaller
Stein et al. (1997)	**21 adult women**, sexual abuse during childhood, **21 healthy controls**	1.5 T MRI Manual hippocampal segmentation Early life adversity	5% smaller left hippocampal volume in abused subjects No significant difference in the right hippocampus	Left hippocampus smaller
Vythilingam et al. (2002)	**21 women with MDD** and a history of prepubertal physical or sexual abuse, **11 women with MDD** without abuse, **14 healthy controls**	1.5 T MRI Manual segmentation of hippocampus Early trauma inventory	Subjects with childhood abuse had 18% smaller mean left hippocampal volumes than the nonabused depressed subjects and 15% smaller mean left hippocampal volume than the healthy subjects No significant difference in the right hippocampus	Left hippocampus smaller
Pedersen et al. (2004)	**17 women with prepubertal abuse** without PTSD, **17 women with prepubertal abuse** with PTSD, **17 controls**	1.5 T MRI Manual segmentation of hippocampus Childhood trauma questionnaire, trauma symptom inventory	No significant difference in hippocampus between groups	n.s.

Golier et al. (2005)	**27 elderly Holocaust survivors, 20 nonexposed healthy controls**	1.5 T MRI Manual segmentation of the hippocampus, superior temporal gyrus, lateral temporal lobe	No significant difference in hippocampus or other regions between groups	n.s.
Cohen, Grieve, et al. (2006) and Cohen, Paul, et al. (2006)	**265 healthy adults,** comparison between more than two events of early life adversity and no event of early life adversity	1.5 T MRI Region of interest-based VBM ACC, the hippocampus, the amygdala, and the caudate nucleus Early life stress questionnaire	Trend for smaller left and right hippocampal volumes in subjects with early life adversity as compared to those without early life adversity Significant smaller ACC and caudate volume in those adults with adverse childhood events No significant difference in the amygdala	Trend for smaller left and right
Weniger et al. (2008)	**10 women with severe childhood abuse** and diagnosis of PTSD, **13 women with severe childhood abuse** and dissociative disorder, **25 healthy controls**	1.5 T MRI Manual segmentation of hippocampal and amygdala volume Traumatic antecedent questionnaire	Smaller hippocampal and amygdala volumes in subjects with PTSD compared to controls No significant difference in hippocampal volumes and larger amygdala volumes between dissociative disorder and controls	Smaller hippocampal volumes
Frodl, Reinhold, Koutsouleris, Reiser, et al. (2010)	**43 patients with MDD and 44 healthy controls,** median split with emotional neglect (low, high)	1.5 T Manual segmentation of hippocampus Childhood trauma questionnaire	Smaller hippocampal volumes in subjects with depression and emotional neglect during childhood as compared to subjects (healthy or patients) without emotional neglect	Smaller hippocampal white matter

Continued

TABLE 1 Structural MRI Studies on the Effect of Childhood Maltreatment in Humans—cont'd

Study	Population	Method	Results	Comment
Gerritsen et al. (2011)	**568 healthy subjects**, 51% in 1.5 -T cohort and 56% in 3 -T cohort reported childhood adversity	1.5 T and 3 T MRI VBM List of threatening life events	History of childhood adversity was associated with smaller volumes in ACC and OFC in the 1.5 -T cohort but not in the 3 -T cohort Hippocampus not altered	n.s. for hippocampus
Van Harmelen et al. (2010)	**84 patients with depression and/or anxiety disorder** who reported childhood emotional maltreatment, **97 patients with depression or anxiety disorder** without childhood maltreatment	3 T MRI VBM Multiple incidents (more than once) of emotional neglect and/or emotional abuse before age 16 years	5.14% reduction in the left dorsal mPFC, no significant differences in hippocampus or amygdala	n.s. for hippocampus

n.s., non significant.

hippocampal volumes in subjects with childhood maltreatment was later replicated by a study in patients with MDD (Vythilingam et al., 2002).

We found that patients with MDD and a history of emotional neglect during childhood had reduced left hippocampal white matter as compared to those without a history of emotional neglect, but no significant differences were detected in the whole hippocampus (Frodl, Reinhold, Koutsouleris, Reiser, & Meisenzahl, 2010). Our finding that childhood maltreatment was also associated with smaller hippocampal head volumes in unaffected first-degree relatives of patients with MDD is interesting (Carballedo et al., 2012). There is an anteroposterior gradient in the proportional volume of each subfield in the head, body, and tail of the hippocampus. Higher proportions of the CA1-3 and subiculum are found in the hippocampal head, whereas the hippocampal body includes the greatest proportion of the dentate gyrus (DG) (Malykhin, Lebel, Coupland, Wilman, & Carter, 2010). The subfields CA1-3 were found to be altered by experimental stress in animal studies (Sapolsky, 2001), explaining why early stress affected the hippocampal head more than other parts of the hippocampus in family-risk subjects.

Subjects with severe childhood maltreatment who developed complex PTSD showed smaller hippocampal volumes (Weniger, Lange, Sachsse, & Irle, 2008). All of these above-mentioned studies had sample sizes of 17–21 cases versus 17–21 controls. In a large study on healthy men and women (N = 265) a trend for smaller hippocampal volumes, and significantly smaller ACC and caudate volume, was reported in adults with at least two adverse child events as compared to those without adverse child events (Cohen, Grieve, et al., 2006).

One longitudinal MRI study in 15 children with childhood maltreatment reported that the presence of early life maltreatment was related to a decrease of hippocampal volumes over time (Carrion, Weems, & Reiss, 2007). This study did not have a comparison group of nonmaltreated individuals, so no conclusion can be drawn about whether the observed changes are pathological. De Bellis, Hall, Boring, Frustaci, and Moritz (2001) did not find significant differences in hippocampal volumes between nine children with early life adversity as compared to those without (N = 9), and no differences were found in hippocampal volume changes over at least 2 years of follow-up. These two studies included a small number of patients, so that these findings should not be over interpreted. De Bellis, Hooper, Woolley, and Shenk (2010) did not find the hippocampus to be related to PTSD symptoms in children, whereas other variables, like socioeconomic status, general maltreatment, and sexual abuse, were predictive for PTSD symptoms. Neither was there a change in hippocampal volumes detected in a study examining 17 children exposed to continuous maternal depressive symptomatology since birth compared to 21 control children who were not exposed to maternal depressive symptomatology (Lupien et al., 2011). In this study, however, the amygdala was larger in children exposed to maternal depression (Lupien et al., 2011).

Interestingly, one paper reported that 61 children with a history of childhood abuse and related PTSD had larger hippocampal volumes as compared to 122 healthy children (Tupler & De Bellis, 2006).

Manual Tracings

There are also two manual tracing studies that have not found effects of childhood maltreatment on the hippocampal volume. Hippocampal volume differences were not seen in 27 survivors of the Nazi Holocaust, with or without PTSD, and who were children during the Holocaust, although memory deficits were present as compared to healthy controls (Golier et al., 2005). Neither were there significant differences in hippocampal volumes in women with prepubertal abuse who developed later PTSD (N = 17), those with abuse who did not develop PTSD (N = 17), and healthy controls (N = 17) (Pederson et al., 2004). Interestingly, the mean age when the trauma occurred was 13.3 years for the group that developed PTSD and 10.5 years for the group that did not develop PTSD in the study of Holocaust survivors (Golier et al., 2005). Thus, the time when trauma, abuse, or neglect occurs may be important in relation to hippocampal damage.

Voxel-Based Morphometry Studies

Studies that used voxel-based morphometry (VBM) to analyze brain structural differences between subjects with childhood maltreatment as compared to those without maltreatment did not report differences between these groups in relation to the hippocampus but did report differences in the OFC and ACC. A large study of 568 healthy subjects found smaller volumes in the ACC and OFC for those subjects measured with 1.5 T MRI, whereas this finding could not be confirmed in the group measured with 3 T MRI (Gerritsen et al., 2011). Eighty-four patients with depression and/or anxiety who reported childhood emotional maltreatment had smaller volumes in the left dorsomedial prefrontal cortex compared to 97 patients with depression or anxiety disorder without childhood maltreatment (Van Harmelen et al., 2010). Analysis methods like VBM have the advantage in the investigation of all cortical brain regions, as this is more difficult because of the lack of clear anatomic boundaries, with manual tracing methods. However, subcortical regions like the hippocampus or the amygdala seem to be more difficult to analyze with automatic VBM methods, and thus the two existing VBM studies in large samples provide strong evidence for the effect of early life adversity on brain development.

A meta-analysis based on 14,250 participants demonstrated that early adverse life events confer significant risk of subsequent MDD (Risch et al., 2009). Studies reveal that certain aspects of stress-related mental disorders result from maladaptive, stress-induced neuroplastic changes in specific neural circuits (Krishnan & Nestler, 2008). Young people and adults exposed to early-life adversity appear to have smaller hippocampus volumes (Woon & Hedges, 2008), although a large population-based study from Australia did not find smaller hippocampal volumes in subjects exposed to adverse childhood events (Cohen, Grieve, et al., 2006). The Woon and Hedges study used VBM to conduct a whole brain analysis. They also corrected for multiple comparisons, decreasing the probability of detecting changes in small structures like the hippocampus.

fMRI Studies

In relation to fMRI studies, childhood maltreatment seems to result in blunted prefrontal cortical activation when processing higher order cognitive functions (Van Harmelen et al., 2014). It was found that childhood maltreatment was associated with increased neural responses in the amygdala to the presentation of both angry and happy emotional expressions (McCrory et al., 2013) and in another study, to the presentation of sad facial expressions only (Dannlowski et al., 2013). In combination with genetic factors, such childhood maltreatment-induced alterations might set the stage for an individual to develop psychopathology.

Thus, while the studies in children are rare, and only a minority of these show an association between childhood maltreatment and smaller hippocampal volumes, studies in adults show evidence for decreased hippocampal and also prefrontal volumes when they have experienced childhood maltreatment. Childhood maltreatment also resulted in a heightened neural response to emotion processing, whereby studies are controversial with regard to their specificity for valences.

Structural Connectivity

With respect to childhood maltreatment, some studies show evidence for decreased FA. Paul et al. (2008) showed a significant reduction in the genu of the corpus callosum in female individuals subjected to high levels of childhood maltreatment as compared to controls, and these changes are also seen in the absence of psychiatric symptoms. A study by Choi, Jeong, Rohan, Polcari, and Teicher (2009) reported a significant decrease in FA in the left superior temporal gyrus in association with parental verbal abuse, whereas Tomoda et al., (2010) reported a 14.1% increase in the gray matter of this area. Despite conflicting evidence, both studies concluded that parental verbal abuse caused alterations in the neural pathways responsible for language processing and development.

A region of interest and tractography study conducted in seven socioeconomically deprived children and matched controls found significant reductions in FA values of the uncinate fasciculus and suggested that these changes may underlie the cognitive and behavioral changes occurring in children exposed to high levels of childhood maltreatment (Eluvathingal et al., 2006).

Combined HPA and Neuroimaging

As described above, there are well-replicated abnormalities in hippocampal structure and in HPA axis function in MDD. Dysregulated feedback mechanisms at the level of the hippocampus appear to drive the HPA axis dynamics observed in depression while the toxic effects of hypercortisolism appear to lead to the cognitive and structural changes of the hippocampus. There has been much speculation about whether cortisol measures, either dynamic or static, correlate with hippocampal neuroimaging measures (Table 2).

The total hippocampal volume of six subjects with increasing or high cortisol levels was significantly reduced by 14% in comparison to that of five subjects with decreasing/moderate cortisol levels, and the degree of hippocampal atrophy correlated strongly with both the degree of cortisol elevation over time and current basal cortisol levels (Lupien et al., 1998). Subsequent studies produced mixed results.

Dexamethasone Suppression Test and Brain Neuroimaging Studies

To date five studies have used the Dexamethasone suppression test (DST) to evaluate the association between HPA axis functioning and hippocampal volumes. In the largest and best-powered study investigating 575 patients with arteriosclerotic disease, participants with higher awakening levels after dexamethasone had smaller hippocampal volumes (Knoops, Gerritsen, Van Der Graaf, Mali, & Geerlings, 2010). However, five independent studies with smaller samples did not show an association between DST and hippocampal volumes. In one of these studies, 10 healthy, elderly nonsuppressors to dexamethasone did not show changes in the hippocampus but had significantly smaller left ACC volumes than 10 suppressors (MacLullich et al., 2006). Another study, examining 41 middle-aged type 2 diabetes mellitus patients, did not report any association between results from the DST test and hippocampal volumes (Bruehl, Wolf, Sweat, et al., 2009). Neither was any association found between post-DST cortisol and hippocampal volume in mild to moderate Alzheimer disease (Elgh et al., 2006).

Likewise, in samples with depression there is not much evidence that the DST status is associated with hippocampal volumes (Table 3). In one study 20 drug-free, first-episode female patients with depression were investigated. The volumes of the left and right hippocampus did not correlate with basal or postdexamethasone cortisol levels, although the depressed sample had smaller hippocampal volumes as compared to 15 healthy controls (Kaymak et al., 2010). No significant association between DST and hippocampal volumes were detected in a depression study (Vythilingam et al., 2004). These studies are limited by being cross-sectional rather than longitudinal. Larger samples, in particular in patients with MDD, and over time, are necessary to thoroughly explore the association between the DST and hippocampal volume.

Awakening and Daily Cortisol Measures and Brain Neuroimaging Studies

Studies in healthy subjects and in patients with arteriosclerotic disease, multiple sclerosis, schizophrenia, and alcohol dependency have reported smaller

TABLE 2 Studies on Cortisol Measurements and Hippocampus in Healthy Controls and Nonpsychiatric/Neurological Samples; Other Brain Region Findings Are Also Described in the Table

Study	Subjects	Method	Result
Studies that Only Used or in Addition Used DST			
MacLullich et al. (2006)	10 **healthy** elderly male nonsupressors and 10 healthy elderly supressors to the DST	1.9 T Semiautomated measure of ACC, superior frontal gyrus, and hippocampus DST (0.25 mg dexamethasone, blood drawing at 9 am)	Nonsuppressors had smaller ACC volumes **No difference in hippocampal volumes** between suppressors and nonsupressors
Knoops et al. (2010)	575 patients with **arteriosclerotic disease**	1.5 T Manual tracing of the hippocampus cortisol measures after awakening and after 30, 45, and 60 min at 10 pm and 11 pm + dexamethasone test	Participants with higher evening levels and higher awakening levels **after dexamethasone** had smaller hippocampal volumes Cortisol awakening response was not significantly associated with hippocampal volumes
Bruehl, Wolf, Sweat, et al. (2009)	41 middle-aged dementia-free volunteers with **type 2 diabetes** mellitus (on average 7 years since diagnosis) with 47 age-, education-, and gender-matched noninsulin-resistant controls	1.5 T Manual segmentation of hippocampus and prefrontal lobe Basal cortisol levels after a 10-h overnight fast HPA axis feedback was assessed on a different day, using the short version of the dex/CRH test	No association was found between cortisol measures and hippocampal volumes Diminished cortisol suppression after dexamethasone and dyslipidemia were associated with decreased cognitive performance, whereas obesity was negatively related to hippocampal volume

Continued

TABLE 2 **Studies on Cortisol Measurements and Hippocampus in Healthy Controls and Nonpsychiatric/Neurological Samples; Other Brain Region Findings Are Also Described in the Table—cont'd**

Study	Subjects	Method	Result
Studies that Used CAR or Daily Cortisol Profiles			
Lupien et al. (1998)	**Healthy participants** Six subjects with a progressive increase in cortisol levels and five subjects with a progressive decrease in cortisol levels	1.5 T MRI Manual hippocampal segmentation 24-h period with sampling each hour of a blood sample For 5 to 6 years, plasma cortisol levels were measured annually over a 24-h period	Longitudinal study: total hippocampal volume of the increasing/high cortisol group was significantly reduced by 14% in comparison to that of the decreasing/moderate cortisol group **The degree of hippocampal atrophy correlated strongly with both the degree of cortisol elevation over time and current basal cortisol levels** No significant effect was seen in the parahippocampal gyrus or fusiforme gyrus
Wiedenmayer et al. (2006)	17 **healthy** children	1.5 T Hippocampal volume and surface morphology Blood samples were obtained in the morning to measure cortisol levels	Cortisol levels were not associated with total hippocampal volumes Positive associations between cortisol and hippocampal surface morphology were found focally for the anterior segment of the hippocampus (CA3 and dentate gyrus); inverse associations were found along the lateral aspects of the anterior, medial, and posterior segments of the hippocampus (CA1)

Pruessner et al. (2007)	13 **healthy** male subjects	1.5 T Manual segmentation of hippocampal volumes CAR Samples were taken at 0, 30, and 60 min after awakening once in four consecutive weeks Trier social stress test (TSST) A total of eight saliva samples for cortisol assessment were taken 45, 15 min, and immediately before the TSST and immediately, 10, 20, 40, and 60 min thereafter.	Larger hippocampal volume was associated with stronger cortisol increase in response to the TSST and a significantly greater CAR
Kremen et al. (2010)	388 middle-aged male **healthy** twins (51–59 years old)	1.5 T Cortical surface reconstruction using freesurfer and hippocampal segmentation Saliva samples were obtained at awakening, 30 min postawakening, 10 am, 3 pm, and 9 pm or bedtime	Higher mean cortisol level and AUC cortisol were significantly associated with thinner cortex in seven prefrontal regions: left and right superior frontal gyrus, left rostral middle frontal gyrus, left pars opercularis, left pars triangularis, left pars orbitalis, and right medial orbital frontal cortex No significant associations were detected between cortisol measures and hippocampal volumes

Continued

TABLE 2 Studies on Cortisol Measurements and Hippocampus in Healthy Controls and Nonpsychiatric/Neurological Samples; Other Brain Region Findings Are Also Described in the Table—cont'd

Study	Subjects	Method	Result
Dedovic et al. (2010)	59 **healthy** young men and women	1.5 T MRI Manual segmentation of the hippocampus Cortisol awakening response, AUC: at the time of awakening, after 30 min, 60 min, at 4 pm, and at 9 pm over a span of three nonconsecutive workdays	No significant association was found in the whole group In men significant positive correlations between hippocampal volume and CAR were seen
Bruehl, Wolf, Convit (2009)	18 **type 2 diabetes mellitus** and 12 healthy controls	1.5 T Volumes of hippocampus, superior temporal gyrus, and frontal lobe at awakening, 15, 30, and 60 min after wake up and at 11 am, 3 pm, and 8 pm	Hippocampal volume was positively correlated with the CAR, independent of diagnosis, no association between AUC and the hippocampus

TABLE 3 Studies on Cortisol Measurements and Hippocampus in Patients with MDD; Other Brain Region Findings are Also Described in the Table

Study	Subjects	Method	Result
Studies that only used or in addition used DST			
Kaymak et al. (2010)	20 drug-free female patients with first-episode **MDD**, 15 healthy female controls	3 T Manual hippocampal volume assessment Blood samples at 8 am and 4 pm to measure baseline cortisol concentrations + DST (1 mg dexamethasone), blood collection at 8 am and 4 pm the following day	**No correlation between** hippocampal volumes and baseline nor postdexamethasone cortisol levels in patients and controls
Vythilingam et al. (2004)	38 subjects with **MDD** and 33 healthy subjects	1.5 T Manual hippocampal segmentation Urine sample over 24 h, DST: (1 mg of dexamethasone at 11 pm) and blood was collected at 4 pm the following day	Baseline plasma or urinary free cortisol (UFC) was not related to either hippocampal volume in patients with major depression A negative correlation between 24-h UFC and both the right and left hippocampus was seen in healthy controls There was **no correlation between hippocampal volume and post-DST** plasma cortisol or baseline DST plasma cortisol
Studies that used CAR or daily cortisol profiles			
Colla et al. (2007)	24 patients with **MDD**, 14 healthy controls	1.5 T Manual segmentation of hippocampal volumes Saliva cortisol was measured at 8 am and 4 pm	Baseline cortisol levels were not related to hippocampal volumes

hippocampal volumes to be associated with increased levels of daytime cortisol (Knoops et al., 2010). In an MRI study of 15 children with a history of maltreatment, mean cortisol levels measured four times a day during three consecutive days were significantly negatively correlated with right hippocampus volumes at follow-up at 12–18 months (Carrion et al., 2007). In 20 healthy subjects, measures of ACTH feedback after cortisol injection were inversely associated with 24-h urinary cortisol (Wolf, Convit, De Leon, Caraos, & Qadri, 2002).

Of interest are also studies examining the relationship between cortisol secretion and hippocampal volume in schizophrenia. The area under the curve (AUC) of diurnal cortisol was found to be correlated negatively with hippocampal volumes in 24 patients with first-episode psychosis (Mondelli et al., 2010). Small but significant inverse associations were also found between cortisol levels and the thickness of the left dorsolateral (superior frontal gyrus, left rostral middle frontal gyrus) and ventrolateral (pars opercularis, pars triangularis, pars orbitalis) prefrontal regions and right dorsolateral (superior frontal gyrus) and medial orbital frontal cortex in 388 middle-aged male twins who were 51–59 years old (Kremen et al., 2010).

The associations between hippocampal volume and HPA axis measures are not consistently found. An association between evening levels of cortisol and hippocampal volumes could not be seen in 24 traumatized police officers (Lindauer, Olff, Van Meijel, Carlier, & Gersons, 2006). In 59 young men and women no significant association was found between hippocampal volumes and AUC either in the whole group or in subsamples at risk for depression based on their current depression ratings (Dedovic et al., 2010). In a study on subjects with first-episode schizophrenia and healthy controls, there was no significant association found between daily cortisol measured as the AUC and hippocampal volumes (Gunduz-Bruce et al., 2007).

With regard to samples with MDD, Vythillingam et al. examined hippocampal-related cognitive tasks, hippocampal volume, and cortisol measures in 38 subjects with depression and no adverse experiences in childhood and 33 healthy subjects and repeated the tests in a subgroup of 22 of the depressed group following treatment with antidepressants. Hippocampal volume was not related to 24-h urinary cortisol levels or post-DST cortisol levels in patients with MDD; however, a negative association was seen between urinary cortisol and hippocampus in healthy controls (Vythilingam et al., 2004). No significant associations between baseline cortisol measures and hippocampal volumes were found in 24 patients with MDD and healthy controls (Colla et al., 2007; Table 3).

Thus, the most consistent findings of an association between higher cortisol levels and smaller hippocampal volumes arise from a continuous measure of the cortisol levels, e.g., as AUC over a day. Here studies in healthy controls, those with alcohol dependency, multiple sclerosis, arteriosclerosis, and schizophrenia indicate inverse associations.

HPA Axis and Brain Function

Since stressful life events and childhood maltreatment have been shown to impact brain function, there should be the expectation that measures of the HPA axis might also be associated with brain function. Studies investigating pre- to poststressor change in cortisol showed variable results, with some studies reporting increases in the prefrontal cortex and OFC related to pre- to poststressor change in cortisol (Wang et al., 2008). Others reported decreases in the hypothalamus, amygdala, hippocampus, orbitofrontal cortex, and ACC in participants reacting with an increase in cortisol following stressors (Pruessner et al., 2008).

Negative associations between cortisol reactivity during passive viewing of negative stimuli and ventromedial prefrontal cortex activity (Root et al., 2009) and negative correlations between cortisol levels and activation of the amygdala were observed (Cunningham-Bussel et al., 2009; Pruessner et al., 2008). On the other hand, a positive association between limbic activation and cortisol change has also been shown (Root et al., 2009; Van Stegeren et al., 2007).

Conversely, exogenous cortisol administration is associated with a decreased activation of the amygdala and hippocampus during rest in healthy controls (Lovallo, Robinson, Glahn, & Fox, 2010). In a study of depression, 15 patients with depression in remission and 15 healthy controls were investigated. A visual stimuli task paradigm was presented to evoke a mild stress response, adapted from the International Affective Picture System. Cortisol and ACTH were measured at baseline and following the onset of the stress paradigm. Interestingly, among remitted MDD patients, the percent signal change in the right amygdala was negatively related to peak cortisol change (Holsen et al., 2013).

Therefore, some studies report that during stress, an increase of cortisol is associated with a decrease of amygdala reactivity; others report the opposite. Thus, no conclusion can be drawn about the direction of association between cortisol reactivity and neural reactivity of limbic brain circuitries.

IMMUNE SYSTEM

The stress hormone system and the immune system are closely linked to each other. Alterations of inflammatory biomarkers have been repeatedly reported in patients with MDD as compared to healthy controls, with indirect evidence that neuroinflammation plays a role in MDD (Eisenberger, Inagaki, Rameson, Mashal, & Irwin, 2009; Miller, Maletic, & Raison, 2009; Reichenberg et al., 2001; Wright, Strike, Brydon, & Steptoe, 2005).

Preclinical and clinical MRI research suggests that neuroinflammation in MDD might be associated with structural and functional anomalies in various regions of the central nervous system. Recently, we reported a significant association of interleukin (IL)-6 and C-reactive protein (CRP) with findings in the left and right hippocampus in 40 patients with MDD, independent of demographic variables (Frodl et al., 2012). Moreover, there was an association between IL-6 blood levels and hippocampal volumes in healthy controls. We found smaller hippocampal volumes in MDD patients who had a lower expression of glucocorticoid-induced leucine zipper (GILZ) mRNA, or SGK-1 mRNA, as markers of reduced activation of the glucocorticoid system, as compared to those with higher serum SGK-1 and GILZ mRNA expression (Frodl et al., 2012). However, these findings need replication in a larger sample size in order to confirm the relationships between inflammation, stress hormone system, and brain structure/function. Although possible antidepressant effects should be taken into account, these findings suggest that reduced glucocorticoid responsiveness and increased inflammation might have a role in the neuroplasticity–neurotoxicity cascade in MDD.

An increase in TNF-α has been found to be associated with more activity in the dorsal ACC and anterior insula in subjects experiencing social rejection (Slavich, Way, Eisenberger, & Taylor, 2010). In another fMRI study it was shown that higher TNF-α plasma concentrations are correlated with increased activation of the right inferior OFC in response to emotional visual stimuli (Kullmann et al., 2012). In another study inflammation was found to alter reward-related neural activity, and this in turn led to a depressed mood (Eisenberger et al., 2010). In a further study, peripheral IL-6 was found to modulate the association between mood states and reduced connectivity of subgenual ACC to the amygdala, medial prefrontal cortex, nucleus accumbens, and superior temporal sulcus (Harrison et al., 2009). Thus, initial research suggests that there is an association between inflammatory markers, brain function, and structure, although more research is needed in this area.

NEUROTROPHIC SYSTEM

Signaling pathways implicated in neuroplasticity, among other downstream targets, involve genes for growth factors such as brain-derived neurotrophic factor (BDNF; Newton & Duman, 2004). BDNF regulates neuronal survival, migration, phenotypic differentiation, axonal and dendritic growth, and synapse formation (Huang & Reichardt, 2001). BDNF is also a key regulator of synaptic plasticity and behavior (Lu, 2003) and may be important for memory acquisition and consolidation (Tyler, Alonso, Bramham, & Pozzo-Miller, 2002). Acute and chronic stress reduces neurogenesis in the rodent hippocampus (reviewed in Dranovsky & Hen, 2006; Duman, 2004). As neurogenesis may be required for the behavioral response to antidepressants in rodents (Santarelli et al., 2003), and impaired neurogenesis may represent a pathophysiological component of MDD (Duman, 2004), suppressed neurogenesis may represent another pathway, whereby the effects of stress contribute to the development of MDD. The specific functional role and relevance of new neurons in the hippocampus is not firmly established, but a link between neurogenesis and the learning-related functions of the hippocampus is an intriguing possibility (Leuner, Gould, & Shors, 2006). Some computational theories of hippocampal function predict a role for new neurons in hippocampal-dependent learning (Becker, MacQueen, & Wojtowicz, 2009).

A single-nucleotide polymorphism in the pro-domain of BDNF converts the 66th amino acid valine into methionine (Val66Met). This Val66Met polymorphism affects dendritic trafficking and synaptic localization of BDNF and impairs its secretion. The Val66Met SNP is also associated with deficits in short-term episodic memory (Egan et al., 2003). In a large population-based study of 1435 participants with lifetime MDD it was shown that BDNF Met-allele carriers had reduced serum BDNF levels if they had experienced childhood abuse; the Val/Val carriers did not have reduced BDNF levels when exposed to childhood abuse (Elzinga et al., 2011). Thus again in the neurotrophic system there is clear evidence that stressful environmental and genetic factors influence the expression of the protein.

Healthy met-BDNF carriers have relatively small hippocampal volumes (Bueller et al., 2006; Pezawas et al., 2004). An effect of the Met-BDNF allele on hippocampal volume is also apparent in patients with MDD (Frodl et al., 2007), although one report did not observe this association but rather found larger differences for patients with the Val phenotype (Gonul et al., 2011). A study in healthy volunteers reported that subjects carrying the Met-BDNF allele have smaller hippocampus volumes when they have more subthreshold symptoms of depression and higher neuroticism scores (Joffe et al., 2009).

In a study of 89 participants, significant interactions between BDNF genotype and early life stress were apparent in hippocampus and amygdala volumes, heart rate, and working memory (Gatt et al., 2009). The investigators used structural equation modeling to investigate the pathways through which the BDNF genotype and early life stress interact to produce effects on brain structure, body arousal, and emotional stability, which in turn predict alterations in symptoms and cognition. Results suggested that the combination of Met carrier status of the BDNF polymorphism and exposure to early life stress predicted reduced hippocampal volumes, associated lateral prefrontal cortex volumes and, in turn, vulnerability to depression.

A study has examined hippocampal dentate gyrus and cornu ammonis subfields in MDD as compared to healthy controls in relation to child maltreatment and which subfields are affected by the BDNF Val66Met polymorphism. Patients with MDD had significantly smaller CA4/DG and CA2/3 volumes as compared to healthy controls. A significant interactive effect of BDNF Val66Met and childhood adversity was found for CA2/3 and CA4/DG volumes. Met allele carriers without childhood adversity had larger, and with childhood adversity smaller, CA4/DG and CA2/3 volumes than Val-allele homozygotes (Frodl et al., 2014).

SEROTONIN AND GENETICS

Serotonin (5-HT) is one of the neurotransmitters most implicated with depression. In addition to being a neurotransmitter, 5-HT plays a major role in the development of the brain. Elimination of 5-HT from the synaptic cleft is mediated by a single protein, the 5-HT transporter (5-HTT), which determines the size and duration of the serotonergic responses (Lesch & Mossner, 1998). Prenatal deletion of the 5-HTT in rodents has been shown to induce various subtle brain alterations (Esaki et al., 2005). In addition, a transient inhibition of the 5-HT transporter in mice during early postnatal development through administration of fluoxetine leads to depressive-anxious behaviors and increases stress sensitivity in adulthood (Ansorge, Zhou, Lira, Hen, & Gingrich, 2004). These effects were not observed following inhibition of the norepinephrine transporter or following brain maturity, suggesting that different neurotransmitters have a different role in neurodevelopment and that the impact of the insult on neurotransmitter function is time specific (Ansorge, Morelli, & Gingrich, 2008).

The promoter region of 5-HTT has a polymorphism that results in allelic variation of functional 5-HTT expression (Lesch et al., 1996). The long (l) allele is associated with the production of more 5-HTT transcription than the short (s) allele (Lesch et al., 1996). The short (s) allele of the 5-HTT polymorphism is associated with anxiety, depression, and aggression-related personality traits in some reports (Lesch & Mossner, 1998; Lesch et al., 1996), but a meta-analysis did not confirm an association between 5-HTT polymorphism and vulnerability to MDD (Risch et al., 2009). The importance of variation in 5-HTTLPR to hippocampal volume, e.g., in patients with MDD, has been noted in three published reports that utilized diallelic analysis and two reports using triallelic analysis (Frodl, Meisenzahl, Zill, et al., 2004; Frodl, Zill, et al., 2008; Taylor et al., 2005).

Early life stress and variations in the 5-HTT gene may interact to predict development of MDD (Caspi et al., 2003). An increased risk of depression was detected in maltreated children homozygous for the *S*-allele (Kaufman et al., 2006). In a study, patients carrying the S-allele had smaller hippocampal volumes when they had a history of emotional neglect as compared to patients who had only one risk factor, either environmental or genetic (Frodl, Reinhold, Koutsouleris, Donohoe, et al., 2010). Childhood stress also predicted hippocampal alterations independent of genotype. Meta-analytic approaches suggest, however, that 5-HTTLPR may not further increase the risk for MDD in subjects who experience critical life events (Risch et al., 2009).

EPIGENETICS

In general, findings are indicating that gene by environment interactions are of central importance in the development of depression. Although the specific mechanisms remain unknown, a number of studies suggest that DNA methylation may be an underlying mechanism mediating the impact of adverse social environments on gene function (Booij, Tremblay, et al., 2015; Booij, Wang, Levesque, Tremblay, & Szyf, 2013). Studies in patients with MDD have shown an association of differential DNA methylation in white blood cells with early maltreatment as well as depressive symptomatology (Booij et al., 2013; Nestler, 2014). The findings are not consistent, but the inconsistent effects might be due to differences in methodology, as different methodologies are being employed in different centers (Booij et al., 2013).

Animal models tracking the trajectory from early life stress to adult depression indicate that sustained stress during development leads to hypermethylation of the GR promotor gene, leading to reduced function of the GR and an inability to shut down stress responses (McGowan et al., 2011). An impact of parental care on epigenetic regulation of hippocampal GR was demonstrated in a study observing that suicide victims with a history of early maltreatment display decreased GR mRNA expression and increased cytosine methylation of a neuron-specific GR (NR3C1) promoter in postmortem hippocampus as compared to either suicide victims with no childhood maltreatment or controls (McGowan et al., 2009).

Peripheral SLC6A4 DNA methylation was found to be related to childhood maltreatment in a sample of pregnant women and a sample of adoptees (Beach, Brody, Todorov, Gunter, & Philibert, 2010, 2011; Devlin, Brain, Austin, & Oberlander, 2010; Kang et al., 2013). In a study on prenatal and postnatal exposure to maternal depression it was detected that increased second trimester maternal depressed mood was associated with decreased maternal and infant SLC6A4 promoter methylation (Devlin et al., 2010). This correlational finding seems to be in the opposite direction than research on childhood adversity, warranting further research to understand the impact of SLC6A4 methylation and its function from a developmental view.

The functional relevance of DNA methylation in SLC6A4 promotor regulation is that it suppresses transcriptional activity (Wang et al., 2012). DNA methylation should be cell-type specific so that DNA methylation changes that are relevant to brain function should be detected only in the brain. Nevertheless, we have previously reported differential methylation of a regulatory region of the SLC6A4 gene in peripheral T-cells that is associated with differences in in vivo measures of lower 5-HT synthesis, measured with positron emission tomography (Wang et al., 2012). This T cell methylation is also associated with hippocampal volume detected by MRI in patients with MDD and healthy controls (Booij et al., 2014). Moreover, an association between peripheral SLC6A4 methylation with several gray matter structures, including the hippocampus, insula, amygdala, and nucleus caudatus, has been reported (Dannlowski et al., 2014).

Previously, we reported an interactive effect between the promotor polymorphism of SLC6A4 and childhood adversity on brain structure—a finding present in patients with depression and not in healthy controls (Booij, Szyf, et al., 2015; Frodl, Reinhold, Koutsouleris, Donohoe, et al., 2010). In a study a significant effect of the state of SLC6A4 methylation in whole blood DNA on fMRI BOLD responses during emotional attention processing was shown

(Frodl et al., 2015). BOLD responses elicited by negative emotional stimuli in the left anterior insula/frontal operculum area were found to be positively associated with methylation of the SLC6A4 regulatory region. Patients with MDD showed increased activation in the insula as well as other emotional brain regions elicited by emotional relevant stimuli and lower activity in the hippocampal area during higher order cognitive processing as compared to controls; therefore, suggesting that the state of SLC6A4 methylation and depression influence brain function in the same direction.

OVERALL CONSIDERATIONS

It is obvious that this is not a simple story. What is emerging is a complex system of interconnected pathways that need to be separated out before the contribution of the component systems can be understood.

In this chapter we focused first on brain imaging in depression and highlighted the evidence of the role of childhood maltreatment and the stress system on neuroimaging findings. The importance of childhood maltreatment in the development of psychiatric diseases like depression is intuitively evident and has now been established scientifically. Environmental factors like childhood maltreatment may modify stress mechanisms via epigenetic processes, and these might in turn lead to structural and functional reorganization of brain systems. Possible systems that have been implicated are the HPA axis and the inflammatory, 5-HT, or neurotropic systems. Alternatively, modification of the HPA axis may mediate the changes in the other systems, given that glucocorticoids have tissue-specific genomic effects.

Despite the experimental findings involving neuroplasticity in the pathophysiology of MDD (Duman, 2002), to date this work has not been translated into the clinical setting in terms of elucidating a causal role for stress and inflammatory cytokines in mediating hippocampal changes. However, it is well established that depressed patients have had higher amounts of early life maltreatment, hypersecrete cortisol (Vreeburg et al., 2009), have impaired GR functioning (Pace & Miller, 2009), have increased circulating concentrations of inflammatory cytokines (Simon et al., 2008) and CRP (Howren, Lamkin, & Suls, 2009), have lower levels of BDNF and reduced hippocampal volumes (MacQueen & Frodl, 2011), and show dysconnectivity between limbic and frontal brain regions, as reviewed in this chapter.

Importantly, the interactions between different systems need to be understood, and this is a major focus of current research. So far, we conclude that a model integrating childhood maltreatment, cortisol abnormalities, a decrease in neurotrophic factors, inflammatory states, and genetics needs to be established to provide a cohesive explanation of all the findings (Figure 2). When one of the components of the integrated systems gets altered, the functional balance of the whole will be disturbed, probably leading to stress-related brain disorders like depression. The hope would be that through disentangling these

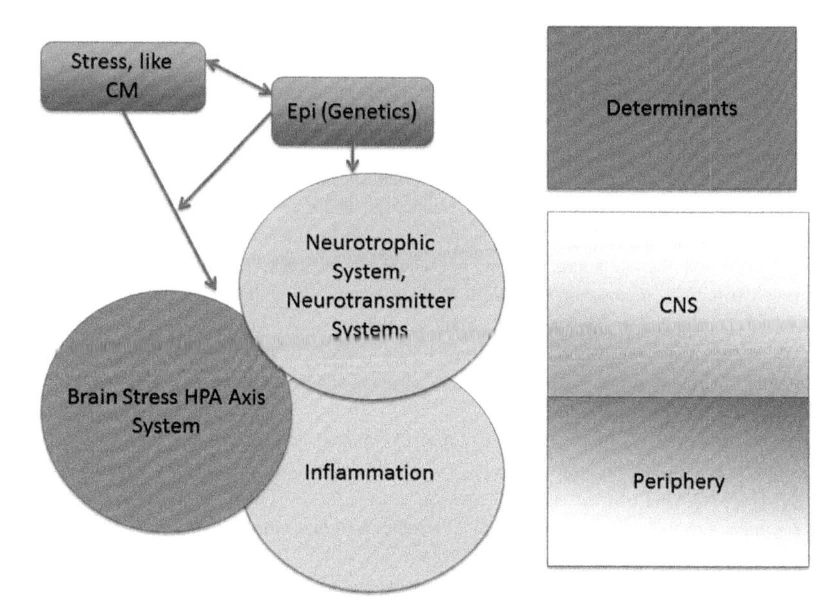

FIGURE 2 A model integrating childhood maltreatment and epigenetics with brain cortisol abnormalities, a decrease in neurotrophic factors, changes in neurotransmitter systems, and inflammatory states might explain abnormalities in behavior as well as brain structure and function.

systems, different subtypes of depression can be defined and diagnosis might be based more closely on what is happening in the brain rather then just on behavioral assessments.

REFERENCES

Ansorge, M. S., Zhou, M., Lira, A., Hen, R., & Gingrich, J. A. (2004). Early-life blockade of the 5-HT transporter alters emotional behavior in adult mice. *Science, 306*, 879–881.

Ansorge, M. S., Morelli, E., & Gingrich, J. A. (2008). Inhibition of serotonin but not norepinephrine transport during development produces delayed, persistent perturbations of emotional behaviors in mice. *Journal of Neuroscience, 28*, 199–207.

Beach, S. R. H., Brody, G. H., Todorov, A. A., Gunter, T. D., & Philibert, R. A. (2010). Methylation at SLC6A4 is linked to family history of child abuse: an examination of the Iowa adoptee sample. *American Journal of Medical Genetics Part B, Neuropsychiatric Genetics, 153B*, 710–713.

Beach, S. R. H., Brody, G. H., Todorov, A. A., Gunter, T. D., & Philibert, R. A. (2011). Methylation at 5HTT mediates the impact of child sex abuse on women's antisocial behavior: an examination of the Iowa adoptee sample. *Psychosomatic Medicine, 73*, 83–87.

Becker, S., MacQueen, G., & Wojtowicz, J. M. (2009). Computational modeling and empirical studies of hippocampal neurogenesis-dependent memory: effects of interference, stress and depression. *Brain Research, 1299*, 45–54.

Booij, L., Wang, D., Levesque, M. L., Tremblay, R. E., & Szyf, M. (2013). Looking beyond the DNA sequence: the relevance of DNA methylation processes for the stress-diathesis model of depression. *Philosophical Transactions of the Royal Society of London Series B: Biological Sciences, 368*, 20120251.

Booij, L., Szyf, M., Carballedo, A., Frey, E. M., Morris, D., Dymov, S., … Frodl, T. (2015). DNA methylation of the serotonin transporter gene in peripheral cells and stress-related changes in hippocampal volume: a study in depressed patients and healthy controls. *PLoS One, 10*, e0119061.

Booij, L., Szyf, M., Carballedo, A., Morris, D., Ly, V., Fahey, C., … Frodl, T. (2014). *The role of SLC6A4 DNA methylation in stress-related changes in hippocampal volume: A study in depressed patients and healthy controls*. Vancouver: CINP.

Booij, L., Tremblay, R. E., Szyf, M., & Benkelfat, C. (2015). Genetic and Early environmental influences on the serotonin system: consequences for brain development and risk for psychopathology. *Journal of Psychiatry & Neuroscience, 40*, 5–18.

Bremner, J. D., Randall, P., Vermetten, E., Staib, L., Bronen, R. A., Mazure, C., … Charney, D. S. (1997). Magnetic resonance imaging-based measurement of hippocampal volume in post-traumatic stress disorder related to childhood physical and sexual abuse—a preliminary report. *Biological Psychiatry, 41*, 23–32.

Bruehl, H., Wolf, O. T., & Convit, A. (2009). A blunted cortisol awakening response and hippocampal atrophy in type 2 diabetes mellitus. *Psychoneuroendocrinology, 34*, 815–821.

Bruehl, H., Wolf, O. T., Sweat, V., Tirsi, A., Richardson, S., & Convit, A. (2009). Modifiers of cognitive function and brain structure in middle-aged and elderly individuals with type 2 diabetes mellitus. *Brain Research, 1280*, 186–194.

Brzezicka, A. (2013). Integrative deficits in depression and in negative mood states as a result of fronto-parietal network dysfunctions. *Acta Neurobiologiae Experimentalis (Wars), 73*, 313–325.

Buckner, R. L., Andrews-Hanna, J. R., & Schacter, D. L. (2008). The brain's default network: anatomy, function, and relevance to disease. *Annals of the New York Academy of Sciences, 1124*, 1–38.

Bueller, J. A., Aftab, M., Sen, S., Gomez-Hassan, D., Burmeister, M., & Zubieta, J. K. (2006). BDNF Val66Met allele is associated with reduced hippocampal volume in healthy subjects. *Biological Psychiatry, 59*, 812–815.

Bush, G., Luu, P., & Posner, M. I. (2000). Cognitive and emotional influences in anterior cingulate cortex. *Trends in Cognitive Sciences, 4*, 215–222.

Campbell, S., Marriott, M., Nahmias, C., & MacQueen, G. M. (2004). Lower hippocampal volume in patients suffering from depression: a meta-analysis. *American Journal of Psychiatry, 161*, 598–607.

Carballedo, A., Lisiecka, D., Fagan, A., Saleh, K., Ferguson, Y., Connolly, G., … Frodl, T. (2012). Early life adversity is associated with brain changes in subjects at family risk for depression. *World Journal of Biological Psychiatry, 13*, 569–578.

Carboni, L., Becchi, S., Piubelli, C., Mallei, A., Giambelli, R., Razzoli, M., … Domenici, E. (2010). Early-life stress and antidepressants modulate peripheral biomarkers in a gene-environment rat model of depression. *Progress in Neuropsycho-pharmacology & Biological Psychiatry.*

Carrion, V. G., Weems, C. F., & Reiss, A. L. (2007). Stress predicts brain changes in children: a pilot longitudinal study on youth stress, posttraumatic stress disorder, and the hippocampus. *Pediatrics, 119*, 509–516.

Caspi, A., Sugden, K., Moffitt, T. E., Taylor, A., Craig, I. W., Harrington, H., … Poulton, R. (2003). Influence of life stress on depression: moderation by a polymorphism in the 5-HTT gene. *Science, 301*, 386–389.

Chen, M. C., Hamilton, J. P., & Gotlib, I. H. (2010). Decreased hippocampal volume in healthy girls at risk of depression. *Archives of General Psychiatry, 67*, 270–276.

Choi, J., Jeong, B., Rohan, M., Polcari, A., & Teicher, M. (2009). Preliminary evidence for white matter tract abnormalities in young adults exposed to parental verbal abuse. *Biological psychiatry, 65*, 227–234.

Claes, S. (2009). Glucocorticoid receptor polymorphisms in major depression. *Annals of the New York Academy of Sciences, 1179*, 216–228.

Cohen, R. A., Grieve, S., Hoth, K. F., Paul, R. H., Sweet, L., Tate, D., … Williams, L. M. (2006). Early life stress and morphometry of the adult anterior cingulate cortex and caudate nuclei. *Biological Psychiatry, 59*, 975–982.

Cohen, R. A., Paul, R. H., Stroud, L., Gunstad, J., Hitsman, B. L., McCaffery, J., … Gordon, E. (2006). Early life stress and adult emotional experience: an international perspective. *International Journal of Psychiatry in Medicine, 36*, 35–52.

Colla, M., Kronenberg, G., Deuschle, M., Meichel, K., Hagen, T., Bohrer, M., & Heuser, I. (2007). Hippocampal volume reduction and HPA-system activity in major depression. *Journal of Psychiatric Research, 41*, 553–560.

Cunningham-Bussel, A. C., Root, J. C., Butler, T., Tuescher, O., Pan, H., Epstein, J., … Silbersweig, D. (2009). Diurnal cortisol amplitude and fronto-limbic activity in response to stressful stimuli. *Psychoneuroendocrinology, 34*, 694–704.

Dannlowski, U., Kugel, H., Huber, F., Stuhrmann, A., Redlich, R., Grotegerd, D., … Suslow, T. (2013). Childhood maltreatment is associated with an automatic negative emotion processing bias in the amygdala. *Human Brain Mapping, 34*, 2899–2909.

Dannlowski, U., Kugel, H., Redlich, R., Halik, A., Schneider, I., Opel, N., … Hohoff, C. (2014). Serotonin transporter gene methylation is associated with hippocampal gray matter volume. *Human Brain Mapping*.

De Bellis, M. D., Hall, J., Boring, A. M., Frustaci, K., & Moritz, G. (2001). A pilot longitudinal study of hippocampal volumes in pediatric maltreatment-related posttraumatic stress disorder. *Biological Psychiatry, 50*, 305–309.

De Bellis, M. D., Hooper, S. R., Woolley, D. P., & Shenk, C. E. (2010). Demographic, maltreatment, and neurobiological correlates of PTSD symptoms in children and adolescents. *Journal of Pediatric Psychology, 35*, 570–577.

De Kloet, E., Joëls, M., & Holsboer, F. (2005). Stress and the brain: from adaptation to disease. *Nature Reviews Neuroscience, 6*, 463–475.

De Quervain, D. J., Roozendaal, B., Nitsch, R. M., McGaugh, J. L., & Hock, C. (2000). Acute cortisone administration impairs retrieval of long-term declarative memory in humans. *Nature Neuroscience, 3*, 313–314.

Dedovic, K., Engert, V., Duchesne, A., Lue, S. D., Andrews, J., Efanov, S. I., … Pruessner, J. C. (2010). Cortisol awakening response and hippocampal volume: vulnerability for major depressive disorder? *Biological Psychiatry, 68*, 847–853.

Devlin, A. M., Brain, U., Austin, J., & Oberlander, T. F. (2010). Prenatal exposure to maternal depressed mood and the *MTHFR C677T* variant affect *SLC6A4* methylation in infants at birth. *PLoS One, 5*, e12201.

Dranovsky, A., & Hen, R. (2006). Hippocampal neurogenesis: regulation by stress and antidepressants. *Biological Psychiatry, 59*, 1136–1143.

Duman, R. S. (2002). Pathophysiology of depression: the concept of synaptic plasticity. *European Psychiatry, 17*(Suppl. 3), 306–310.

Duman, R. S. (2004). Depression: a case of neuronal life and death? *Biological Psychiatry, 56*, 140–145.

Egan, M. F., Kojima, M., Callicott, J. H., Goldberg, T. E., Kolachana, B. S., Bertolino, A., … Weinberger, D. R. (2003). The BDNF val66met polymorphism affects activity-dependent secretion of BDNF and human memory and hippocampal function. *Cell, 112*, 257–269.

Eisenberger, N. I., Inagaki, T. K., Rameson, L. T., Mashal, N. M., & Irwin, M. R. (2009). An fMRI study of cytokine-induced depressed mood and social pain: the role of sex differences. *Neuroimage, 47*, 881–890.

Eisenberger, N. I., Berkman, E. T., Inagaki, T. K., Rameson, L. T., Mashal, N. M., & Irwin, M. R. (2010). Inflammation-induced anhedonia: endotoxin reduces ventral striatum responses to reward. *Biological Psychiatry, 68*, 748–754.

Elgh, E., Lindqvist Astot, A., Fagerlund, M., Eriksson, S., Olsson, T., & Nasman, B. (2006). Cognitive dysfunction, hippocampal atrophy and glucocorticoid feedback in Alzheimer's disease. *Biological Psychiatry, 59*, 155–161.

Eluvathingal, T. J., Chugani, H. T., Behen, M. E., Juhasz, C., Muzik, O., Maqbool, M., & Makki, M. (2006). Abnormal brain connectivity in children after early severe socioemotional deprivation: a diffusion tensor imaging study. *Pediatrics, 117*, 2093.

Elzinga, B. M., Molendijk, M. L., Oude Voshaar, R. C., Bus, B. A., Prickaerts, J., Spinhoven, P., & Penninx, B. J. (2011). The impact of childhood abuse and recent stress on serum brain-derived neurotrophic factor and the moderating role of BDNF Val66Met. *Psychopharmacology (Berl), 214*, 319–328.

Ende, G., Demirakca, T., & Tost, H. (2006). The biochemistry of dysfunctional emotions: proton MR spectroscopic findings in major depressive disorder. *Progress in Brain Research, 156*, 481–501.

Esaki, T., Cook, M., Shimoji, K., Murphy, D. L., Sokoloff, L., & Holmes, A. (2005). Developmental disruption of serotonin transporter function impairs cerebral responses to whisker stimulation in mice. *Proceedings of the National Academy of Sciences of the United States of America, 102*, 5582–5587.

Fox, M. D., Corbetta, M., Snyder, A. Z., Vincent, J. L., & Raichle, M. E. (2006). Spontaneous neuronal activity distinguishes human dorsal and ventral attention systems. *Proceedings of the National Academy of Sciences, 103*, 10046–10051.

Frodl, T., Meisenzahl, E. M., Zetzsche, T., Hohne, T., Banac, S., Schorr, C., ... Moller, H. J. (2004). Hippocampal and amygdala changes in patients with major depressive disorder and healthy controls during a 1-year follow-up. *Journal of Clinical Psychiatry, 65*, 492–499.

Frodl, T., Meisenzahl, E. M., Zill, P., Baghai, T., Rujescu, D., Leinsinger, G., ... Moller, H. J. (2004). Reduced hippocampal volumes associated with the long variant of the serotonin transporter polymorphism in major depression. *Archives of General Psychiatry, 61*, 177–183.

Frodl, T., Schule, C., Schmitt, G., Born, C., Baghai, T., Zill, P., ... Meisenzahl, E. M. (2007). Association of the brain-derived neurotrophic factor Val66Met polymorphism with reduced hippocampal volumes in major depression. *Archives of General Psychiatry, 64*, 410–416.

Frodl, T., Szyf, M., Carballedo, A., Ly, V., Dymov, S., Vaisheva, F., ... Booij, L. (2015). DNA methylation of the serotonin transporter gene (SLC6A4) is associated with brain function involved in processing emotional stimuli. *Journal of Psychiatry & Neuroscience, 40*, 296–305.

Frodl, T., Zill, P., Baghai, T., Schule, C., Rupprecht, R., Zetzsche, T., ... Meisenzahl, E. M. (2008). Reduced hippocampal volumes associated with the long variant of the tri- and diallelic serotonin transporter polymorphism in major depression. *American Journal of Medical Genetics Part B, Neuropsychiatric Genetics, 147B*, 1003–1007.

Frodl, T. S., Koutsouleris, N., Bottlender, R., Born, C., Jager, M., Scupin, I., ... Meisenzahl, E. M (2008). Depression-related variation in brain morphology over 3 years: effects of stress? *Archives of General Psychiatry, 65*, 1156–1165.

Frodl, T., Bokde, A. L., Scheuerecker, J., Lisiecka, D., Schoepf, V., Hampel, H., ... Meisenzahl, E. (2010). Functional connectivity bias of the orbitofrontal cortex in drug-free patients with major depression. *Biological Psychiatry, 67*, 161–167.

Frodl, T., Reinhold, E., Koutsouleris, N., Donohoe, G., Bondy, B., Reiser, M., ... Meisenzahl, E. M. (2010). Childhood stress, serotonin transporter gene and brain structures in major depression. *Neuropsychopharmacology, 35*, 1383–1390.

Frodl, T., Reinhold, E., Koutsouleris, N., Reiser, M., & Meisenzahl, E. M. (2010). Interaction of childhood stress with hippocampus and prefrontal cortex volume reduction in major depression. *Journal of Psychiatric Research, 44*, 799–807.

Frodl, T., Carballedo, A., Hughes, M. M., Saleh, K., Fagan, A., Skokauskas, N., ... Connor, T. J. (2012). Reduced expression of glucocorticoid-inducible genes GILZ and SGK-1: high IL-6 levels are associated with reduced hippocampal volumes in major depressive disorder. *Translational Psychiatry, 2*, e88.

Frodl, T., Skokauskas, N., Frey, E. M., Morris, D., Gill, M., & Carballedo, A. (2014). BDNF Val66Met genotype interacts with childhood adversity and influences the formation of hippocampal subfields. *Human Brain Mapping, 35*, 5776–5783.

Gatt, J. M., Nemeroff, C. B., Dobson-Stone, C., Paul, R. H., Bryant, R. A., Schofield, P. R., ... Williams, L. M. (2009). Interactions between BDNF Val66Met polymorphism and early life stress predict brain and arousal pathways to syndromal depression and anxiety. *Molecular Psychiatry, 14*, 681–695.

Gerritsen, L., Tendolkar, I., Franke, B., Vasquez, A. A., Kooijman, S., Buitelaar, J., ... Rijpkema, M. (2011). BDNF Val66Met genotype modulates the effect of childhood adversity on subgenual anterior cingulate cortex volume in healthy subjects. *Molecular Psychiatry.*

Golier, J. A., Yehuda, R., De Santi, S., Segal, S., Dolan, S., & De Leon, M. J. (2005). Absence of hippocampal volume differences in survivors of the Nazi Holocaust with and without post-traumatic stress disorder. *Psychiatry Research, 139*, 53–64.

Gonul, A. S., Kitis, O., Eker, M. C., Eker, O. D., Ozan, E., & Coburn, K. (2011). Association of the brain-derived neurotrophic factor Val66Met polymorphism with hippocampus volumes in drug-free depressed patients. *The World Journal of Biological Psychiatry, 12*, 110–118.

Goosens, K. A., & Sapolsky, R. M. (2007). Stress and glucocorticoid contributions to normal and pathological aging. In D. R. Riddle (Ed.), Brain Aging: *Models, Methods, and Mechanisms*. Boca Raton (FL): CRC Press/Taylor & Francis; 2007. Chapter 13.

Gunduz-Bruce, H., Szeszko, P. R., Gueorguieva, R., Ashtari, M., Robinson, D. G., Kane, J. M., & Bilder, R. M. (2007). Cortisol levels in relation to hippocampal sub-regions in subjects with first episode schizophrenia. *Schizophrenia Research, 94*, 281–287.

Gusnard, D. A., Akbudak, E., Shulman, G. L., & Raichle, M. E. (2001). Medial prefrontal cortex and self-referential mental activity: relation to a default mode of brain function. *Proceedings of the National Academy of Sciences of the United States of America, 98*, 4259–4264.

Harrison, N. A., Brydon, L., Walker, C., Gray, M. A., Steptoe, A., & Critchley, H. D. (2009). Inflammation causes mood changes through alterations in subgenual cingulate activity and mesolimbic connectivity. *Biological Psychiatry, 66*, 407–414.

Hasler, G., Van Der Veen, J. W., Tumonis, T., Meyers, N., Shen, J., & Drevets, W. C. (2007). Reduced prefrontal glutamate/glutamine and gamma-aminobutyric acid levels in major depression determined using proton magnetic resonance spectroscopy. *Archives of General Psychiatry, 64*, 193–200.

Heim, C., & Nemeroff, C. (2001). The role of childhood trauma in the neurobiology of mood and anxiety disorders: preclinical and clinical studies. *Biological Psychiatry, 49*(12), 1023–1039.

Holsboer, F. (2000). The corticosteroid receptor hypothesis of depression. *Neuropsychopharmacology, 23*, 477–501.

Holsen, L. M., Lancaster, K., Klibanski, A., Whitfield-Gabrieli, S., Cherkerzian, S., Buka, S., & Goldstein, J. M. (2013). HPA-axis hormone modulation of stress response circuitry activity in women with remitted major depression. *Neuroscience, 250*, 733–742.

Horn, D. I., Yu, C., Steiner, J., Buchmann, J., kaufmann, J., Osoba, A., ... Walter, M. (2010). Glutamatergic and resting-state functional connectivity correlates of severity in major depression - the role of pregenual anterior cingulate cortex and anterior insula. *Frontiers in Systems Neuroscience, 4*.

Howren, M. B., Lamkin, D. M., & Suls, J. (2009). Associations of depression with C-reactive protein, IL-1, and IL-6: a meta-analysis. *Psychosomatic Medicine, 71*, 171–186.

Huang, E. J., & Reichardt, L. F. (2001). Neurotrophins: roles in neuronal development and function. *Annual Review of Neuroscience, 24*, 677–736.

Ising, M., Horstmann, S., Kloiber, S., Lucae, S., Binder, E. B.Kern, N., ... (2007). Combined dexamethasone/corticotropin releasing hormone test predicts treatment response in major depression - a potential biomarker? *Biological Psychiatry, 62*, 47–54.

Joffe, R. T., Gatt, J. M., Kemp, A. H., Grieve, S., Dobson-Stone, C., Kuan, S. A., ... Williams, L. M. (2009). Brain derived neurotrophic factor Val66Met polymorphism, the five factor model of personality and hippocampal volume: Implications for depressive illness. *Human Brain Mapping, 30*, 1246–1256.

Juruena, M. F., Pariante, C. M., Papadopoulos, A. S., Poon, L., Lightman, S., & Cleare, A. J. (2009). Prednisolone suppression test in depression: prospective study of the role of HPA axis dysfunction in treatment resistance. *British Journal of Psychiatry, 194*, 342–349.

Juster, R. P., McEwen, B. S., & Lupien, S. J. (2010). Allostatic load biomarkers of chronic stress and impact on health and cognition. *Neuroscience and Biobehavioral Reviews, 35*, 2–16.

Kang, H. J., Kim, J. M., Stewart, R., Kim, S. Y., Bae, K. Y., Kim, S. W., ... Yoon, J. S. (2013). Association of SLC6A4 methylation with early adversity, characteristics and outcomes in depression. *Progress in Neuro-psychopharmacology & Biological Psychiatry, 44*, 23–28.

Kaufman, J., Yang, B. Z., Douglas-Palumberi, H., Grasso, D., Lipschitz, D., Houshyar, S., ... Gelernter, J. (2006). Brain-derived neurotrophic factor-5-HTTLPR gene interactions and environmental modifiers of depression in children. *Biological Psychiatry, 59*, 673–680.

Kaymak, S. U., Demir, B., Senturk, S., Tatar, I., Aldur, M. M., & Ulug, B. (2010). Hippocampus, glucocorticoids and neurocognitive functions in patients with first-episode major depressive disorders. *European Archives of Psychiatry and Clinical Neuroscience, 260*, 217–223.

Kelly, W. F., Kelly, M. J., & Faragher, B. (1996). A prospective study of psychiatric and psychological aspects of Cushing's syndrome. *Clinical Endocrinology (Oxf), 45*, 715–720.

Kempton, M. J., Salvador, Z., Munafo, M. R., Geddes, J. R., Simmons, A., Frangou, S., & Williams, S. C. (2011). Structural neuroimaging studies in major depressive disorder. Meta-analysis and comparison with bipolar disorder. *Archives of General Psychiatry, 68*, 675–690.

Kessler, R. C. (1997). The effects of stressful life events on depression. *Annual Review of Psychology, 48*, 191–214.

Knoops, A. J., Gerritsen, L., Van Der Graaf, Y., Mali, W. P., & Geerlings, M. I. (2010). Basal hypothalamic pituitary adrenal axis activity and hippocampal volumes: the SMART-Medea study. *Biological Psychiatry, 67*, 1191–1198.

Kremen, W. S., O'brien, R. C., Panizzon, M. S., Prom-Wormley, E., Eaves, L. J., Eisen, S. A., ... Franz, C. E. (2010). Salivary cortisol and prefrontal cortical thickness in middle-aged men: a twin study. *Neuroimage, 53*, 1093–1102.

Krishnan, V., & Nestler, E. J. (2008). The molecular neurobiology of depression. *Nature, 455*, 894–902.

Kullmann, J. S., Grigoleit, J. S., lichte, P., Kobbe, P., Rosenberger, C., Banner, C., … Schedlowski, M. (2012). Neural response to emotional stimuli during experimental human endotoxemia. *Human Brain Mapping*.

Lahti, J., Raikkonen, K., Bruce, S., Heinonen, K., Pesonen, A. K., Rautanen, A., … Eriksson, J. G. (2011). Glucocorticoid receptor gene haplotype predicts increased risk of hospital admission for depressive disorders in the Helsinki birth cohort study. *Journal of Psychiatric Research*.

Lai, C. H., & Hsu, Y. Y. (2011). A subtle grey-matter increase in first-episode, drug-naive major depressive disorder with panic disorder after 6 weeks' duloxetine therapy. *The International Journal of Neuropsychopharmacology, 14*, 225–235.

Lee, T., Jarome, T., Li, S. J., Kim, J. J., & Helmstetter, F. J. (2009). Chronic stress selectively reduces hippocampal volume in rats: a longitudinal magnetic resonance imaging study. *Neuroreport, 20*, 1554–1558.

Lesch, K. P., Bengel, D., Heils, A., Sabol, S. Z., Greenberg, B. D., Petri, S., … Murphy, D. L. (1996). Association of anxiety-related traits with a polymorphism in the serotonin transporter gene regulatory region. *Science, 274*, 1527–1531.

Lesch, K. P., & Mossner, R. (1998). Genetically driven variation in serotonin uptake: is there a link to affective spectrum, neurodevelopmental, and neurodegenerative disorders? *Biological Psychiatry, 44*, 179–192.

Leuner, B., Gould, E., & Shors, T. J. (2006). Is there a link between adult neurogenesis and learning? *Hippocampus, 16*, 216–224.

Liao, Y., Huang, X., Wu, Q., Yang, C., Kuang, W., Du, M., … Gong, Q. (2013). Is depression a disconnection syndrome? Meta-analysis of diffusion tensor imaging studies in patients with MDD. *Journal of Psychiatry & Neuroscience, 38*, 49–56.

Lindauer, R. J., Olff, M., Van Meijel, E. P., Carlier, I. V., & Gersons, B. P. (2006). Cortisol, learning, memory, and attention in relation to smaller hippocampal volume in police officers with posttraumatic stress disorder. *Biological Psychiatry, 59*, 171–177.

Lisiecka, D., Meisenzahl, E., Scheuerecker, J., Schoepf, V., Whitty, P., Chaney, A., … Frodl, T. (2011). Neural correlates of treatment outcome in major depression. *International Journal of Neuropsychopharmacology, 14*, 521–534.

Lovallo, W. R., Robinson, J. L., Glahn, D. C., & Fox, P. T. (2010). Acute effects of hydrocortisone on the human brain: an fMRI study. *Psychoneuroendocrinology, 35*, 15–20.

Lu, B. (2003). Pro-region of neurotrophins: role in synaptic modulation. *Neuron, 39*, 735–738.

Luine, V., Martinez, C., Villegas, M., Magarinos, A. M., & McEwen, B. S. (1996). Restraint stress reversibly enhances spatial memory performance. *Physiology & Behavior, 59*, 27–32.

Lupien, S. J., De Leon, M., De Santi, S., Convit, A., Tarshish, C., Nair, N. P., … Meaney, M. J. (1998). Cortisol levels during human aging predict hippocampal atrophy and memory deficits. *Nature Neuroscience, 1*, 69–73.

Lupien, S. J., Wilkinson, C. W., Briere, S., Menard, C., Ng Ying Kin, N. M., & Nair, N. P. (2002). The modulatory effects of corticosteroids on cognition: studies in young human populations. *Psychoneuroendocrinology, 27*, 401–416.

Lupien, S. J., Parent, S., Evans, A. C., Tremblay, R. E., Zelazo, P. D., Corbo, V., … Seguin, J. R. (2011). Larger amygdala but no change in hippocampal volume in 10-year-old children exposed to maternal depressive symptomatology since birth. *Proceedings of the National Academy of Sciences of the United States of America, 108*, 14324–14329.

Lyons, D. M., Yang, C., Sawyer-Glover, A. M., Moseley, M. E., & Schatzberg, A. F. (2001). Early life stress and inherited variation in monkey hippocampal volumes. *Archives of General Psychiatry, 58*, 1145–1151.

MacLullich, A. M., Ferguson, K. J., Wardlaw, J. M., Starr, J. M., Deary, I. J., & Seckl, J. R. (2006). Smaller left anterior cingulate cortex volumes are associated with impaired hypothalamic-pituitary-adrenal axis regulation in healthy elderly men. *Journal of Clinical Endocrinology and Metabolism, 91*, 1591–1594.

MacQueen, G., & Frodl, T. (2011). The hippocampus in major depression: evidence for the convergence of the bench and bedside in psychiatric research? *Molecular Psychiatry, 16*, 252–264.

Malykhin, N. V., Lebel, R. M., Coupland, N. J., Wilman, A. H., & Carter, R. (2010). In vivo quantification of hippocampal subfields using 4.7 T fast spin echo imaging. *Neuroimage, 49*, 1224–1230.

Mann, J. J., & Currier, D. M. (2010). Stress, genetics and epigenetic effects on the neurobiology of suicidal behavior and depression. *European Psychiatry, 25*, 268–271.

McCrory, E. J., De Brito, S. A., Kelly, P. A., Bird, G., Sebastian, C. L., Mechelli, A., ... Viding, E. (2013). Amygdala activation in maltreated children during pre-attentive emotional processing. *British Journal of Psychiatry, 202*, 269–276.

McGowan, P. O., Sasaki, A., D'alessio, A. C., Dymov, S., Labonte, B., Szyf, M., ... Meaney, M. J. (2009). Epigenetic regulation of the glucocorticoid receptor in human brain associates with childhood abuse. *Nature Neuroscience, 12*, 342–348.

McGowan, P. O., Suderman, M., Sasaki, A., Huang, T. C., Hallett, M., Meaney, M. J., & Szyf, M. (2011). Broad epigenetic signature of maternal care in the brain of adult rats. *PLoS One, 6*, e14739.

McKinnon, M. C., Yucel, K., Nazarov, A., & MacQueen, G. M. (2009). A meta-analysis examining clinical predictors of hippocampal volume in patients with major depressive disorder. *Journal of Psychiatry & Neuroscience, 34*, 41–54.

Miller, A. H., Maletic, V., & Raison, C. L. (2009). Inflammation and its discontents: the role of cytokines in the pathophysiology of major depression. *Biological Psychiatry, 65*, 732–741.

Miller, E. J., Saint Marie, L. R., Breier, M. R., & Swerdlow, N. R. (2010). Pathways from the ventral hippocampus and caudal amygdala to forebrain regions that regulate sensorimotor gating in the rat. *Neuroscience, 165*, 601–611.

Modell, S., Lauer, C., Schreiber, W., Huber, J., Krieg, J., & Holsboer, F. (1998). Hormonal response pattern in the combined DEX-CRH test is stable over time in subjects at high familial risk for affective disorders. *Neuropsychopharmacology, 18*, 253–262.

Mondelli, V., Pariante, C. M., Navari, S., Aas, M., D'albenzio, A., Di Forti, M., ... Dazzan, P. (2010). Higher cortisol levels are associated with smaller left hippocampal volume in first-episode psychosis. *Schizophrenia Research, 119*, 75–78.

Murphy, M. L., & Frodl, T. (2011). Meta-analysis of diffusion tensor imaging studies shows altered fractional anisotropy occurring in distinct brain areas in association with depression. *Biology of Mood &Anxiety Disorders, 1*, 3.

Nestler, E. J. (2014). Epigenetic mechanisms of depression. *JAMA Psychiatry.*

Newcomer, J. W., Selke, G., Melson, A. K., Hershey, T., Craft, S., Richards, K., & Alderson, A. L. (1999). Decreased memory performance in healthy humans induced by stress-level cortisol treatment. *Archives of General Psychiatry, 56*, 527–533.

Newton, S. S., & Duman, R. S. (2004). Regulation of neurogenesis and angiogenesis in depression. *Current Neurovascular Research, 1*, 261–267.

O'Mara, S. M., Sanchez-Vives, M. V., Brotons-Mas, J. R., & O'Hare, E. (2009). Roles for the subiculum in spatial information processing, memory, motivation and the temporal control of behaviour. *Progress in Neuropsychopharmacology & Biological Psychiatry, 33*, 782–790.

Oberlander, T., Weinberg, J., Papsdorf, M., Grunau, R., Misri, S., & Devlin, A. (2008). Prenatal exposure to maternal depression, neonatal methylation of human glucocorticoid receptor gene (NR3C1) and infant cortisol stress responses. *Epigenetics, 3*, 97.

Ochsner, K. N., Ray, R. D., Cooper, J. C., Robertson, E. R., Chopra, S., Gabrieli, J. D., & Gross, J. J. (2004). For better or for worse: neural systems supporting the cognitive down- and up-regulation of negative emotion. *Neuroimage, 23*, 483–499.

Ohl, F., Michaelis, T., Vollmann-Honsdorf, G. K., Kirschbaum, C., & Fuchs, E. (2000). Effect of chronic psychosocial stress and long-term cortisol treatment on hippocampus-mediated memory and hippocampal volume: a pilot-study in tree shrews. *Psychoneuroendocrinology, 25*, 357–363.

Öngür, D., Ferry, A. T., & Price, J. L. (2003). Architectonic subdivision of the human orbital and medial prefrontal cortex. *Journal of Comparative Neurology, 460*, 425–449.

Pace, T. W., & Miller, A. H. (2009). Cytokines and glucocorticoid receptor signaling. Relevance to major depression. *Annals of the New York Academy of Sciences, 1179*, 86–105.

Pariante, C. M., & Lightman, S. L. (2008). The HPA axis in major depression: classical theories and new developments. *Trends in Neurosciences, 31*, 464–468.

Paslakis, G., Krumm, B., Gilles, M., Schweiger, U., Heuser, I., Richter, I., & Deuschle, M. (2011). Discrimination between patients with melancholic depression and healthy controls: comparison between 24-h cortisol profiles, the DST and the Dex/CRH test. *Psychoneuroendocrinology, 36*, 691–698.

Paul, R., Henry, L., Grieve, S., Guilmette, T., Niaura, R., Bryant, R., … Cohen, R. (2008). The relationship between early life stress and microstructural integrity of the corpus callosum in a non-clinical population. *Neuropsychiatric Disease and Treatment, 4*, 193.

Pederson, C. L., Maurer, S. H., Kaminski, P. L., Zander, K. A., Peters, C. M., Stokes-Crowe, L. A., & Osborn, R. E. (2004). Hippocampal volume and memory performance in a community-based sample of women with posttraumatic stress disorder secondary to child abuse. *Journal of Traumatic Stress, 17*, 37–40.

Peterson, A., Thome, J., Frewen, P., & Lanius, R. A. (2014). Resting-state neuroimaging studies: a new way of identifying differences and similarities among the anxiety disorders? *Canadian Journal of Psychiatry, 59*, 294–300.

Pezawas, L., Verchinski, B. A., Mattay, V. S., Callicott, J. H., Kolachana, B. S., Straub, R. E., … Weinberger, D. R. (2004). The brain-derived neurotrophic factor val66met polymorphism and variation in human cortical morphology. *Journal of Neuroscience, 24*, 10099–10102.

Price, J. L., & Drevets, W. C. (2012). Neural circuits underlying the pathophysiology of mood disorders. *Trends in Cognitive Sciences, 16*, 61–71.

Pruessner, M., Pruessner, J. C., Hellhammer, D. H., Bruce Pike, G., & Lupien, S. J. (2007). The associations among hippocampal volume, cortisol reactivity, and memory performance in healthy young men. *Psychiatry Research, 155*, 1–10.

Pruessner, J. C., Dedovic, K., Khalili-Mahani, N., Engert, V., Pruessner, M., Buss, C., … Lupien, S. (2008). Deactivation of the limbic system during acute psychosocial stress: evidence from positron emission tomography and functional magnetic resonance imaging studies. *Biological Psychiatry, 63*, 234–240.

Reichenberg, A., Yirmiya, R., Schuld, A., Kraus, T., Haack, M., Morag, A., & Pollmacher, T. (2001). Cytokine-associated emotional and cognitive disturbances in humans. *Archives of General Psychiatry, 58*, 445–452.

Risch, N., Herrell, R., Lehner, T., Liang, K. Y., Eaves, L., Hoh, J., … Merikangas, K. R. (2009). Interaction between the serotonin transporter gene (5-HTTLPR), stressful life events, and risk of depression: a meta-analysis. *JAMA, 301*, 2462–2471.

Root, J. C., Tuescher, O., Cunningham-Bussel, A., Pan, H., Epstein, J., Altemus, M., … Silbersweig, D. (2009). Frontolimbic function and cortisol reactivity in response to emotional stimuli. *Neuroreport, 20*, 429–434.

Rush, A. J., Giles, D. E., Schlesser, M. A., Orsulak, P. J., Parker, C. R., Jr., Weissenburger, J. E., … Vasavada, N. (1996). The dexamethasone suppression test in patients with mood disorders. *Journal of Clinical Psychiatry, 57*, 470–484.

Santarelli, L., Saxe, M., Gross, C., Surget, A., Battaglia, F., Dulawa, S., … Hen, R. (2003). Requirement of hippocampal neurogenesis for the behavioral effects of antidepressants. *Science, 301*, 805–809.

Sapolsky, R. M. (2001). Atrophy of the hippocampus in posttraumatic stress disorder: how and when? *Hippocampus, 11*, 90–91.

Sapolsky, R. M. (2003). Stress and plasticity in the limbic system. *Neurochemical Research, 28*, 1735–1742.

Sapolsky, R. M., Krey, L. C., & McEwen, B. S. (1986). The neuroendocrinology of stress and aging: the glucocorticoid cascade hypothesis. *Endocrine Reviews, 7*, 284–301.

Seeley, W. W., Menon, V., Schatzberg, A. F., Keller, J., Glover, G. H., Kenna, H., … Greicius, M. D. (2007). Dissociable intrinsic connectivity networks for salience processing and executive control. *Journal of Neuroscience, 27*, 2349–2356.

Sexton, C., MacKay, C., & Ebmeier, K. (2009). A systematic review of diffusion tensor imaging studies in affective disorders. *Biological Psychiatry, 66*, 814–823.

Sheline, Y. I., Price, J. L., Yan, Z., & Mintun, M. A. (2010). Resting-state functional MRI in depression unmasks increased connectivity between networks via the dorsal nexus. *Proceedings of the National Academy of Sciences, 107*, 11020–11025.

Simmons, N. E., Do, H. M., Lipper, M. H., & Laws, E. R., Jr. (2000). Cerebral atrophy in Cushing's disease. *Surgical Neurology, 53*, 72–76.

Simon, N. M., McNamara, K., Chow, C. W., Maser, R. S., Papakostas, G. I., Pollack, M. H., … Wong, K. K. (2008). A detailed examination of cytokine abnormalities in Major Depressive Disorder. *European Neuropsychopharmacology, 18*, 230–233.

Slavich, G. M., Way, B. M., Eisenberger, N. I., & Taylor, S. E. (2010). Neural sensitivity to social rejection is associated with inflammatory responses to social stress. *Proceedings of the National Academy of Sciences of the United States of America, 107*, 14817–14822.

Spinelli, S., Chefer, S., Suomi, S. J., Higley, J. D., Barr, C. S., & Stein, E. (2009). Early-life stress induces long-term morphologic changes in primate brain. *Archives of General Psychiatry, 66*, 658–665.

Stein, M. B., Koverola, C., Hanna, C., Torchia, M. G., & McClarty, B. (1997). Hippocampal volume in women victimized by childhood sexual abuse. *Psychological Medicine, 27*, 951–959.

Stockmeier, C. A., Mahajan, G. J., Konick, L. C., Overholser, J. C., Jurjus, G. J., Meltzer, H. Y., … Rajkowska, G. (2004). Cellular changes in the postmortem hippocampus in major depression. *Biological Psychiatry, 56*, 640–650.

Sundermann, B., Olde Lutke Beverborg, M., & Pfleiderer, B. (2014). Toward literature-based feature selection for diagnostic classification: a meta-analysis of resting-state fMRI in depression. *Frontiers in Human Neuroscience, 8*, 692.

Sylvester, C. M., Barch, D. M., Corbetta, M., Power, J. D., Schlaggar, B. L., & Luby, J. L. (2013). Resting state functional connectivity of the ventral attention network in children with a history of depression or anxiety. *Journal of the American Academy of Child and Adolescent Psychiatry, 52*, 1326–1336 e5.

Taylor, M. J. (2014). Could glutamate spectroscopy differentiate bipolar depression from unipolar? *Journal of Affective Disorders, 167*, 80–84.

Taylor, W. D., Steffens, D. C., Payne, M. E., MacFall, J. R., Marchuk, D. A., Svenson, I. K., & Krishnan, K. R. (2005). Influence of serotonin transporter promoter region polymorphisms on hippocampal volumes in late-life depression. *Archives of General Psychiatry, 62*, 537–544.

Teicher, M. H., Andersen, S. L., Polcari, A., Anderson, C. M., Navalta, C. P., & Kim, D. M. (2003). The neurobiological consequences of early stress and childhood maltreatment. *Neuroscience and Biobehavioral Reviews, 27*, 33–44.

Thomas, R., Hotsenpiller, G., & Peterson, D. (2007). Acute psychosocial stress reduces cell survival in adult hippocampal neurogenesis without altering proliferation. *Journal of Neuroscience, 27*, 2734.

Tomoda, A., Sheu, Y. S., Rabi, K., Suzuki, H., Navalta, C. P., Polcari, A., & Teicher, M. H. (2010). Exposure to parental verbal abuse is associated with increased gray matter volume in superior temporal gyrus. *Neuroimage*.

Tsankova, N., Berton, O., Renthal, W., Kumar, A., Neve, R., & Nestler, E. (2006). Sustained hippocampal chromatin regulation in a mouse model of depression and antidepressant action. *Nature Neuroscience, 9*, 519–525.

Tupler, L. A., & De Bellis, M. D. (2006). Segmented hippocampal volume in children and adolescents with posttraumatic stress disorder. *Biological Psychiatry, 59*, 523–529.

Tyler, W. J., Alonso, M., Bramham, C. R., & Pozzo-Miller, L. D. (2002). From acquisition to consolidation: on the role of brain-derived neurotrophic factor signaling in hippocampal-dependent learning. *Learning & Memory, 9*, 224–237.

Ugwu, I. D., Amico, F., Carballedo, A., Fagan, A. J., & Frodl, T. (2014). Childhood adversity, depression, age and gender effects on white matter microstructure: a DTI study. *Brain Structure & Function*.

Van Den Heuvel, M., Mandl, R., Luigjes, J., & Hulshoff Pol, H. (2008). Microstructural organization of the cingulum tract and the level of default mode functional connectivity. *Journal of Neuroscience, 28*, 10844–10851.

Van Harmelen, A. L., Van Tol, M. J., Van Der Wee, N. J., Veltman, D. J., Aleman, A., Spinhoven, P., … Elzinga, B. M. (2010). Reduced medial prefrontal cortex volume in adults reporting childhood emotional maltreatment. *Biological Psychiatry, 68*, 832–838.

Van Harmelen, A. L., Van Tol, M. J., Dalgleish, T., Van Der Wee, N. J., Veltman, D. J., Aleman, A., … Elzinga, B. M. (2014). Hypoactive medial prefrontal cortex functioning in adults reporting childhood emotional maltreatment. *Social Cognitive and Affective Neuroscience*.

Van Stegeren, A. H., Wolf, O. T., Everaerd, W., Scheltens, P., Barkhof, F., & Rombouts, S. A. (2007). Endogenous cortisol level interacts with noradrenergic activation in the human amygdala. *Neurobiology of Learning and Memory, 87*, 57–66.

Videbech, P., & Ravnkilde, B. (2004). Hippocampal volume and depression: a meta-analysis of MRI studies. *American Journal of Psychiatry, 161*, 1957–1966.

Vreeburg, S. A., Hoogendijk, W. J., Van Pelt, J., Derijk, R. H., Verhagen, J. C., Van Dyck, R., … Penninx, B. W. (2009). Major depressive disorder and hypothalamic-pituitary-adrenal axis activity: results from a large cohort study. *Archives of General Psychiatry, 66*, 617–626.

Vreeburg, S. A., Hartman, C. A., Hoogendijk, W. J., Van Dyck, R., Zitman, F. G., Ormel, J., & Penninx, B. W. (2010). Parental history of depression or anxiety and the cortisol awakening response. *British Journal of Psychiatry, 197*, 180–185.

Vythilingam, M., Heim, C., Newport, J., Miller, A. H., Anderson, E., Bronen, R., ... Bremner, J. D. (2002). Childhood trauma associated with smaller hippocampal volume in women with major depression. *American Journal of Psychiatry, 159*, 2072–2080.

Vythilingam, M., Vermetten, E., Anderson, G. M., Luckenbaugh, D., Anderson, E. R., Snow, J., ... Bremner, J. D. (2004). Hippocampal volume, memory, and cortisol status in major depressive disorder: effects of treatment. *Biological Psychiatry, 56*, 101–112.

Walter, M., Henning, A., Grimm, S., Schulte, R. F., Beck, J., Dydak, U., ... Northoff, G. (2009). The relationship between aberrant neuronal activation in the pregenual anterior cingulate, altered glutamatergic metabolism, and anhedonia in major depression. *Archives of General Psychiatry, 66*, 478–486.

Wang, L., Krishnan, K. R., Steffens, D. C., Potter, G. G., Dolcos, F., & McCarthy, G. (2008). Depressive state- and disease-related alterations in neural responses to affective and executive challenges in geriatric depression. *American Journal of Psychiatry, 165*, 863–871.

Wang, D., Szyf, M., Benkelfat, C., Provencal, N., Turecki, G., Caramaschi, D., ... Booij, L. (2012). Peripheral SLC6A4 DNA methylation is associated with in vivo measures of human brain serotonin synthesis and childhood physical aggression. *PLoS One, 7*, e39501.

Weaver, I., Cervoni, N., Champagne, F., D'alessio, A., Sharma, S., Seckl, J., ... Meaney, M. (2004). Epigenetic programming by maternal behavior. *Nature Neuroscience, 7*, 847–854.

Weniger, G., Lange, C., Sachsse, U., & Irle, E. (2008). Amygdala and hippocampal volumes and cognition in adult survivors of childhood abuse with dissociative disorders. *Acta Psychiatrica Scandinavica, 118*, 281–290.

Wiedenmayer, C. P., Bansal, R., Anderson, G. M., Zhu, H., Amat, J., Whiteman, R., & Peterson, B. S. (2006). Cortisol levels and hippocampus volumes in healthy preadolescent children. *Biological Psychiatry, 60*, 856–861.

Wolf, O. T., Convit, A., De Leon, M. J., Caraos, C., & Qadri, S. F. (2002). Basal hypothalamo-pituitary-adrenal axis activity and corticotropin feedback in young and older men: relationships to magnetic resonance imaging-derived hippocampus and cingulate gyrus volumes. *Neuroendocrinology, 75*, 241–249.

Woon, F. L., & Hedges, D. W. (2008). Hippocampal and amygdala volumes in children and adults with childhood maltreatment-related posttraumatic stress disorder: a meta-analysis. *Hippocampus, 18*, 729–736.

Wright, C. E., Strike, P. C., Brydon, L., & Steptoe, A. (2005). Acute inflammation and negative mood: mediation by cytokine activation. *Brain, Behavior, and Immunity, 19*, 345–350.

Young, A. H., Sahakian, B. J., Robbins, T. W., & Cowen, P. J. (1999). The effects of chronic administration of hydrocortisone on cognitive function in normal male volunteers. *Psychopharmacology (Berl), 145*, 260–266.

Zakzanis, K. K., Leach, L., & Kaplan, E. (1998). On the nature and pattern of neurocognitive function in major depressive disorder. *Neuropsychiatry, Neuropsychology, and Behavioral Neurology, 11*, 111–119.

Chapter 10

Depression in Children and Adolescents

Tomoya Hirota[1], Gordana Milavić[2], Fiona McNicholas[3,4,5], Thomas Frodl[6,7], Norbert Skokauskas[6,8]

[1]*Department of Psychiatry, Division of Child and Adolescent Psychiatry, University of California San Francisco, San Francisco, CA, USA;* [2]*National and Specialist Services, Michael Rutter Centre, Maudsley Hospital, London, UK;* [3]*Department of Psychiatry, University College Dublin, Dublin, Ireland;* [4]*Lucena Clinic, Dublin, Ireland;* [5]*Department of Child Psychiatry, Our Lady's Children Hospital Crumlin, Dublin, Ireland;* [6]*Department of Psychiatry, Trinity College Dublin, Ireland;* [7]*Department of Psychiatry and Psychotherapy, Otto von Guericke University, Magdeburg, Germany;* [8]*Centre for Child and Adolescent Mental Health and Child Protection, Department of Neuroscience, Norges teknisk-naturvitenskapelige universitet NTNU, Trondheim, Norway*

BACKGROUND

For over half a century clinicians have been aware that children and adolescents can become depressed and require treatment (Frommer, 1967; Pearce, 1978; Puig-Antich et al., 1987). There has, however, been a considerable delay before depressive disorders in children and adolescents became the focus of renewed academic study (Birmaher, Ryan, et al., 1996) with further delay in recognizing and treating depressive disorders in routine clinical practice (Coyle et al., 2003). It is now widely accepted that depression is one of the most common psychological disorders in children and adolescents, often presenting as a severe, chronic, and recurring condition with a high risk of self-harm and suicide. Depressive disorders are associated with adverse psychosocial circumstances and adverse life events and may lead to serious impairment in the child's and adolescent's current functioning, and usher a poor prognosis and chronic course that extend well into adulthood (Birmaher, Ryan, et al., 1996).

The clinical picture of depression in children and adolescents is, in the main, similar to that found in adults, but some differences stemming from biological, developmental, and social attributes will influence the way depression presents in children, particularly in younger children. For instance, irritability, low frustration levels, temper tantrums, somatic complaints, and social withdrawal are found more commonly in children (Birmaher et al., 2007;

Systems Neuroscience in Depression. http://dx.doi.org/10.1016/B978-0-12-802456-0.00010-8

Goodyer & Cooper, 1993). Disinterest in general appearance, changes in peer relationships and social withdrawal, a drop in academic performance, self-harming, and other risk-taking behaviors may precede a more typical presentation of depression in adolescence. According to the World Health Organization (WHO), clinically significant depression is diagnosed when a child or adolescent presents with at least two weeks' duration of depressed mood or irritability and either loss of interest or pleasure and/or tiredness as well as a number of other symptoms, including poor or increased sleep, poor concentration or indecisiveness, low self-confidence, poor or increased appetite, suicidal thoughts or acts, agitation or slowing of movements, guilt, or self-blame (WHO, 1992). The degree of severity of the depressive disorder is determined by the number of associated symptoms (WHO, 1992).

Prevalence rates of depressive disorders in children and adolescents vary across studies. The most quoted rates are Angold and Costello's 12-month period prevalence rates (Angold & Costello, 2001) for major depression of 1% for prepubertal children and 3% for postpubertal children and adolescents. Merikangas et al. (2010), looking at the lifetime prevalence of adolescent depression in the National Comorbidity Survey, Adolescent Supplement, a face-to-face study of 10,123 US adolescents aged 13–18 years, used the modified version of the WHO composite international diagnostic interview to establish rates of major depressive disorder (MDD) or dysthymia. The authors of the study found a total of 11.7% rate of depression with 8.7% having suffered severe impairment. Merikangas et al. noticed a female-to-male ratio of 15.9 to 7.7 in this age group as well and showed how the rates of depressive disorder increased as the young people grew older: 13- to 14-year-olds had rates of MDDs and dysthymia of 8.4%, 15- to 16-year olds 12.6%, and 17- to 18-year-olds 15.4%.

The transition to adulthood is a critical developmental period, marked by a series of tasks that include strengthening and expanding self-concept, forming stable intimate relationships, making career decisions, and achieving independence from parents. Homotypic and heterotypic continuities of depressive disorders from childhood to adulthood have been the subject of an increasing number of studies (Birmaher et al., 2004; Copeland, Shanahan, Costello, & Angold, 2009; Fombonne, Wostear, Cooper, Harrington, & Rutter, 2001; Harrington, Fudge, Rutter, Pickles, & Hill, 1990; Kovacs et al., 1984; Luby, Si, Belden, Tandon, & Spitznagel, 2009; Reinke & Ostrander, 2008; Rutter, Kim-Cohen, & Maughan, 2006). Heterotypical continuity means that an underlying (developmental) process or impairment stays the same, but the manifestations (i.e., depressive symptoms) do not. Homeotypic continuity is seen as the opposite of heterotypic continuity. Carballo et al. (2011) found that homotypic continuity of depressive disorder into adulthood was associated with an absence of psychiatric comorbidity prior to age 18 years. Also, adolescent-onset depression, by contrast to childhood-onset depression, showed greater homotypic continuity. Luby et al. (2009) established homotypic continuities from preschool to early school years, but other studies point to the "complexity" underlying heterotypic outcomes.

Depression in children or adolescents can be difficult to diagnose, and a clinician may need to see a patient over time to determine the appropriate diagnosis. A thorough mental health assessment, including joint and separate interviews with the child and his or her parents/carers, is a prerequisite to a comprehensive evaluation aimed at establishing the presence or otherwise of a depressive disorder. Psychoeducation, self-help, and family support techniques are recommend at the outset of any intervention (National Institute for Health and Clinical Excellence NICE, 2015). Watchful waiting, nondirective supportive therapy, group cognitive behavioral therapy (CBT), and guided self-help are recommended for managing mild depression. Should the disorder become more severe or not respond to interventions at the primary care level, evidence-based psychological treatments, including CBT, interpersonal psychotherapy, and short-term family therapy, should be introduced (Birmaher et al., 2007; Brent et al., 1997; Mufson et al., 2004; National Institute for Health and Clinical Excellence NICE, 2015). Should these interventions not be effective, fluoxetine, a selective serotonin reuptake inhibitor (SSRI) antidepressant, is recommended as the first-line medication (National Institute for Health and Clinical Excellence NICE, 2015). However, it should be prescribed in conjunction with psychological treatments and only following an assessment and diagnosis by a child and adolescent psychiatrist (National Institute for Health and Clinical Excellence NICE, 2015). SSRIs other than fluoxetine, such as sertraline or citalopram, should only be used as a second-line treatment. A meta-analysis of studies demonstrated the overall benefit of the use of fluoxetine medication, showing that there is consistent evidence for a statistically significant benefit but pointing to modest effect sizes. The response rates in the Fluoxetine trial groups varied between 41% and 61% and in the placebo groups between 20% and 35% (Hetrick, Merry, McKenzie, Sindahl, & Proctor, 2007).

The administration of antidepressants at a much earlier stage of treatment is indicated in severe depression, as has been shown in clinical trials (Dubicka et al., 2010; Goodyer et al., 2007; March et al., 2004, 2007). The study of the Treatment of Resistant Depression in Adolescents showed that in those young people where there was no response to one SSRI, a switch to another antidepressant and the addition of CBT showed better results than a switch to another antidepressant without CBT (Brent & Birmaher, 2006).

The safety and effectiveness debate with respect to the use of antidepressants continued after the publication of the NICE Guideline (2005) in the United Kingdom and the Practice Parameters in the United States (Birmaher et al., 2007). The Treatment for Adolescents with Depression Study (TADS) offered extended findings, which established that all treatment arms were effective in reducing suicidal ideation but that the combined treatment of fluoxetine and

CBT reduced the risk of suicidal events in contrast to fluoxetine-treated patients who had more suicide-related events (Emslie et al., 2006). The TADS predictors of suicidal events included higher levels of self-reported suicidal ideation and depression at baseline, only slight improvement in depression, and acute interpersonal conflict (73% cases) (Vitiello et al., 2009).

Overall, "adverse-events" analyses of clinical trials show that one to three individuals in 100 taking antidepressants will have an onset or worsening of suicidality. It should be emphasized that there were no completed suicides in any of the quoted clinical trials.

Electroconvulsive therapy (ECT) is rarely used in the treatment of depression in children and adolescents, and apart from some anecdotal cases, there are no randomized controlled trials using ECT in children and adolescents. Children and young people have a lower seizure threshold, and the risk of prolonged seizures is increased. However, procedures and practices have been progressively refined, including modern anesthetic methods. Nevertheless, fewer than 1% of all patients treated with ECT are children and adolescents (Ghaziuddin et al., 2004).

NORMAL BRAIN STRUCTURAL AND FUNCTIONAL DEVELOPMENT

The minimal increase in brain size from childhood to adolescence clouds the significant remodeling and restructuring of both white and gray matter. Significant increases in neuronal production and the size of neurons (white matter) are matched with subsequent selective pruning, which is genetically preprogrammed and subject to modification by experience, following "use-it-or-lose-it" principles. Around adolescence there is a peak in gray matter volume, only to reduce again in adulthood, following synaptic pruning. Myelination of neurons allows for faster and more efficient transmission with reduced energy expenditure involving fewer structural areas alongside increased connectivity between the various neural systems. There is an uneven timeline for development going from "bottom up" with the subcortical limbic areas developing first, followed by the prefrontal cortical (PFC) areas, which do not reach full functional maturity until the mid-twenties. This uneven developmental timing exposes the adolescent to more influence by the earlier maturing subcortical system, which is reward-driven, over the later developing "top-down" systems, which influence the executive functions of planning and impulse control (Casey & Jones, 2010). This unbalance has been suggested for the higher risk of risk-taking behaviors and substance misuse problems in adolescents as compared to adulthood. Disturbance in the developmental progression can have a wide-ranging influence on subsequent behavior, emotions, and cognitive development. The depressant effect of substance misuses or secondary effects will no doubt confuse the biological picture found in adolescent depression, making neurobiological comparisons with the adult depressed cohorts difficult. Neuroimaging studies have found a correlation between PFC maturation and cognitive mastery and control and behavioral inhibition (Casey, Tottenham, Liston, & Durston, 2005), and therefore, atypical neurodevelopmental trajectories may be harder to interpret in clinical groups.

In addition to structural findings, functional brain imagining studies have exposed differences in an adolescent's response when compared to either children or adults. When presented with emotionally laden images, adolescents show more pronounced amygdala activation, with a relative underuse of PFC areas (Hare et al., 2008). Similarly, adolescences recruit different and more numerous brain areas than adults when faced with tasks requiring calculation and impulse control, with a preference to use more caudal than frontal areas. This typical developmental difference needs to be considered when interpreting similarities or differences between adolescent or adult depression.

The neurotransmitter system also follows a distinctive development trajectory. The cholinergic and serotinergic systems are thought to be significant in facilitating neuronal maturation, plasticity, and connectivity, particularly in the corticolimbic pathways, and develop earlier, declining during adolescence. Development of the dopamine-rich mesostriatal and mesocorticolimbic systems, responsible for reward-directed behaviors, are more pronounced and occur later with increased dopamine projections to the PFC as a result of subcortical dopamine pruning. The reduced serotinergic, GABAnergic, and glutaminergic inputs to the PFC in adolescence allow for a relative dopamine-dominated state, which might explain the increased sensitivity to rewards and risk-taking behavior in adolescence (Spear, 2009) and may also explain the relative lack of efficacy of tricyclic antidepressant medications in youths versus adults. The age-related variations in neurotransmitter system may also explain the lack of consistent findings when comparing blood or cerebrospinal fluid (CSF) levels between pediatric and adult cohorts. For example, lower levels of serotonin and its metabolite, 5-HIAA, in either blood or CSF have been found in adults and have been linked with aggression and suicidal behavior (Lidberg, Belfrage, Bertilsson, Evenden, & Asberg, 2000). Postmortem studies in adults have found abnormalities in the brain serotonin system and in the few adolescent brains (Pandey et al., 2002), but none were published specifically on depression. Studies of CSF 5-HIAA levels in children suggest similar inverse correlations between 5-HIAA levels and aggression (Kruesi et al., 1990), whereas platelet studies have been contradictory (Zalsman, Huang, et al., 2006).

NEUROENDOCRINE STUDIES

Many studies of adult patients have implicated the hypothalamic–pituitary–axis in depression, with a blunted response following stimulation and higher levels in the CSF in depressed patients as compared to controls, with some evidence of normalization posttreatment. This has been less consistently found in children or adolescents (Birmaher, Dahl, et al., 1996). Nonsuppression of cortisol release by dexamethasone was almost considered diagnostic, occurring in 50–70% of depressed adults, especially associated with psychotic subtypes and a past history of depression (Zalsman, Brent, & Weersing, 2006). Although some studies have reported similar findings in children and adolescence (Casat, Arana, & Powell, 1989), this is by no means typical. The higher levels of cortisol found among adult depressed cohorts have not been identified in children (Puig-Antich et al., 1989). Similarly, confusing findings have been reported with regard to growth hormone (GH) levels. In depressed adults, higher daytime and lower nocturnal secretion has been reported as compared to nondepressed samples, yet in children (Puig-Antich et al., 1984) and adolescents (Kutcher et al., 1991), almost the reverse has been reported, with increased nocturnal GH release.

STRUCTURAL BRAIN CHANGES

The most consistent evidence from neuroimaging studies suggests smaller PFC and basal ganglia in depressed adults with corresponding decreased blood flow and metabolism (Soares & Mann, 1997). Pediatric studies examining the amygdala and striatum have also reported reduced volumes in depressed youth as compared to controls (MacMillan et al., 2003), but the findings are difficult to interpret given the lack of association with depressive severity, age of onset, or chronicity along with the variability of findings to adult studies. In contrast, in a meta-analysis of magnetic resonance imaging studies, volume reduction in PFC regions in depressed adult patients was associated with illness severity (Koolschijn, van Haren, Lensvelt-Mulders, Hulshoff Pol, & Kahn, 2009).

Studies of both children and adults have found reduced hippocampal volumes in depressed individuals (McKinnon, Yucel, Nazarov, & MacQueen, 2009) and in healthy adolescents with a family history of depression, with volumes correlating with a subsequent risk of depression (Rao et al., 2010). Given the role that the hippocampus plays in mood regulation, sensitivity to stress and, evidence of impaired neurogenesis and atrophy in chronically stressed animal models (Campbell & Macqueen, 2004; Samuels & Hen, 2011), these findings are of interest and may be linked with both heritability to depression along with the exposure to stress if raised by a parent with depression.

FUNCTIONAL BRAIN CHANGES

Functional brain changes have been identified in adults but less consistently in adolescent depression. Following negative emotional stimuli, individuals with depression, or at risk of depression, have increased neuronal activity in the corticolimbic and corticostriatal circuits and reduced activation following positive cues, at least in adults (Leppänen, 2006). Findings in adolescences are less consistent. It is not clear whether these findings are state-dependent, a risk state for developing depression, or a result of the illness or its treatment (Weir, Zakama, & Rao, 2012).

It is difficult to disentangle the significant but normal neurodevelopment in adolescences from a more pathological divergence of maturational trajectories, particularly given the need to consider individual variation, gene–environmental interaction, and state–trait phenomena. Substance misuse in common at the time of adolescence also creates its own deviations, with a reduced white matter integrity and density of gray matter. Taken together, this makes the study of neurobiology of pediatric depression exciting and frustrating, yielding as much difference as similarity.

GENETICS

Relatively little is known in pediatric populations as to how genes contribute to the development of depressive symptoms and how they interact with the environment, despite a high prevalence of pediatric depression. To investigate these enigmas, large amounts of genetic data as well as a sophisticated study design and analytical techniques are required. In this section we will first review genetic epidemiology, including family, twin, and adoption studies, to shed the light on familial aggregation, heritability, environmental influences, and correlation and interaction between gene and environment in MDD. Scientific achievements in molecular genetics, including linkage analysis and candidate gene studies, will be highlighted.

Family Studies

Several family studies have demonstrated higher levels of depression in the offspring of depressed parents in comparison to the offspring of parents with no psychopathology (Rice, Harold, & Thapar, 2002a). In a systematic literature review and meta-analysis of genetic epidemiology in adult populations with MDD, five family studies that met the inclusion criteria suggested that there was strong evidence for an association between MDD in the probands and MDD in first-degree relatives (Sullivan, Neale, & Kendler, 2000). According to the study, a person with first-degree relatives with MDD has a 2.84 increased likelihood of developing MDD. Rice et al. (2002a) performed a systematic review and meta-analysis of family studies of MDD to investigate the familial risk to become depressed in childhood. Four parent-to-offspring studies (i.e., studies of children of depressed parents) and six "bottom-up" studies (i.e., studies of the relatives of child probands with MDD) were included in a meta-analysis, revealing odds ratio (OR) of 3.98 and 2.30, respectively, in comparison with the normal control group. This study suggested a strong familial aggregation in pediatric depression. However, its familial risk was not substantially greater than that computed from an adult study (OR 2.89). Retrospective studies indicate a stronger familiarity in populations with early-onset MDD than in ones with late-onset MDD, leading to the hypothesis that childhood-onset (prepubertal) MDD can harbor a stronger familial aggregation in comparison to adolescent- (postpubertal) and adult-onset MDD (Weissman et al., 1984; Wickramaratne & Weissman, 1998). However, one study using samples from a longitudinal British cohort failed to prove this hypothesis, with a relative risk of 1.0 and 0.8 in first- and second-degree relatives of probands with pre- versus postpubertal MDD, respectively, and 0.8 and 1.0 in first- and second-degree relatives of probands with adult versus postpubertal MDD (Harrington et al., 1997).

Twin Studies

Twin and adoption studies are the two principal approaches for distinguishing genetic and environmental contributions to MDD. The twin study requires affected probands from both pairs of monozygotic (MD) and dizygotic twins, and the presence (concordance) or absence (discordance) of the disorder in the cotwins of these affected probands is assessed. Concordance and discordance proportions are used to determine the relative contributions of genetic and environmental factors in the development of illness, which are conceptualized as genes, shared environment (that is, environmental factors that make twins more alike—socioeconomic status, for example), and unique environment (that is, nonshared environmental factors that makes twins dissimilar).

In an adult study, heritability computed from data from five twin studies (three studies conducted in the community, one study done in a clinical setting, and one study using population-based samples both in the community and a clinical setting) was 37% (95% confidence interval (CI) = 31−42%) (Sullivan et al., 2000). On the contrary, heritability of depressive symptoms in youth varies across studies (Rice et al., 2002a; Thapar & McGuffin, 1994), depending on who rated the symptoms and what measurements were selected. In the review by Rice et al. (2002a), heritability estimates for parent-rated depressive symptoms varied from 30% to 80%, while lower heritability estimates were found for self-reported depressive symptoms. Despite these inconsistent results, the only consistent finding from several twin studies is that the genetic factors becoming more important from childhood to adolescence (Rice, Harold, & Thapar, 2002b; Thapar & McGuffin, 1994), which might be due to uncovered gene−environment correlation, gene−environment interaction, or additive genetic influence in adolescence (Eley & Stevenson, 1999). Additionally, two studies from the national Swedish Twin Registry suggested higher heritability in women than in men (40% versus 30%: PMID11352363 and 42% versus 29%; Kendler, Gatz, Gardner, & Pedersen, 2006), and clear evidence was found for sex-specific genetic effects with genetic correlations estimated at +0.55 and +0.63.

Adoption Studies

Despite the significant importance in genetic epidemiology of MDD, the fact that even MD twins can share much of the same environment poses a challenge to us in differentiating the relative contribution of genetic and environmental factors to the development of illness in twin studies. Adoption study is utilized to overcome this difficulty.

Three adult adoption studies on MDD were included in a systematic review by Sullivan et al. (2000), although results of them were not rigorous due to methodological issues (the diagnosis of MDD was based on indirect sources rather than personal interviews) and limited statistical power. In the younger generation, three adoption studies investigated childhood depression (two studies examined internalizing problems, including depression and anxiety, and one study examined the diagnosis of MDD), none of which provided rigorous evidence (Eley, Deater-Deckard, Fombonne, Fulker, & Plomin, 1998; van den Oord, Boomsma, & Verhulst, 1994; Tully, Iacono, & McGue, 2008).

In summary, early-onset depression may be more strongly genetically influenced than adult-onset MDD, suggesting strong familial aggregation (with less environmental contribution) and heritability.

Molecular Genetics

Historically, the process of gene discovery required genetic linkage studies, following the confirmation of familial aggregation described above. In the genomic era, however, we see different approaches for detecting candidate genes using rapid scientific advancements.

In a review article by Flint and Kendler (2014), 26 previously detected candidate genes were meta-analyzed, among which only the following seven genes yielded significant ($p < 0.05$) results: 5-HTTP/SLC6A4, APOE, DRD4, GNB3, HTR1A, MTHFR, and SLC6A3. Despite these results, the authors reported that little conclusive evidence was yielded from the candidate genes to support each gene's involvement in depression, based on the fact that the odds ratio that estimated overall candidate gene meta-analysis was 1.15.

In genetic linkage studies, two loci (the disease locus and the marker locus) are considered to be linked and inherited together more often than expected by chance within families if they are on the same chromosome and close enough together on it. In MDD, like other diseases, data from genetic linkage studies tend to generate a series of publications because the authors acquire additional data. Among those studies, obvious heterogeneity and poor internal consistency are frequently noted, raising a concern for the existence of false-positive or negative findings.

The Genetics of Recurrent Early-Onset Depression (GenRED) study is one example that highlights these problems in genetic linkage studies in MDD. This study was conducted in six facilities in the United States, using a large sample of families with recurrent MDD (Levinson et al., 2003). The authors used genome scan methods and tried to shed light on the idea that early onset and the recurrence of depressive episodes might predict a substantially increased risk of depression in first-degree relatives as compared with the general population. Initial reports from this study revealed 15q linkage, although this was not corroborated by the following paper that included additional samples, which provided suggestive linkage results on chromosome (Holmans et al., 2004) 17p12 and chromosome 8p22-p21.3 (Holmans et al., 2004, 2007).

Most candidate genes are studies using a case–control association study design. In this study design, genetic variants are usually selected based on a priori hypothesis, such as neurobiological plausibility (serotonin transporter for antidepressants, for example) or genomic location of a candidate gene (i.e., linkage peak). One small study consisting of 68 unrelated patients with *Diagnostic and Statistical Manual of Mental Disorders, Fourth Edition* depressive disorders and 68 unrelated healthy control subjects reported a significant association between a short variant of the serotonin transporter gene-linked polymorphic region (5-HTTLPR) and pediatric-onset depressive disorders, including both MDD and dysthymia, using a case–control design and a family-based association design (Nobile et al., 2004).

With evidence of candidate genes, further attempts have been made to elucidate whether or not genetic vulnerability to MDD can be affected by environments. Using longitudinal data of representative samples from a birth cohort in New Zealand, Capsi et al. (2003) concluded that individuals with a genetic variation of the short allele of 5-HTTLPR were at an elevated risk for the development of major depression in adult life in interaction with stressful life experiences in childhood, suggesting a potential gene–environment interaction in this illness. Although a significant association was found between the number of stressful events and depression (OR 1.41, 95% CI 1.25–1.57) in the meta-analysis of published data conducted by Risch et al. (2009), the effect of the 5-HTTLPR genotype alone failed to reach statistical significance on the development of depression. Furthermore, aggregate data from this meta-analysis did not suggest that the 5-HTTLPR genotype interacts with childhood stressful experiences on the risk of depression in later life (OR 1.01; CI 0.94–1.10). From a different perspective, the study by Kaufman et al. (2004) revealed that the risk of developing MDD in participants with a short allele of 5-HTTLPR was removed in the presence of social support.

Genetic variants of the Val66Met polymorphism of the brain-derived neutrophic factor (BDNF) gene were found to have a potential interaction with environmental factors on the risk of major depression (Aquirela et al., 2009). The three-way interaction among the Val66Met polymorphism of the BDNF gene and the short variant of 5-HTTLPR has been reported to be associated with a higher risk of pediatric depression in children with maltreatment history in comparison to a control group (Kaufman et al., 2006). In this study, social support was also protective factor against the higher rate of depression following childhood maltreatment in children with the Met allele of the BDNP gene. A similar moderating effect on childhood abuse and subsequent adult depression was found with regard to the corticotrophin-releasing hormone receptor gene but in males only (Heim et al., 2009).

Polymorphism in 5-HTTLPR genes and variants in BDNF polymorphisms have been associated with both structural and functional changes in the corticolimbic areas, including hippocampal size and amygdala activation in children with stress who are exposed to emotionally laden cues (Chien, Chau, & Lu, 1991; Frodl, Möller, & Meisenzahl, 2008).

Finally, as described above, genome-wide association studies (GWAS) are available due to rapid technological advances. This design enables us to scan markers across the complete set of genomes to detect small variations, which are called single nucleotide polymorphisms, that occur more frequently in probands with a particular disease than in people without the disease. Flint and Kendler (2014) summarized nine previously published studies in MDD, revealing that no significant findings have been reported in this illness. The authors assumed that the failure of GWAS studies to detect candidate loci was most likely attributable to an underpowered sample size of each study. To the best of the authors' knowledge, no such studies exclusive in pediatric populations with MDD have been conducted.

FINAL REMARKS

Despite a growing body of research into the neurobiological correlates of depression, the evidence in children and adolescents is limited to the extent that no firm conclusions may yet be drawn. Studies that do exist are not always consistent and are often hampered by ethical and methodological difficulties, particularly in the choice of controls, given the salience of age, gender, previous childhood experience, and motivation along with the definition and duration of depression. Equally, it is recognized that adolescence, a developmental phase characterized by a significant increase in the prevalence of depression, is also a time of marked physical, cognitive, and social change, occurring at different chronological and developmental ages. Thus standardizing for this in studies is difficult and might explain the inconsistent findings in the few studies conducted. A lack of evidence does not mean that similarities do not exist—only that more robustly conducted and replicated studies are needed. An understanding of normal brain development during adolescence is necessary to put into context possible variations found in studies of youth with or at risk for depression.

REFERENCES

Angold, A., & Costello, E. J. (2001). *The epidemiology of depression in children and adolescents.* Cambridge University Press.

Aquirela, M., Arias, B., Wichers, M., Barrantes-Vidal, N., Moya, J., Villa, V., ... Fañanás, L. (2009). Early adversity and 5-HTT/BDNF genes: new evidence of gene-environment interactions on depressive symptoms in a general population. *Psychological Medicine, 239*(9), 1425–1432.

Birmaher, B., Brent, D., AACAP Work Group on Quality Issues, Bernet, W., Bukstein, O., Walter, H., ... Medicus, J. (2007). Practice parameter for the assessment and treatment of children and adolescents with depressive disorders. *Journal of the American Academy of Child and Adolescent Psychiatry, 46*(11), 1503–1526.

Birmaher, B., Dahl, R. E., Perel, J., Williamson, D. E., Nelson, B., Stull, S., ... Ryan, N. D. (1996). Corticotropin-releasing hormone challenge in prepubertal major depression. *Biological Psychiatry, 39*(4), 267–277.

Birmaher, B., Ryan, N. D., Williamson, D. E., Brent, D. A., Kaufman, J., Dahl, R. E., ... Nelson, B. (1996). Childhood and adolescent depression: a review of the past 10 years. Part I. *Journal of the American Academy of Child and Adolescent Psychiatry, 35*(11), 1427–1439.

Birmaher, B., Williamson, D. E., Dahl, R. E., Axelson, D. A., Kaufman, J., Dorn, L. D., & Ryan, N. D. (2004). Clinical presentation and course of depression in youth: does onset in childhood differ from onset in adolescence? *Journal of the American Academy of Child and Adolescent Psychiatry, 43*(1), 63–70.

Brent, D. A., & Birmaher, B. (2006). Treatment-resistant depression in adolescents: recognition and management. *Child and Adolescent Psychiatric Clinics of North America, 15*(4), 1015–1034.

Brent, D. A., Holder, D., Kolko, D., Birmaher, B., Baugher, M., Roth, C., ... Johnson, B. A. (1997). A clinical psychotherapy trial for adolescent depression comparing cognitive, family and supportive therapy. *Archives of General Psychiatry, 54*(9), 877–885.

Campbell, S., & Macqueen, G. (2004). The role of the hippocampus in the pathophysiology of major depression. *Journal of Psychiatry and Neuroscience, 29*(6), 417–426.

Capsi, A., Sugden, K., Moffitt, T. E., Taylor, A., Craig, I. W., Harrington, H., ... Poulton, R. (2003). Influence of life stress on depression: moderation by a polymorphism in the 5-HTT gene. *Science, 301*(5631), 386–389.

Carballo, J. J., MuÒoz-Lorenzo, L., Blasco-Fontecilla, H., Lopez-Castroman, J., García-Nieto, R., Dervic, K., ... Baca-García, E. (2011). Continuity of depressive disorders from childhood and adolescence to adulthood: a naturalistic study in community mental health centers. *Primary Care Companion for CNS Disorders, 13*(5).

Casat, C. D., Arana, G. W., & Powell, K. (1989). The DST in children and adolescents with major depressive disorder. *American Journal of Psychiatry, 146*(4), 503–507.

Casey, B. J., & Jones, R. M. (2010). Neurobiology of the adolescent brain and behavior: implications for substance use disorders. *Journal of the American Academy of Child and Adolescent Psychiatry, 49*(12), 1189–1201. quiz 1285.

Casey, B. J., Tottenham, N., Liston, C., & Durston, S. (2005). Imaging the developing brain: what have we learned about cognitive development? *Trends in Cognitive Sciences, 9*(3), 104–110.

Chien, C. L., Chau, Y. P., & Lu, K. S. (1991). Ultrastructural studies on the barrier properties of the paraganglia in the rat recurrent laryngeal nerve. *Acta Anatomica (Basel), 141*(3), 262–268.

Copeland, W. E., Shanahan, L., Costello, E. J., & Angold, A. (2009). Childhood and adolescent psychiatric disorders as predictors of young adult disorders. *Archives of General Psychiatry, 66*(7), 764–772.

Coyle, J. T., Pine, D. S., Charney, D. S., Lewis, L., Nemeroff, C. B., Carlson, G. A., ... Depression and Bipolar Support Alliance Consensus Development Panel. (2003). Depression and bipolar support alliance consensus statement on the unmet needs in diagnosis and treatment of mood disorders in children and adolescents. *Journal of the American Academy of Child and Adolescent Psychiatry, 42*(12), 1494–1503.

Dubicka, B., Elvins, R., Roberts, C., Chick, G., Wilkinson, P., & Goodyer, I. M. (2010). Combined treatment with cognitive-behavioural therapy in adolescent depression: meta-analysis. *British Journal of Psychiatry, 197*(6), 433–440.

Eley, T. C., Deater-Deckard, K., Fombonne, E., Fulker, D. W., & Plomin, R. (1998). An adoption study of depressive symptoms in middle childhood. *Journal of Child Psychology and Psychiatry, 39*(3), 337–345.

Eley, T. C., & Stevenson, J. (1999). Exploring the covariation between anxiety and depression symptoms: a genetic analysis of the effects of age and sex. *Journal of Child Psychology and Psychiatry, 40*(8), 1273–1282.

Emslie, G. J., Kratochvil, C., Vitiello, B., Silva, S., Mayes, T., McNulty, S., ... TADS Team. (2006). Treatment for adolescents with depression study (TADS): safety results. *Journal of the American Academy of Child and Adolescent Psychiatry, 45*(12), 1440–1455.

Flint, J., & Kendler, K. S. (2014). The genetics of major depression. *Neuron, 81*(3), 484–503.

Fombonne, E., Wostear, G., Cooper, V., Harrington, R., & Rutter, M. (2001). The Maudsley long-term follow-up of child and adolescent depression, 1: psychiatric outcomes in adulthood. *British Journal of Psychiatry, 179*, 210–217.

Frodl, T., Möller, H. J., & Meisenzahl, E. (2008). Neuroimaging genetics: new perspectives in research on major depression? *Acta Psychiatrica Scandinavica, 118*(5), 363–372.

Frommer, E. A. (March 25, 1967). Treatment of childhood depression with antidepressant drugs. *British Medical Journal, 1*(5542), 729–732.

Ghaziuddin, N., Kutcher, S. P., Knapp, P., Bernet, W., Arnold, V., Beitchman, J., ... AACAP. (2004). Practice parameter for use of electroconvulsive therapy with adolescents. *Journal of the American Academy of Child and Adolescent Psychiatry, 43*(12), 1521–1539.

Goodyer, I. M., & Cooper, P. J. (1993). A community study of depression in adolescent girls II: the clinical features of identified disorder. *British Journal of Psychiatry, 163*, 374–380.

Goodyer, I. M., Dubicka, B., Wilkinson, P., Kelvin, R., Roberts, C., Byford, S., ... Harrington, R. (2007). Selective serotonin reuptake inhibitors (SSRIs) and routine specialist care with and without cognitive behaviour therapy in adolescents with major depression: randomised controlled trial. *BMJ, 335*(7611), 142.

Hare, T. A., Tottenham, N., Galvan, A., Voss, H. U., Glover, G. H., & Casey, B. J. (2008). Biological substrates of emotional reactivity and regulation in adolescence during an emotional go-nogo task. *Biological Psychiatry, 63*(10), 927–934.

Harrington, R., Fudge, H., Rutter, M., Pickles, A., & Hill, J. (1990). Adult outcomes of childhood and adolescent depression, I: psychiatric status. *Archives of General Psychiatry, 47*(5), 465–473.

Harrington, R., Rutter, M., Weissman, M., Fudge, H., Groothues, C., Bredenkamp, D., ... Wickramaratne, P. (1997). Psychiatric disorders in the relatives of depressed probands. I. Comparison of prepubertal, adolescent and early adult onset cases. *Journal of Affective Disorders, 42*(1), 9–22.

Heim, C., Bradley, B., Mletzko, T. C., Deveau, T. C., Musselman, D. L., Nemeroff, C. B., ... Binder, E. B. (2009). Effect of childhood trauma on adult depression and neuroendocrine function: sex-specific moderation by CRH receptor 1 gene. *Frontiers in Behavioral Neuroscience, 3*, 41.

Hetrick, S., Merry, S., McKenzie, J., Sindahl, P., & Proctor, M. (2007). Selective serotonin reuptake inhibitors (SSRIs) for depressive disorders in children and adolescents. *Cochrane Database of Systematic Reviews, 3*, CD004851.

Holmans, P., Weissman, M. M., Zubenko, G. S., Scheftner, W. A., Crowe, R. R., Depaulo, J. R., Jr., ... Levinson, D. F. (2007). Genetics of recurrent early-onset major depression (GenRED): final genome scan report. *American Journal of Psychiatry, 164*(2), 248–258.

Holmans, P., Zubenko, G. S., Weissman, M. M., Crowe, R. R., DePaulo, J. R., Jr., Scheftner, W. A., ... Levinson, D. F. (2004). Genomewide significant linkage to recurrent, early-onset major depressive disorder on chromosome 15q. *American Journal of Human Genetics, 74*(6), 1154–1167.

Kaufman, J., Yang, B. Z., Douglas-Palumberi, H., Grasso, D., Lipschitz, D., Houshyar, S., ... Gelernter, J. (2006). Brain-derived neurotrophic factor-5-HTTLPR gene interactions and environmental modifiers of depression in children. *Biological Psychiatry, 59*(8), 673–680.

Kaufman, J., Yang, B. Z., Douglas-Palumberi, H., Houshyar, S., Lipschitz, D., Krystal, J. H., & Gelernter, J. (2004). Social supports and serotonin transporter gene moderate depression in maltreated children. *Proceedings of the National Academy of Sciences of the United States of America, 101*(49), 17316–17321.

Kendler, K. S., Gatz, M., Gardner, C. O., & Pedersen, N. L. (2006). A Swedish national twin study of lifetime major depression. *American Journal of Psychiatry, 163*(1), 109–114.

Koolschijn, P. C., van Haren, N. E., Lensvelt-Mulders, G. J., Hulshoff Pol, H. E., & Kahn, R. S. (2009). Brain volume abnormalities in major depressive disorder: a meta-analysis of magnetic resonance imaging studies. *Human Brain Mapping, 30*(11), 3719–3735.

Kovacs, M., Feinberg, T. L., Crouse-Novak, M., Paulauskas, S. L., Pollock, M., & Finkelstein, R. (1984). Depressive disorders in childhood. II. A longitudinal study of the risk for a subsequent major depression. *Archives of General Psychiatry, 41*(7), 643–649.

Kruesi, M. J., Rapoport, J. L., Hamburger, S., Hibbs, E., Potter, W. Z., & Lenane, M. (1990). Cerebrospinal fluid monoamine metabolites, aggression, and impulsivity in disruptive behavior disorders of children and adolescents. *Archives of General Psychiatry, 47*(5), 419–426.

Kutcher, S., Malkin, D., Silverberg, J., Marton, P., Williamson, P., Malkin, A., ... Katic, M. (1991). Nocturnal cortisol, thyroid stimulating hormone, and growth hormone secretory profiles in depressed adolescents. *Journal of the American Academy of Child and Adolescent Psychiatry, 30*(3), 407–414.

Leppänen, J. M. (2006). Emotional information processing in mood disorders: a review of behavioral and neuroimaging findings. *Current Opinion in Psychiatry, 19*(1), 34–39.

Levinson, D. F., Zubenko, G. S., Crowe, R. R., DePaulo, R. J., Scheftner, W. S., Weissman, M. M., ... Chellis, J. (May 15, 2003). Genetics of recurrent early-onset depression (GenRED): design and preliminary clinical characteristics of a repository sample for genetic linkage studies. *American Journal of Medical Genetics Part B: Neuropsychiatric Genetics, 119B*(1), 118–130.

Lidberg, L., Belfrage, H., Bertilsson, L., Evenden, M. M., & Asberg, M. (2000). Suicide attempts and impulse control disorder are related to low cerebrospinal fluid 5-HIAA in mentally disordered violent offenders. *Acta Psychiatrica Scandinavica, 101*(5), 395–402.

Luby, J. L., Si, X., Belden, A. C., Tandon, M., & Spitznagel, E. (2009). Preschool depression: homotypic continuity and course over 24 months. *Archives of General Psychiatry, 66*(8), 897–905.

MacMillan, S., Szeszko, P. R., Moore, G. J., Madden, R., Lorch, E., Ivey, J., ... Rosenberg, D. R. (2003). Increased amygdala: hippocampal volume ratios associated with severity of anxiety in pediatric major depression. *Journal of Child and Adolescent Psychopharmacology, 13*(1), 65–73.

March, J., Silva, S., Petrycki, S., Curry, J., Wells, K., Fairbank, J., ... Severe, J. (2007). The treatment for adolescents with depression study (TADS): long-term effectiveness and safety outcomes. *Archives of General Psychiatry, 64*(10), 1132–1143.

March, J., Silva, S., Petrycki, S., Curry, J., Wells, K., Fairbank, J., ... Treatment for Adolescents With Depression Study (TADS) Team. (2004). Fluoxetine, cognitive-behavioral therapy, and their combination for adolescents with depression: treatment for adolescents with depression study (TADS) randomized controlled trial. *Journal of the American Medical Association, 292*(7), 807–820.

McKinnon, M. C., Yucel, K., Nazarov, A., & MacQueen, G. M. (2009). A meta-analysis examining clinical predictors of hippocampal volume in patients with major depressive disorder. *Journal of Psychiatry and Neuroscience, 34*(1), 41–54.

Merikangas, K. R., He, J. P., Burstein, M., Swanson, S. A., Avenevoli, S., Cui, L., ... Swendsen, J. (2010). Lifetime prevalence of mental disorders in U.S. adolescents: results from the National Comorbidity Survey Replication–Adolescent Supplement (NCS-A). *Journal of the American Academy of Child and Adolescent Psychiatry, 49*(10), 980–989.

Mufson, L., Dorta, K. P., Wickramaratne, P., Nomura, Y., Olfson, M., & Weissman, M. M. (2004). A randomized effectiveness trial of interpersonal psychotherapy for depressed adolescents. *Archives of General Psychiatry, 61*(6), 577–584.

National Institute for Health and Clinical Excellence NICE. (2015). *Depression in children and young people: Identification and management in primary, community and secondary care.* National Practice Guideline No 28. London, UK: The British Psychological Society and the Royal College of Psychiatrists.

NICE guideline CG28. (2005). *Depression in children and young people: identification and management in primary, community and secondary care.* London: British Psychological Society.

Nobile, M., Cataldo, M. G., Giorda, R., Battaglia, M., Baschirotto, C., Bellina, M., ... Molteni, M. (2004). A case-control and family-based association study of the 5-HTTLPR in pediatric-onset depressive disorders. *Biological Psychiatry, 56*(4), 292–295.

van den Oord, E. J., Boomsma, D. I., & Verhulst, F. C. (1994). A study of problem behaviors in 10- to 15-year-old biologically related and unrelated international adoptees. *Behavior Genetics, 24*(3), 193–205.

Pandey, G. N., Dwivedi, Y., Rizavi, H. S., Ren, X., Pandey, S. C., Pesold, C., ... Tamminga, C. A. (2002). Higher expression of serotonin 5-HT(2A) receptors in the postmortem brains of teenage suicide victims. *American Journal of Psychiatry, 159*(3), 419–429.

Pearce, J. B. (1978). The recognition of depressive disorder in children. *Journal of Royal Society of Medicine, 71*(7), 494–500.

Puig-Antich, J., Dahl, R., Ryan, N., Novacenko, H., Goetz, D., Goetz, R., ... Klepper, T. (1989). Cortisol secretion in prepubertal children with major depressive disorder. Episode and recovery. *Archives of General Psychiatry, 46*(9), 801–809.

Puig-Antich, J., Goetz, R., Davies, M., Fein, M., Hanlon, C., Chambers, W. J., ... Weitzman, E. D. (1984). Growth hormone secretion in prepubertal children with major depression. II. Sleep-related plasma concentrations during a depressive episode. *Archives of General Psychiatry, 41*(5), 463–466.

Puig-Antich, J., Perel, J., Lupatkin, W., Chambers, W. J., Tabrizi, M. A., King, J., ... Stiller, R. L. (1987). Imipramine in prepubertal major depressive disorders. *Archives of General Psychiatry, 44*(1), 81–89.

Rao, U., Chen, L. A., Bidesi, A. S., Shad, M. U., Thomas, M. A., & Hammen, C. L. (2010). Hippocampal changes associated with early-life adversity and vulnerability to depression. *Biological Psychiatry, 67*(4), 357–364.

Reinke, W. M., & Ostrander, R. (2008). Heterotyic and homotypic continuity: the moderating effects of age and gender. *Journal of Abnormal Child Psychology, 36*(7), 1109–1121.

Rice, F., Harold, G. T., & Thapar, A. (2002a). The genetic aetiology of childhood depression: a review. *Journal of Child Psychology and Psychiatry, 43*(1), 65–79.

Rice, F., Harold, G. T., & Thapar, A. (2002b). Assessing the effects of age, sex and shared environment on the genetic aetiology of depression in childhood and adolescence. *Journal of Child Psychology and Psychiatry, 43*(8), 1039–1051.

Risch, N., Herrell, R., Lehner, T., Liang, K. Y., Eaves, L., Hoh, J., ... Merikangas, K. R. (2009). Interaction between the serotonin transporter gene (5-HTTLPR), stressful life events, and risk of depression: a meta-analysis. *Journal of the American Medical Association, 301*(23), 2462–2471.

Rutter, M., Kim-Cohen, J., & Maughan, B. (2006). Continuities and discontinuities in psychopathology between childhood and adult life. *Journal of Child Psychology and Psychiatry, 47*(3–4), 276–295.

Samuels, B. A., & Hen, R. (2011). Neurogenesis and affective disorders. *European Journal of Neuroscience, 33*(6), 1152–1159.

Soares, J. C., & Mann, J. J. (1997). The anatomy of mood disorders–review of structural neuroimaging studies. *Biological Psychiatry, 41*(1), 86–106.

Spear, L. P. (2009). Heightened stress responsivity and emotional reactivity during pubertal maturation: Implications for psychopathology. *Development and Psychopathology, 21*(1), 87–97.

Sullivan, P. F., Neale, M. C., & Kendler, K. S. (2000). Genetic epidemiology of major depression: review and meta-analysis. *American Journal of Psychiatry, 157*(10), 1552–1562.

Thapar, A., & McGuffin, P. (1994). A twin study of depressive symptoms in childhood. *British Journal of Psychiatry, 165*(2), 259–265.

Tully, E. C., Iacono, W. G., & McGue, M. (2008). An adoption study of parental depression as an environmental liability for adolescent depression and childhood disruptive disorders. *American Journal of Psychiatry, 165*(9), 1148–1154.

Vitiello, B., Silva, S. G., Rohde, P., Kratochvil, C. J., Kennard, B. D., Reinecke, M. A., ... March, J. S. (2009). Suicidal events in the treatment for adolescents with depression study (TADS). *Journal of Clinical Psychiatry, 70*(5), 741–747.

Weir, J. M., Zakama, A., & Rao, U. (2012). Developmental risk I: depression and the developing brain. *Child and Adolescent Psychiatric Clinics of North America, 21*(2), 237–259. vii.

Weissman, M. M., Wickramaratne, P., Merikangas, K. R., Leckman, J. F., Prusoff, B. A., Caruso, K. A., ... Gammon, G. D. (1984). Onset of major depression in early adulthood. Increased familial loading and specificity. *Archives of General Psychiatry, 41*(12), 1136–1143.

WHO. (1992). *The ICD-10 classification of mental and behavioural disorders: Clinical descriptions and diagnostic guidelines*. WHO.

Wickramaratne, P. J., & Weissman, M. M. (1998). Onset of psychopathology in offspring by developmental phase and parental depression. *Journal of the American Academy of Child and Adolescent Psychiatry, 37*(9), 933–942.

Zalsman, G., Brent, D. A., & Weersing, V. R. (2006). Depressive disorders in childhood and adolescence: an overview: epidemiology, clinical manifestation and risk factors. *Child and Adolescent Psychiatric Clinics of North America, 15*(4), 827–841. vii.

Zalsman, G., Huang, Y. Y., Oquendo, M. A., Burke, A. K., Hu, X. Z., Brent, D. A., ... Mann, J. J. (2006). Association of a triallelic serotonin transporter gene promoter region (5-HTTLPR) polymorphism with stressful life events and severity of depression. *American Journal of Psychiatry, 163*(9), 1588–1593.

Chapter 11

Systems Neuroscience in Late-Life Depression

Kevin J. Manning, David C. Steffens
Department of Psychiatry, University of Connecticut Health Center, Farmington, CT, USA

INTRODUCTION

Major depression in the elderly is a heterogeneous disorder. Depression in older adults can have a variety of presentations, including recurrent disease stemming from earlier life (i.e., early-onset depression) or new-onset depression in late life (i.e., late-onset depression). Here we define the term late-life depression (LLD) as any depression occurring in older adults, regardless of the age of onset. Prevalence estimates for LLD have been reported to be 4.4% in women and 2.7% in men (Steffens et al., 2000) with a total prevalence of 16.52% for lifetime major depressive disorder (MDD); (Volkert, Schulz, Harter, Wlodarczyk, & Andreas, 2013). Clinically, LLD presents with a constellation of cognitive and affective symptoms. Neuroanatomical models describe altered patterns of regulation between the dorsal-cognitive and ventral-affective components of the brain in major depression (Mayberg, 1997; Phillips, Drevets, Rauch, & Lane, 2003b). Neuroimaging, including structural and functional magnetic resonance imaging (fMRI), has helped elucidate the contributions of these brain regions to the pathophysiology of LLD.

Investigators have explored the intrinsic functional connectivity of the brain and have concentrated on the effort to understand the inner workings of several regions or networks in tandem. Functional connectivity is based on the observation that spontaneous blood oxygen-dependent (BOLD) signal fluctuations among interconnected brain regions tend to be temporally correlated. It is thought that functional connectivity during rest (where subjects are instructed to lie still with their eyes closed) or when subjects complete experimental tasks reflects important interrelationships among brain structures with related functions. Functional connectivity is commonly assessed with (1) seed-based correlation techniques, whereby a primary region of interest (ROI) is chosen a priori and brain regions that show correlated activity over time with the ROI are identified, and (2) independent

Systems Neuroscience in Depression. http://dx.doi.org/10.1016/B978-0-12-802456-0.00011-X

components analysis (ICA), where several regions thought to comprise a neural network are statistically extracted based upon similar BOLD activity levels. A more detailed description about different connectivity methods is given in the Chapter 7 by Beckmann et al.

Converging evidence using both seed-based correlation and ICA has identified three prominent functional networks in LLD and other psychiatric illnesses (Sheline, Price, Yan, & Mintun, 2010): (1) a cognitive control network (CCN) active during cognitively demanding tasks, (2) a default mode network (DMN) active during resting state and inhibited during cognitively demanding activity that is involved in internal mentation, and (3) a salience network (SN) relevant to attending to survival-relevant events in the environment. Our understanding of the underlying function of these networks in LLD has advanced by means of studies that employ various methodologies. For example, some research contrasts network functional connectivity in geriatric depressed adults and age-matched controls, and other studies explore the functional neuroanatomical correlates of behavioral syndromes observed in LLD. In this chapter, we review the research on the prominent functional networks apparent in LLD. We begin with a succinct review of the neuroanatomy relevant to LLD before discussing special considerations in the pathophysiology of depression in older adults. We then provide an individual overview of the cognitive control, default mode, and SNs as observed in LLD. That is not to say that these networks function independently of one another; instead, emerging evidence suggests that the three networks work in unison, so we end our discussion on how dysfunction in multiple networks contributes to common behavioral syndromes in LLD, including apathy, anxious-depression, and suicidality.

FUNCTIONAL NEUROANATOMY RELEVANT TO LATE-LIFE DEPRESSION

Major depression generally involves hypometabolism of cognitive-dorsal and hypermetabolism of affective-limbic regions. The primary cognitive functions of the dorsolateral prefrontal cortex (DLPFC; Brodmann areas [BA] 9, 10) include the regulation of effortful attention and working memory. With input from the superior parietal cortex (BA 7), the DLPFC maintains the attentional demands of tasks, while the dorsal anterior cingulate cortex (dACC; BA 24, 32) monitors ongoing performance (MacDonald, Cohen, Stenger, & Carter, 2000). Ventromedial prefrontal regions, including the orbitofrontal cortex (OFC; BA 10, 11, 47), ventral-rostral anterior cingulate (BA areas 25, 24, 32, 33), and their connections with the amygdala, mediate divergent functions such as the processing of emotional information, utilization of cues in the environment to predict rewarding or aversive events, and the regulation of behavioral responses, particularly in the context of changing reinforcement contingencies (Bonelli & Cummings, 2007).

The ventral-rostral anterior cingulate is the distribution center of emotional information (Alexopoulos, Gunning-Dixon, Latoussakis, Kanellopoulos, & Murphy, 2008). It has extensive connections with the amygdala, thalamic nuclei, nucleus accumbens, and ventral striatum and has been associated with autonomic function and emotional behavior (Phillips, Drevets, Rauch, & Lane, 2003a). Lesions to this region, including the subgenual anterior cingulate cortex (sgACC; BA 24a, 24b, 25), result in abnormal autonomic responses to emotional experiences, an inability to experience emotion related to concepts that ordinarily evoke emotion, and an inability to use information regarding the likelihood of punishment versus reward in guiding social behavior (Drevets, 2001). Volume reduction in sgACC is prevalent early in the course of unipolar and bipolar depression and is also observed in young adults at risk of major depression. Moreover, glucose metabolism in the sgACC is abnormally decreased in the depressed but increased in the manic phase of bipolar disorder, suggesting that sgACC activity is correlated with dopamine release and subsequent negative (apathy, anhedonia) and elevated (hedonic response, increased motivation) mood states (Drevets, 2001).

CONTRIBUTIONS SPECIFIC TO DEPRESSION IN LATE-LIFE: VASCULAR DISEASE

The vascular depression hypothesis posits that cerebrovascular disease may predispose, precipitate, or perpetuate a depressive syndrome in elderly patients (Taylor, Aizenstein, & Alexopoulos, 2013). The hallmark of MRI-defined vascular depression is the presence of white matter lesions identified as white matter hyperintensities (WMH) on T2-weighted or fluid-attenuated inversion recovery MRI. WMHs are associated with advanced age and cerebrovascular risk factors, including diabetes, cardiac disease, and hypertension (Abraham et al., 2015; Wolfson et al., 2013). The presence of WMH and white matter microstructural abnormalities in the dACC/DLPFC is associated with a worse illness course and cognitive performance in LLD (Alexopoulos, Murphy, et al., 2008; Gunning et al., 2009; Gunning-Dixon et al., 2010). White matter abnormalities are also prevalent in the OFC (Taylor et al., 2007) and sgACC (Aizenstein et al., 2011) in older adults with major depression. Thus, cerebrovascular disease may contribute to dysfunction of frontolimbic systems responsible for cognitive and emotional processing. Greater WMH volume was associated with an increased BOLD response in the rostral cingulate (including the sgACC) and insula in older adults with major depression when completing an emotional regulation task. This association was specific to major depression, as a similar relationship was not observed between WMH and cingulate activity in healthy older adults (Aizenstein et al., 2011). Moreover, decreased functional connectivity between the posterior cingulate and medial prefrontal cortex was highly associated with the presence of WMH ($r = -0.72$, $p < 0.05$) (Wu et al., 2011). Overall, while this and additional evidence (Sheline, Pieper, et al., 2010) lends support to the vascular depression hypothesis, other mechanisms do contribute to LLD. The same factors that predispose to depression in younger adults continue to confer vulnerability to depression in later life, including genetic predisposition, early adverse events, limited social support, interpersonal stressors, and personality characteristics (e.g., neuroticism). Descriptions of these factors are beyond the scope of this chapter but are described in detail elsewhere (Aziz & Steffens, 2013).

SYSTEMS NEUROSCIENCE IN LATE-LIFE DEPRESSION

The Cognitive Control Network

Disruption of the CCN, which encompasses the DLPFC, dorsal and rostral regions of the anterior cingulate, and parietal association regions, is prominent in LLD (Alexopoulos et al., 2012, 2015). Clinically, disruption of cognitive control results in symptoms of executive dysfunction, including a tendency to attend to irrelevant information, impaired concentration, disorganization, and difficulty shifting attention (Manning et al., 2014). Cognitive control is thus a central aspect of voluntary behavior and is necessary for solving novel or complex tasks, inhibiting automatic responses, monitoring behavior for errors, and sustaining attention (MacDonald et al., 2000).

Executive dysfunction is common in LLD. Roughly 30–40% of non-demented older adults with major depression exhibit signs of executive dysfunction on cognitive examination (Lockwood, Alexopoulos, & van Gorp, 2002). Depressed older adults often perform poorly on tests of verbal fluency, the Wisconsin Card Sorting Test (a measure of problem-solving and cognitive flexibility), the Tower of London (a test of planning), and the Stroop Color–Word Interference (a test of response inhibition) (Lim et al., 2013). Performance on these tests correlates with structural and electrophysiological abnormalities in cognitive control regions (Murphy et al., 2007). Executive dysfunction is also predictive of poor clinical outcomes in LLD (Morimoto, Kanellopoulos, Manning, & Alexopoulos, 2015). Deficits in verbal fluency and response inhibition predict poor and slow antidepressant response, relapse, and greater levels of functional disability (Morimoto et al., 2012, 2011; Potter, Kittinger, Wagner, Steffens, & Krishnan, 2004).

Functional neuroimaging illustrates the role of the CCN in LLD. Using a seed-based correlation approach to explore resting state functional connectivity, Alexopoulos et al. (2012) found that 16 older adults with nonpsychotic unipolar depression, who were without mild cognitive impairment or dementia (mean age $= 69.0 \pm 5.5$; mean MMSE $= 29.1 \pm 0.8$), exhibited decreased functional connectivity between the left DLPFC (BA 9) and left dACC (BA 24/32) and the left DLPFC and bilateral inferior parietal cortices (BA 40) when compared to nondepressed healthy older adults (mean age $= 68 \pm 7.0$; mean MMSE $= 28.5 \pm 0.9$). Moreover, the authors found that decreased CCN connectivity was associated with a worse clinical outcome. Specifically, decreased functional connectivity of the CCN at baseline correlated with behavioral measures of executive functioning and distinguished older depressed adults who remitted following 12 weeks of escitalopram treatment from those who did not (Alexopoulos et al., 2012). Compared to remitters, nonremitters demonstrated decreased functional connectivity at baseline between the bilateral DLPFC, dACC, and inferior parietal cortices.

The Default Mode Network

The DMN was originally described by Shulman et al. (1997) and subsequently Raichle et al. (2001), who observed that select brain regions experienced increased metabolic activity during rest and decreased activity when engaged in goal-directed (i.e., cognitively demanding) behavior. Several subsequent investigations support the notion that the DMN is normally engaged during internally focused tasks yet is reduced during cognitive demands. Brain regions activated within the DMN include the ventral and dorsal medial prefrontal cortices (BA 24, 10m, 10r, 10p, 32a, 32c, 9), posterior cingulate/retrosplenial cortex (BA 29/30, 23/31), inferior parietal lobule (BA 39, 40), lateral temporal cortex (BA 21), and hippocampal formation.

Evidence suggests that the DMN contributes to the mental exploration of social and emotional content. DMN activity increases during perspective-taking of the desires, beliefs, and intentions of others (i.e., theory of mind), in remembering the past (e.g., autobiographical memory), and in planning the future (Buckner, Andrews-Hanna, & Schacter, 2008). These functions inherently involve the self as a reference point (e.g., one must imagine how events would impact oneself in order to take another's perspective). The self-referential properties of these functions suggest that the DMN may contribute to adaptive behavior by allowing scenarios to be constructed, replayed, and explored in the mind, both to ponder past events and to derive expectations about the future (Buckner et al., 2008). Reduction of activity in the DMN during effortful cognitive processing may reflect the need to reduce the brain's self-referential activity in order to focus on the (external) task at hand (Raichle, 2015). Failure to reduce DMN activity may result in interference from internal mentation or emotional processing (Sheline et al., 2009).

Overactivity of the DMN may help to explain common symptoms in major depression. Major depression is characterized by increased rumination or the recurrent, reflective, and uncontrollable focus on the depressed mood and its causes and consequences. Major depression in middle-aged adults has been repeatedly associated with increased activity within the DMN (Greicius et al., 2007), which in turn is associated with negative rumination (Hamilton, Farmer, Fogelman, & Gotlib, 2015). Similar to their younger depressed counterparts, older adults with major depression exhibit high resting state connectivity within the DMN when contrasted with age-matched nondepressed controls, and greater DMN functional connectivity among depressed subjects is correlated with more abundant negative thinking (Alexopoulos et al., 2012)

The Salience Network

In contrast to the DMN that assigns valence to internally represented stimuli, the SN assesses the significance of external stimuli. The SN has prominent nodes in the anterior insula, ventrolateral prefrontal cortex, and anterior cingulate cortex, which in turn have robust connections to several limbic areas and subcortical structures, including the amygdala, ventral striatum/nucleus accumbens, hypothalamus, dorsomedial thalamus, periaqueductal gray, and substantia nigra/ventral tegmental area. The SN is important in fear, vigilance, and the relationship of emotion to visceral function (Yuen, Gunning-Dixon, et al., 2014). It is conceptualized as a bottom-up processor of salient experiences—whether cognitive, homeostatic, or emotional—that subsequently recruits other large-scale networks to influence ensuing behavior in response to the salient stimulus (Menon & Uddin, 2010; Seeley et al., 2007). For example, the SN may receive emotional information from limbic inputs, suggesting a need for an adaptive response, wherein the SN disengages DMN and engages CEN in mediating attention, working memory, and other higher cognitive processes in problem-solving. The right anterior insula (rAI) is the primary SN node and is deployed first upon the detection of salient stimuli (Downar, Crawley, Mikulis, & Davis, 2000), and the insula, often together with amygdala activity, represents an individual's subjective and conscious emotional state as well as the emotional value of external stimuli (Seeley et al., 2007). The insula acts as the major causal hub that triggers subsequent CEN versus DMN node activation (Hamilton et al., 2011). Increased functional connectivity of the SN with the CCN and DMN has been observed in midlife depression (Manoliu et al., 2013; Sheline, Price, et al., 2010). Whereas the SN has been relatively unexplored in LLD, decreased functional connectivity is associated with increased apathy in LLD (Yuen, Gunning-Dixon, et al., 2014), and increased connectivity between the SN and CCN is associated with behavioral measures of executive dysfunction in elderly depressed subjects (Wang, Chou, Potter, & Steffens, 2015). These studies are detailed below.

SIMULTANEOUS NETWORK ACTIVITY

Dysfunction among several neural systems very likely produces the constellation of cognitive and affective symptoms that constitutes major depression. Wang et al. (2015) examined the functional connectivity among the cognitive control, default mode, and SNs and contrasted correlation patterns between elderly subjects with major depression and age-matched controls. The CCN and SN were significantly correlated with an affective network that overlapped with the dorsomedial prefrontal cortex of the DMN and also included the sgACC and rostral anterior cingulate, but this association was only apparent in healthy older adults. No significant correlation was observed among elderly depressed subjects, suggesting that "dysynchronization" among these networks predisposes or perpetuates major depression. The exact mechanisms by which neural systems synergistically contribute to major depression are unclear. One hypothesis is that heightened responses to negative emotional internal (DMN) and external (SN) stimuli may provide a bottom-up source of input that can serve to dysregulate cognitive control systems that might normally suppress such affective responses (Carballedo et al., 2011; Sheline, Price, et al., 2010). For example, Fales et al. (2008) found that control subjects recruited the DLPFC in response to the processing of fearful faces (activating both DMN and SN regions), whereas middle-aged major depression subjects did not, suggesting that cognitive control recruitment is normally increased in response to emotional distractions, perhaps as a means to suppress amygdala activity. Decreased CCN and increased DMN activation may explain why elderly depressed subjects are more likely to benefit from distraction rather than cognitive reappraisal following negative rumination (Smoski, LaBar, & Steffens, 2014). Wang et al. (2015) observed that older adults who exhibited stronger associations between the SN and affective networks were the same subjects who performed worse on a behavioral measure of cognitive control (Stroop Color and Word Test). Overall, these findings suggest that increased synchronization between salience and emotional regulation networks may represent an increased attention to emotional events, which then could distract them from ongoing cognitive tasks (Wang et al., 2015). Understanding about the interconnectivity among various networks can also be gleaned from exploring the functional neuroanatomical correlates of behavioral syndromes observed in LLD, as described below.

Network Activity Differences in Older and Younger Adults with Major Depression

Younger and older adults with major depression may engage different aspects of the same network when engaged in cognitive and emotional processing. Rao et al. (2015) compared the effects of age and illness on cognitive processing by contrasting the network activity of older adults with major depression (age 65+) engaged in a cognitive control test to age-matched controls and young adults with major depression (age range 18–33). Older adults with LLD exhibited increased activity in several regions (e.g., anterior cingulate and caudate) as compared to both age-matched controls and younger depression subjects, indicating that hyperactivity was not due to either age or depression alone but rather a synergy of the two. Rao et al. (2015) thus suggest that CCN dysfunction in LLD may represent a type of accelerated aging process, consistent with the notion that depression precedes or places older adults at an increased risk of cognitive decline and neurodegeneration (Rushing, Sachs-Ericsson, & Steffens, 2014; Taylor, McQuoid, et al., 2013). Research on functional connectivity differences between young and older adults with MDD is a topic of ongoing research.

FUNCTIONAL CONNECTIVITY UNDERLYING BEHAVIOR IN LATE-LIFE DEPRESSION

The heterogeneous nature of major depression gives rise to various behavioral syndromes. Major depression in the elderly is very commonly accompanied by prominent apathy, anxiety, or suicidality. Whereas once conceived as distinct clinical entities, investigation of the prominent networks reveals that shared or overlapping neuroanatomical substrates may explain seemingly diverse symptom presentations.

APATHY

Apathy is defined as a primary motivational impairment that influences emotional, cognitive, and behavioral functioning (Alexopoulos et al., 2013). Apathetic and depressed older adults present with indifference and flattened affect and a lack of intellectual interest. They also exhibit a lack of initiative and goal-directed behavior. Approximately 30–40% of older depressed adults are described as apathetic by caregivers or acknowledge significant symptoms of apathy on questionnaires (e.g., Apathy Evaluation Scale) (Yuen, Bhutani, et al., 2014). Apathy in LLD is associated with a worse clinical outcome, including poor treatment response, greater burden to caregivers, and increased levels of disability (Steffens, Hays, & Krishnan, 1999; Yuen, Bhutani, et al., 2014).

Older adults with apathetic LLD exhibit behavioral abnormalities in reward-related decision making. Examples of reward-related decision-making tests include the Iowa Gambling Test. When performing this task, individuals select cards, one at a time, from one of four decks. Half of the decks are disadvantageous (i.e., higher immediate rewards but long-term negative outcomes) and the other half are advantageous (i.e., lower immediate rewards but long-term positive outcomes). Overall, older adults with major depression do not differ from age-matched controls on the Iowa or other gambling paradigms. However, when contrasting apathetic and nonapathetic depressed elderly subjects, apathetic subjects demonstrate an advantageous strategy on the Iowa Gambling Task (selecting cards from the conservative decks), suggesting that they are not influenced by immediate reinforcers (the high rewards of disadvantageous desks) (McGovern, Alexopoulos, Yuen, Morimoto, & Gunning-Dixon, 2014).

Dysfunction in reward centers and salience and CCNs contributes to apathy in LLD. The anterior insula and cingulate appear to be prominent in the behavioral manifestation of apathy. Apathy in LLD is correlated with a reduced volume of the anterior cingulate (Lavretsky, Ballmaier, Pham, Toga, & Kumar, 2007; Yuen, Gunning, et al., 2014). Depressed elders who exhibit persistent apathy following 3 months of escitalopram have smaller volumes of the sgACC than depressed patients whose apathy improved, consistent with evidence that this region is in involved in responses to emotional experiences. Increased functional connectivity between the insula and reward centers (seed placement in the nucleus accumbens), cognitive control regions (seed placement in the dACC), and the ventromedial prefrontal cortex regions was found in apathetic older depressed patients ($n = 7$; mean age $= 69.9 \pm 4.9$; mean MMSE $= 28.9 \pm 1.1$) as compared to nonapathetic depressed patients ($n = 9$; mean age $= 68.3 \pm 6.1$; mean MMSE $= 29.3 \pm 0.7$) (Alexopoulos et al., 2013). The insula, as the focal point of the SN, connects to a broad region overlapping between the DMN and CCN, suggesting that the SN acts as a transition center, shifting attention from internal (DMN) to external stimuli (CCN) (Hamilton et al., 2011). Using the right anterior insula as the focal region of interest, Yuen, Gunning-Dixon, et al. (2014) found that, relative to nonapathetic depressed elderly patients, apathetic depressed elders had a lower resting state functional connectivity of the right anterior insula with the right dACC (BA 32), right nucleus accumbens, left caudate, left thalamus, bilateral amygdala, bilateral hippocampus, and bilateral posterior parietal cortex. Relative to nonapathetic depressed elderly patients, apathetic depressed patients had a higher resting state functional connectivity of the right anterior insula with the right DLPFC (BA 9), right posterior cingulate cortex, left posterior cingulate, and precuneus. Overall, as Yuen, Gunning-Dixon, et al. (2014) suggest, an aberrant increased resting state functional connectivity between the SN and DMN may predispose individuals to depression, core features of which may include difficulty in engaging in cognitively demanding tasks while ignoring irrelevant, negatively valenced stimuli, and impaired functional connectivity of the SN, when added to abnormalities in other networks related to depression, can manifest in a picture of depression marked by high comorbid apathy.

ANXIOUS-DEPRESSION

Generalized anxiety and MDDs commonly co-occur. Estimates suggest that roughly 50% of older adults with major depression either meet current criteria or lifetime criteria for an anxiety disorder (Jeste, Hays, & Steffens, 2006). Elderly depressives with comorbid anxiety symptoms present with more severe depression, greater suicidal ideation, and more impairment on subjective social support, and experience a more difficult illness course, including decreased or delayed treatment response (Jeste et al., 2006; Lenze et al., 2001).

Anxious-depressed older adults may also exhibit worse cognitive functioning as compared to older adults with depression alone (DeLuca et al., 2005).

Evidence suggests that late-life anxious depression is associated with abnormalities in both the CCN and DMN. Andreescu et al. (2009) compared the fMRI activation patterns of depressed elders and anxious-depressed elders on a cognitive control task where subjects had to respond to commonly presented stimuli (75% of the time) by pressing a button and then switch button responses for less commonly presented stimuli (25% of the time). Findings revealed that older adults with comorbid depression and anxiety, as compared to elders with depression alone, experienced longer and greater activation in the dorsal and posterior cingulate during task completion. Cingulate overactivity in the context of anxious-depression could represent CCN dysfunction, such as a pathological bias to threat (perceived conflict in the absence of actual conflict) and/or interference from introspective ruminative worry associated with DMN abnormalities (Andreescu et al., 2009). Prominent DMN dysfunction was further suggested in a subsequent seed-based (ROI posterior cingulate) functional connectivity study where elderly depressed subjects with high anxiety had increased connectivity in the posterior regions of the DMN (including the parietal and occipital areas) and decreased connectivity in the anterior regions of the DMN (the rostral ACC, medial prefrontal, and OFC) (Andreescu et al., 2011). The authors speculated that the increased connectivity in the posterior areas of the DMN suggests that subjects with increased anxiety maintain a "higher alert," scanning both the environment (occipital areas) and themselves (parietal areas) excessively, in an effort to detect external or internal potential sources of threat. Based upon work with younger depressed adults, examination of the SN network in LLD may prove to be useful in clarifying the nature of anxious-depression (Hamilton et al., 2011).

SUICIDALITY

Suicide rates in older adults are high as compared to other age groups. The highest risk group for suicide is older white males 85 years or older, and the rate of suicide among older adults aged 50–74 years has almost doubled from 2000 to 2011 (Kiosses, Szanto, & Alexopoulos, 2014). Mood disorders, especially major depression, are the most consistently found psychiatric disorders in suicide completers. Notably, one study found that 42% of older adults with major depression felt that they wanted to die, 25% thought a lot about suicide, and 4.7% attempted suicide (Chou & Cheung, 2013). Multiple factors are associated with suicidal ideation in older adults, including a higher severity of depression, prior suicide attempts, cognitive deficits, white matter disease, lower subjective social support, hopelessness, and impulsive-aggressive traits (Alexopoulos et al., 2009; Dombrovski, Szanto, Clark, Reynolds, & Siegle, 2013; Sachs-Ericsson et al., 2014).

Behavioral deficits in cognitive control represent the most consistent finding in both middle-aged (Keilp et al., 2001, 2014) and older (Gujral et al., 2014; McGirr, Jollant, & Turecki, 2013; Richard-Devantoy, Szanto, Butters, Kalkus, & Dombrovski, 2015) suicide attempters. An emerging literature suggests that the tendency to make disadvantageous decisions is also associated with suicidal behavior. Older suicide attempters exhibit an impaired performance on gambling paradigms (disadvantageous selections) as compared to depressed nonsuicidal and healthy controls (Clark et al., 2011). Poor self-monitoring and impulsivity may drive impaired decision making. Dombrovski et al. (2013) found that suicide attempts in LLD were correlated with impulsivity and monitoring errors on a gambling test. Older adults who have attempted suicide exhibit abnormalities in the ventromedial prefrontal cortex and the basal ganglia as compared to elderly depressed subjects without a history of suicide attempts (Dombrovski et al., 2012, 2013). Neuroanatomical models of suicide suggest that ventral prefrontal cortex is implicated in decision-making impairments in suicide attempters, while more dorsal parts of the prefrontal cortex (ACC) may explain deficits in cognitive control and emotion regulation processes (Ding et al., 2015; Richard-Devantoy et al., 2013). Ongoing research using functional connectivity may provide corroborating evidence in support of cognitive control and limbic dysfunction in elderly depressed subjects with a history of suicide.

CONCLUSION

Multiple neural systems explain the constellation of cognitive and affective symptoms observed in LLD. Understanding abnormalities within and between the cognitive control, default mode, and salience network has helped elucidate the neuroanatomical substrates of major depression in the elderly. Moreover, knowledge of interconnected network circuitry has clarified the association between major depression and heterogeneous symptom presentations, such as apathy, rumination, comorbid anxiety, and suicidality. Ongoing research using functional connectivity holds promise in unlocking the pathophysiology of major depression and identifying the neural patterns that might successfully predict treatment response and illness course in LLD.

REFERENCES

Abraham, H. M., Wolfson, L., Moscufo, N., Guttmann, C. R., Kaplan, R. F., & White, W. B. (2015). Cardiovascular risk factors and small vessel disease of the brain: blood pressure, white matter lesions, and functional decline in older persons. *Journal of Cerebral Blood Flow and Metabolism*, 1–7.

Aizenstein, H. J., Andreescu, C., Edelman, K. L., Cochran, J. L., Price, J., Butters, M. A., … Reynolds, C. F., 3rd (2011). fMRI correlates of white matter hyperintensities in late-life depression. *American Journal of Psychiatry, 168*(10), 1075–1082.

Alexopoulos, G. S., Gunning-Dixon, F. M., Latoussakis, V., Kanellopoulos, D., & Murphy, C. F. (2008). Anterior cingulate dysfunction in geriatric depression. *International Journal of Geriatric Psychiatry, 23*(4), 347–355.

Alexopoulos, G. S., Murphy, C. F., Gunning-Dixon, F. M., Latoussakis, V., Kanellopoulos, D., Klimstra, S., … Hoptman, M. J. (2008). Microstructural white matter abnormalities and remission of geriatric depression. *American Journal of Psychiatry, 165*(2), 238–244.

Alexopoulos, G. S., Reynolds, C. F., 3rd, Bruce, M. L., Katz, I. R., Raue, P. J., Mulsant, B. H., … Group, P. (2009). Reducing suicidal ideation and depression in older primary care patients: 24-month outcomes of the PROSPECT study. *American Journal of Psychiatry, 166*(8), 882–890.

Alexopoulos, G. S., Hoptman, M. J., Kanellopoulos, D., Murphy, C. F., Lim, K. O., & Gunning, F. M. (2012). Functional connectivity in the cognitive control network and the default mode network in late-life depression. *Journal of Affective Disorders, 139*(1), 56–65.

Alexopoulos, G. S., Hoptman, M. J., Yuen, G., Kanellopoulos, D., Seirup, J. K., Lim, K. O., & Gunning, F. M. (2013). Functional connectivity in apathy of late-life depression: a preliminary study. *Journal of Affective Disorders, 149*(1–3), 398–405.

Alexopoulos, G. S., Manning, K. J., Kanellopoulos, D., McGovern, A. R., Seirup, J. K., Banerjee, S., & Gunning, F. M. (2015). Cognitive control, reward related decision making and response of late-life depression to an antidepressant. *Psychologie Medicale*, 1–10.

Andreescu, C., Butters, M., Lenze, E. J., Venkatraman, V. K., Nable, M., Reynolds, C. F., 3rd, & Aizenstein, H. J. (2009). fMRI activation in late-life anxious depression: a potential biomarker. *International Journal of Geriatric Psychiatry, 24*(8), 820–828.

Andreescu, C., Wu, M., Butters, M. A., Figurski, J., Reynolds, C. F., 3rd, & Aizenstein, H. J. (2011). The default mode network in late-life anxious depression. *American Journal of Geriatric Psychiatry, 19*(11), 980–983.

Aziz, R., & Steffens, D. C. (2013). What are the causes of late-life depression? *Psychiatric Clinics of North America, 36*(4), 497–516.

Bonelli, R. M., & Cummings, J. L. (2007). Frontal-subcortical circuitry and behavior. *Dialogues in Clinical Neurosciences, 9*(2), 141–151.

Buckner, R. L., Andrews-Hanna, J. R., & Schacter, D. L. (2008). The brain's default network: anatomy, function, and relevance to disease. *Annals of the New York Academy of Sciences, 1124*, 1–38.

Carballedo, A., Scheuerecker, J., Meisenzahl, E., Schoepf, V., Bokde, A., Moller, H. J., … Frodl, T. (2011). Functional connectivity of emotional processing in depression. *Journal of Affective Disorders, 134*(1–3), 272–279.

Chou, K. L., & Cheung, K. C. (2013). Major depressive disorder in vulnerable groups of older adults, their course and treatment, and psychiatric comorbidity. *Depression and Anxiety, 30*(6), 528–537.

Clark, L., Dombrovski, A. Y., Siegle, G. J., Butters, M. A., Shollenberger, C. L., Sahakian, B. J., & Szanto, K. (2011). Impairment in risk-sensitive decision-making in older suicide attempters with depression. *Psychology and Aging, 26*(2), 321–330.

DeLuca, A. K., Lenze, E. J., Mulsant, B. H., Butters, M. A., Karp, J. F., Dew, M. A., … Reynolds, C. F., 3rd (2005). Comorbid anxiety disorder in late life depression: association with memory decline over four years. *International Journal of Geriatric Psychiatry, 20*(9), 848–854.

Ding, Y., Lawrence, N., Olie, E., Cyprien, F., le Bars, E., Bonafe, A., … Jollant, F. (2015). Prefrontal cortex markers of suicidal vulnerability in mood disorders: a model-based structural neuroimaging study with a translational perspective. *Translational Psychiatry, 5*, e516.

Dombrovski, A. Y., Siegle, G. J., Szanto, K., Clark, L., Reynolds, C. F., & Aizenstein, H. (2012). The temptation of suicide: striatal gray matter, discounting of delayed rewards, and suicide attempts in late-life depression. *Psychologie Medicale, 42*(6), 1203–1215.

Dombrovski, A. Y., Szanto, K., Clark, L., Reynolds, C. F., & Siegle, G. J. (2013). Reward signals, attempted suicide, and impulsivity in late-life depression. *JAMA Psychiatry, 70*(10), 1.

Downar, J., Crawley, A. P., Mikulis, D. J., & Davis, K. D. (2000). A multimodal cortical network for the detection of changes in the sensory environment. *Nature Neuroscience, 3*(3), 277–283.

Drevets, W. C. (2001). Neuroimaging and neuropathological studies of depression: implications for the cognitive-emotional features of mood disorders. *Current Opinion in Neurobiology, 11*(2), 240–249.

Fales, C. L., Barch, D. M., Rundle, M. M., Mintun, M. A., Snyder, A. Z., Cohen, J. D., … Sheline, Y. I. (2008). Altered emotional interference processing in affective and cognitive-control brain circuitry in major depression. *Biological Psychiatry, 63*(4), 377–384.

Greicius, M. D., Flores, B. H., Menon, V., Glover, G. H., Solvason, H. B., Kenna, H., … Schatzberg, A. F. (2007). Resting-state functional connectivity in major depression: abnormally increased contributions from subgenual cingulate cortex and thalamus. *Biological Psychiatry, 62*(5), 429–437.

Gujral, S., Dombrovski, A. Y., Butters, M., Clark, L., Reynolds, C. F., 3rd, & Szanto, K. (August 2014). Impaired executive function in contemplated and attempted suicide in late life. *American Journal of Geriatric Psychiatry, 22*(8), 811–819.

Gunning, F. M., Cheng, J., Murphy, C. F., Kanellopoulos, D., Acuna, J., Hoptman, M. J., … Alexopoulos, G. S. (2009). Anterior cingulate cortical volumes and treatment remission of geriatric depression. *International Journal of Geriatric Psychiatry, 24*(8), 829–836.

Gunning-Dixon, F. M., Walton, M., Cheng, J., Acuna, J., Klimstra, S., Zimmerman, M. E., … Alexopoulos, G. S. (2010). MRI signal hyperintensities and treatment remission of geriatric depression. *Journal of Affective Disorders, 126*(3), 395–401.

Hamilton, J. P., Furman, D. J., Chang, C., Thomason, M. E., Dennis, E., & Gotlib, I. H. (2011). Default-mode and task-positive network activity in major depressive disorder: implications for adaptive and maladaptive rumination. *Biological Psychiatry, 70*(4), 327–333.

Hamilton, J. P., Farmer, M., Fogelman, P., & Gotlib, I. H. (2015). Depressive rumination, the default-mode network, and the dark matter of clinical neuroscience. *Biological Psychiatry, 78*(4), 224–230.

Jeste, N. D., Hays, J. C., & Steffens, D. C. (2006). Clinical correlates of anxious depression among elderly patients with depression. *Journal of Affective Disorders, 90*(1), 37–41.

Keilp, J. G., Sackeim, H. A., Brodsky, B. S., Oquendo, M. A., Malone, K. M., & Mann, J. J. (2001). Neuropsychological dysfunction in depressed suicide attempters. *American Journal of Psychiatry, 158*(5), 735–741.

Keilp, J. G., Beers, S. R., Burke, A. K., Melhem, N. M., Oquendo, M. A., Brent, D. A., & Mann, J. J. (2014). Neuropsychological deficits in past suicide attempters with varying levels of depression severity. *Psychologie Medicale, 44*(14), 2965–2974.

Kiosses, D. N., Szanto, K., & Alexopoulos, G. S. (2014). Suicide in older adults: the role of emotions and cognition. *Current Psychiatry Reports, 16*(11), 495.

Lavretsky, H., Ballmaier, M., Pham, D., Toga, A., & Kumar, A. (2007). Neuroanatomical characteristics of geriatric apathy and depression: a magnetic resonance imaging study. *American Journal of Geriatric Psychiatry, 15*(5), 386–394.

Lenze, E. J., Mulsant, B. H., Shear, M. K., Alexopoulos, G. S., Frank, E., & Reynolds, C. F., 3rd. (2001). Comorbidity of depression and anxiety disorders in later life. *Depression and Anxiety, 14*(2), 86–93.

Lim, J., Oh, I. K., Han, C., Huh, Y. J., Jung, I. K., Patkar, A. A., … Jang, B. H. (2013). Sensitivity of cognitive tests in four cognitive domains in discriminating MDD patients from healthy controls: a meta-analysis. *International Psychogeriatric, 25*(9), 1543–1557.

Lockwood, K. A., Alexopoulos, G. S., & van Gorp, W. G. (2002). Executive dysfunction in geriatric depression. *American Journal of Psychiatry, 159*(7), 1119–1126.

MacDonald, A. W., 3rd, Cohen, J. D., Stenger, V. A., & Carter, C. S. (2000). Dissociating the role of the dorsolateral prefrontal and anterior cingulate cortex in cognitive control. *Science, 288*(5472), 1835–1838.

Manning, K. J., Alexopoulos, G. S., McGovern, A. R., Morimoto, S., Yuen, G. S., Kanellopoulos, D., & Gunning, F. M. (2014). Executive functioning in late-life depression. *Psychiatric Annals, 44*, 143–146.

Manoliu, A., Meng, C., Brandl, F., Doll, A., Tahmasian, M., Scherr, M., … Sorg, C. (2013). Insular dysfunction within the salience network is associated with severity of symptoms and aberrant inter-network connectivity in major depressive disorder. *Frontiers in Human Neuroscience, 7*, 930.

Mayberg, H. S. (1997). Limbic-cortical dysregulation: a proposed model of depression. *Journal of Neuropsychiatry and the Clinical Neurosciences, 9*(3), 471–481.

McGirr, A., Jollant, F., & Turecki, G. (2013). Neurocognitive alterations in first degree relatives of suicide completers. *Journal of Affective Disorders, 145*(2), 264–269.

McGovern, A. R., Alexopoulos, G. S., Yuen, G. S., Morimoto, S. S., & Gunning-Dixon, F. M. (2014). Reward-related decision making in older adults: relationship to clinical presentation of depression. *International Journal of Geriatric Psychiatry, 29*(11), 1125–1131.

Menon, V., & Uddin, L. Q. (2010). Saliency, switching, attention and control: a network model of insula function. *Brain Structure and Function, 214*(5–6), 655–667.

Morimoto, S. S., Gunning, F. M., Kanellopoulos, D., Murphy, C. F., Klimstra, S. A., Kelly, R. E., Jr., & Alexopoulos, G. S. (2012). Semantic organizational strategy predicts verbal memory and remission rate of geriatric depression. *International Journal of Geriatric Psychiatry, 27*(5), 506–512.

Morimoto, S. S., Gunning, F. M., Murphy, C. F., Kanellopoulos, D., Kelly, R. E., & Alexopoulos, G. S. (2011). Executive function and short-term remission of geriatric depression: the role of semantic strategy. *American Journal of Geriatric Psychiatry, 19*(2), 115–122.

Morimoto, S. S., Kanellopoulos, D., Manning, K. J., & Alexopoulos, G. S. (2015). Diagnosis and treatment of depression and cognitive impairment in late life. *Annals of the New York Academy of Sciences, 1345*(1), 36–46.

Murphy, C. F., Gunning-Dixon, F. M., Hoptman, M. J., Lim, K. O., Ardekani, B., Shields, J. K., … Alexopoulos, G. S. (2007). White-matter integrity predicts stroop performance in patients with geriatric depression. *Biological Psychiatry, 61*(8), 1007–1010.

Phillips, M. L., Drevets, W. C., Rauch, S. L., & Lane, R. (2003a). Neurobiology of emotion perception I: the neural basis of normal emotion perception. *Biological Psychiatry, 54*(5), 504–514.

Phillips, M. L., Drevets, W. C., Rauch, S. L., & Lane, R. (2003b). Neurobiology of emotion perception II: implications for major psychiatric disorders. *Biological Psychiatry, 54*(5), 515–528.

Potter, G. G., Kittinger, J. D., Wagner, H. R., Steffens, D. C., & Krishnan, K. R. (2004). Prefrontal neuropsychological predictors of treatment remission in late-life depression. *Neuropsychopharmacology, 29*(12), 2266–2271.

Raichle, M. E., MacLeod, A. M., Snyder, A. Z., Powers, W. J., Gusnard, D. A., & Shulman, G. L. (2001). A default mode of brain function. *Proceedings of the National Academy of Sciences of the United States of America, 98*(2), 676–682.

Raichle, M. E. (2015). The brain's default mode network. *Annual Review of Neuroscience, 38*, 433–447.

Rao, J. A., Kassel, M. T., Weldon, A. L., Avery, E. T., Briceno, E. M., Mann, M., … Weisenbach, S. L. (2015). The double burden of age and major depressive disorder on the cognitive control network. *Psychology and Aging, 30*(2), 475–485.

Richard-Devantoy, S., Olie, E., Guillaume, S., Bechara, A., Courtet, P., & Jollant, F. (2013). Distinct alterations in value-based decision-making and cognitive control in suicide attempters: toward a dual neurocognitive model. *Journal of Affective Disorders, 151*(3), 1120–1124.

Richard-Devantoy, S., Szanto, K., Butters, M. A., Kalkus, J., & Dombrovski, A. Y. (2015). Cognitive inhibition in older high-lethality suicide attempters. *International Journal of Geriatric Psychiatry, 30*(3), 274–283.

Rushing, N. C., Sachs-Ericsson, N., & Steffens, D. C. (2014). Neuropsychological indicators of preclinical Alzheimer's disease among depressed older adults. *Neuropsychology, Development, and Cognition. Section B, Aging, Neuropsychology and Cognition, 21*(1), 99–128.

Sachs-Ericsson, N., Hames, J. L., Joiner, T. E., Corsentino, E., Rushing, N. C., Palmer, E., … Steffens, D. C. (2014). Differences between suicide attempters and nonattempters in depressed older patients: depression severity, white-matter lesions, and cognitive functioning. *American Journal of Geriatric Psychiatry, 22*(1), 75–85.

Seeley, W. W., Menon, V., Schatzberg, A. F., Keller, J., Glover, G. H., Kenna, H., … Greicius, M. D. (2007). Dissociable intrinsic connectivity networks for salience processing and executive control. *Journal of Neuroscience, 27*(9), 2349–2356.

Sheline, Y. I., Barch, D. M., Price, J. L., Rundle, M. M., Vaishnavi, S. N., Snyder, A. Z., … Raichle, M. E. (2009). The default mode network and self-referential processes in depression. *Proceedings of the National Academy of Sciences of the United States of America, 106*(6), 1942–1947.

Sheline, Y. I., Pieper, C. F., Barch, D. M., Welsh-Bohmer, K., McKinstry, R. C., MacFall, J. R., … Doraiswamy, P. M. (2010). Support for the vascular depression hypothesis in late-life depression: results of a 2-site, prospective, antidepressant treatment trial. *Archives of General Psychiatry, 67*(3), 277–285.

Sheline, Y. I., Price, J. L., Yan, Z., & Mintun, M. A. (2010). Resting-state functional MRI in depression unmasks increased connectivity between networks via the dorsal nexus. *Proceedings of the National Academy of Sciences of the United States of America, 107*(24), 11020–11025.

Shulman, G. L., Fiez, J. A., Corbetta, M., Buckner, R. L., Miezin, F. M., Raichle, M. E., & Petersen, S. E. (1997). Common blood flow changes across visual tasks: II. Decreases in cerebral cortex. *Journal of Cognitive Neuroscience, 9*(5), 648–663.

Smoski, M. J., LaBar, K. S., & Steffens, D. C. (2014). Relative effectiveness of reappraisal and distraction in regulating emotion in late-life depression. *American Journal of Geriatric Psychiatry, 22*(9), 898–907.

Steffens, D. C., Hays, J. C., & Krishnan, K. R. (1999). Disability in geriatric depression. *American Journal of Geriatric Psychiatry, 7*(1), 34–40.

Steffens, D. C., Skoog, I., Norton, M. C., Hart, A. D., Tschanz, J. T., Plassman, B. L., ... Breitner, J. C. (2000). Prevalence of depression and its treatment in an elderly population: the cache county study. *Archives of General Psychiatry, 57*(6), 601–607.

Taylor, W. D., Macfall, J. R., Payne, M. E., McQuoid, D. R., Steffens, D. C., Provenzale, J. M., & Krishnan, K. R. (2007). Orbitofrontal cortex volume in late life depression: influence of hyperintense lesions and genetic polymorphisms. *Psychologie Medicale, 37*(12), 1763–1773.

Taylor, W. D., Aizenstein, H. J., & Alexopoulos, G. S. (2013). The vascular depression hypothesis: mechanisms linking vascular disease with depression. *Molecular Psychiatry, 18*(9), 963–974.

Taylor, W. D., McQuoid, D. R., Payne, M. E., Zannas, A. S., Macfall, J. R., & Steffens, D. C. (2013). Hippocampus atrophy and the longitudinal course of late-life depression. *American Journal of Geriatric Psychiatry, 22*(12), 1504–1512.

Volkert, J., Schulz, H., Harter, M., Wlodarczyk, O., & Andreas, S. (2013). The prevalence of mental disorders in older people in Western countries - a meta-analysis. *Ageing Research Review, 12*(1), 339–353.

Wang, L., Chou, Y. H., Potter, G. G., & Steffens, D. C. (2015). Altered synchronizations among neural networks in geriatric depression. *BioMed Research International, 2015*, 343720.

Wolfson, L., Wakefield, D. B., Moscufo, N., Kaplan, R. F., Hall, C. B., Schmidt, J. A., ... White, W. B. (2013). Rapid buildup of brain white matter hyperintensities over 4 years linked to ambulatory blood pressure, mobility, cognition, and depression in old persons. *Journal of Gerontology Series A: Biological Sciences and Medical Sciences, 68*(11), 1387–1394.

Wu, M., Andreescu, C., Butters, M. A., Tamburo, R., Reynolds, C. F., 3rd, & Aizenstein, H. (2011). Default-mode network connectivity and white matter burden in late-life depression. *Psychiatry Research, 194*(1), 39–46.

Yuen, G. S., Bhutani, S., Lucas, B. J., Gunning, F. M., AbdelMalak, B., Seirup, J. K., ... Alexopoulos, G. S. (2014). Apathy in late-life depression: common, persistent, and disabling. *American Journal of Geriatric Psychiatry, 23*(5), 488–494.

Yuen, G. S., Gunning-Dixon, F. M., Hoptman, M. J., AbdelMalak, B., McGovern, A. R., Seirup, J. K., & Alexopoulos, G. S. (2014). The salience network in the apathy of late-life depression. *International Journal of Geriatric Psychiatry, 29*(11), 1116–1124.

Yuen, G. S., Gunning, F. M., Woods, E., Klimstra, S. A., Hoptman, M. J., & Alexopoulos, G. S. (2014). Neuroanatomical correlates of apathy in late-life depression and antidepressant treatment response. *Journal of Affective Disorders, 166*, 179–186.

Chapter 12

Arousal Regulation in Affective Disorders

Ulrich Hegerl[1,2], Christian Sander[1,2], Tilman Hensch[1]
[1]*Department of Psychiatry and Psychotherapy, University of Leipzig, Leipzig, Germany;*
[2]*Research Centre of the German Depression Foundation, Leipzig, Germany*

INTRODUCTION

Arousal, specifically brain arousal, fundamentally impacts all human behaviors (NIMH, 2012; Pfaff, Ribeiro, Matthews, & Kow, 2008). The long tradition of analyzing the role of general arousal in normal and abnormal behavior and cognition (Eysenck, 1990; Yerkes & Dodson, 1908; Zuckerman, 1979) has been renewed by the Research Domain Criteria project of the National Institute of Mental Health, which implemented arousal as a basic dimension of mental diseases (Cuthbert & Insel, 2013). In daily life, brain arousal has to be precisely regulated to fulfill situational requirements. For example, brain arousal must be heightened in case of potential danger or maintained during cognitive tasks and reduced at bedtime.

In this chapter a new concept will be introduced that links affective disorders and other psychiatric conditions to a disturbed regulation of brain arousal. After a short overview on terminological difficulties, theoretical models, and common means of assessments an electroencephalography (EEG)-based assessment approach, the Vigilance Algorithm Leipzig (VIGALL), will be described, facilitating research on brain arousal regulation. Afterward, the arousal model of affective disorders will be described with respect to depression, mania, and attention deficit/hyperactivity disorder (ADHD).

DISTURBED AROUSAL REGULATION IN AFFECTIVE DISORDERS

Studies on disturbance of brain arousal in affective disorders have previously focused on disturbed sleep. Although sleep disturbances are a very common complaint in many psychiatric disorders, they are of special prominence in affective disorders. Most patients suffering from depression experience some

Systems Neuroscience in Depression. http://dx.doi.org/10.1016/B978-0-12-802456-0.00012-1

kind of sleep disorder. Insomnia is typical for cases of unipolar depression, where pathological sleep patterns with prolonged sleep latencies (Armitage, 2007; Kayumov et al., 2000; Tsuno, Besset, & Ritchie, 2005), disturbed sleep continuity, and early awakenings are often seen, paralleled by an altered sleep architecture with decreased slow-wave sleep and increased rapid eye movement density in the first sleep cycle (Riemann, Berger, & Voderholzer, 2001; Wichniak, Wierzbicka, & Jernajczyk, 2012). Hypersomnia, however, is a symptom of atypical depression. In the past, sleep disorders have been considered to be a mere symptom of depression, but evidence now suggests that disrupted sleep plays a more central role in the pathophysiology of depression (Baglioni et al., 2011; Riemann & Voderholzer, 2003) and should be considered as another core symptom of depression (Nutt, Wilson, & Paterson, 2008).

Furthermore, an association between sleep duration and mood has been observed: prolonged time in bed and long sleep are associated with a decline in mood, whereas short sleep duration and sleep deprivation have mood-enhancing properties and in vulnerable persons may even trigger manic episodes (Bauer et al., 2006; Wehr, 1989). This effect is further illustrated by the effectiveness of therapeutic sleep deprivation, which in about 60% of patients quickly reduces depressive symptoms (Giedke & Schwarzler, 2002). However, this antidepressive effect often only lasts until the next sleep episode is initiated, after which the depressive symptomatology resurfaces (Riemann, Wiegand, Lauer, & Berger, 1993). Some evidence suggests that chronically restricting sleep or time in bed might improve mood and depressive symptoms (Dirksen & Epstein, 2008; Manber et al., 2008).

It would be unjustified, however, to focus on disrupted sleep alone, as the wake state comprises the largest part of the day, and subjects suffering from sleep problems may also exhibit dysregulation of arousal during wakefulness. Tiredness and feelings of fatigue or weariness are typically reported by depressed patients (Shen et al., 2011) and, especially when severe sleep problems are experienced at night, patients consider themselves in grave need of sleep. However, in contrast to their subjective feeling of exhaustion, patients with typical depression do not show increased daytime sleepiness as assessed by sleep onset latencies during the day. Their sleep onset latencies are prolonged, and patients report difficulties in relaxing (Kayumov et al., 2000; Reynolds, Coble, Kupfer, & Holzer, 1982). Furthermore, they often carry signs of higher noradrenergic and hypothalamic–pituitary–adrenal (HPA) axis activity (Pariante & Lightman, 2008; Wong et al., 2000) and subjectively report high inner tension. Hypersomnia, although often reported subjectively by patients, is in most cases not verifiable with objective assessments (Dauvilliers, Lopez, Ohayon, & Bayard, 2013; Nofzinger et al., 1991).

This points to an important terminological problem (Hegerl, 2014). Expressions such as tiredness and fatigue are, in many cases, used to describe completely distinct phenomena (Hegerl et al., 2013; Neu et al., 2008):

(a) tiredness/fatigue in the sense of sleepiness, i.e., increased tendency to get drowsy or fall asleep and (b) tiredness/fatigue in the sense of exhaustion with a tonically high inner tension and physiological arousal. It is the latter syndrome that is typically found in patients with unipolar depression. Patients are convinced that they could improve their condition by extended bed rest, yet in many cases this only aggravates the underlying problem (as will be described below).

Terminological blurs are also one reason for the lesser acknowledgment of the wake stage differentiation within the research community. Different levels of wakefulness have been conceptualized independently within several research fields, e.g., psychophysiology, psychology, or cognitive neuroscience, so that several terms and concepts exist, which are in parts synonymous, contradictory, or simply referring to specific aspects of wakefulness. Some terms are used to describe what is going on within the organism (physiological phenomena) and others to describe patterns of nonverbal and verbal behavior (behavioral phenomena). Brain arousal describes different states of physiological activation, and a generalized central nervous system arousal is considered to underlie all motivated behavior (Pfaff et al., 2008). Alertness describes different behavioral patterns, especially different states of responsiveness and watchfulness to mostly external stimuli. The term vigilance itself is used both at the behavioral and physiological levels (Oken, Salinsky, & Elsas, 2006). Originally it was coined to describe a state of maximal physiologic efficiency (Head, 1923) and later to describe the brain arousal levels assessable with EEG during the transition period between sleep and wakefulness (vigilance levels; see below). Within cognitive psychology, however, vigilance is frequently used to describe behavioral patterns, especially the ability to maintain attention and alertness over long durations, and has become a synonym for sustained attention in psychology. Accordingly, different assessments have been put forward (see below).

In the context of this chapter, the term vigilance is used to describe different states of global brain function, indicating different brain arousal states. It is common knowledge that during sleep distinct sleep stages can be separated using EEG. However, in sleep research the wake state is often described simply as the non-sleeping state and is not further considered. Still, the wake state can also be subdivided into several vigilance stages. Just as sleep stages, these vigilance stages can best be classified using EEG (see below). This knowledge has existed since the 1960s, when these substages were first described (Bente, 1964; Roth, 1961); however, it has received much less research attention as compared to the sleep stages. Yet, sleep and vigilance stages are both global functional physiological states of the organism. Within these different levels an organism is more or less sensitive to internal or external stimuli, and therefore performance on the behavioral level differs according to the global functional state. Besides the different vigilance stages the regulation of vigilance (or arousal respectively) is of special importance. Life is a constant interaction with the environment, and an organism needs to adapt to the environmental needs and challenges. On one hand, it is important to adapt the degree of brain arousal to the specific needs and challenges of the current environment to achieve goals and avoid harm and potential death (e.g., avoiding or coping with dangerous situations); on the other hand, it is important to actively shape or seek an environment fitting to the current arousal level (e.g., find a safe place to sleep). If the internal arousal regulation is disturbed, health, functioning, and well-being are severely threatened.

MODELS OF AROUSAL, SLEEP, AND WAKEFULNESS REGULATION

According to the Two-Process Model of Sleep Regulation (Borbely, 1982; Daan, Beersma, & Borbely, 1984), the timing of sleep and wakefulness is the result of two interacting processes: a circadian process C and a homeostatic process S. The homeostatic process S rises during wakefulness, resulting in a growing sleep propensity, and degrades exponentially during sleep. The circadian process C rises and declines in a periodical manner, which has been shown to follow an about 24-h cycle. The transition from wakefulness to sleep occurs when the increase in process S reaches a certain threshold, whereas the sleep–wake transition sets in after the degrading process S reaches another threshold during sleep.

Based on this model, a Three-Process Model of Alertness and Sleepiness has been formulated (Akerstedt & Folkard, 1996, 1997), in which process C describes sleepiness due to circadian influences and process S is an exponential function of time since awaking. Accordingly, process S is meant to decline during wakefulness and exponentially rises during sleep. The time course of daytime alertness is a result of the interaction between processes S and C with a third component, process W, which describes sleep inertia after waking. Inertia components (at the transition points from wake to sleep and sleep to wakefulness) have also been applied to the original two-process model to simulate daytime vigilance, alertness, and sleepiness (Achermann, 2004).

Furthermore, it has been suggested that solely focusing on a sleep-promoting drive is not sufficient to explain many phenomena surrounding sleep–wake regulation. Thus a Four-Process Model of Sleep and Wakefulness has been introduced (Johns, 1998), which postulates a sleep and an antagonizing wake drive that are considered mutually inhibitory. Here, a person's current arousal state depends on the preponderance of the relative strength of the two drives but not the absolute strength of one. Both drives result from the additive effects of a primary and a secondary component, the primary one being derived from activity of neuronal groups and the secondary one influenced by homeostatic aspects or behavior (i.e., the secondary sleep drive corresponds to process S of the two-process model).

ASSESSMENT OF WAKEFULNESS LEVEL AND SLEEPINESS

There are several available means to assess arousal and wakefulness (for reviews see Cluydts, De Valck, Verstraeten, & Theys, 2002; Mathis & Hess, 2009). For practical purposes, questionnaires are a convenient and easily applied means of assessment. The most broadly used instruments for the assessment of daytime sleepiness or arousal state are:

- the Epworth Sleepiness Scale (ESS; Johns, 1991);
- the Stanford Sleepiness Scale (SSS; Hoddes, Dement, & Zarcone, 1972); and
- the Karolinska Sleepiness Scale (KSS; Akerstedt & Gillberg, 1990).

With the ESS, a subject is requested to rate the likelihood of falling asleep within typical daytime activities and situations; therefore the scale quantifies the overall amount of sleepiness and its impact on functioning within a certain period of time. Clinically, the questionnaire is mostly used as a screening instrument for excessive daytime sleepiness but can neither be used to assess the acute level of sleepiness nor minor fluctuations in wakefulness within short intervals. For these purposes, the SSS and KSS were developed, which are short rating scales on which a subject is asked to rate the current level of wakefulness. This can be repeated frequently; however, answers reflect the subjective estimation of the subject, which can differ from the physiological sleep propensity or level of wakefulness. Therefore, objective assessments have been developed; the most widely used being:

- the Multiple Sleep Latency Test (MSLT; Carskadon et al., 1986); and
- the Maintenance of Wakefulness Test (MWT; Mitler, Gujavarty, & Browman, 1982).

Both tests need to be performed in a sleep laboratory as they require a polysomnography setup and have comparable implementation requirements (e.g., several trials repeated every 2 h). However, they assess different aspects. The MSLT measures the propensity of falling asleep, whereas the MWT measures the ability to resist falling asleep, two skills that are not necessarily associated (Sangal, Thomas, & Mitler, 1992). Within the MSLT, subjects are placed in a comfortable position (lying in bed in a dark and quiet room) and are instructed to try to fall asleep. It is recorded whether or not they do so within a 20-min trial and after what amount of time (sleep onset latency, SOL). Normally, four to five trials are performed every 2 h. An average SOL of 10 min and more is considered as normal sleepiness while an average SOL of 5 min or less is interpreted as abnormal sleepiness. Due to its setup and instruction, the MSLT cannot be used to assess the ability to stay awake, which may be the more relevant skill for daily functioning. For this end, the MWT is the more suitable test. Subjects are usually seated in a chair in a dark room and are instructed to stay awake during a 20- or 40-min trial, which are also repeated (normally four trials every 2 h).

Apart from the MSLT and MWT, other objective but less resource-demanding assessments of wakefulness and sleepiness are available. One class is composed of performance tests, such as the Psychomotor Vigilance Task (Dinges & Powell, 1985). In these tests, subjects are required to perform an easy and monotonous task for an extended period of time, and it is recorded whether or not they can successfully carry out that task. The performance decrement is used as an indicator of sleepiness. Other tests are based on electrophysiological measures, such as the Karolinska drowsiness test (Akerstedt & Gillberg, 1990) or the alpha attenuation test (Stampi, Stone, & Michimori, 1995). These tests estimate the current level of wakefulness by comparing EEG activity between eyes open and eyes closed conditions. This takes into account the distinct changes in EEG activity in both conditions with increasing sleepiness (increase of alpha activity in eyes open condition versus decreased alpha activity with eyes closed). Such assessment can be repeated several times but are still not useful to continuously monitor fluctuations in arousal. One approach to overcome this limitation has been the pupillographic sleepiness test (Wilhelm et al., 2001), where the diameter of the pupil is continuously monitored. Pupil diameter is inversely related to sleepiness, and its variability over time is used as an indicator for arousal changes. Still, EEG recordings provide the best temporal resolution and therefore remain the gold standard to objectively assess sleep stages. They should also be the method of choice to assess wakefulness fluctuations.

EEG VIGILANCE AS A MARKER OF BRAIN AROUSAL

As described above, different EEG-vigilance stages can be discerned not only during sleep but also during wakefulness. According to original conceptions from the 1960s (Bente, 1964; Roth, 1961), the following EEG-vigilance stages can be observed during the transition from high alertness to relaxed wakefulness to drowsiness and finally sleep onset:

- Stage 0 is characterized by a desynchronized nonalpha EEG in the absence of slow horizontal eye movements. This stage is typically seen during activated states (e.g., reflecting mental effort).
- Stage A (with substages A1, A2, and A3) is characterized by dominant alpha activity in the EEG trace and corresponds to relaxed wakefulness. With decreasing vigilance there is a slight slowing of alpha activity and a shift from occipital to more anterior regions.
- Stage B1 is again characterized by desynchronized non-alpha EEG with low amplitude (similar spectral composition as stage 0) but often (yet not necessarily) in the presence of slow horizontal eye movements. This stage corresponds to drowsiness.

- Stage B2/3 is characterized by a non-alpha EEG with predominant theta/delta activity and occasional occurrence of vertex waves. It reflects a state of more severe drowsiness and marks the transition to sleep onset.
- Stage C is reached when sleep spindles or K-complexes are seen, which are signs of sleep onset.

Studies on changes of EEG activity during the transition from active wakefulness to sleep onset endorse these classifications (Benca et al., 1999; Cantero, Atienza, & Salas, 2002; Corsi-Cabrera, Guevara, Del Rio-Portilla, Arce, & Villanueva-Hernandez, 2000; De Gennaro, Ferrara, & Bertini, 2001; De Gennaro, Ferrara, Curcio, et al., 2001, 2004; Kaida et al., 2006; Marzano et al., 2007; Strijkstra, Beersma, Drayer, Halbesma, & Daan, 2003; Tsuno et al., 2002). Research on wakefulness regulation has been hindered by the absence of explicit scoring rules, which have long been established for the scoring of sleep stages (Iber, Ancoli-Israel, Chessonn, & Quan, 2007; Rechtschaffen & Kales, 1968). Furthermore, changes in wakefulness are not as uniform as the typical changes in sleep stages, as subjects go back and forth between vigilance stages with sometimes very short-lasting switches. Therefore, a segmentation of the resting EEG into 30-s epochs, as is the consensus in sleep medicine, is not feasible for scoring vigilance changes in a resting EEG, where much shorter periods have to be considered. Visual classification of vigilance stages in a resting EEG has therefore been an arduous and time-consuming task, and the problem of inter- and intrarater reliability has always been a crucial issue. Therefore, the development of computer-assisted scoring algorithms has been essential for rejuvenating research interest in wakefulness regulation. Several algorithms have been put forward (Khushaba, Kodagoda, Lal, & Dissanayake, 2011; Sauvet et al., 2014; Shi, Duan, & Lu, 2013), in most cases with the aim of detecting drowsiness/sleep lapses during task performance, e.g., for driver safety. A new algorithm (Vigilance Algorithm Leipzig, VIGALL) has been introduced, which was developed with the specific aim of facilitating research on brain arousal in psychiatric conditions and will be described in detail in the next paragraph. In parallel to the development of the VIGALL algorithm a pathogenetic concept has been formulated, linking affective disorders to disturbances of brain arousal regulation. This brain arousal model of affective disorders will be presented and discussed below.

THE VIGALL

The VIGALL (for manual and download go to http://research.uni-leipzig.de/vigall/) is an EEG- and EOG-based tool that allows for objectively classifying vigilance levels within multichannel EEG recordings by automatically attributing one of the above-mentioned vigilance stages to a certain EEG segment of preferably 1 s duration (Hegerl et al., 2008; Olbrich et al., 2009, 2011). The VIGALL algorithm takes into account EEG activity in different frequency bands (delta/theta, alpha) and the cortical distribution of EEG activity using EEG source localization approaches (Pascual-Marqui, Esslen, Kochi, & Lehmann, 2002; Pascual-Marqui, Michel, & Lehmann, 1994). However, EEG activity is characterized by high intraindividual stability and large interindividual variability; therefore, VIGALL has adaptive features concerning individual alpha peaks and amplitude levels (see Figure 1). Before the vigilance classification is performed, VIGALL automatically detects the individual alpha frequency and power from a representative epoch of alpha activity. Some of the parameters (e.g., upper and lower border of the alpha band) and decision criteria of the VIGALL (e.g., absolute alpha power cutoff to classify an A-stage) are then adapted accordingly. It should be taken into account that VIGALL should not be applied in cases of alpha variant rhythms, major modifications due to drugs (e.g., anticholinergic drugs), or diseases (e.g., severe Alzheimer disease). EEGs of children under the age of 10 (or older in the case of delayed maturation) should also be assessed with caution.

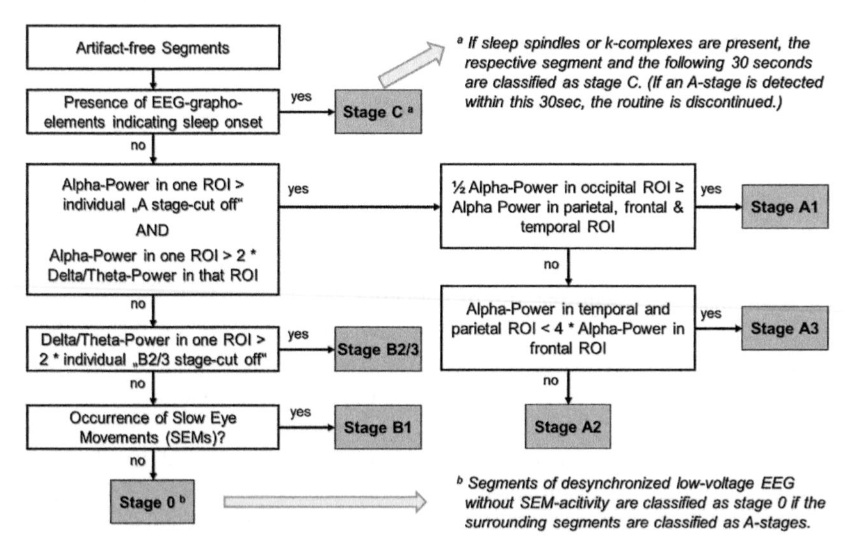

FIGURE 1 Operational sequence and decision criteria of the VIGALL 2.0 algorithm. VIGALL first screens the EEG trace for a 10-s epoch with prominent alpha activity (default range 7.5–12.5 Hz). For the respective epoch, the Alpha center of gravity frequency (ACF) and mean alpha power at occipital sites (OAP) are calculated. ACF is then used to set the individual alpha range (ACF ±2 Hz); the delta/theta range is fixed to 2–7 Hz. Afterward, VIGALL calculates spectral power (alpha versus delta/theta) in four regions of interest (ROI, i.e., the frontal, parietal, temporal, and occipital lobe) using LORETA. For classification of vigilance stages, the OAP is used to determine an individual alpha threshold, which is then used as a cutoff value in the classification of A- and B2/3-stages. Segments not classified as A- or B2/3 stages are classified as B1- or 0 stages, according to the presence of slow horizontal eye movements. If graphoelements indicating sleep onset (sleep spindles or K-complexes) are present, segments are classified as stage C.

The very high temporal resolution of 1 s allows investigations on brain arousal regulation by assessing the time course of vigilance levels during the recording period. This regulation shows considerable interindividual differences (Huang et al., 2015). During eyes closed resting conditions of 15–20 min duration, most subjects show progressive declines to lower vigilance stages (adaptive vigilance regulation). However, whereas some subjects steadily remain in stages of high vigilance, others exhibit rapid declines within only a few seconds. These patterns are called hyperstable or unstable vigilance regulation, respectively (see Figure 2). This regulation of EEG vigilance has been found to be intraindividually stable (Huang et al., 2015) with, at the same

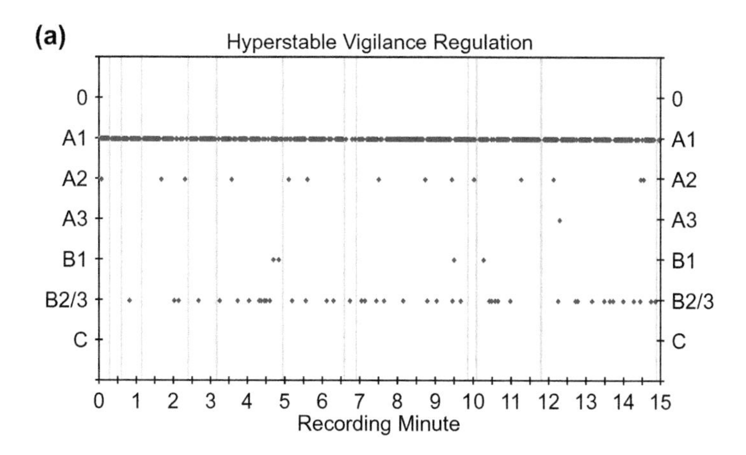

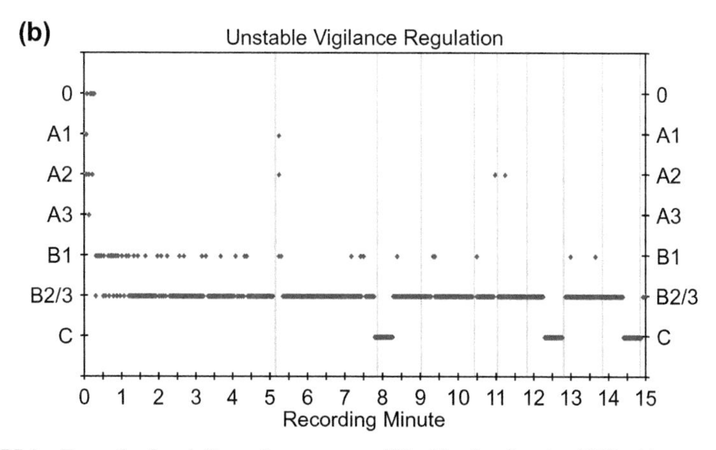

FIGURE 2 Examples for vigilance time courses within 15 min of resting EEG with eyes closed. (a) A subject with a hyperstable EEG-vigilance pattern, i.e., remaining continuously in A1-stages. (b) A subject with an unstable EEG-vigilance pattern, i.e., immediate decline to drowsiness (B2/3-stages) and sleep onset (stage C) within the seventh minute of recording. Red dots indicate the respective vigilance stage that was assigned to each segment (resolution 1 s); gray vertical lines mark segments containing artifacts.

time, considerable interindividual differences. This trait is modulated by many individual and environmental factors such as sleep deficits, arousal-enhancing substances, effort, motivation, and disease-related factors.

VIGALL 2.0 improves upon earlier versions of the algorithm, which have been validated performing simultaneous EEG–functional magnetic resonance imaging (fMRI; Olbrich et al., 2009) as well as simultaneous EEG–*positron emission tomography* (PET) studies (Guenther et al., 2011). These studies showed that decreases in EEG-vigilance levels are associated with increased metabolic activity in cortical but decreased activity in subcortical areas. A further study relating vigilance stages to autonomous functions found evidence that during lower vigilance stages heart rates and skin conductance levels also decline (Olbrich et al., 2011). Further studies related the vigilance stages to different behavioral parameters (Bekhtereva et al., 2014; Minkwitz et al., 2011). Finally, the influence of different vigilance stages on evoked potentials and reaction times has been demonstrated in an oddball task (Huang, Spada, Sander, Hegerl, & Hensch, 2014). These basic research studies also imply clinical relevance given the importance of cognitive tests, MRI, and PET in diagnostic procedures, where VIGALL might contribute to improve diagnostic accuracy by assessing arousal-induced error variance.

THE AROUSAL REGULATION MODEL OF AFFECTIVE DISORDERS

The arousal regulation model of affective disorders (Hegerl & Hensch, 2014) suggests that the level and the regulation of brain arousal are not only affected by the environment but also that humans can create a more or less "arousing" environment by their own behavior. In an autoregulatory manner, a more or less stimulating environment can be actively created in order to reduce or increase brain arousal levels. An everyday life example for such an autoregulatory attempt to increase arousal by enhancing external stimulation may be overtired children, who often develop hyperactive, sensation seeking, and talkative behavior. In contrast, states of tonic hyperarousal are often associated with the tendency to withdraw and to avoid external stimulations, such as loud music or social interactions.

The arousal regulation model builds on earlier concepts of the autoregulatory function of behavioral syndromes (Bente, 1964; Ulrich, 1994); related concepts have also been suggested for ADHD (Weinberg & Brumback, 1990; Zentall & Zentall, 1983). In biological personality research, traits such as extraversion (Eysenck, 1990) and sensation seeking (Zuckerman, 1979) were interpreted as autoregulatory behavior in order to achieve an optimal level of arousal. These personality traits were also suggested to reflect some vulnerability to affective disorders and ADHD (Hensch, Herold, & Brocke, 2007; White, 1999).

The arousal regulation model assumes that in vulnerable subjects such autoregulatory mechanisms can result in clinically relevant behavioral syndromes. Several lines of arguments indicate that this is the case for manic and depressed states as well as for ADHD.

Arousal Regulation in Mania

The suggested pathogenetic mechanisms in mania are illustrated in Figure 3. In vulnerable subjects an unstable arousal regulation can induce an exaggerated autoregulatory behavior as an attempt to stabilize brain arousal. This autoregulatory syndrome includes sensation and novelty seeking, hyperactivity, talkativeness, distractibility, and impulsivity. It can override the physiological tendency to seek sleep (e.g., by partying), thus aggravating the sleep deficits and as a consequence the instability of brain arousal. A vicious circle can be started, contributing to full-blown mania.

This pathogenetic model is supported by findings that during manic episodes many patients show an unstable arousal regulation, which at a first glance seems to be in striking contrast to their highly energetic behavior (Van Sweden, 1986). However, when studied in a quiet environment with eyes closed and low external stimulation, rapid declines to low-vigilance stages and microsleeps with sleep spindles can often be seen within the first seconds of EEG recording (Small, Milstein, Malloy, Medlock, & Klapper, 1999; Ulrich, 1994; Van Sweden, 1986).

Different findings suggest that this unstable arousal regulation in mania should not only be seen as a consequence of mania-induced sleep deficits but also seems to play a causal pathogenetic role:

- Several factors, which are associated with sleep deficits, are among the strongest triggers of mania and/or worsen manic behavior (Harvey, 2008; Wehr, 1992). Sleep deprivation was suggested as an animal model of mania (Gessa, Pani, Fadda, & Fratta, 1995) and can induce a switch into (hypo) mania in bipolar patients (Colombo, Benedetti, Barbini, Campori, & Smeraldi, 1999; Kasper & Wehr, 1992; Wu, & Bunney, 1990). The causal

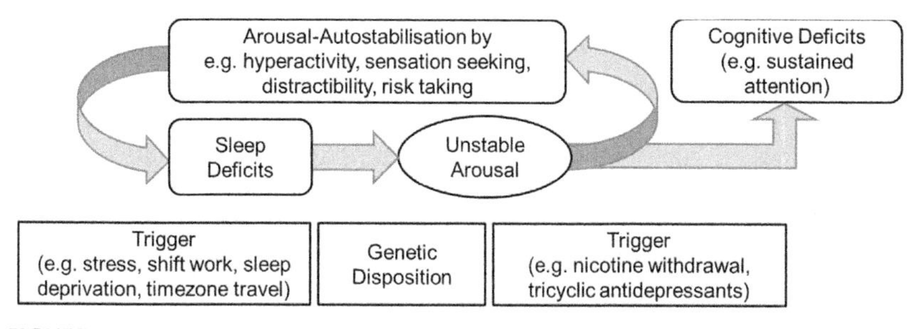

FIGURE 3 Model of a pathogenetic circle of an arousal stabilization syndrome contributing to full-blown mania.

relevance of sleep reduction (e.g., as a consequence of obstructive sleep apnea, bereavement, newborn infants, travel, or shift work) for triggering mania is reviewed in Plante and Winkelman (2008). Furthermore, sleep disturbances are by far the most robust early symptom of mania (median prevalence of 77%; Jackson, Cavanagh, & Scott, 2003), and it has been found that life events disturbing sleep—wake rhythms can trigger or aggravate (hypo)manic syndromes (Barbini, Bertelli, Colombo, & Smeraldi, 1996; Plante & Winkelman, 2008; Wehr, 1991).

- Stabilization of sleep—wake rhythms is an established and important element in behavioral therapies for bipolar affective disorders (Frank et al., 2005; Leibenluft & Suppes, 1999; Riemann, Voderholzer, & Berger, 2002). Additionally, extended bed rest and darkness as an add-on to the usual treatment of acute mania resulted in a faster decrease of manic symptoms in those patients with a recent (within 2 weeks) onset of mania (Barbini et al., 2005; for similar results see also Nowlin-Finch, Altshuler, Szuba, & Mintz, 1994; Wehr et al., 1998). These interventions can be expected to stop the pathogenetic circle described in Figure 3 by stabilizing arousal regulation.
- All standard antidepressants reduce the firing rate of locus coeruleus (LC; see Arousal Regulation in Depression) and are often associated with drowsiness as a side effect (Hensch et al., 2015). It might be that this arousal reduction contributes to the antidepressants' potential to induce manic episodes. It is interesting that this switch risk has been found, in some studies, to be higher in sedating antidepressants with anticholinergic and antihistaminic properties such as tricyclic antidepressants than in less drowsiness-inducing selective serotonin reuptake inhibitors (Gijsman, Geddes, Rendell, Nolen, & Goodwin, 2004; Peet, 1994).
- Arousal-enhancing psychostimulants are considered to be contraindicated in mania by many clinicians. However, following the model presented here, brain arousal-stabilizing drugs could be able to stop the manic vicious circle. Indeed, when reviewing the literature, there is a lack of empirical evidence for detrimental effects of psychostimulants in mania (Hegerl, Sander, Olbrich, & Schoenknecht, 2009). If psychostimulants had a high risk to induce or worsen mania, then the broad description of stimulants in ADHD would result in considerable problems due to the high comorbidity between ADHD and bipolar affective disorders (Singh, DelBello, Kowatch, & Strakowski, 2006). Given the differential diagnostic difficulties in distinguishing both diseases, especially in children, one can assume that many unrecognized or misdiagnosed pediatric manic patients have already received stimulants. Therefore, a reanalysis of randomized trials with stimulants in ADHD was carried out by the Food and Drug Administration, demonstrating that psychotic or maniclike reactions occurred rarely (in about 1 of 400 treated patients), and in the majority of cases (55 of 60), the

symptoms resolved within 2 days (Gelperin & Phelan, 2006; Mosholder, 2006; Phelan, 2006a, 2006b; Ross, 2006). Additionally, in a controlled trial in children with ADHD and severe mood dysregulation, an improvement in manic symptoms was observed under methylphenidate treatment (Waxmonsky et al., 2008). Stimulants have already been prescribed to bipolar patients as an add-on to mood stabilizers. In children and adolescents, open trials (Kowatch, Sethuraman, Hume, Kromelis, & Weinberg, 2003; Kummer & Teixeira, 2008) and controlled trials (Findling et al., 2007; Scheffer, Kowatch, Carmody, & Rush, 2005; Zeni, Tramontina, Ketzer, Pheula, & Rohde, 2009) showed that adding a psychostimulant did not worsen but often improved manic symptomatology. Mirroring these findings, in adults neither uncontrolled studies (Carlson, Merlock, & Suppes, 2004; El-Mallakh, 2000; Fernandes & Petty, 2003; Lydon & El-Mallakh, 2006; Nasr, Wendt, & Steiner, 2006) nor controlled trials (Calabrese, Frye, Yang, & Ketter, 2014; Calabrese et al., 2010; Frye et al., 2007; Ketter, Yang, & Frye, 2015) could detect a greater risk for (hypo)manic symptoms in bipolar depressed patients treated with stimulants as an add-on to mood stabilizers.

- In conclusion, stimulants in bipolar disorder seem to be relatively safe, and there are even several case reports suggesting rapid antimanic effects of psychostimulants (Beckmann & Heinemann, 1976; Garvey, Hwang, Teubner-Rhodes, Zander, & Rhem, 1987; Max, Richards, & Hamdanallen, 1995). In a study by Bschor, Müller-Oerlinghausen, and Ulrich (2001), improvement of manic symptoms occurred about 2 h after oral intake of methylphenidate in a manic patient with signs of unstable EEG-vigilance regulation. Three months later, when the patient was admitted anew, a rapid antimanic effect was again shown after re-exposition to methylphenidate. In contrast, no improvement was found in another manic patient without this EEG pattern. Schoenknecht, Olbrich, Sander, Spindler, and Hegerl (2010) reported a rapid response of an acutely manic patient to monotherapy with the arousal-stabilizing drug modafinil. After 5 days the patient had clearly improved and after stopping modafinil, treatment was continued with lithium. Clinical improvement went along with a stabilization of arousal regulation. Based on these findings, an international randomized placebo-controlled clinical trial was started, analyzing the effect of acute treatment with methylphenidate in mania (Kluge et al., 2013; NCT01541605).

The arousal regulation model can also explain symptomatology and effects of stimulants in ADHD. At the symptom level, manic episodes show remarkable similarities to ADHD (Hegerl, Himmerich, Engmann, & Hensch, 2010), in line with the high comorbidity of bipolar disorder and ADHD. Therefore, in the following ADHD will be discussed in the context of the arousal regulation model.

Based on studies with skin conductance level and quantitative EEG, a chronic hypoarousal had been postulated in ADHD for many years (reviewed in Geissler, Romanos, Hegerl, & Hensch, 2014). Furthermore, unstable arousal regulation has also been found in ADHD using MSLT (Geissler et al., 2014) and VIGALL (Sander, Arns, Olbrich, & Hegerl, 2010). Additionally, a higher subjective sleepiness was reported in ADHD and associated with symptom severity (Cortese, Faraone, Konofal, & Lecendreux, 2009; Gamble, May, Besing, Tankersly, & Fargason, 2013; Yoon, Jain, & Shapiro, 2012).

An unstable arousal regulation provides an explanation for the attention deficits in ADHD, especially the well-documented deficits in continuous performance tasks (Nichols & Waschbusch, 2004), and can also explain the ADHD presentation specifiers according to the Diagnostic and Statistical Manual of Mental Disorders, fifth edition (DSM-V) (and the ADHD subtypes as their predecessors in the DSM-IV-TR). In the predominantly inattentive presentation (formerly named predominantly inattentive subtype), the deficits are explained by the instability of the arousal regulation. In the combined presentation (subtype) with attention deficits and hyperactivity, additional autoregulatory aspects come into play with hyperactivity, sensation, and novelty seeking as an attempt to stabilize brain arousal. The arousal regulation model is also able to explain why studies reported low prevalence rates for the predominantly hyperactive-impulsive subtype or even called into question the general validity of this subtype (Hurtig et al., 2007; Willcutt et al., 2012): The model suggests that the unstable brain arousal is a core pathogenetic factor in ADHD, which results in attention deficits. Hyperactivity, in contrast, does not represent a primary disorder per se but rather an autoregulatory response, which may or may not be present. Thus, a "pure" hyperactive subgroup should not exist (for a more detailed discussion see Geissler et al., 2014).

In accordance with the model proposed for mania, the well-established therapeutic effects of stimulants in ADHD can be explained by their arousal-stabilizing effects, which interrupt the autoregulatory hyperactivity and sensation-seeking behavior. Stimulants reduce attention deficits, sensation-seeking behavior, and hyperactivity in patients with ADHD (Pietrzak, Mollica, Maruff, & Snyder, 2006; Riccio, Waldrop, Reynolds, & Lowe, 2001; Spencer et al., 2005), and symptom improvement is usually rapid (Greydanus, Pratt, & Patel, 2007), similar to the quick antimanic effects observed in case reports.

Arousal Regulation in Depression

While the behavioral syndrome in mania and ADHD is suggested to stabilize an unstable brain arousal by creating a stimulating environment, the opposite is suggested to be the case in depression. Depressed patients are characterized by withdrawal from all interactions and sensation avoidance, possibly as a reaction to a hyperstable arousal regulation. Furthermore, other symptoms of

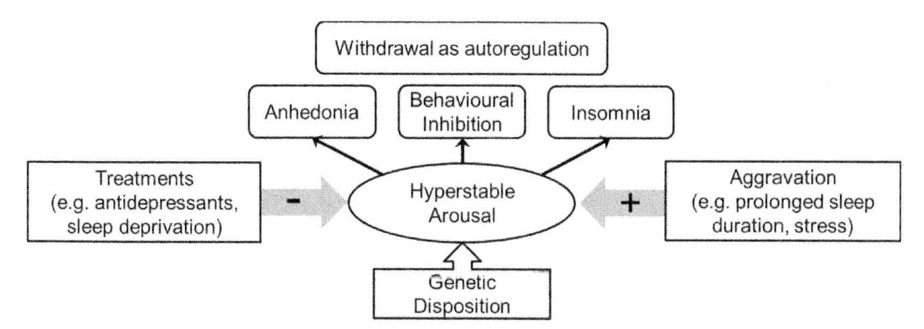

FIGURE 4 Hyperstable arousal regulation as a core pathogenetic factor in typical depression. The withdrawal behavior of depressed patients is interpreted as a compensatory reaction to reduce brain arousal. Other symptoms of depression and effects of treatments or aggravating factors can also be explained by the model (see text).

depression, such as insomnia, anxiety, and anhedonia as well as treatment effects of antidepressants can be explained by assuming hyperarousal as a core feature of unipolar depression, as delineated in Figure 4 and outlined in the following:

- During depression a hyperstable arousal regulation can often be found (Ulrich, 1994; Ulrich & Fuerstenberg, 1999). When arousal regulation was assessed using the VIGALL algorithm, a hyperstable vigilance regulation pattern was found more frequently in unmedicated, depressed patients as compared to healthy controls (Hegerl, Wilk, Olbrich, Schoenknecht, & Sander, 2012; Olbrich et al., 2012). This hyperstable vigilance is in line with the delayed sleep onset latency observed in depressive patients (Armitage, 2007; Tsuno et al., 2005), their inner restlessness and tension, and the hyperactivity of the HPA axis in depression (Pariante & Lightman, 2008). The high arousal level in depression explains the seemingly paradoxical finding that the worse the nighttime sleep, the longer the daytime sleep latency on the MSLT (Kayumov et al., 2000).
- Normalizing the hyperarousal in depression might also be one mode of action of antidepressants. Somnolence is a well-known frequent side effect of all common antidepressants (Bull et al., 2002; Cascade, Kalali, & Kennedy, 2009; Fava et al., 2006; Papakostas, 2008). In line with this, most antidepressants, including those that are often labeled as "activating" drugs, reduce the firing rate of neurons in the noradrenergic LC, which plays an important role in brain arousal. Preclinical studies found this firing rate reduction for acute and 2-week applications of different serotonin, serotonin–norepinephrine, norepinephrine, and norepinephrine–dopamine reuptake inhibitors; tricyclic antidepressants; and monoamine oxidase inhibitors (West, Ritchie, Boss-Williams, & Weiss, 2009). It was suggested that this reduction might be a common pathway of antidepressant action, as electroconvulsive shocks also reduce the firing rate of neurons in the LC (Grant & Weiss, 2001). The importance of heightened noradrenergic

activity for depressive symptomatology is further supported by research, suggesting symptoms of anhedonia and behavioral inhibition as a consequence of noradrenergic hyperactivity (Stone, Lin, Sarfraz, & Quartermain, 2011; West & Weiss, 2011).

- The antidepressant or depressiogenic effects of other drugs might also be partly explained by their arousal-modulating effects—for example, the depressiogenic effects of cholinesterase inhibitors in manic patients and healthy subjects (Burt, Sachs, & Demopulos, 1999; Dagyte, Den Boer, & Trentani, 2011; Janowsky, el-Yousef, Davis, & Sekerke, 1972) as well as the antidepressant and promanic effects of anticholinergics (Drevets & Furey, 2010; Fleischhacker et al., 1987; Furey & Drevets, 2006; Knable, 1989) or ketamine (Coyle & Laws, 2015). For more details see Hegerl and Hensch (2014).

Sleep deprivation is another well-established treatment in depressive episodes. Sleep deprivation in the second half of the night results in a pronounced reduction of the depressive symptoms in more than half of the patients (Benedetti & Colombo, 2011). Unfortunately, even a short recovery nap can be followed by the immediate recurrence of depressive symptomatology (Berger, van Calker, & Riemann, 2003). The arousal model of affective disorders provides a straightforward explanation for these effects: sleep deprivation might increase sleep propensity and reduce the hyperstable arousal regulation found in depressed patients and by that reduce the autoregulatory behavior with withdrawal and sensation avoidance. Further tentative support for this hypothesis comes from studies showing that patients with higher subjective arousal ratings or higher sustained attention benefit more from sleep deprivation (Bouhuys, Vandenburg, & Vandenhoofdakker, 1995; Wu et al., 1992).

- While sleep deprivation has antidepressant effects and can trigger mania, sleep can have depressiogenic effects in vulnerable subjects. During depressive episodes within major depressive and bipolar disorder, many patients clearly describe their depression as most severe in the morning, becoming less severe during the course of the day and the late evening. Sleep may reduce sleep propensity and aggravate the hyperstable arousal regulation, whereas being awake may increase sleep propensity and by that reduce this arousal dysregulation during the course of the day. In line with this reasoning, switches from mania to depression tend to occur during the second half of the night, whereas switches from depression to mania tend to occur during the afternoon and evening (Feldman-Naim, Turner, & Leibenluft, 1997; Wilk & Hegerl, 2010).
- The relationship between changes in sleep and changes in mood has been longitudinally analyzed in patients with bipolar affective disorder (Bauer et al., 2006). Cross-correlations between self-reported sleep or bed rest and mood demonstrated that in the majority of patients with a significant

cross-correlation, an increase in sleep or bed rest was followed by an increase in depression, whereas a reduction of sleep or bed rest was followed by hypomania or mania. Relationships between reduced sleep and (hypo) mania as well as increased sleep and depressive symptoms have also been reported by other research groups (Leibenluft, Albert, Rosenthal, & Wehr, 1996; Wehr, Goodwin, Wirz-Justice, Breitmaier, & Craig, 1982).

- Whereas the pathogenetic arousal model outlined here suggests stimulants as a possible acute antimanic treatment, it would in consequence not suggest stimulants in typical depression. Nonetheless, due to the energizing properties of psychostimulants seen in healthy subjects, stimulants have been tried as antidepressants in numerous studies with limited success. This failure can be explained, because depressive symptoms only superficially suggest sleepiness and a lack of drive, symptoms that might respond to psychostimulants. As mentioned before, many patients with typical depression do not suffer from sleepiness (tendency to fall asleep) but rather from insomnia and decreased sleep drive (prolonged sleep latencies) despite feelings of exhaustion and weariness. They also do not suffer from lack of drive but rather from inhibition of drive (retardation) combined with high inner tension. Thus, given the increased brain arousal during depressive episodes, psychostimulants are unlikely to be helpful in general. In line with this assumption, evidence for an antidepressant effect of stimulants in patients with typical major depressive disorder is indeed lacking. A Cochrane review (Candy, Jones, Williams, Tookman, & King, 2008) analyzed randomized, controlled trials from the past six decades, testing antidepressant effects of stimulants as monotherapy or an add-on in depression. With respect to clinical response, no significant effects could be shown. Concerning the second outcome variable (reduction in depression symptoms), only one of the subanalyses by Candy et al. (2008) showed a significant effect based on three studies. However, two of the trials were on patients with serious comorbid diseases (HIV with hypersomnia and traumatic brain injury). The third trial included 20 outpatients with "moderate depression" and "apathy, fatigue, [and] lack of energy" (Elizur, Wintner, & Davidson, 1979). No diagnostic details are specified, but considering the symptom of apathy, one might suspect atypical depression in this group. In the years following the meta-analysis by Candy et al. (2008) several studies on stimulants as an add-on to antidepressants or mood stabilizers were published. In unipolar depression, an add-on of methylphenidate (Ravindran et al., 2008), atomoxetine (Michelson et al., 2007), or modafinil (Beck et al., 2010; Dunlop et al., 2007) failed, but two further studies showed promising augmentation effects: an underpowered, short-term trial with a modafinil add-on (Abolfazli et al., 2011; $N = 46$; 6 weeks) and a study with lisdexamfetamine augmentation in partial escitalopram responders with residual symptoms (Trivedi et al., 2013; $N = 129$; $p = 0.09$, significant at prespecified alpha level of 0.1). However, two phase 3 studies with a lisdexamfetamine add-on again failed to meet the primary efficacy end point, and the manufacturer accordingly stopped the clinical development program (Shire, 2014). In bipolar depression, three trials showed some add-on effect of (ar)modafinil (Calabrese et al., 201, 2010; Frye et al., 2007), whereas two following multicenter studies failed. Again, due to efficacy concerning the primary end point, the manufacturer announced that the company would not proceed with regulatory filings for armodafinil for depression in bipolar I patients (Teva, 2013). Two other studies with lisdexamfetamine augmentation in bipolar disorder (NCT01093963, NCT01131559) were also stopped by the sponsor.

- To conclude, in uni- and bipolar depression there is no evidence for specific antidepressant effects of stimulants as a monotherapy or add-on. However, depression is a heterogeneous condition (Baumeister & Parker, 2012), and the arousal regulation model may help to identify subgroups, who nonetheless might respond to stimulants. Uncontrolled studies point to a possible antidepressant effect of stimulants in secondary depression (Masand, Pickett, & Murray, 1991). Such secondary depressive syndromes may be characterized by sleepiness and a lack of drive. Similarly, in atypical depression, which is likely characterized by unstable arousal regulation, stimulants might show some possible benefit.

CONCLUSION

Sufficient brain arousal is an important prerequisite for higher cognitive functions, and a successful regulation of brain arousal according to environmental and physiological needs is crucial for any organism. Disturbances in brain arousal regulation have been linked to psychiatric conditions, especially affective disorders. The outlined arousal regulation model gives an explanation for different clinical phenomena, such as response to psychostimulants in ADHD (established) and mania (currently being tested in an international controlled trial; Kluge et al., 2013; NCT01541605), the antidepressant but potentially mania-triggering effects of sleep deprivation, and the antimanic but depressiogenic effects of increased sleep duration. The model allows for deriving testable hypotheses concerning treatment response, thereby possibly contributing to personalized treatment. The biomarker EEG-vigilance regulation is suggested to stratify the highly heterogeneous category of major depressive disorder into biologically more homogenous subgroups.

Clinicians should differentiate between exhaustion and weariness in the context of chronic hyperarousal (as typically observed during a depressive episode) and actual sleepiness. According to the arousal regulation model, it is not surprising that current empirical studies have consistently shown that treatments with stimulants are not effective in depression, as they would further increase the hyperarousal. However, first evidence exists that stimulants could be effective in mania or secondary depression where they would improve the unstable brain arousal. In contrast, depressed patients who show no signs of increased sleepiness when assessed with VIGALL might not only profit from antidepressants, which reduce noradrenergic activity and thereby arousal, but also from monitoring their sleeping behavior. Many depressed patients have long bedtimes, trying to relax and to get as much sleep as possible. However, in the case of a chronic hyperarousal, this behavior will perpetuate symptomatology. Instead, patients without objectively confirmed increased sleepiness should carefully monitor their bedtimes, reduce their sleep duration, and avoid additional daytime sleep in order to counter their hyperstable arousal regulation.

REFERENCES

Abolfazli, R., Hosseini, M., Ghanizadeh, A., Ghaleiha, A., Tabrizi, M., Raznahan, M., ... Akhondzadeh, S. (2011). Double-blind randomized parallel-group clinical trial of efficacy of the combination fluoxetine plus modafinil versus fluoxetine plus placebo in the treatment of major depression. *Depression and Anxiety, 28*(4), 297–302. http://dx.doi.org/10.1002/da.20801.

Achermann, P. (2004). The two-process model of sleep regulation revisited. *Aviation Space and Environmental Medicine, 75*(3 Suppl.), A37–A43.

Akerstedt, T., & Folkard, S. (1996). Predicting duration of sleep from the three process model of regulation of alertness. *Occupational and Environmental Medicine, 53*(2), 136–141.

Akerstedt, T., & Folkard, S. (1997). The three-process model of alertness and its extension to performance, sleep latency, and sleep length. *Chronobiology International, 14*(2), 115–123.

Akerstedt, T., & Gillberg, M. (1990). Subjective and objective sleepiness in the active individual. *International Journal of Neuroscience, 52*(1–2), 29–37.

Armitage, R. (2007). Sleep and circadian rhythms in mood disorders. *Acta Psychiatrica Scandinavica, 115*(Suppl. 433), 104–115. http://dx.doi.org/10.1111/j.1600-0447.2007.00968.x.

Baglioni, C., Battagliese, G., Feige, B., Spiegelhalder, K., Nissen, C., Voderholzer, U., ... Riemann, D. (2011). Insomnia as a predictor of depression: a meta-analytic evaluation of longitudinal epidemiological studies. *Journal of Affective Disorders, 135*(1–3), 10–19. http://dx.doi.org/10.1016/j.jad.2011.01.011.

Barbini, B., Benedetti, F., Colombo, C., Dotoli, D., Bernasconi, A., Cigala-Fulgosi, M., ... Smeraldi, E. (2005). Dark therapy for mania: a pilot study. *Bipolar Disorders, 7*(1), 98–101. http://dx.doi.org/10.1111/j.1399-5618.2004.00166.x.

Barbini, B., Bertelli, S., Colombo, C., & Smeraldi, E. (1996). Sleep loss, a possible factor in augmenting manic episode. *Psychiatry Research, 65*(2), 121–125. http://dx.doi.org/10.1016/S0165-1781(96)02909-5.

Bauer, M., Grof, P., Rasgon, N., Bschor, T., Glenn, T., & Whybrow, P. C. (2006). Temporal relation between sleep and mood in patients with bipolar disorder. *Bipolar Disorders, 8*(2), 160–167. http://dx.doi.org/10.1111/j.1399-5618.2006.00294.x.

Baumeister, H., & Parker, G. (2012). Meta-review of depressive subtyping models. *Journal of Affective Disorders, 139*(2), 126–140. http://dx.doi.org/10.1016/j.jad.2011.07.015.

Beck, J., Hemmeter, U., Brand, S., Muheim, F., Hatzinger, M., & Holsboer-Trachsler, E. (2010). Modafinil reduces microsleep during partial sleep deprivation in depressed patients. *Journal of Psychiatric Research, 44*(13), 853–864. http://dx.doi.org/10.1016/j.jpsychires.2010.01.008. pii:S0022-3956(10)00017-8.

Beckmann, H., & Heinemann, H. (1976). D-Amphetamine in manic syndrome. *Arznelmittel-Forschung, 26*(6), 1185–1186.

Bekhtereva, V., Sander, C., Forschack, N., Olbrich, S., Hegerl, U., & Muller, M. M. (2014). Effects of EEG-vigilance regulation patterns on early perceptual processes in human visual cortex. *Clinical Neurophysiology, 125*(1), 98–107. http://dx.doi.org/10.1016/j.clinph.2013.06.019.

Benca, R. M., Obermeyer, W. H., Larson, C. L., Yun, B., Dolski, I., Kleist, K. D., … Davidson, R. J. (1999). EEG alpha power and alpha power asymmetry in sleep and wakefulness. *Psychophysiology, 36*(4), 430–436.

Benedetti, F., & Colombo, C. (2011). Sleep deprivation in mood disorders. *Neuropsychobiology, 64*(3), 141–151. http://dx.doi.org/10.1159/000328947.

Bente, D. (1964). Vigilanz, dissoziative Vigilanzverschiebung und Insuffizienz des Vigilitätstonus. In H. Kranz, & K. Heinrich (Eds.), *Begleitwirkungen und Mißerfolge der psychiatrischen Pharmakotherapie*. Stuttgart: Georg Thieme Verlag. pp. 13–28.

Berger, M., van Calker, D., & Riemann, D. (2003). Sleep and manipulations of the sleep-wake rhythm in depression. *Acta Psychiatrica Scandinavica Supplementum, 418*, 83–91.

Borbely, A. A. (1982). A two process model of sleep regulation. *Human Neurobiology, 1*(3), 195–204.

Bouhuys, A. L., Vandenburg, W., & Vandenhoofdakker, R. H. (1995). The relationship between tiredness prior to sleep-deprivation and the antidepressant response to sleep-deprivation in depression. *Biological Psychiatry, 37*(7), 457–461.

Bschor, T., Müller-Oerlinghausen, B., & Ulrich, G. (2001). Decreased level of EEG-vigilance in acute mania as a possible predictor for a rapid effect of methylphenidate: a case study. *Clinical Electroencephalography, 32*(1), 36–39.

Bull, S. A., Hunkeler, E. M., Lee, J. Y., Rowland, C. R., Williamson, T. E., Schwab, J. R., & Hurt, S. W. (2002). Discontinuing or switching selective serotonin-reuptake inhibitors. *Annals of Pharmacotherapy, 36*(4), 578–584.

Burt, T., Sachs, G. S., & Demopulos, C. (1999). Donepezil in treatment-resistant bipolar disorder. *Biological Psychiatry, 45*(8), 959–964. pii:S0006-3223(98)00320-5.

Calabrese, J. R., Frye, M. A., Yang, R., & Ketter, T. A. (2014). Efficacy and safety of adjunctive armodafinil in adults with major depressive episodes associated with bipolar I disorder: a randomized, double-blind, placebo-controlled, multicenter trial. *Journal of Clinical Psychiatry, 75*(10), 1054–1061. http://dx.doi.org/10.4088/JCP.13m08951.

Calabrese, J. R., Ketter, T. A., Youakim, J. M., Tiller, J. M., Yang, R., & Frye, M. A. (2010). Adjunctive armodafinil for major depressive episodes associated with bipolar I disorder: a randomized, multicenter, double-blind, placebo-controlled, proof-of-concept study. *Journal of Clinical Psychiatry, 71*(10), 1363–1370. http://dx.doi.org/10.4088/JCP.09m05900gry.

Candy, M., Jones, L., Williams, R., Tookman, A., & King, M. (2008). Psychostimulants for depression. *Cochrane Database of Systematic Reviews, 2*, CD006722. http://dx.doi.org/10.1002/14651858.CD006722.pub2.

Cantero, J. L., Atienza, M., & Salas, R. M. (2002). Human alpha oscillations in wakefulness, drowsiness period, and REM sleep: different electroencephalographic phenomena within the alpha band. *Neurophysiologie Clinique, 32*(1), 54–71.

Carlson, P. J., Merlock, M. C., & Suppes, T. (2004). Adjunctive stimulant use in patients with bipolar disorder: treatment of residual depression and sedation. *Bipolar Disorders, 6*(5), 416–420. http://dx.doi.org/10.1111/j.1399-5618.2004.00132.x.

Carskadon, M. A., Dement, W. C., Mitler, M. M., Roth, T., Westbrook, P. R., & Keenan, S. (1986). Guidelines for the multiple sleep latency test (MSLT): a standard measure of sleepiness. *Sleep, 9*(4), 519–524.

Cascade, E., Kalali, A. H., & Kennedy, S. H. (2009). Real-world data on SSRI antidepressant side effects. *Psychiatry (Edgmont), 6*(2), 16–18.

Cluydts, R., De Valck, E., Verstraeten, E., & Theys, P. (2002). Daytime sleepiness and its evaluation. *Sleep Medicine Reviews, 6*(2), 83–96.

Colombo, C., Benedetti, F., Barbini, B., Campori, E., & Smeraldi, E. (1999). Rate of switch from depression into mania after therapeutic sleep deprivation in bipolar depression. *Psychiatry Research, 86*(3), 267–270. pii:S0165-1781(99)00036-0.

Corsi-Cabrera, M., Guevara, M. A., Del Rio-Portilla, Y., Arce, C., & Villanueva-Hernandez, Y. (2000). EEG bands during wakefulness, slow-wave and paradoxical sleep as a result of principal component analysis in man. *Sleep, 23*(6), 738–744.

Cortese, S., Faraone, S. V., Konofal, E., & Lecendreux, M. (2009). Sleep in children with attention-deficit/hyperactivity disorder: meta-analysis of subjective and objective studies. *Journal of the American Academy of Child and Adolescent Psychiatry, 48*(9), 894–908.

Coyle, C. M., & Laws, K. R. (2015). The use of ketamine as an antidepressant: a systematic review and meta-analysis. *Human Psychopharmacology, 30*(3), 152–163. http://dx.doi.org/10.1002/hup.2475.

Cuthbert, B. N., & Insel, T. R. (2013). Toward the future of psychiatric diagnosis: the seven pillars of RDoC. *BMC Medicine, 11*, 126. http://dx.doi.org/10.1186/1741-7015-11-126.

Daan, S., Beersma, D. G., & Borbely, A. A. (1984). Timing of human sleep: recovery process gated by a circadian pacemaker. *American Journal of Physiology, 246*(2 Pt 2), R161–R183.

Dagyte, G., Den Boer, J. A., & Trentani, A. (2011). The cholinergic system and depression. *Behavioural Brain Research, 221*(2), 574–582. http://dx.doi.org/10.1016/j.bbr.2010.02.023. pii:S0166-4328(10)00123-3.

Dauvilliers, Y., Lopez, R., Ohayon, M., & Bayard, S. (2013). Hypersomnia and depressive symptoms: methodological and clinical aspects. *BMC Medicine, 11*, 78. http://dx.doi.org/10.1186/1741-7015-11-78.

De Gennaro, L., Ferrara, M., & Bertini, M. (2001). The boundary between wakefulness and sleep: quantitative electroencephalographic changes during the sleep onset period. *Neuroscience, 107*(1), 1–11.

De Gennaro, L., Ferrara, M., Curcio, G., & Cristiani, R. (2001). Antero-posterior EEG changes during the wakefulness-sleep transition. *Clinical Neurophysiology, 112*(10), 1901–1911.

De Gennaro, L., Vecchio, F., Ferrara, M., Curcio, G., Rossini, P. M., & Babiloni, C. (2004). Changes in fronto-posterior functional coupling at sleep onset in humans. *Journal of Sleep Research, 13*(3), 209–217. http://dx.doi.org/10.1111/j.1365-2869.2004.00406.x.

Dinges, D. F., & Powell, J. W. (1985). Microcomputer analyses of performance on a portable, simple visual RT task during sustained operations. *Behavior Research Methods Instruments and Computers, 17*(6), 652–655.

Dirksen, S. R., & Epstein, D. R. (2008). Efficacy of an insomnia intervention on fatigue, mood and quality of life in breast cancer survivors. *Journal of Advanced Nursing, 61*(6), 664–675. http://dx.doi.org/10.1111/j.1365-2648.2007.04560.x.

Drevets, W. C., & Furey, M. L. (2010). Replication of scopolamine's antidepressant efficacy in major depressive disorder: a randomized, placebo-controlled clinical trial. *Biological Psychiatry, 67*(5), 432–438. http://dx.doi.org/10.1016/j.biopsych.2009.11.021. pii:S0006-3223(09)01414-0.

Dunlop, B. W., Crits-Christoph, P., Evans, D. L., Hirschowitz, J., Solvason, H. B., Rickels, K., ... Ninan, P. T. (2007). Coadministration of modafinil and a selective serotonin reuptake inhibitor from the initiation of treatment of major depressive disorder with fatigue and sleepiness: a double-blind, placebo-controlled study. *Journal of Clinical Psychopharmacology, 27*(6), 614–619. http://dx.doi.org/10.1097/jcp.0b013e31815abefb.

El-Mallakh, R. S. (2000). An open study of methylphenidate in bipolar depression. *Bipolar Disorders, 2*(1), 56–59.

Elizur, A., Wintner, I., & Davidson, S. (1979). The clinical and psychological effects of pemoline in depressed patients—a controlled study. *International Pharmacopsychiatry, 14*(3), 127–134.

Eysenck, H. J. (1990). Biological dimensions of personality. In L. A. Pervin (Ed.), *Handbook of personality: Theory and research* (pp. 244–276). New York: Guilford Press (Reprinted from: Not in File).

Fava, M., Graves, L. M., Benazzi, F., Scalia, M. J., Iosifescu, D. V., Alpert, J. E., & Papakostas, G. I. (2006). A cross-sectional study of the prevalence of cognitive and physical symptoms during long-term antidepressant treatment. *Journal of Clinical Psychiatry, 67*(11), 1754–1759.

Feldman-Naim, S., Turner, E. H., & Leibenluft, E. (1997). Diurnal variation in the direction of mood switches in patients with rapid-cycling bipolar disorder. *Journal of Clinical Psychiatry, 58*(2), 79–84.

Fernandes, P. P., & Petty, F. (2003). Modafinil for remitted bipolar depression with hypersomnia. *Annals of Pharmacotherapy, 37*(12), 1807–1809. http://dx.doi.org/10.1345/aph.1D226.

Findling, R. L., Short, E. J., McNamara, N. K., Demeter, C. A., Stansbrey, R. J., Gracious, B. L., ... Calabrese, J. R. (2007). Methylphenidate in the treatment of children and adolescents with bipolar disorder and attention-deficit/hyperactivity disorder. *Journal of the American Academy of Child and Adolescent Psychiatry, 46*(11), 1445–1453. http://dx.doi.org/10.1097/chi.0b013e31814b8d3b.

Fleischhacker, W. W., Barnas, C., Gunther, V., Meise, U., Stuppack, C., & Unterweger, B. (1987). Mood-altering effects of biperiden in healthy volunteers. *Journal of Affective Disorders, 12*(2), 153–157.

Frank, E., Kupfer, D. J., Thase, M. E., Mallinger, A. G., Swartz, H. A., Fagiolini, A. M., ... Monk, T. (2005). Two-year outcomes for interpersonal and social rhythm therapy in individuals with bipolar I disorder. *Archives of General Psychiatry, 62*(9), 996–1004. http://dx.doi.org/10.1001/archpsyc.62.9.996.

Frye, M. A., Grunze, H., Suppes, T., McElroy, S. L., Keck, P. E., Jr., Walden, J., ... Post, R. M. (2007). A placebo-controlled evaluation of adjunctive modafinil in the treatment of bipolar depression. *American Journal of Psychiatry, 164*(8), 1242–1249. http://dx.doi.org/10.1176/appi.ajp.2007.06060981.

Furey, M. L., & Drevets, W. C. (2006). Antidepressant efficacy of the antimuscarinic drug scopolamine: a randomized, placebo-controlled clinical trial. *Archives of General Psychiatry, 63*(10), 1121–1129. http://dx.doi.org/10.1001/archpsyc.63.10.1121.

Gamble, K. L., May, R. S., Besing, R. C., Tankersly, A. P., & Fargason, R. E. (2013). Delayed sleep timing and symptoms in adults with attention-deficit/hyperactivity disorder: a controlled actigraphy study. *Chronobiology International, 30*(4), 598–606. http://dx.doi.org/10.3109/07420528.2012.754454.

Garvey, M. J., Hwang, S., Teubner-Rhodes, D., Zander, J., & Rhem, C. (1987). Dextroamphetamine treatment of mania. *Journal of Clinical Psychiatry, 48*(10), 412–413.

Geissler, J., Romanos, M., Hegerl, U., & Hensch, T. (2014). Hyperactivity and sensation seeking as autoregulatory attempts to stabilize brain arousal in ADHD and mania? *Attention Deficit and Hyperactive Disorders, 6*(3), 159–173. http://dx.doi.org/10.1007/s12402-014-0144-z.

Gelperin, K., & Phelan, K. (2006). *Psychiatric adverse events associated with drug treatment of ADHD: Review of postmarketing safety data.* FDA Report PID D050243. US Food and Drug Administration. March 3, 2006. Available at www.fda.gov/ohrms/dockets/ac/06/briefing/2006-4210b_11_01_AdverseEvents.pdf.

Gessa, G. L., Pani, L., Fadda, P., & Fratta, W. (1995). Sleep deprivation in the rat: an animal model of mania. *European Neuropsychopharmacology, 5*(Suppl), 89–93. 10.1016/0924-977X(95)00023-I.

Giedke, H., & Schwarzler, F. (2002). Therapeutic use of sleep deprivation in depression. *Sleep Medicine Reviews, 6*(5), 361–377.

Gijsman, H. J., Geddes, J. R., Rendell, J. M., Nolen, W. A., & Goodwin, G. M. (2004). Antidepressants for bipolar depression: a systematic review of randomized, controlled trials. *American Journal of Psychiatry, 161*(9), 1537–1547. http://dx.doi.org/10.1176/appi.ajp.161.9.1537.

Grant, M. M., & Weiss, J. M. (2001). Effects of chronic antidepressant drug administration and electroconvulsive shock on locus coeruleus electrophysiologic activity. *Biological Psychiatry, 49*(2), 117–129. pii:S0006-3223(00)00936-7.

Greydanus, D. E., Pratt, H. D., & Patel, D. R. (2007). Attention deficit hyperactivity disorder across the lifespan: the child, adolescent, and adult. *Disease-a-month, 53*(2), 70–131.

Guenther, T., Schonknecht, P., Becker, G., Olbrich, S., Sander, C., Hesse, S., ... Sabri, O. (2011). Impact of EEG-vigilance on brain glucose uptake measured with [(18)F]FDG and PET in patients with depressive episode or mild cognitive impairment. *NeuroImage, 56*(1), 93–101. http://dx.doi.org/10.1016/j.neuroimage.2011.01.059. pii:S1053-8119(11)00095-4.

Harvey, A. G. (2008). Sleep and circadian rhythms in bipolar disorder: seeking synchrony, harmony, and regulation. *American Journal of Psychiatry, 165*(7), 820–829. http://dx.doi.org/10.1176/appi.ajp.2008.08010098.

Head, H. (1923). The conception of nervous and mental energy: II. vigilance: a physiological state of the nervous system. *British Journal of Psychology, 14*, 126–147.

Hegerl, U. (2014). Largely unnoticed flaws in the fundamentals of depression diagnosis: the semantics of core symptoms. *Australian and New Zealand Journal of Psychiatry, 48*(12), 1166. http://dx.doi.org/10.1177/0004867414559550.

Hegerl, U., & Hensch, T. (2014). The vigilance regulation model of affective disorders and ADHD. *Neuroscience and Biobehavioral Reviews, 44*, 45–57. http://dx.doi.org/10.1016/j.neubiorev.2012.10.008.

Hegerl, U., Himmerich, H., Engmann, B., & Hensch, T. (2010). Mania and attention-deficit/hyperactivity disorder: common symptomatology, common pathophysiology and common treatment? *Current Opinion in Psychiatry, 23*(1), 1–7. http://dx.doi.org/10.1097/YCO.0b013e328331f694.

Hegerl, U., Lam, R. W., Malhi, G. S., McIntyre, R. S., Demyttenaere, K., Mergl, R., & Gorwood, P. (2013). Conceptualising the neurobiology of fatigue. *Australian and New Zealand Journal of Psychiatry, 47*(4), 312–316. http://dx.doi.org/10.1177/0004867413481505.

Hegerl, U., Sander, C., Olbrich, S., & Schoenknecht, P. (2009). Are psychostimulants a treatment option in mania? *Pharmacopsychiatry, 42*(5), 169–174. http://dx.doi.org/10.1055/s-0029-1220888.

Hegerl, U., Stein, M., Mulert, C., Mergl, R., Olbrich, S., Dichgans, E., ... Pogarell, O. (2008). EEG-vigilance differences between patients with borderline personality disorder, patients with obsessive-compulsive disorder and healthy controls. *European Archives of Psychiatry and Clinical Neuroscience, 258*(3), 137–143. http://dx.doi.org/10.1007/s00406-007-0765-8.

Hegerl, U., Wilk, K., Olbrich, S., Schoenknecht, P., & Sander, C. (2012). Hyperstable regulation of vigilance in patients with major depressive disorder. *World Journal of Biological Psychiatry, 13*(6), 436–446.

Hensch, T., Blume, A., Bottger, D., Sander, C., Niedermeier, N., & Hegerl, U. (2015). Yawning in depression: Worth looking into. *Pharmacopsychiatry, 48*(3), 118–120. http://dx.doi.org/10.1055/s-0035-1545332.

Hensch, T., Herold, U., & Brocke, B. (2007). An electrophysiological endophenotype of hypomanic and hyperthymic personality. *Journal of Affective Disorders, 101*(1–3), 13–26. http://dx.doi.org/10.1016/j.jad.2006.11.018. pii:S0165-0327(06)00500-3.

Hoddes, E., Dement, W., & Zarcone, V. (1972). The development and use of the Stanford sleepiness scale (SSS). *Psychophysiology, 9*, 150–151.

Huang, J., Sander, C., Jawinski, P., Ulke, C., Spada, J., Hegerl, U., & Hensch, T. (2015). Stability of brain arousal regulation as assessed with EEG vigilance. *Neuropsychiatric Electrophysiology, 1*, 13. http://dx.doi.org/10.1186/s40810-015-0013-9.

Huang, J., Spada, J., Sander, C., Hegerl, U., & Hensch, T. (2014). *Auditory event related potentials during different vigilance stages*. Paper presented at the Leipzig Research Festival for Life Sciences 2014, Leipzig.

Hurtig, T., Ebeling, H., Taanila, A., Miettunen, J., Smalley, S. L., McGough, J. J., ... Moilanen, I. K. (2007). ADHD symptoms and subtypes: relationship between childhood and adolescent symptoms. *Journal of American Academy of Child and Adolescent Psychiatry, 46*(12), 1605–1613. http://dx.doi.org/10.1097/chi.0b013e318157517a. pii:S0890-8567(09)61872-4.

Iber, C., Ancoli-Israel, S., Chessonn, A., & Quan, S. F. (2007). *The AASM manual for the scoring of sleep and associated events: rules, terminology and technical specifications* (1st ed.). Westchester, IL: American Academy of Sleep Medicine.

Jackson, A., Cavanagh, J., & Scott, J. (2003). A systematic review of manic and depressive prodromes. *Journal of Affective Disorders, 74*(3), 209–217.

Janowsky, D. S., el-Yousef, M. K., Davis, J. M., & Sekerke, H. J. (1972). A cholinergic-adrenergic hypothesis of mania and depression. *Lancet, 2*(7778), 632–635.

Johns, M. W. (1991). A new method for measuring daytime sleepiness: the Epworth sleepiness scale. *Sleep, 14*(6), 540–545.

Johns, M. (1998). Rethinking the assessment of sleepiness. *Sleep Medicine Reviews, 2*(1), 3–15.

Kaida, K., Takahashi, M., Akerstedt, T., Nakata, A., Otsuka, Y., Haratani, T., & Fukasawa, K. (2006). Validation of the Karolinska sleepiness scale against performance and EEG variables. *Clinical Neurophysiology, 117*(7), 1574–1581.

Kasper, S., & Wehr, T. A. (1992). The role of sleep and wakefulness in the genesis of depression and mania. *L'Encéphale, 18*(Spec No 1), 45–50.

Kayumov, L., Rotenberg, V., Buttoo, K., Auch, C., Pandi-Perumal, S. R., & Shapiro, C. M. (2000). Interrelationships between nocturnal sleep, daytime alertness, and sleepiness: two types of alertness proposed. *Journal of Neuropsychiatry and Clinical Neuroscience, 12*(1), 86–90.

Ketter, T. A., Yang, R., & Frye, M. A. (2015). Adjunctive armodafinil for major depressive episodes associated with bipolar I disorder. *Journal of Affective Disorders, 181*, 87–91. http://dx.doi.org/10.1016/j.jad.2015.04.012.

Khushaba, R. N., Kodagoda, S., Lal, S., & Dissanayake, G. (2011). Driver drowsiness classification using fuzzy wavelet-packet-based feature-extraction algorithm. *IEEE Transactions on Biomedical Engineering, 58*(1), 121–131. http://dx.doi.org/10.1109/tbme.2010.2077291.

Kluge, M., Hegerl, U., Sander, C., Dietzel, J., Mergl, R., Bitter, I., … Lopez-Garcia, P. (2013). Methylphenidate in mania project (MEMAP): study protocol of an international randomised double-blind placebo-controlled study on the initial treatment of acute mania with methylphenidate. *BMC Psychiatry, 13*, 71. http://dx.doi.org/10.1186/1471-244x-13-71.

Knable, M. B. (1989). Euphorigenic properties of anticholinergics. *Journal of Clinical Psychiatry, 50*(5), 186.

Kowatch, R. A., Sethuraman, G., Hume, J. H., Kromelis, M., & Weinberg, W. A. (2003). Combination pharmacotherapy in children and adolescents with bipolar disorder. *Biological Psychiatry, 53*(11), 978–984. pii:S0006322303000672.

Kummer, A., & Teixeira, A. (2008). Methylphenidate in attention deficit hyperactivity disorder and bipolar disorder. *Australasian Psychiatry, 16*(6), 458–459.

Leibenluft, E., Albert, P. S., Rosenthal, N. E., & Wehr, T. A. (1996). Relationship between sleep and mood in patients with rapid-cycling bipolar disorder. *Psychiatry Research, 63*(2–3), 161–168.

Leibenluft, E., & Suppes, T. (1999). Treating bipolar illness: focus on treatment algorithms and management of the sleep-wake cycle. *American Journal of Psychiatry, 156*(12), 1976–1981.

Lydon, E., & El-Mallakh, R. S. (2006). Naturalistic long-term use of methylphenidate in bipolar disorder. *Journal of Clinical Psychopharmacology, 26*(5), 516–518. http://dx.doi.org/10.1097/01.jcp.0000236655.62920.dc.

Manber, R., Edinger, J. D., Gress, J. L., San Pedro-Salcedo, M. G., Kuo, T. F., & Kalista, T. (2008). Cognitive behavioral therapy for insomnia enhances depression outcome in patients with comorbid major depressive disorder and insomnia. *Sleep, 31*(4), 489–495.

Marzano, C., Fratello, F., Moroni, F., Pellicciari, M. C., Curcio, G., Ferrara, M., … De Gennaro, L. (2007). Slow eye movements and subjective estimates of sleepiness predict EEG power changes during sleep deprivation. *Sleep, 30*(5), 610–616.

Masand, P., Pickett, P., & Murray, G. B. (1991). Psychostimulants for secondary depression in medical illness. *Psychosomatics, 32*(2), 203–208.

Mathis, J., & Hess, C. W. (2009). Sleepiness and vigilance tests. *Swiss Medical Weekly, 139*(15–16), 214–219.

Max, J. E., Richards, L., & Hamdanallen, G. (1995). Antimanic effectiveness of dextroamphetamine in a brain-injured adolescent. *Journal of the American Academy of Child and Adolescent Psychiatry, 34*(4), 472–476.

Michelson, D., Adler, L. A., Amsterdam, J. D., Dunner, D. L., Nierenberg, A. A., Reimherr, F. W., … Williams, D. W. (2007). Addition of atomoxetine for depression incompletely responsive to sertraline: a randomized, double-blind, placebo-controlled study. *Journal of Clinical Psychiatry, 68*(4), 582–587.

Minkwitz, J., Trenner, M. U., Sander, C., Olbrich, S., Sheldrick, A. J., Schoenknecht, P., … Himmerich, H. (2011). Prestimulus vigilance predicts response speed in an easy visual discrimination task. *Behavioral and Brain Functions, 7*(1), 31. http://dx.doi.org/10.1186/1744-9081-7-31.

Mitler, M. M., Gujavarty, K. S., & Browman, C. P. (1982). Maintenance of wakefulness test: a polysomnographic technique for evaluation treatment efficacy in patients with excessive somnolence. *Electroencephalography and Clinical Neurophysiology, 53*(6), 658–661.

Mosholder, A. (2006). *Psychiatric adverse events in clinical trials of drugs for ADHD*. FDA Report PID D060163. US Food and Drug Administration. March 3, 2006. Available at http://www.fda.gov/ohrms/dockets/ac/06/briefing/2006-4210b_10_01_Mosholder.pdf.

Nasr, S., Wendt, B., & Steiner, K. (2006). Absence of mood switch with and tolerance to modafinil: a replication study from a large private practice. *Journal of Affective Disorders, 95*(1−3), 111−114. http://dx.doi.org/10.1016/j.jad.2006.01.010. pii:S0165-0327(06)00011-5.

Neu, D., Hoffmann, G., Moutrier, R., Verbanck, P., Linkowski, P., & Le Bon, O. (2008). Are patients with chronic fatigue syndrome just 'tired' or also 'sleepy'? *Journal of Sleep Research, 17*(4), 427−431. http://dx.doi.org/10.1111/j.1365-2869.2008.00679.x.

Nichols, S. L., & Waschbusch, D. A. (2004). A review of the validity of laboratory cognitive tasks used to assess symptoms of ADHD. *Child Psychiatry Hum.Dev, 34*(4), 297−315. http://dx.doi.org/10.1023/B:CHUD.0000020681.06865.97.

NIMH. (2012). *Arousal and regulatory systems: Workshop proceedings*. Retrieved 18.04.15, from: http://www.nimh.nih.gov/research-priorities/rdoc/arousal-and-regulatory-systems-workshop-proceedings.shtml.

Nofzinger, E. A., Thase, M. E., Reynolds, C. F., 3rd, Himmelhoch, J. M., Mallinger, A., Houck, P., & Kupfer, D. J. (1991). Hypersomnia in bipolar depression: a comparison with narcolepsy using the multiple sleep latency test. *American Journal of Psychiatry, 148*(9), 1177−1181.

Nowlin-Finch, N. L., Altshuler, L. L., Szuba, M. P., & Mintz, J. (1994). Rapid resolution of first episodes of mania: sleep related? *Journal of Clinical Psychiatry, 55*(1), 26−29.

Nutt, D., Wilson, S., & Paterson, L. (2008). Sleep disorders as core symptoms of depression. *Dialogues in Clinical Neurosciences, 10*(3), 329−336.

Oken, B. S., Salinsky, M. C., & Elsas, S. M. (2006). Vigilance, alertness, or sustained attention: physiological basis and measurement. *Clinical Neurophysiology, 117*(9), 1885−1901. http://dx.doi.org/10.1016/j.clinph.2006.01.017. pii:S1388-2457(06)00049-6.

Olbrich, S., Mulert, C., Karch, S., Trenner, M., Leicht, G., Pogarell, O., & Hegerl, U. (2009). EEG-vigilance and BOLD effect during simultaneous EEG/fMRI measurement. *Neuro-Image, 45*(2), 319−332. http://dx.doi.org/10.1016/j.neuroimage.2008.11.014. pii:S1053-8119(08)01213-5.

Olbrich, S., Sander, C., Matschinger, H., Mergl, R., Trenner, M., Schoenknecht, P., & Hegerl, U. (2011). Brain and body: associations between EEG-vigilance and the autonomous nervous system activity during rest. *Journal of Psychophysiology, 25*(4), 190−200. http://dx.doi.org/10.1027/0269-8803/a000061.

Olbrich, S., Sander, C., Minkwitz, J., Chittka, T., Mergl, R., Hegerl, U., & Himmerich, H. (2012). EEG vigilance regulation patterns and their discriminative power to separate patients with major depression from healthy controls. *Neuropsychobiology, 65*(4), 188−194. http://dx.doi.org/10.1159/000337000.

Papakostas, G. I. (2008). Tolerability of modern antidepressants. *Journal of Clinical Psychiatry, 69*(Suppl. E1), 8−13.

Pariante, C. M., & Lightman, S. L. (2008). The HPA axis in major depression: classical theories and new developments. *Trends in Neurosciences, 31*(9), 464−468. http://dx.doi.org/10.1016/j.tins.2008.06.006. pii:S0166-2236(08)00164-1.

Pascual-Marqui, R. D., Esslen, M., Kochi, K., & Lehmann, D. (2002). Functional imaging with low-resolution brain electromagnetic tomography (LORETA): a review. *Methods and Findings in Experimental and Clinical Pharmacology, 24*(Suppl. C), 91−95.

Pascual-Marqui, R. D., Michel, C. M., & Lehmann, D. (1994). Low resolution electromagnetic tomography: a new method for localizing electrical activity in the brain. *International Journal of Psychophysiology, 18*(1), 49−65.

Peet, M. (1994). Induction of mania with selective serotonin re-uptake inhibitors and tricyclic antidepressants. *British Journal of Psychiatry, 164*(4), 549–550.

Pfaff, D., Ribeiro, A., Matthews, J., & Kow, L. M. (2008). Concepts and mechanisms of generalized central nervous system arousal. *Annals of the New York Academy of Sciences, 1129*, 11–25. http://dx.doi.org/10.1196/annals.1417.019.

Phelan, K. (2006a). *One year post-pediatric exclusivity postmarketing adverse event review: Adderall XR*. FDA Report PID D040761. US Food and Drug Administration. January 5, 2006. Available at http://www.fda.gov/ohrms/dockets/ac/06/briefing/2006-4210b_05_02_AdderallSafetyReview.pdf.

Phelan, K. (2006b). *Summary of psychiatric and neurological adverse events from June 2005 1-year post pediatric exclusivity reviews of Concerta and other methylphenidate products*. US Food and Drug Administration. Available at http://www.fda.gov/ohrms/dockets/ac/06/briefing/2006-4210b_09_01_Methsummary.pdf.

Pietrzak, R. H., Mollica, C. M., Maruff, P., & Snyder, P. J. (2006). Cognitive effects of immediate-release methylphenidate in children with attention-deficit/hyperactivity disorder. *Neuroscience and Biobehavioral Reviews, 30*(8), 1225–1245.

Plante, D. T., & Winkelman, J. W. (2008). Sleep disturbance in bipolar disorder: therapeutic implications. *American Journal of Psychiatry, 165*(7), 830–843. http://dx.doi.org/10.1176/appi.ajp.2008.08010077.

Ravindran, A. V., Kennedy, S. H., O'Donovan, M. C., Fallu, A., Camacho, F., & Binder, C. E. (2008). Osmotic-release oral system methylphenidate augmentation of antidepressant monotherapy in major depressive disorder: results of a double-blind, randomized, placebo-controlled trial. *Journal of Clinical Psychiatry, 69*(1), 87–94.

Rechtschaffen, A., & Kales, A. (1968). *A manual of standardized terminology techniques and scoring systems for sleep of human subjects*. Washington DC: US Government Printing Office.

Reynolds, C. F., 3rd, Coble, P. A., Kupfer, D. J., & Holzer, B. C. (1982). Application of the multiple sleep latency test in disorders of excessive sleepiness. *Electroencephalography and Clinical Neurophysiology, 53*(4), 443–452.

Riccio, C. A., Waldrop, J. J. M., Reynolds, C. R., & Lowe, P. (2001). Effects of stimulants on the continuous performance test (CPT): implications for CPT use and interpretation. *Journal of Neuropsychiatry and Clinical Neurosciences, 13*(3), 326–335.

Riemann, D., Berger, M., & Voderholzer, U. (2001). Sleep and depression—results from psychobiological studies: an overview. *Biological Psychology, 57*(1–3), 67–103.

Riemann, D., & Voderholzer, U. (2003). Primary insomnia: a risk factor to develop depression? *Journal of Affective Disorders, 76*(1–3), 255–259.

Riemann, D., Voderholzer, U., & Berger, M. (2002). Sleep and sleep-wake manipulations in bipolar depression. *Neuropsychobiology, 45*(Suppl. 1), 7–12.

Riemann, D., Wiegand, M., Lauer, C. J., & Berger, M. (1993). Naps after total sleep deprivation in depressed patients: are they depressiogenic? *Psychiatry Research, 49*(2), 109–120.

Ross, R. G. (2006). Psychotic and manic-like symptoms during stimulant treatment of attention deficit hyperactivity disorder. *American Journal of Psychiatry, 163*(7), 1149–1152.

Roth, B. (1961). The clinical and theoretical importance of EEG rhythms corresponding to states of lowered vigilance. *Electroencephalography and Clinical Neurophysiology, 13*, 395–399.

Sander, C., Arns, M., Olbrich, S., & Hegerl, U. (2010). EEG-vigilance and response to stimulants in paediatric patients with attention deficit/hyperactivity disorder. *Clinical Neurophysiology, 121*(9), 1511–1518. http://dx.doi.org/10.1016/j.clinph.2010.03.021. pii:S1388-2457(10)00309-3.

Sangal, R. B., Thomas, L., & Mitler, M. M. (1992). Maintenance of wakefulness test and multiple sleep latency test. Measurement of different abilities in patients with sleep disorders. *Chest, 101*(4), 898–902.

Sauvet, F., Bougard, C., Coroenne, M., Lely, L., Van Beers, P., Elbaz, M., ... Chennaoui, M. (2014). In-flight automatic detection of vigilance states using a single EEG channel. *IEEE Transactions on Biomedical Engineering, 61*(12), 2840–2847. http://dx.doi.org/10.1109/tbme.2014.2331189.

Scheffer, R. E., Kowatch, R. A., Carmody, T., & Rush, A. J. (2005). Randomized, placebo-controlled trial of mixed amphetamine salts for symptoms of comorbid ADHD in pediatric bipolar disorder after mood stabilization with divalproex sodium. *American Journal of Psychiatry, 162*(1), 58–64.

Schoenknecht, P., Olbrich, S., Sander, C., Spindler, P., & Hegerl, U. (2010). Treatment of acute mania with modafinil monotherapy. *Biological Psychiatry, 67*(11), e55–e57. http://dx.doi.org/10.1016/j.biopsych.2009.12.021. pii:S0006-3223(09)01482-6.

Shen, J., Hossain, N., Streiner, D. L., Ravindran, A. V., Wang, X., Deb, P., ... Shapiro, C. M. (2011). Excessive daytime sleepiness and fatigue in depressed patients and therapeutic response of a sedating antidepressant. *Journal of Affective Disorders, 134*(1–3), 421–426. http://dx.doi.org/10.1016/j.jad.2011.04.047.

Shi, L. C., Duan, R. N., & Lu, B. L. (2013). A robust principal component analysis algorithm for EEG-based vigilance estimation. *Conference Proceedings: IEEE Engineering in Medicine and Biology Society, 2013*, 6623–6626. http://dx.doi.org/10.1109/embc.2013.6611074.

Shire. (2014). *Shire reports top-line results from two phase 3 studies for Vyvanse® (lisdexamfetamine dimesylate) capsules (CII) as an adjunctive treatment for adults with major depressive disorder [Press release].* Retrieved from: http://www.shire.com/shireplc/uploads/press/MDDPhase3release06Feb2014.pdf.

Singh, M. K., DelBello, M. P., Kowatch, R. A., & Strakowski, S. M. (2006). Co-occurrence of bipolar and attention-deficit hyperactivity disorders in children. *Bipolar Disorders, 8*(6), 710–720.

Small, J. G., Milstein, V., Malloy, F. W., Medlock, C. E., & Klapper, M. H. (1999). Clinical and quantitative EEG studies of mania. *Journal of Affective Disorders, 53*(3), 217–224.

Spencer, T., Biederman, J., Wilens, T., Doyle, R., Surman, C., Prince, J., ... Faraone, S. (2005). A large, double-blind, randomized clinical trial of methylphenidate in the treatment of adults with attention-deficit/hyperactivity disorder. *Biological Psychiatry, 57*(5), 456–463.

Stampi, C., Stone, P., & Michimori, A. (1995). A new quantitative method for assessing sleepiness. *Work Stress, 9*, 368–376.

Stone, E. A., Lin, Y., Sarfraz, Y., & Quartermain, D. (2011). The role of the central noradrenergic system in behavioral inhibition. *Brain Research Reviews, 67*(1–2), 193–208. http://dx.doi.org/10.1016/j.brainresrev.2011.02.002. pii:S0165-0173(11)00015-4.

Strijkstra, A. M., Beersma, D. G., Drayer, B., Halbesma, N., & Daan, S. (2003). Subjective sleepiness correlates negatively with global alpha (8–12 Hz) and positively with central frontal theta (4–8 Hz) frequencies in the human resting awake electroencephalogram. *Neuroscience Letters, 340*(1), 17–20.

Teva. (2013). *Teva reports top-line results from final phase III study of armodafinil (NUVIGIL®) in patients with major depression associated with bipolar I disorder [Press release].* Retrieved from: http://ir.tevapharm.com/phoenix.zhtml?c=73925&p=irol-newsArticle&ID=1851019.

Trivedi, M. H., Cutler, A. J., Richards, C., Lasser, R., Geibel, B. B., Gao, J., ... Patkar, A. A. (2013). A randomized controlled trial of the efficacy and safety of lisdexamfetamine dimesylate as augmentation therapy in adults with residual symptoms of major depressive disorder after treatment with escitalopram. *Journal of Clinical Psychiatry, 74*(8), 802–809. http://dx.doi.org/10.4088/JCP.13m08360.

Tsuno, N., Besset, A., & Ritchie, K. (2005). Sleep and depression. *Journal of Clinical Psychiatry, 66*(10), 1254–1269.

Tsuno, N., Shigeta, M., Hyoki, K., Kinoshita, T., Ushijima, S., Faber, P. L., & Lehmann, D. (2002). Spatial organization of EEG activity from alertness to sleep stage 2 in old and younger subjects. *Journal of Sleep Research, 11*(1), 43–51.

Ulrich, G. (1994). *Psychiatrische Elektroenzephalographie*. Jena: Gustav Fischer Verlag.

Ulrich, G., & Fuerstenberg, U. (1999). Quantitative assessment of dynamic electroencephalogram (EEG) organization as a tool for subtyping depressive syndromes. *European Psychiatry, 14*, 217–229

Van Sweden, B. (1986). Disturbed vigilance in mania. *Biological Psychiatry, 21*(3), 311–313.

Waxmonsky, J., Pelham, W. E., Gnagy, E., Cummings, M. R., O'Connor, B., Majumdar, A., … Robb, J. A. (2008). The efficacy and tolerability of methylphenidate and behavior modification in children with attention-deficit/hyperactivity disorder and severe mood dysregulation. *Journal of Child and Adolescent Psychopharmacology, 18*(6), 573–588.

Wehr, T. A. (1989). Sleep loss: a preventable cause of mania and other excited states. *Journal of Clinical Psychiatry, 50*(Suppl.), 8–16. discussion 45–17.

Wehr, T. A. (1991). Sleep-loss as a possible mediator of diverse causes of mania. *British Journal of Psychiatry, 159*, 576–578.

Wehr, T. A. (1992). Improvement of depression and triggering of mania by sleep deprivation. *Journal of American Medical Association, 267*(4), 548–551.

Wehr, T. A., Goodwin, F. K., Wirz-Justice, A., Breitmaier, J., & Craig, C. (1982). 48-hour sleep-wake cycles in manic-depressive illness: naturalistic observations and sleep deprivation experiments. *Archives of General Psychiatry, 39*(5), 559–565.

Wehr, T. A., Turner, E. H., Shimada, J. M., Lowe, C. H., Barker, C., & Leibenluft, E. (1998). Treatment of rapidly cycling bipolar patient by using extended bed rest and darkness to stabilize the timing and duration of sleep. *Biological Psychiatry, 43*(11), 822–828. pii:S0006322397005428.

Weinberg, W. A., & Brumback, R. A. (1990). Primary disorder of vigilance: a novel explanation of inattentiveness, daydreaming, boredom, restlessness, and sleepiness. *Journal of Pediatrics, 116*(5), 720–725.

West, C. H., Ritchie, J. C., Boss-Williams, K. A., & Weiss, J. M. (2009). Antidepressant drugs with differing pharmacological actions decrease activity of locus coeruleus neurons. *International Journal of Neuropsychopharmacology, 12*(5), 627–641. http://dx.doi.org/10.1017/S1461145708009474. pii:S1461145708009474.

West, C. H., & Weiss, J. M. (2011). Effects of chronic antidepressant drug administration and electroconvulsive shock on activity of dopaminergic neurons in the ventral tegmentum. *International Journal of Neuropsychopharmacology, 14*(2), 201–210. http://dx.doi.org/10.1017/s1461145710000489.

White, J. D. (1999). Personality, temperament and ADHD: a review of the literature. *Personality and Individual Differences, 27*, 589–598.

Wichniak, A., Wierzbicka, A., & Jernajczyk, W. (2012). Sleep and antidepressant treatment. *Current Pharmaceutical Design, 18*(36), 5802–5817.

Wilhelm, B., Giedke, H., Ludtke, H., Bittner, E., Hofmann, A., & Wilhelm, H. (2001). Daytime variations in central nervous system activation measured by a pupillographic sleepiness test. *Journal of Sleep Research, 10*(1), 1–7.

Wilk, K., & Hegerl, U. (2010). Time of mood switches in ultra-rapid cycling disorder: a brief review. *Psychiatry Research, 180*(1), 1–4. http://dx.doi.org/10.1016/j.psychres.2009.08.011. pii:S0165-1781(09)00318-7.

Willcutt, E. G., Nigg, J. T., Pennington, B. F., Solanto, M. V., Rohde, L. A., Tannock, R., ... Lahey, B. B. (2012). Validity of DSM-IV attention deficit/hyperactivity disorder symptom dimensions and subtypes. *Journal of Abnormal Psychology, 121*(4), 991–1010. http://dx.doi.org/10.1037/a0027347.

Wong, M. L., Kling, M. A., Munson, P. J., Listwak, S., Licinio, J., Prolo, P., ... Gold, P. W. (2000). Pronounced and sustained central hypernoradrenergic function in major depression with melancholic features: relation to hypercortisolism and corticotropin-releasing hormone. *Proceedings of the National Academy of Sciences of the United States of America, 97*(1), 325–330.

Wu, J. C., & Bunney, W. E. (1990). The biological basis of an antidepressant response to sleep deprivation and relapse: review and hypothesis. *American Journal of Psychiatry, 147*(1), 14–21.

Wu, J. C., Gillin, J. C., Buchsbaum, M. S., Hershey, T., Johnson, J. C., & Bunney, W. E., Jr. (1992). Effect of sleep deprivation on brain metabolism of depressed patients. *American Journal of Psychiatry, 149*(4), 538–543.

Yerkes, R. M., & Dodson, J. D. (1908). The relation of strength of stimulus to rapidity of habit-formation. *Journal of Comparative Neurology and Psychology, 18*, 459–482. http://dx.doi.org/10.1002/cne.920180503.

Yoon, S. Y., Jain, U., & Shapiro, C. (2012). Sleep in attention-deficit/hyperactivity disorder in children and adults: past, present, and future. *Sleep Medicine Reviews, 16*(4), 371–388. http://dx.doi.org/10.1016/j.smrv.2011.07.001.

Zeni, C. P., Tramontina, S., Ketzer, C. R., Pheula, G. F., & Rohde, L. A. (2009). Methylphenidate combined with aripiprazole in children and adolescents with bipolar disorder and attention-deficit/hyperactivity disorder: a randomized crossover trial. *Journal of Child and Adolescent Psychopharmacology, 19*(5), 553–561. http://dx.doi.org/10.1089/cap.2009.0037.

Zentall, S. S., & Zentall, T. R. (1983). Optimal stimulation: a model of disordered activity and performance in normal and deviant children. *Psychological Bulletin, 94*(3), 446–471.

Zuckerman, M. (1979). *Sensation-seeking: Beyond the optimal level of arousal*. Hillsdale, NJ: Erlbaum.

Index

'*Note*: Page numbers followed by "f" indicate figures, "t" indicate tables, and "b" indicate boxes.'

D

E

F

G

N

O

P

R

S

T

U

V

W

Z

Printed and bound by CPI Group (UK) Ltd, Croydon, CR0 4YY
08/05/2025
01865022-0003